CALGARY PUBLIC LIBRARY

MAR — 2008

Hydroecology and Ecohydrology

Hydroecology and Ecohydrology: Past, Present and Future

Edited by

PAUL J. WOOD

Department of Geography, Loughborough University, UK

DAVID M. HANNAH

School of Geography, Earth and Environmental Sciences, University of Birmingham, UK

and

JONATHAN P. SADLER

School of Geography, Earth and Environmental Sciences, University of Birmingham, UK

John Wiley & Sons, Ltd

Copyright © 2007   John Wiley & Sons Ltd, The Atrium, Southern Gate, Chichester,
West Sussex PO19 8SQ, England

Telephone   (+44) 1243 779777

Email (for orders and customer service enquiries): cs-books@wiley.co.uk
Visit our Home Page on www.wileyeurope.com or www.wiley.com

All Rights Reserved. No part of this publication may be reproduced, stored in a retrieval system or transmitted in any form or by any means, electronic, mechanical, photocopying, recording, scanning or otherwise, except under the terms of the Copyright, Designs and Patents Act 1988 or under the terms of a licence issued by the Copyright Licensing Agency Ltd, 90 Tottenham Court Road, London W1T 4LP, UK, without the permission in writing of the Publisher. Requests to the Publisher should be addressed to the Permissions Department, John Wiley & Sons Ltd, The Atrium, Southern Gate, Chichester, West Sussex PO19 8SQ, England, or emailed to permreq@wiley.co.uk, or faxed to (+44) 1243 770620.

Designations used by companies to distinguish their products are often claimed as trademarks. All brand names and product names used in this book are trade names, service marks, trademarks or registered trademarks of their respective owners. The Publisher is not associated with any product or vendor mentioned in this book.

This publication is designed to provide accurate and authoritative information in regard to the subject matter covered. It is sold on the understanding that the Publisher is not engaged in rendering professional services. If professional advice or other expert assistance is required, the services of a competent professional should be sought.

The Publisher and the Author make no representations or warranties with respect to the accuracy or completeness of the contents of this work and specifically disclaim all warranties, including without limitation any implied warranties of fitness for a particular purpose. The advice and strategies contained herein may not be suitable for every situation. In view of ongoing research, equipment modifications, changes in governmental regulations, and the constant flow of information relating to the use of experimental reagents, equipment, and devices, the reader is urged to review and evaluate the information provided in the package insert or instructions for each chemical, piece of equipment, reagent, or device for, among other things, any changes in the instructions or indication of usage and for added warnings and precautions. The fact that an organization or Website is referred to in this work as a citation and/or a potential source of further information does not mean that the author or the publisher endorses the information the organization or Website may provide or recommendations it may make. Further, readers should be aware that Internet Websites listed in this work may have changed or disappeared between when this work was written and when it is read. No warranty may be created or extended by any promotional statements for this work. Neither the Publisher nor the Author shall be liable for any damages arising herefrom.

Other Wiley Editorial Offices

John Wiley & Sons Inc., 111 River Street, Hoboken, NJ 07030, USA

Jossey-Bass, 989 Market Street, San Francisco, CA 94103-1741, USA

Wiley-VCH Verlag GmbH, Boschstr. 12, D-69469 Weinheim, Germany

John Wiley & Sons Australia Ltd, 42 McDougall Street, Milton, Queensland 4064, Australia

John Wiley & Sons (Asia) Pte Ltd, 2 Clementi Loop # 02-01, Jin Xing Distripark, Singapore 129809

John Wiley & Sons Ltd, 6045 Freemont Blvd, Mississauga, Ontario L5R 4J3, Canada

Wiley also publishes its books in a variety of electronic formats. Some content that appears in print may not be available in electronic books.

Anniversary Logo Design: Richard J. Pacifico

Library of Congress Cataloging-in-Publication Data

Hydroecology and ecohydrology : past, present, and future / edited by Paul
J. Wood, David M. Hannah, and Jonathan P. Sadler.
    p. cm.
Includes bibliographical references and index.
ISBN 978-0-470-01017-4 (cloth : alk. paper)
 1. Hydrology.   2. Groundwater ecology.   3. Ecohydrology.   I. Wood, Paul J.
II. Hannah, David M.   III. Sadler, J. P.
GB653.H83 2007
551.48—dc22

2007041703

British Library Cataloguing in Publication Data

A catalogue record for this book is available from the British Library

ISBN 978-0-470-01017-4 (HB)

Typeset in 10/12 pt Times by SNP Best-set Typesetter Ltd., Hong Kong
Printed and bound in Great Britain by Antony Rowe Ltd, Chippenham, Wiltshire
This book is printed on acid-free paper responsibly manufactured from sustainable forestry
in which at least two trees are planted for each one used for paper production.

Paul Wood for Maureen, Connor and Ryan
David Hannah for Angela and Ellie
Jonathan Sadler for Elizabeth, Matthew, Thomas, Rachel and Rebecca

Contents

List of Contributors		xix
Preface		xxiii

1 Ecohydrology and Hydroecology: An Introduction   1
Paul J. Wood, David M. Hannah and Jonathan P. Sadler
    1.1   Wider Context   1
    1.2   Hydroecology and Ecohydrology: A Brief Retrospective   2
    1.3   A Focus   3
    1.4   This Book   4
    1.5   Final Opening Remarks   4
    References   5

PART I   PROCESSES AND RESPONSES

2 How Trees Influence the Hydrological Cycle in Forest Ecosystems   7
Barbara J. Bond, Frederick C. Meinzer and J. Renée Brooks
    2.1   Introduction   7
    2.2   Key Processes and Concepts in Evapotranspiration –
          Their Historical Development and Current Status   8
          2.2.1   The SPAC   8
          2.2.2   Transpiration   9
          2.2.3   Liquid Water Transport through Trees and the Role
                  of Hydraulic Architecture   14
          2.2.4   Water Uptake by Roots   19
    2.3   Evapotranspiration in Forest Ecosystems   21
          2.3.1   Evaporation and Transpiration   21
          2.3.2   Transpiration from the Understory   22
    2.4   Applying Concepts: Changes in Hydrologic Processes through
          the Life Cycle of Forests   22
          2.4.1   A Summary of Age-related Changes in Forest
                  Composition, Structure, and Function   23

		2.4.2	Impacts of Tree Size on Stomatal Conductance and Whole-tree Water Use	23
		2.4.3	Age-related Change in Transpiration, Interception and Water Storage on the Forest Stand Level	25
		2.4.4	Impacts of Change in Species Composition on Transpiration in Aging Forests	27
		2.4.5	Implications for Predictive Models	27
	Acknowledgments			28
	References			28

3 The Ecohydrology of Invertebrates Associated with Exposed Riverine Sediments — 37
Jonathan P. Sadler and Adam J. Bates

	3.1	Introduction		37
	3.2	ERS Habitats		38
	3.3	Invertebrate Conservation and ERS Habitats		38
	3.4	Flow Disturbance in ERS Habitats		39
	3.5	The Importance of Flow Disturbance for ERS Invertebrate Ecology		41
		3.5.1	Principle (i): Physical Variability and ERS Invertebrates	41
		3.5.2	Principle (ii): Life History Patterns and Function Ecology	46
		3.5.3	Principle (iii): Lateral and Longitudinal Connectivity and Population Viability	47
	3.6	How Much Disturbance is Needed to Sustain ERS Diversity?		48
	3.7	Threats to ERS Invertebrate Biodiversity		50
	3.8	Conclusions		52
	References			52

4 Aquatic–Terrestrial Subsidies Along River Corridors — 57
Achim Paetzold, John L. Sabo, Jonathan P. Sadler, Stuart E.G. Findlay and Klement Tockner

	4.1	Introduction		57
	4.2	What Controls Aquatic–Terrestrial Flows?		58
		4.2.1	Subsidies from Land to Water	59
		4.2.2	Subsidies from Water to Land	59
	4.3	Aquatic–Terrestrial Flows Along River Corridors		61
		4.3.1	Aquatic–Terrestrial Subsidies in Forested Headwater Streams	61
		4.3.2	Aquatic–Terrestrial Subsidies in a Braided River Reach	63
		4.3.3	Aquatic–Terrestrial Subsidies in Temperate Lowland Rivers	66
	4.4	Influence of Human Impacts on Aquatic–Terrestrial Subsidies		67
		4.4.1	Riparian Deforestation	67
		4.4.2	River Channelization and Regulation	67
	4.5	Conclusions		68
	4.6	Future Research		68
	References			69

5 Flow-generated Disturbances and Ecological Responses: Floods and Droughts   75
Philip S. Lake
- 5.1 Introduction   75
- 5.2 Definition of Disturbance   75
- 5.3 Disturbances and Responses   76
- 5.4 Disturbance and Refugia   77
- 5.5 Floods   78
  - 5.5.1 The Disturbance   78
- 5.6 Droughts   79
  - 5.6.1 The Disturbance   79
- 5.7 The Responses to Floods   80
  - 5.7.1 Constrained Streams   80
  - 5.7.2 Floodplain Rivers   82
- 5.8 Responses to Drought   82
  - 5.8.1 Impacts   82
  - 5.8.2 Recovery from Drought   85
- 5.9 Summary   86
- 5.10 Hydrological Disturbances and Future Challenges   87
- Acknowledgements   88
- References   88

6 Surface Water–Groundwater Exchange Processes and Fluvial Ecosystem Function: An Analysis of Temporal and Spatial Scale Dependency   93
Pascal Breil, Nancy B. Grimm and Philippe Vervier
- 6.1 Introduction   93
- 6.2 Fluvial Ecosystems: The Hydrogeomorphic Template and Ecosystem Function   94
  - 6.2.1 Fluvial Ecosystem Function: Biogeochemical Dynamics   94
  - 6.2.2 Fluvial Ecosystem Structure: Biotic Communities   95
- 6.3 Flow Variability and SGW Water Movements   96
  - 6.3.1 In Space   96
  - 6.3.2 In Time   99
  - 6.3.3 An Analysis of Flow Variability Dependency with Basin Area   99
  - 6.3.4 Linkage Between SGW and Flow Dynamics   103
- 6.4 Implications of Flow Variability for SGW Exchange and Fluvial Ecosystem Structure and Function   103
  - 6.4.1 Material Delivery to and within Fluvial Ecosystems   103
  - 6.4.2 Modulation of Nutrient and Organic Matter Delivery by the Riparian Interface Zone   105
  - 6.4.3 In-stream Biogeochemical Function and Flow Variability   106
- 6.5 Conclusion   107
- Acknowledgments   108
- References   108

7 Ecohydrology and Climate Change — 113
Wendy S. Gordon and Travis E. Huxman

- 7.1 Introduction — 113
- 7.2 Ecohydrological Controls on Streamflow — 114
- 7.3 Simulation Studies of Ecohydrological Effects of Climate Change — 116
- 7.4 Experimental Studies of Ecohydrological Effects of Climate Change — 117
- 7.5 Differing Perspectives of Hydrologists and Ecologists — 121
- 7.6 Future Research Needs — 122
- 7.7 Postscript — 123
- References — 124

PART II METHODS AND CRITIQUES

8 The Value of Long-term (Palaeo) Records in Hydroecology and Ecohydrology — 129
Tony G. Brown

- 8.1 River–Floodplain–Lake Systems and the Limits of Monitoring — 129
- 8.2 Key Concepts — 130
- 8.3 Palaeoecology and Palaeohydrology: Proxies and Transfer Functions — 132
  - 8.3.1 Dendrohydrology — 132
  - 8.3.2 Coleoptera (Beetles) — 133
  - 8.3.3 Chironomids (Non-biting Midges) — 133
  - 8.3.4 Cladocera (Water Fleas) — 134
  - 8.3.5 Diatoms — 134
  - 8.3.6 Pollen and Spores — 135
- 8.4 Palaeoecology, Restoration and Enhancement — 136
- 8.5 Case Study I. The River Culm in South-west England — 137
- 8.6 Case Study II. The Changing Status of Danish Lakes — 138
- 8.7 Conclusions — 141
- Acknowledgements — 142
- References — 142

9 Field Methods for Monitoring Surface/Groundwater Hydroecological Interactions in Aquatic Ecosystems — 147
Andrew J. Boulton

- 9.1 Introduction — 147
- 9.2 Research Contexts: Questions, Scales, Accuracy and Precision — 148
- 9.3 Direct Hydrological Methods for Assessing SGW Interactions — 150
  - 9.3.1 Seepage Meters — 151
  - 9.3.2 Mini-piezometers and Groundwater Mapping — 151
  - 9.3.3 Synoptic Surveys of Surface Discharge or Lake Levels — 157
- 9.4 Indirect Hydrological Methods for Assessing SGW Interactions — 157
  - 9.4.1 Water Temperature and Thermal Patterns — 157
  - 9.4.2 Water Chemistry and Chemical Signatures — 158
  - 9.4.3 Dyes and Added Tracers — 159
- 9.5 Future Technical Challenges and Opportunities — 160
- Acknowledgements — 161
- References — 161

10 Examining the Influence of Flow Regime Variability on Instream Ecology — 165
Wendy A. Monk, Paul J. Wood and David. M. Hannah

- 10.1 Introduction — 165
- 10.2 The Requirement for Hydroecological Data — 166
- 10.3 Bibliographic Analysis — 167
- 10.4 Importance of Scale — 167
- 10.5 River Flow Data: Collection and Analysis — 171
- 10.6 Ecological Data: Collection and Analysis — 172
- 10.7 Integration of Hydrological and Ecological Data for Hydroecolical Analysis — 175
- 10.8 River Flow Variability and Ecological Response: Future Directions and Challenges — 176
- References — 178

11 High Resolution Remote Sensing for Understanding Instream Habitat — 185
Stuart N. Lane and Patrice E. Carbonneau

- 11.1 Introduction — 185
- 11.2 Scale, the Grain of Instream Habitat and the Need for Remotely Sensed Data — 185
- 11.3 Depth and Morphology — 188
  - 11.3.1 Image Processing — 188
  - 11.3.2 Photogrammetry — 190
  - 11.3.3 Laser Scanning — 194
- 11.4 Substrate — 196
- 11.5 Discrete Grain Identification — 196
  - 11.5.1 Principles — 196
  - 11.5.2 Example Application: Exposed Gravel Grain-size Distributions — 197
- 11.6 Ensemble Grain Size Parameter Determination — 197
  - 11.6.1 Principles — 197
- 11.7 Example Application: Substrate Mapping in a Salmon River — 198
- 11.8 Future Developments — 200
- References — 200

12 A Mathematical and Conceptual Framework for Ecohydraulics — 205
John M. Nestler, R. Andrew Goodwin, David L. Smith and James J. Anderson

- 12.1 Introduction — 205
- 12.2 Ecohydraulics: Where Do the Ideas Come From? — 207
- 12.3 Reference Frameworks of Engineering and Ecology — 208
  - 12.3.1 Eulerian Reference Framework — 209
  - 12.3.2 Lagrangian Reference Framework — 209
  - 12.3.3 Agent Reference Framework — 209
- 12.4 Concepts for Ecohydraulics — 211

12.5 Two Examples of Ecohydraulics ... 212
    12.5.1 Example 1: Semi-quantitatively Describing Habitat of Drift Feeding Salmonids ... 212
    12.5.2 Example 2: Quantitatively Describing Fish Swim Path Selection in Complex Flow Fields ... 215
12.6 Discussion ... 219
    12.6.1 An Opportunity for Engineers and Ecologists ... 219
    12.6.2 Challenges and Limits for Ecohydraulics ... 220
12.7 Conclusions ... 221
Acknowledgements ... 221
References ... 221

13 Hydroecology: The Scientific Basis for Water Resources Management and River Regulation ... 225
Geoffrey E. Petts

13.1 Introduction ... 225
13.2 A Scientific Basis for Water Resources Management ... 227
    13.2.1 Principles for Sustainable River Regulation ... 230
13.3 Hydroecology in Water Management ... 231
    13.3.1 Water Allocation to Protect Riverine Systems ... 232
    13.3.2 Defining Ecologically Acceptable Flow Regimes ... 232
    13.3.3 Determining Environmental Flows ... 234
13.4 Applications to Water Resource Problems ... 242
    13.4.1 Communication and Policy Development ... 244
13.5 Conclusions ... 245
References ... 246

PART III CASE STUDIES

14 The Role of Floodplains in Mitigating Diffuse Nitrate Pollution ... 253
Tim Burt, Mariet M. Hefting, Gilles Pinay and Sergi Sabater

14.1 Context ... 253
14.2 Nitrogen Removal by Riparian Buffers: Results of a Pan-European Experiment ... 254
    14.2.1 The NICOLAS Experiment ... 255
    14.2.2 Climatic and Hydrological Controls on the Efficiency of Riparian Buffers ... 255
    14.2.3 The Effect of the Riparian Vegetation Type on Nitrate Removal ... 258
    14.2.4 Nitrogen Saturation Effect ... 259
    14.2.5 N_2O Emissions ... 260
14.3 Landscape Perspectives ... 261
    14.3.1 Upslope–Riparian Zone–Channel Linkage ... 262
    14.3.2 Catchment-scale Considerations ... 263
    14.3.3 N Loading ... 263
14.4 Future Perspectives ... 264
References ... 265

15 Flow–Vegetation Interactions in Restored Floodplain Environments 269
Rachel Horn and Keith Richards
- 15.1 The Need for Ecohydraulics 269
- 15.2 The Basic Hydraulics of Flow–Vegetation Interaction 271
  - 15.2.1 Roughness Properties of Vegetation 272
  - 15.2.2 Nonlinearities 276
- 15.3 Drag Coefficients and Vegetation 277
- 15.4 Velocity, Velocity Profiles and Vegetation Character 278
- 15.5 Dimensionality: Flow Velocity in Compound Channels with Vegetation 281
- 15.6 Some Empirical Illustrations of Flow–Vegetation Interactions 283
  - 15.6.1 Velocity Profiles in Submerged Rigid Vegetation 283
  - 15.6.2 Velocity Profiles in Emergent Rigid Vegetation 284
  - 15.6.3 Velocity Profiles in a Mixture of Submerged and Emergent Vegetation 284
  - 15.6.4 Velocity Profiles in Submerged, Flexible Vegetation 286
  - 15.6.5 Complex Velocity Patterns in Staggered Arrays 286
  - 15.6.6 Velocity Variation Across a Partially Vegetated Channel 287
  - 15.6.7 An Alternative Means of Assessing Vegetative Roughness: The Water Surface Slope 289
  - 15.6.8 An Alternative Means of Measuring Velocity in the Field 289
  - 15.6.9 Modelling the Wienflüss Flows 290
- 15.7 Conclusions 291
- References 292

16 Hydrogeomorphological and Ecological Interactions in Tropical Floodplains: The Significance of Confluence Zones in the Orinoco Basin, Venezuela 295
Judith Rosales, Ligia Blanco-Belmonte and Chris Bradley
- 16.1 Introduction 295
- 16.2 Hydrogeomorphological Dynamics 296
- 16.3 The Riparian Ecosystem 303
- 16.4 Longitudinal Gradients at Confluence Zones 305
  - 16.4.1 Sediment Gradient 305
  - 16.4.2 Biological Gradient 306
- 16.5 Synthesis and Conclusions 311
- Acknowledgements 313
- References 314

17 Hydroecological Patterns of Change in Riverine Plant Communities 317
Birgitta M. Renöfält and Christer Nilsson
- 17.1 Introduction 317
- 17.2 Vegetation in Riverine Habitats 318

17.3	Hydrological–Ecological Interactions	319
	17.3.1 Hydrological Drivers	319
	17.3.2 Ecological Drivers	321
17.4	Natural Patterns of Change	321
	17.4.1 Lateral Dimension	322
	17.4.2 Vertical Dimension	323
	17.4.3 Longitudinal Dimension	325
	17.4.4 Temporal Dimension	326
17.5	Human Impacts	327
17.6	Ways Forward	330
Acknowledgements	332
References	332

18 Hydroecology of Alpine Rivers   339
Lee E. Brown, Alexander M. Milner and David M. Hannah

18.1	Introduction	339
18.2	Water Sources Dynamics in Alpine River Systems	341
18.3	Physicochemical Properties of Alpine Rivers	342
	18.3.1 Stream Discharge	342
	18.3.2 Stream Temperature	343
	18.3.3 Suspended Sediment Concentration	344
	18.3.4 Hydrochemistry	346
18.4	Biota of Alpine Rivers	346
	18.4.1 Biota of Alpine Glacier-fed Rivers	346
	18.4.2 Biota of Other (Nonglacier-fed) Alpine Rivers	348
	18.4.3 Temporal Variability of Biota in Alpine Rivers	349
18.5	Towards an Integrated Hydroecological Understanding of Alpine River Systems	351
18.6	Conclusions and Future Research Directions	353
Acknowledgements	356
References	356

19 Fluvial Sedimentology: Implications for Riverine Ecosystems   361
Gregory H. Sambrook Smith

19.1	Introduction	361
19.2	The Sedimentology of Barforms	362
	19.2.1 Ecological Implications	362
	19.2.2 Grain Size and Sorting	363
	19.2.3 Bar Surface Structure and Hydraulics	371
19.3	The Evolution of Barforms	373
	19.3.1 Ecological Implications	373
	19.3.2 Bar Migration	374
	19.3.3 Avulsion	376
19.4	Discussion and Conclusion	377
References	380

20	**Physical–Ecological Interactions in a Lowland River System: Large Wood, Hydraulic Complexity and Native Fish Associations in the River Murray, Australia**	**387**
	Victor Hughes, Martin C. Thoms, Simon J. Nicol and John D. Koehn	
	20.1 Introduction	387
	20.2 Study Area	389
	20.3 Methods	391
	20.4 Results	394
	20.4.1 Physical and Hydraulic Characteristics of Large Wood	394
	20.4.2 Fish Capture	396
	20.4.3 Analysis of Fish Abundance in Areas of Low, Medium and High Velocities	397
	20.5 Discussion	397
	20.6 Conclusions	399
	Acknowledgements	400
	References	401
21	**The Ecological Significance of Hydraulic Retention Zones**	**405**
	Friedrich Schiemer and Thomas Hein	
	21.1 Introduction	405
	21.2 Geomorphology and Patch Dynamics Creating Retention Zones	406
	21.3 Retention, Hydraulics and Physiographic Conditions	407
	21.4 Habitat Conditions for Characteristic Biota	409
	21.5 Retention and Water Column Processes	411
	21.6 The Significance of Retention Zones for the River Network	414
	21.7 Implications for River Management	416
	References	417
22	**Hydroecology and Ecohydrology: Challenges and Future Prospects**	**421**
	David M. Hannah, Jonathan P. Sadler and Paul J. Wood	
	22.1 Introduction	421
	22.2 The Need for an Interdisciplinary Approach	422
	22.3 Future Research Themes	423
	22.3.1 Ecosystem Sensitivity to Hydrological Change	424
	22.3.2 Disturbance: Water and Ecological Stress	424
	22.3.3 Aquatic–Terrestrial Linkages	424
	22.3.4 Modern and Palaeo-analogue Studies	425
	22.3.5 Applied Hydroecology	425
	References	426
Index		**431**

Paul J. Wood is a Senior Lecturer in the Department of Geography at Loughborough University, UK. His interdisciplinary research centres on three themes at the interface of ecology and hydrology: (1) the interactions between hydrological variability and aquatic invertebrate populations and communities within headwater streams, springs and hypogean systems (including groundwater, caves and the hyporheic zone); (2) the influence of river flow regime variability upon lotic invertebrate communities; and (3) the role of hydrological disturbances (droughts and floods) in structuring aquatic faunal communities and the processes involved in change over varying time-scales.

David M. Hannah is a Senior Lecturer in Physical Geography at the University of Birmingham, UK. His research is interdisciplinary, focusing on three complementary themes within hydroclimatology (interface between hydrology-climatology): (1) hydroclimatological processes within alpine, mountain and glacierized river basins; (2) climate and river flow regimes; and (3) river energy budget and thermal dynamics. He has a strong crosscutting interest in hydroecology, specifically ecological response to hydroclimatological and physicochemical habitat variability/change. He has also developed new methods for monitoring, analysing and modelling environmental dynamics at a range of space-time scales.

Jonathan P. Sadler is a Reader in Biogeography and Ecology at the University of Birmingham, UK. His research activity is split between urban ecology, hydroecology, and the use and reconfiguration of knowledges associated with biodiversity. The central focus of his hydroecological work examines: (1) how the interaction of flow variability and fluvial geomorphology affects riparian animal communities; (2) how hydrological disturbance affects aquatic riparian animal community structure, food-web fluxes and subsidies, and species populations, and; (3) how riparian management and aquatic pollution affects both lotic and riparian animal communities.

List of Contributors

James J. Anderson University of Washington, Columbia Basin Research

Adam J. Bates School of Geography, Earth and Environmental Sciences, University of Birmingham, UK

Ligia Blanco-Belmonte Fundación La Salle de Ciencias Naturales, Estación de Investigaciones Hidrobiológica de Guayana, Venezuela

Barbara J. Bond Department of Forest Science, Oregon State University, USA

Andrew J. Boulton University of New England, New South Wales, Australia

Chris Bradley School of Geography, Earth and Envirommental Sciences, University of Birmingham, UK

Pascal Breil *Cemagref*, Agricultural and Environmental Engineering Research, Hydrology-hydraulics research unit, Lyon, France

J. Renée Brooks US Environmental Protection Agency, Western Ecology Division, Corvallis, OR, USA

Lee E. Brown School of Geography, University of Leeds, UK

Tony G. Brown School of Archaeology, Geography and Earth Resources, The University of Exeter, UK

Tim Burt Department of Geography, Durham University, UK

Patrice E. Carbonneau Department of Geography, Durham University, UK

Stuart E.G. Findlay Institute of Ecosystem Studies, Millbrook, USA

R. Andrew Goodwin U.S. Army Engineer Research and Development Center

Wendy S. Gordon The University of Texas at Austin, TX, USA

Nancy B. Grimm School of Life Sciences, Arizona State University, USA

David M. Hannah School of Geography, Earth and Environmental Sciences, University of Birmingham, UK

Mariet M. Hefting Department of Geobiology, Utrecht University, The Netherlands

Thomas Hein Dept. of Limnology and Hydrobotany, University of Vienna, Austria

Rachel Horn Department of Geography, University of Cambridge, UK

Victor Hughes Riverine Landscapes Research Laboratory, University of Canberra, Australia

Travis E. Huxman University of Arizona, AZ, USA

John D. Koehn Arthur Rylah Institute, Department of Sustainability and Environment, Heidelberg, VIC, Australia

Philip S. Lake School of Biological Sciences, Monash University, Victoria, Australia

Stuart N. Lane Department of Geography, Durham University, UK

Frederick C. Meinzer USDA Forest Service, Corvallis, OR, USA

Alexander M. Milner School of Geography, Earth and Environmental Sciences, University of Birmingham and Institute of Arctic Biology, University of Alaska, USA

Wendy A. Monk Department of Geography, Loughborough University, UK and Canadian Rivers Institute, Department of Biology, University of New Brunswick, Canada

John M. Nestler U.S. Army Engineer Research and Development Center

Simon J. Nicol Arthur Rylah Institute, Department of Sustainability and Environment, Heidelberg, VIC, Australia

Christer Nilsson Department of Ecology and Environmental Science, Umeå University, Sweden

Achim Paetzold Catchment Science Centre, The University of Sheffield, UK and Department of Aquatic Ecology, eawag, Duebendorf, Switzerland

List of Contributors   xxi

Geoffrey E. Petts   University of Westminster, London, UK

Gilles Pinay   Centre d'Ecologie Fonctionnelle et Evolutive, CNRS, France

Birgitta M. Renöfält   Department of Ecology and Environmental Science, Umeå University, Sweden

Keith Richards   Department of Geography, University of Cambridge, UK

Judith Rosales   Universidad Nacional Experimental de Guayana, Centro de Investigaciones Ecológicas, Venezuela

Sergi Sabater   Institute of Aquatic Ecology and Department of Environmental Sciences, University of Girona, Spain

John L. Sabo   School of Life Sciences, Arizona State University, USA

Jonathan P. Sadler   School of Geography, Earth and Environmental Sciences, University of Birmingham, UK

Gregory H. Sambrook Smith   School of Geography, Earth and Environmental Sciences, University of Birmingham, UK

Friedrich Schiemer   Department of Limnology and Hydrobotany, University of Vienna, Austria

David L. Smith   U.S. Army Engineer Research and Development Center and formerly of S.P. Cramer and Associates, Gresham

Martin C. Thoms   Riverine Landscapes Research Laboratory, University of Canberra, Australia

Klement Tockner   Department of Aquatic Ecology, eawag, Duebendorf, Switzerland and Institute of Freshwater Ecology and Inland Fisheries (IGB), Berlin, Germany

Philippe Vervier   UMR CNRS-UPS, Paul Sabatier University, France

Paul J. Wood   Department of Geography, Loughborough University, UK

Preface

Water-dependent habitats are extremely diverse in terms of their nature (e.g., drylands, wetlands, streams/rivers and ponds/lakes), geography (poles to equator, low to high latitude) and many support communities and species of high conservation value, some of which are under threat from extinction. As pressure is increasing on water-dependent habitats due to global change and ever growing anthropogenic impacts, it is clear that balancing the water needs of people against those of ecosystems (terrestrial and aquatic) is, and will increasingly become, a premier environmental issue. This crucial, often precarious 'balancing act' involves some highly complex issues and, thus, it has compelled recent workers to identify the need, not only for new integrative science (between traditional fields of hydrology–ecology) and analytical approaches, but for truly interdisciplinary research. In this context, it has been argued that the 'new' discipline(s) of hydroecology/ecohydrology has the potential not only to unlock elements of this complexity, but also to provide a foundation for the sustainable management of water resources.

The terms ecohydrology and hydroecology have been used increasingly in the international scientific literature over the last decade, and this emerging interdisciplinary subject area has gathered considerable momentum as evidenced by the publication of textbooks, special issues of journals, over 150 peer-reviewed scientific journal papers, and the imminent launch of a dedicated John Wiley & Sons journal, *Ecohydrology*. Although there is a growing volume of research output at the interface between the hydrological and biological sciences, the terms hydroecology and ecohydrology and the scientific remit of the field remain remarkably poorly defined. Hence, this book aims to address this research gap and capture the vitality of this current scientific hot-topic in a cutting-edge research text that: (i) reviews the evolution of the discipline (past); (ii) provides detailed coverage of the present state of the art, and (iii) looks to the horizon for the ecohydrology/hydroecology of the future. To achieve this goal, we invited international leaders within their respective fields to author individual chapters. The resultant chapters present significant new results and methodological developments within the field of ecohydrology/hydroecology, while outlining key historical developments and identifying future research needs. The chapters are positioned at the forefront of their fields and draw together individuals from hydrological and biological/ecological sciences and engineering disciplines to 'bridge the gap' between traditional academic disciplines and to ensure that the book is inclusive and truly interdisciplinary.

We have made every effort to encapsulate the variety of ecohydrological/hydroecological research currently being conducted across the globe and suggest that this book is significantly different from previous texts in providing coverage of: (i) a range of organisms (plants, invertebrates and fish), (ii) physical processes within terrestrial, riparian (aquatic–terrestrial ecotones) and aquatic habitats, and (iii) palaeo-ecological/hydrological perspectives. We have endeavoured to provide a comprehensive overview of the research conducted under the banner of ecohydrology/hydroecology. However, we acknowledge that, due to the rapidly developing nature of the subject, there are inevitably omissions. We hope that in capturing the state of the art, this book will provide a catalyst for future interdisciplinary research and a starting point for scientists, practitioners and end-users with an interest in hydrology–ecology interactions.

There are number of people that require acknowledgement for their contributions to this book. We would like to express our sincere appreciation to the teachers, mentors and colleagues that kindled our interest in hydroecology, and opened opportunities (and our eyes) to develop our work in this interdisciplinary research field. We would like to thank the following chapter reviewers for their important work in enhancing the quality and rigour of this book: Maureen Agnew, Patrick Armitage, Martin Baptist, Valerie Black, Chris Bradley, Lee Brown, Paul Buckland, Leopold Füreder, Jane Fisher, Rob Francis, Alan Hill, Etienne Muller, Pierre Marmonier, Yenory Morales-Chaves, Eric Pattee, Ian Reid, Christopher Robinson, Geoffrey Petts, Gregory Sambrook Smith, Barnaby Smith, Chris Soulsby, Klement Tockner and Larry Weber. We are also very grateful to several anonymous reviewers. We appreciate the support and efforts of John Wiley & Sons' staff at all stages in the preparation of this book, particularly Richard Davies, Colleen Goldring Richard Lawrence and Fiona Murphy.

Paul J. Wood
David M. Hannah
Jonathan P. Sadler

1
Ecohydrology and Hydroecology: An Introduction

Paul J. Wood, David M. Hannah and Jonathan P. Sadler

1.1 Wider Context

Water is essential for life on our 'blue planet' but just 2.5% of all the Earth's water comprises freshwater. Of this precious freshwater resource, 0.3% is estimated to be surface water, held in rivers and streams (2%), wetlands (11%) and lakes (87%) (see, for example, Oki and Kanae, 2006). This tiny proportion of global water supports at least 6% (>100 000) of all described species (Dudgeon *et al.*, 2006). Water-dependent habitats are extremely diverse, located from poles to equator and low to high altitude, and comprise dynamic systems that vary in scale from individual plants to large complex vegetation communities, from the smallest headwater stream to the largest lowland river, complex floodplains, a myriad of wetland types, and lentic ecosystems ranging from ponds to vast lakes. Many of these habitats support communities and species of high conservation value, some of which are under threat from extinction (e.g., Grootjans *et al.*, 2006; Hannah *et al.*, 2007; Ricciardi and Rasmussen 1999; Sadler *et al.*, 2004; Wilcox and Thurow, 2006). Because of water's many uses, humans have fundamentally altered natural hydrological processes and conditions in many areas and, consequently, freshwater ecosystems have experienced reductions in biodiversity at least as great as the most impacted terrestrial ecosystems (Dudgeon *et al.*, 2006). When viewed against a backdrop of rapidly rising global population (predicted to reach about 8 billion by 2025; United Nations, 2000), it is clear that balancing the water needs of people against those of ecosystems (terrestrial and aquatic) is, and will increasing become, a premier environmental issue (see, for example, Petts *et al.*, 2006a). This crucial, precarious 'balancing act'

Hydroecology and Ecohydrology: Past, Present and Future, Edited by Paul J. Wood, David M. Hannah and Jonathan P. Sadler
© 2007 John Wiley & Sons, Ltd.

involves some highly complex issues and, thus, it has compelled recent workers to identify the need, not only for new integrative science and new analytical approaches (e.g. Newman *et al.*, 2006; Petts *et al.*, 2006b), but for truly interdisciplinary research (e.g. Hannah *et al.*, 2004). In this context, it has been argued that the 'new' discipline(s) of hydroecology/ecohydrology has the potential not only to unlock elements of this complexity, but also to provide a foundation for the sustainable management of water resources (e.g. Zalewski, 2000; Zalewski *et al.*, 1997).

1.2 Hydroecology and Ecohydrology: A Brief Retrospective

The terms ecohydrology (eco-hydrology) and hydroecology (hydro-ecology), which include the subdiscipline of ecohydraulics, are being used increasingly by the international scientific community (see e.g. Janauer, 2000; Wilcox and Newman 2003, Wood *et al.*, 2001). Although there is a growing volume of research output at the interface between the hydrological and life (biological) sciences (e.g. Gurnell *et al.*, 2000; Zalewski, 2000), the terms hydroecology and ecohydrology and the scientific remit of the field are remarkably poorly defined, with limited consensus between many of the published definitions (e.g. Wassen and Grootjans, 1996; Zalewski *et al.*, 1997; Baird and Wilby, 1999; Dunbar and Acreman, 2001; Eagleson, 2002; Nuttle, 2002; Bond, 2003; Rodriguez-Iturbe, 2005). Furthermore, it has been suggested that while physical scientists, and hydrologists in particular, are embracing the new 'hot topic', ecologists and biologists are less aware or less inclined to engage in the debate regarding its status as a 'new paradigm' or subdiscipline (Bond, 2003). Bibliographic analysis (for details refer to Hannah *et al.*, 2004) has demonstrated the proliferation of the terms ecohydrology and hydroecology, and the breadth of subject matter; however, it has also identified a larger body of 'hidden' ecohydrological and hydroecological literature that does not flag itself as such (also see Bond, 2003; Bonell, 2002; Kundzewicz, 2002). Perhaps most significantly, the review of hydroecological/ecohydrological literature showed clearly that the majority of the research was undertaken within traditional subject boundaries (i.e., groups dominated by either physical or biological scientists) rather than interdisciplinary teams, uniting researchers from both traditions (Hannah *et al.*, 2004). Whether or not the terms are widely recognised or accepted as something 'new' (or as a subdiscipline) by scientists, the fact remains that the terms are increasingly making an impact within the hydrological and ecological literature. Indeed, the yearly citation rate of the terms has more than doubled since 2004 (Figure 1.1).

1.3 A Focus

It is apparent from our previous literature review and bibliographic analysis (Hannah *et al.*, 2004) that a definition identifying a theoretical core is needed before hydroecology and ecohydrology become established paradigms or disciplines. A definition including the discipline's aim and subject scope would serve as a focal point to help unite the research community. In this regard, a single definition that applies equally to hydroecology and ecohydrology is essential. At present, there is arguably no single accepted definition of either term, never mind a joint definition.

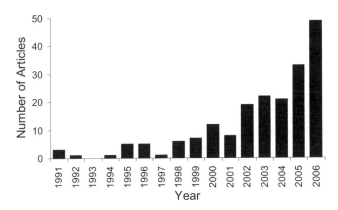

Figure 1.1 Number of peer-reviewed journal articles using the terms ecohydrology, ecohydrology, hydroecology and hydro-ecology (1991–2006)

There has been specific use of the term ecohydrology to refer to plant–water interactions both in the past (e.g., Baird and Wilby, 1999; Eagleson, 2002) and increasingly more recently, with special reference to semi-arid/dryland/rangeland environments (e.g. Newman et al., 2006; Wilcox and Thurow, 2006). However, ecohydrology has also been employed to describe wider hydrology–ecology linkages (i.e. all biota and environments, e.g., Kundzewicz, 2002; Zalewski, 2002). Arguably, this specific versus generalist usage of ecohydrology could lead to confusion and misunderstanding. Therefore, we propose the use of the term hydroecology to refer to hydrology–ecology interactions in the broadest sense (cf. Dunbar and Acreman, 2001) and so provide an umbrella under which ecohydrology in its stricter form can be included.

We recognise the potential danger that definitions can become either too restrictive or nebulous to be effective and/or applicable. Like hydroecology, other 'new' scientific paradigms have begun life as 'hot topics' but they have faded away due to a problem of identity. If hydroecology is to avoid a similar fate, we must ensure that it is an identifiable and constructive discipline, and not a deconstructed version of existing paradigms or academic disciplines. A clear and inclusive definition of hydroecology should help to this end. Rather than debating and deliberating over the appropriate form of words in a revised definition, we suggest that it may be more instructive to provide a list of 'target elements' that outline the theoretical core and range of process interactions and scales that should encapsulate hydroecological research (Hannah et al., 2004):

(i) the bi-directional nature of hydrological–ecological interactions and importance of feedback mechanisms;
(ii) the requirement for fundamental process understanding, rather than simply establishment of functional (statistical) links without a probable chain of causality;
(iii) the subject scope to encompass: (a) the full range of (natural and human-impacted) water-dependent habitats/environments, and (b) flora, fauna and whole ecosystems;

(iv) the need to consider process interactions operating at a range of spatial and temporal scales (including palaeohydrological and palaeoecological viewpoints); and
(v) the interdisciplinary nature of the research philosophy (see Chapter 22).

1.4 This Book

Given the current upsurge in hydroecology and ecohydrology, we thought it important and timely to capture the current state-of-the art in one place as a research-level text. Our intention was to create a cutting-edge volume, which presents new results and methodological developments within the rapidly evolving field of hydroecology/ecohydrology. To achieve this goal, recognised international leaders in their subjects have written the individual chapters and have aimed to position their contributions at the research forefront. In terms of content, the book covers a range of hydrological and ecological processes, methodological approaches and hydroecologically sensitive habitats from an array of geographical locations (e.g., Australia, Europe, and North and South America). The book differs from others currently available not only in terms of its environmental breadth but it also covers a wide range of organisms (plants, invertebrates and fish) and their interactions with water.

The book is structured in three sections: Part 1 considers fundamental ecohydrological/hydroecological process understanding and how floral and faunal communities and ecosystem functions (e.g. nutrient cycling) are influenced and respond to water and its availability (Chapter 2 to Chapter 7). Part 2 of the book draws together up-to-date methodological approaches and critiques of how ecohydrological/hydroecological patterns and processes can (may) be monitored/modelled to maintain and protect the natural environment, and be managed to ensure the continued supply of water for human uses (Chapters 8–13). Part 3 comprises detailed ecohydrological and hydroecological case-studies of research undertaken on different floral and faunal groups in different environments across the globe (Chapter 14 to Chapter 21). The final chapter (Chapter 22) identifies some challenges and future prospects for hydroecology/ecohydrology.

We do not claim that this volume is all encompassing in its coverage of research that could be deemed to be hydroecology or ecohydrology. Indeed, we are conscious that this volume only begins to address palaeohydrology and palaeoecology. Palaeoecohydrology studies may provide valuable baseline information regarding pre-human influences on the environment and for climatic change/variability investigations (Lytle, 2005; Prebble *et al.*, 2005). In addition, the chapters almost exclusively deal with freshwater and do not consider marine/brackish water ecosystems or terrestrial environments subject to salinisation (e.g. Williams and Williams, 1998; Brown *et al.*, 2006). Saline water represents a challenging environment for floral and faunal communities and, as yet, these avenues of research have not been explored fully by ecohydrologists and hydroecologists.

1.5 Final Opening Remarks

It is clear that researchers actively involved in ecohydrological and hydroecological studies are increasingly aware of the need for a truly interdisciplinary philosophical framework, as opposed to a multidisciplinary approach within clearly defined traditional subject boundaries (Hannah *et al.*, 2004; Newman *et al.*, 2006; Petts *et al.*, 2006a). We believe this volume indicates that while these barriers have not necessarily been removed, they may be overcome. The book addresses a range of methodological developments, habitats and floral/faunal groups, and illustrates the diversity of approaches to hydro-ecological/ecohydrological study. The chapters highlight significant advances in our knowledge, as well as providing thorough reviews of the subject matter, and crucially they identify the gaps in our knowledge that remain to be addressed. We hope that this book will provide a catalyst and starting point for others to become involved in dynamic interdisciplinary research at the hydrology–ecology interface. Perhaps, this volume may even provide a landmark in helping to set the foundation for the subdiscipline of hydro-ecology/ecohydrology to emerge from its larger 'parent' subjects and evolve into an accepted discipline in its own right.

References

Baird AJ and Wilby RL. 1999. *Eco-Hydrology: Plants and Water in Terrestrial and Aquatic Environments*. Routledge: London.

Bond B. 2003. Hydrology and ecology meet – and the meeting is good. *Hydrological Processes*, **17**, 2087–2089.

Bonell M. 2002. Ecohydrology – a completely new idea? *Hydrological Sciences Journal*, **47**, 809–810.

Brown CE, Pezeshki SR and DeLaune RD. 2006. The effects of salinity and soil drying on nutrient uptake and growth of *Spartina alterniflora* in a simulated tidal system. *Environmental and Experimental Botany*, **58**, 140–148.

Dudgeon D, Arthington AH, Gessner MO, Kawabata ZI, Knowler DJ, Leveque C, Naiman RJ, Prieur-Richard AH, Soto D, Stiassny MLJ and Sullivan CA. 2006. Freshwater biodiversity: importance, threats, status and conservation challenges. *Biological Reviews*, **81**, 163–182.

Dunbar MJ and Acreman MC. 2001. Applied hydro-ecological sciences for the twenty-first century. In Acreman, MC. (ed.) *Hydro-Ecology: Linking Hydrology and Aquatic Ecology*. Proceedings of Birmingham Workshop, July 1999. IAHS Publication No. 266, pp. 1–17.

Eagleson PS. 2002. *Ecohydrology: Darwinian Expression of Vegetation Form and Function*. Cambridge University Press: Cambridge.

Grootjans AP, Adema EB, Bleuten W, Joostn H, Madaras M and Janakova M. 2006. Hydrological landscape setting of base-rich fen mires and fen meadows: an overview. *Applied Vegetation Science*, **9**, 175–184.

Gurnell AM, Hupp CR and Gregory SV. 2000. Preface. Linking hydrology and ecology. *Hydrological Processes*, **14**, 2813–2815.

Hannah DM, Wood PJ and Sadler JP. 2004 Ecohydrology and hydroecology: A 'new paradigm'? *Hydrological Processes*, **18**, 3439–3445.

Hannah DM, Brown LE, Milner AM, Gurnell AM, McGregor GR, Petts GE, Smith BPG and Snook DL. 2007, Integrating climate-hydrology-ecology for alpine river systems, *Aquatic Conservation: Marine and Freshwater Ecosystems*, **17**, 636–656 doi: 10.1002/aqc.800.

Janauer GA. 2000. Ecohydrology: fusing concepts and scales. *Ecological Engineering*, **16**, 9–16.

Kundzewicz ZW. 2002. Ecohydrology – seeking consensus on interpretation of the notion. *Hydrological Sciences Journal*, **47**, 799–804.

Lytle DE. 2005. Palaeoecological evidence of state shifts between forest and barrens on a Michigan sand plain, USA. *Holocene*, **15**, 821–836.

Newman BD, Wilcox BP, Archer AR, Breshears DD, Dahm CN, Duffy CJ, McDowell NG, Phillips FM, Scanlon BR and Vivoni ER. 2006. Ecohydrology of water-limited environments: A scientific vision. *Water Resources Research*, **42**, W06302.

Nuttle WK. 2002. Is ecohydrology one idea or many? *Hydrological Sciences Journal*, **47**, 805–807.
Oki T and Kanae S. 2006. Global hydrological cycles and world water resources. *Science*, **313**, 1068–1072.
Petts G, Morales Y and Sadler J. 2006a. Linking hydrology and biology to assess the water needs of river ecosystems. *Hydrological Processes*, **20**, 2247–2251.
Petts GE, Nestler J and Kennedy R. 2006b. Advancing science for water resources management. *Hydrobiologia*, **565**, 277–288.
Prebble M, Sim R, Finn J and Fink D. 2005. A holocene pollen and diatom record from Vanderlin Island, Gulf of Carpentaria, lowland tropical Australia. *Quaternary Research*, **64**, 357–371.
Ricciardi A and Rasmussen JB. 1999. Extinction rates of North American freshwater fauna. *Conservation Biology*, **13**, 1220–1222.
Rodriguez-Iturbe I. 2005. *Plants and Soil Moisture Dynamics: A Theoretical Approach to the Ecohydrology of Water-controlled Ecosystems*. Cambridge University Press: Cambridge.
Sadler JP, Bell D and Fowles AP. 2004. The hydroecological controls and conservation value of beetles on exposed riverine sediments in England and Wales. *Biological Conservation*, **118**, 41–56.
United Nations. 2000. *World Population Prospects: The 2000 Revision – Highlights*. UN Population Division, Department of Economic and Social Affairs, New York, USA.
Wassen MJ and Grootjans AP. 1996. Ecohydrology: an interdisciplinary approach for wetland management and restoration. *Vegetatio*, **126**, 1–4.
Wilcox BP and Newman DB. 2003. Ecohydrology of semiarid landscapes. *Ecology*, **86**, 275–276.
Wilcox BP and Thurow TL. 2006. Emerging issues in rangeland ecohydrology. *Hydrological Processes*, **20**, 3159–3178.
Williams DD and Williams NE. 1998. Aquatic insects in an estuarine environment: densities, distribution and salinity tolerance. *Freshwater Biology*, **39**, 411–421.
Wood PJ, Hannah DM, Agnew MD and Petts GE. 2001. Scales of hydroecological variability within a groundwater-dominated chalk stream. *Regulated Rivers: Research and Management*, **17**, 347–367.
Zalewski M, Janauer GA and Jolankaj G. 1997. Ecohydrology: a new paradigm for the sustainable use of aquatic resources. UNESCO IHP Technical Documents in Hydrology No. 7. IHP-V Projects 2.3\2.4, UNESCO, Paris, France.
Zalewski M. 2000. Ecohydrology – the scientific background to use ecosystem properties as management tools toward sustainability of water resources. *Ecological Engineering*, **16**, 1–8.
Zalewski M. 2002. Ecohydrology – the use of ecological and hydrological processes for sustainable management of water resources. *Hydrological Sciences Journal*, **47**, 823–832.

2
How Trees Influence the Hydrological Cycle in Forest Ecosystems

Barbara J. Bond, Frederick C. Meinzer and J. Renée Brooks

2.1 Introduction

Ultimately, the quest of ecohydrology (or hydroecology) is to apply fundamental knowledge from hydrology, ecology, atmospheric science, and related disciplines to solve real world problems involving biological systems and hydrologic cycles. Achieving this goal requires sharing information across disciplines, and this chapter is structured toward that end. Our aim is to present current ecological concepts concerning the ways that the structure and function of forest vegetation influence hydrologic processes. To cover this topic in a single chapter, we emphasize some aspects of the interactions between forest trees and hydrology, especially transpiration, over others, such as moisture interception by forest canopies. Other important topics are not covered at all, such as the influence of forest trees and the myriad flora and fauna associated with them on soil hydraulic properties, and root channels as preferential water flow paths in soils. Research is needed to develop a broader conceptual understanding of these belowground processes, especially over long time periods.

Forests occupy approximately one-third of the Earth's land area, accounting for over two-thirds of the leaf area of land plants, and thus play a very important role in terrestrial hydrology. Our discussion emphasizes temperate coniferous trees in North America because that is where we have the most experience, but the processes discussed are generally applicable to all forest trees, and the tables and figures include information about rates and processes for a variety of species and ecosystems in order to provide perspective on the upper and lower boundaries. Section 2.2 explores transpiration from top (leaves) to bottom (roots), emphasizing the importance of tree hydraulic architecture to transpiration. Section 2.3 expands consideration of evapotranspiration from trees to forest ecosystems. The chapter concludes (Section 2.4) by applying concepts presented in earlier sections to the question of how hydrological processes in forests change as they age – a topic of great relevance as humans alter the age class distribution of forests around the world through land management activities.

2.2 Key Processes and Concepts in Evapotranspiration – Their Historical Development and Current Status

2.2.1 The SPAC

The 'Soil-Plant-Atmosphere Continuum', or SPAC, is a key concept in studies of plant water use. The notion of the SPAC emerges from the cohesion–tension (CT) theory of water movement through plants (Dixon and Joly, 1894), and the recognition that water moves from soil into roots, through plants and into the atmosphere along thermodynamic gradients in water potential (see van den Honert, 1948); these processes are described in detail later in this section. Although the CT theory has been disputed (e.g., Canny, 1995; 1998), it has held up to robust examination (Holbrook *et al.*, 1995; Pockman *et al.*, 1995; Sperry *et al.*, 1996) and is now widely accepted (Angeles *et al.*, 2004).

An electric circuit analogy is often used to characterize physical controls on the movement of water into and through plants and to the atmosphere (van den Honert, 1948). In its simplest form, the pathway can be visualized as a chain of resistances connected in series. The total hydraulic resistance, therefore, is the sum of the individual resistances along the path, including the aerodynamic boundary layer resistances associated with canopy elements, the boundary layer at the leaf surface, stomatal pores, through the xylem pathway of the plant, across root membranes to the soil, and through the soil. Whereas micrometeorologists prefer to view the SPAC in terms of resistances, plant physiologists typically use the inverse of resistance, or conductance, because transpiration increases linearly with conductance at a constant vapor pressure gradient.

While the SPAC model provides a powerful conceptual basis for understanding plant–water relations, it also tends to constrain ecological concepts and models of hydrological cycles to a one-dimensional perspective, limited to vertical fluxes. In this respect, most ecological models and analyses of water balance differ fundamentally from hydrological models and analyses, which typically consider three-dimensional flows of liquid water over and through a landscape. On the other hand, hydrological models are often limited to gravity-driven flowpaths of liquid water, often ignoring or oversimplifying the influences of vegetation on the water cycle. An especially fruitful arena for ecologists and hydrologists to work together is in merging modern, mechanistic models of plant water use, which are almost always one dimensional, with three-dimensional hydrological models (Bond, 2003).

2.2.2 Transpiration

The ratio of transpiration to biomass accumulation varies across plant growth forms, but forest trees typically lose 170 to 340 kg of water vapor for every kg of biomass accumulated (Larcher, 1975). Extensive research has established that the air saturation deficit (D) and net radiation (R_n) are the principal drivers of transpiration (symbols used in this chapter are listed in Table 2.1). Air saturation deficit directly affects transpiration by establishing the vapor pressure gradient between the vapor-saturated leaf interior and the surrounding air. Net radiation indirectly affects transpiration through heating of the canopy, which causes the leaf-to-air vapor pressure gradient to increase as the vapor pressure in the air spaces within leaves increases exponentially with leaf temperature.

Nearly all transpirational vapor loss occurs through the stomatal pores – water losses through leaf cuticles and stems are typically negligible except in unusual circumstances. Over short time periods, plants control transpiration by regulating the size of stomatal pores, while over longer time periods water balance is regulated largely by changes in the amount of leaf area and species composition.

Table 2.1 Terms and symbols

Symbol	Description	Typical units
A_L	Surface area of foliage (projected or total surface)	m^2
A_r	Surface area of roots	m^2
A_s	Surface area of sapwood, measured at breast height (1.37 m) unless specified otherwise	m^2
C	Capacitance (defined as the change in water content of plant tissue per unit change in bulk water potential of the tissue, or $dV/d\Psi$)	$m^3 kPa^{-1}$
D	Air saturation deficit	kPa
g_s	Stomatal conductance	$mol\,m^{-2}\,s^{-1}$
G_c	Canopy conductance	$mol\,m^{-2}\,s^{-1}$
L	Length of stem or hydraulic transport path	m
η	Viscosity	Pa s
k	Permeability; specific conductivity	m^2
k_L	Leaf-specific hydraulic conductivity ($= k\,A_L^{-1}$)	$m^2\,m^{-2}$
K	Hydraulic conductance ($= Q\,\Delta\Psi^{-1}$)	$m^3\,Pa^{-1}\,s^{-1}$
LAI	Leaf area index	Dimensionless ($m^2\,m^{-2}$)
Q	Volume flow per unit time	$m^3\,s^{-1}$
R	Hydraulic resistance	$Pa\,s\,m^{-3}$
R_n	Net radiation	Watts
Ω	Decoupling coefficient	dimensionless
Ψ (Ψ_{soil}, Ψ_{leaf}, $\Delta\Psi$)	Water potential (soil water potential, leaf water potential, difference in water potential at either end of a hydraulic path)	MPa

Maximum and mean stomatal conductances (g_s) vary widely among species and forest types (Table 2.2). Stomatal pore size, and therefore g_s, is dynamic and has been shown to respond rapidly to numerous environmental and physiological variables. Light (especially in blue wavelengths) and D are key components of the aerial environment that exert opposing effects on g_s. Stomatal conductance exhibits a characteristic saturating or asymptotic response to increasing light. Light-saturation points for g_s of different types of forest trees vary considerably, with g_s of coniferous forest trees typically saturating at photosynthetic photon flux densities (PFD) substantially lower than those of temperate and tropical broadleaf trees. Both the light saturation of g_s and maximum g_s are highly variable due to adaptation of foliage to the local light environment. Although it is widely assumed that stomata of woody species are tightly closed at night, resulting in negligible nocturnal transpiration rates, a number of reports indicate that nocturnal transpiration can be substantial, often contributing 25% or more to the daily total (Green *et al.*, 1989; Benyon, 1999; Donovan *et al.*, 1999; Oren *et al.*, 1999a; Sellin, 1999; Feild and Holbrook, 2000),

Table 2.2 *Examples of maximum stomatal (g_s) and canopy (G_c) conductance of different types of forest vegetation*

Forest/vegetation type	Species	LAI	g_s (mmol m^{-2} s^{-1})	G_c (mmol m^{-2} s^{-1})	Reference
Conifer					
boreal	*Picea mariana*	4.4	25	98	Rayment et al., 2000
temperate	*Pinus pinaster*	2.7	150	320	Loustau et al., 1996
Mediterranean temperate mesic	*Pseudotsuga menziesii/ Tsuga heterophylla*	9.0	50–70	480	Phillips et al., 2002; Meinzer et al., 2004c; Unsworth et al., 2004
temperate semiarid	*Pinus ponderosa*	2.1	166	287	Ryan et al., 2000; Anthoni et al., 2002
Angiosperm					
boreal	*Populus tremuloides*	5.6	490	1200	Blanken et al., 1997
temperate deciduous	*Fagus sylvatica*	4.5	250	900	Herbst, 1998
temperate evergreen	*Nothofagus menziesii/ N. fusca*	7.0	160	440	Köstner et al., 1992
tropical plantation	*Goupia glabra*	3.7	180	600	Granier et al., 1992
Amazonian rainforest	mixed	6.6	200	420	Shuttleworth et al., 1984; Roberts et al., 1990

especially in environments where nighttime relative humidity remains relatively low (Bucci *et al.*, 2004).

The response of transpiration to increasing D is regulated (Schulze *et al.*, 1972) through partial stomatal closure (Figure 2.1). Thus, when light is adequate and D is low (i.e., high humidity), stomata are maximally open and transpiration increases linearly with increasing D. For many species, beyond a critical level of D, stomatal conductance declines exponentially with increasing D, causing transpiration to level off at a maximum rate. In some cases, transpiration actually decreases at very high evaporative demand once a maximum value has been attained (Farquhar, 1978; Mott and Parkhurst, 1991; Franks *et al.*, 1997). The responses of both stomatal conductance and transpiration to D change depending on the availability of soil moisture (Figure 2.1).

The apparent sensitivity of g_s to D varies widely among tree species (Figure 2.2A), largely because of differences in their hydraulic characteristics (discussed in next section) and species-specific differences in leaf anatomical traits such as stomatal pore depth and density (which determine maximum g_s at low D). The characteristic exponential decline in g_s with increasing D has been exploited in a model that demonstrates that the sensitivity of g_s to D is proportional to the magnitude of g_s at low D in the same manner across a

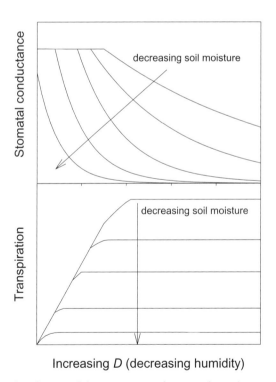

Figure 2.1 *A generalized view of the responses of stomatal conductance and transpiration to soil and atmospheric water deficits for* isohydric *plants (see Section 2.2.3 for a discussion of isohydric and anisohydric behavior in plants)*

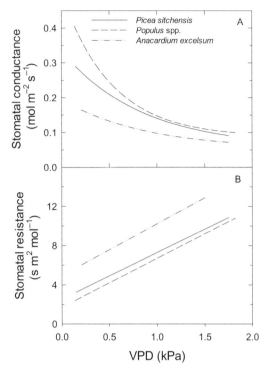

Figure 2.2 Stomatal conductance (A) and resistance (B) in relation to the leaf-to-air vapor pressure deficit (VPD) for a temperate conifer (Picea sitchensis; Schulze and Hall, 1982), temperate broadleaf (Populus spp.; Meinzer et al., 1997), and a tropical broadleaf (Anacardium excelsum; Meinzer et al., 1993) species

broad range of species and conditions (Oren *et al.*, 1999b). Apparent species-specific differences in stomatal sensitivity to D are diminished when the reciprocal of g_s, stomatal resistance, is plotted against D (Figure 2.2B). Because stomatal resistance changes linearly with pore radius, nearly parallel linear responses of resistance to D imply similar sensitivity of stomatal aperture to variation in D, whereas different y-intercepts (minimum resistance) imply differences in leaf anatomy and hydraulic properties of the species.

There is still no comprehensive understanding of how stomata integrate external and internal signals to regulate transpiration. Although stomata are obviously autonomous to some degree in responding directly to external variables such as *PFD* and *D*, the amplitude of stomatal responses is constrained by internal variables. Stomatal responses to increasing soil water deficits exemplify the complex regulatory interactions that ultimately limit forest transpiration during periods of drought. As the soil dries, two types of signal, hydraulic and chemical, may be generated and transmitted to the leaves. Hydraulic signals consist of increases in xylem tension that are rapidly propagated to the leaves as a result of changes in root/soil hydraulic resistance (Section 2.2.4) or soil water potential. Chemical signals may consist of changes in levels of plant growth regulators such as abscisic acid that are exported in the transpiration stream from the roots to the leaves, where they cause partial stomatal closure (Davies and Zhang, 1991). Chemical signals may be generated during incipient soil drying well in advance of hydraulic signals (Gollan et al., 1986), and the magnitude of the hydraulic signal (xylem tension) may determine stomatal responsiveness to chemical signals (Tardieu and Davies, 1993). The role of chemical signals in stomatal regulation in tall trees is uncertain because of the slow propagation of chemical signals – in tall coniferous trees it may take two weeks or more for chemical signals to move from roots to leaves (Meinzer et al., 2006) – relative to nearly instantaneous hydraulic signals. Regardless of the signals or response mechanisms involved, it appears that under a broad range of conditions stomata regulate transpiration to prevent leaf water potentials from dropping below some species-specific minimum (Bond and Kavanagh, 1999), although that minimum may vary slightly with tree size or age within a species (McDowell et al., 2002b). This behavior balances vapor and liquid phase water transport (Meinzer, 2002) and appears to ensure integrity of the liquid water transport pathway in the plant (see next section). Even in deeply rooted tree species that are able partially to sustain transpiration during periods of drought by tapping soil layers that never undergo pronounced drying, conditions sensed by roots in the dry upper soil generate signals that cause partial stomatal closure, leading to relative seasonal homeostasis of maximum leaf water deficits in the canopy (Domec et al., 2004; Warren et al., 2005).

The extent to which transpiration is passively driven by environmental variables such as R_n or is under physiological control by g_s has been debated. Differences in interpretation of the role played by stomata in limiting transpiration are related to the nature of the pathway of water movement in the vapor phase. Closer inspection of the vapor pathway and its associated resistances shows that stomatal control of transpiration is strongest when boundary layer resistance is small in relation to stomatal resistance. Vapor diffusion through stomata would thus represent the controlling resistance. High boundary layer resistance associated with low wind speed, short stature, large leaves, or dense canopies (as is often the case for crop plants and grasslands), will promote local equilibration of humidity near the leaf surfaces, thereby uncoupling the vapor pressure and evaporative demand at the leaf surface from that in the bulk air. Under these circumstances, transpiration is partly uncoupled from g_s, making it appear to be driven largely by R_n. This combination of conditions has sometimes led to the characterization of well-watered vegetation as a wick that passively conducts water from the soil to the atmosphere. However, this apparently passive behavior may conceal pronounced stomatal regulation of transpiration that leads to similar responses of transpiration to environmental drivers across different types of vegetation. The degree of decoupling between stomatal conductance and transpiration has been quantified with a dimensionless decoupling coefficient (Ω) ranging from zero to one (Jarvis and McNaughton, 1986). Stomatal control of transpiration diminishes as Ω approaches 1.0 because the vapor pressure at the leaf surface becomes increasingly decoupled from that in the bulk air. Typical values of Ω range from near 0.1 in needle-leaved coniferous trees with low stomatal and high boundary layer conductance to 0.5 or greater in broadleaf trees; they are higher in dense, herbaceous vegetation. Regardless of the degree of decoupling of transpiration from g_s when soil water is abundant, stomata increasingly limit transpiration as soil water deficits develop.

At the canopy level, transpiration is influenced by additional variables that include leaf area (often described in terms of leaf area index, or LAI, the ratio of leaf area to ground area), canopy structure and aerodynamic properties that determine canopy boundary layer properties. At this scale, controls on transpiration are typically represented by canopy conductance (G_c), a term that combines stomatal and boundary layer conductances. Variation in canopy conductance among forest types thus reflects both leaf level and higher order properties (Table 2.2), and these properties do not necessarily vary across ecosystem types in a consistent way. For example, the canopy conductance of Douglas fir (*Psuedotsuga menziesii*) is nearly double that of ponderosa pine (*Pinus ponderosa*). This is consistent with the environments they grow in – Douglas fir grows in temperate mesic regions, whereas ponderosa pine grows in much drier areas. However, maximum g_s of ponderosa pine is significantly greater than that of Douglas fir; the greater stomatal conductance of the pines is more than offset by lower leaf area of pine forests.

2.2.3 Liquid Water Transport through Trees and the Role of Hydraulic Architecture

Canopy conductance controls transpiration; however, canopy conductance is itself strongly influenced by the hydraulic architecture of trees and forests. Atmospheric conditions create a demand for water, and hydraulic architecture influences the supply of water from the soil. Ultimately, stomata regulate transpiration to ensure that losses do not exceed the supply capacity. In order to understand how vegetation controls transpiration, and to predict how alterations to vegetation will alter evapotranspiration, it is necessary to understand how hydraulic properties of trees influence their use of water.

According to the CT theory, the volume flow per unit time (Q) of liquid water through plants (the 'supply' for transpiration) is directly proportional to difference in water potential between leaves and soil ($\Delta\Psi$; or $\Psi_{leaf} - \Psi_{soil}$) and to whole-tree hydraulic conductance (K); it is therefore inversely proportional to whole-tree hydraulic resistance (R):

$$Q = \Delta\Psi * K \tag{2.1a}$$

$$Q = \Delta\Psi / R \tag{2.1b}$$

Application of Equation (2.1) can be misleading about causes and effects. Does transpiration control $\Delta\Psi$, or does $\Delta\Psi$ control transpiration? In fact, the causality works both ways. The driving force ($\Delta\Psi$) for liquid water movement from soil through the xylem is generated by the transpirational loss of water vapor, which lowers Ψ_{leaf} and transmits tension, or negative pressure, through continuous water columns running from the evaporative surfaces in the leaves to the soil (Tyree and Zimmermann, 2002). However, as will be demonstrated below, when Ψ_{leaf} drops to a critical level, partial stomatal closure occurs, limiting transpiration.

In the absence of transpiration, gravitational forces result in a minimum tension gradient of -0.01 MPa m^{-1} through the vertical dimension of trees. When transpiration occurs, frictional resistances make the vertical tension gradient considerably steeper (Tyree and Zimmermann, 2002). Following Equation (2.1), the magnitude of tension at a given point in the xylem depends upon the water potential of the soil from which the water has been taken up, the cumulative hydraulic resistance to that point, the flow rate, and the height above the ground (for the purposes of this illustration, gravitational forces can be included with R or K).

The development of substantial tension makes xylem water transport potentially vulnerable to disruption by cavitation and embolism (Tyree and Sperry, 1989). Cavitation is the separation of the water column within a xylem conduit (tracheid or vessel) forming a vapor-filled partial vacuum, whereas embolism results from the entry of air into a xylem conduit with or without prior cavitation. Both phenomena block water transport in the affected conduit. The vulnerability of xylem to loss of conductivity from cavitation and embolism is a key component of tree hydraulic architecture. Vulnerability curves relating percent loss of hydraulic conductivity to negative pressure in the xylem have been determined for numerous tree species. Within individual trees, xylem vulnerability typically decreases along a gradient from roots to trunk to terminal branches (Figure 2.3A), corresponding to gradients of increasing tension from roots to branch tips. Not surprisingly, xylem vulnerability varies widely among tree species growing under different ecological conditions (Figure 2.3B), and it is an important determinant of the limits to species distribution (embolism can occur as a result of freezing as well as water stress). Xylem embolism was formerly thought to be largely irreversible over the short term; however,

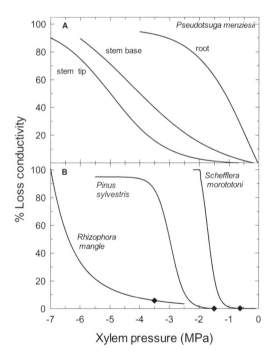

Figure 2.3 *Xylem vulnerability curves showing loss of hydraulic conductivity as a function of xylem pressure (tension). (A) Axial gradient of decreasing vulnerability from roots to terminal branches in a temperate conifer. Data from Domec and Gartner (2001) and Domec et al. (2004). (B) Examples of species showing highly vulnerable, moderately vulnerable, and highly resistant stem xylem. In all of the examples, stomata regulate minimum leaf water potential (diamonds) to prevent excessive loss of conductivity. Data from Cochard (1992), Melcher et al. (2001), and Meinzer et al. unpublished observations*

increasing evidence is emerging to show that it is rapidly reversible in some plant organs (Zwieniecki and Holbrook, 1998; Tyree *et al.*, 1999; Melcher *et al.*, 2001; Bucci *et al.*, 2003).

Plants can be aggregated into two groups based on the relationship between water potential and g_s. In *isohydric* species, which include most temperate forest trees, g_s is regulated to prevent the water potential of xylem from dropping to levels that would provoke excessive loss of conductivity as soil water deficits develop. Thus, isohydric species have a minimum midday water potential that remains more or less constant as soils dry (Figure 2.3B, diamonds). In *anisohydric* species there is no threshold minimum water potential, and transpiration is not as tightly regulated by stomatal closure. Isohydric species may tend to be more vulnerable to embolism and have greater capacity for embolism repair than do anisohydric species (Vogt, 2001). In Section 2.4.2 we discussed species-specific relationships between stomatal conductance, soil water deficits and atmospheric vapor pressure deficit. These environmental controls are usually presented as empirically derived characteristics of species, but in fact they are strongly associated with plant hydraulic architecture as they regulate the transpirational flux of water so that water

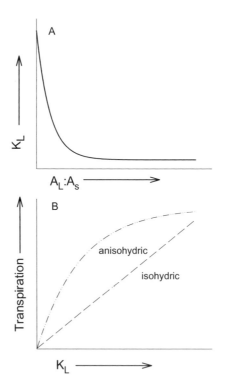

Figure 2.4 *(A) Typical relationship between leaf-specific conductivity (k_L) and the leaf area:sapwood area ratio ($A_L:A_S$), an index of transpirational demand in relation to water supply capacity. (B) Stomatal regulation causes transpiration to increase with k_L in a predictable manner in isohydric and anisohydric species (see text for details)*

potentials do not fall to a damaging level. The upper limit to transpiration from a community of isohydric plants is therefore strongly influenced by their vulnerability to cavitation as well as g_s and LAI.

The hydraulic resistance (or its inverse, conductance) of stems is determined in part by the permeability (k) of wood (many authors use *specific conductivity* for this property), which is primarily determined by the length and diameter of xylem cells. Leaf-specific conductivity (k_L) is k normalized by leaf area distal to the stem ($k\,A_L^{-1}$). As the ratio between leaf area and sapwood area ($A_L:A_s$) increases, (k_L) typically declines exponentially (Figure 2.4A). $A_L:A_s$ is a fundamental allometric trait that reflects the balance between transpirational demand (A_L) and water supply capacity (A_s). Both k_L and $A_L:A_s$ can be expressed at multiple scales from terminal branches to entire trees. Co-occurring tree species often share a common relationship between k_L and $A_L:A_s$ (Bucci *et al.*, 2005). Because k_L represents the balance between the demand for and efficiency of water supply, it constrains stomatal regulation of transpiration within limits that avoid catastrophic loss of xylem function from cavitation and embolism. Stomatal conductance and transpiration thus co-vary with k_L in a coordinated manner (Figure 2.4B). In isohydric species, transpiration exhibits a linear dependence on k_L. Transpiration increases asymptotically with k_L in anisohydric species, causing minimum leaf water potential to vary with k_L. As with the relationship between k_L and $A_L:A_s$, co-occurring tree species often share common relationships between k_L, g_s and transpiration (Meinzer *et al.*, 1995; Andrade *et al.*, 1998; Bucci *et al.*, 2005).

It is important to note that k_L and xylem vulnerability are not static properties within species or individual trees. In many trees, k_L decreases from the base of the stem to the apex, from larger to smaller diameter branches, and with increasing tree age and size (Tyree and Ewers, 1991; Ryan *et al.*, 2000; McDowell *et al.*, 2002a). In addition, seasonal variation in leaf area may partially conserve k_L during dry periods that cause reduced hydraulic conductivity due to cavitation (Bucci *et al.*, 2005). Xylem vulnerability to cavitation can vary dramatically among populations of the same species growing under different environmental conditions (Tognetti *et al.*, 1997; Kavanagh *et al.*, 1999; Sparks and Black, 1999; Melcher *et al.*, 2001), and even within the same growth ring, with latewood being more vulnerable than earlywood (Domec and Gartner, 2002).

Leaf level responses of g_s to the combination of architectural and environmental variables can be summarized in the following equation (Whitehead, 1998):

$$g_s = \frac{kA_s \Delta \Psi}{L \eta A_L D_s} \qquad (2.2)$$

k, A_s, A_L, and $\Delta\Psi$ are defined in Table 2.1; L is the length of the stem or the hydraulic path, η is the temperature-dependent viscosity of water, and D_s is the air saturation deficit at the leaf surface. The equation is typically applied to individual leaves or plants, but can also be applied on the stand level by substituting basal sapwood area for A_s, *LAI* for A_L, and stand-average metrics for the other variables. Although this equation is 'inexact' (it involves assumptions about steady-state processes that are not strictly true, and root resistances – see below – are difficult to incorporate), it yields many insights into the relationships between plant canopies, their environment and transpiration. Across a range of environments with different humidities, for example, g_s may be conserved through adjustment in A_L:A_s via change in allocation patterns. Such adjustments have been measured in Scots pine (*Pinus sylvestris*) (Mencuccini and Grace, 1995). Likewise, pruning part of a canopy decreases A_L:A_s, and often results in increased g_s in remaining foliage. Thus, in response to partial defoliation, leaf-level transpiration rates increased in loblolly pine (*Pinus taeda*), resulting in more or less constant rates of water flow through sapwood (Pataki *et al.*, 1998). This also helps to explain the good relationships often found between transpiration and the sapwood conducting area. In another application, McDowell *et al.* (2002b) found that tall Douglas fir trees have higher wood permeability (k) and lower A_L:A_s compared with smaller trees, partially compensating for the impact that increased L would otherwise have on g_s.

The sapwood of large trees may serve as a storage reservoir for water as well as a conduit. The hydraulic capacitance of sapwood and other plant tissues can be thought of as a component of hydraulic architecture in that it plays an important role in determining the dynamics of water movement through trees. Following the Ohm's law analogue for water movement along the SPAC, the capacitance (C) of a tissue is defined as:

$$C = dV / d\Psi \qquad (2.3)$$

where $dV/d\Psi$ is the volume of water released per change in water potential of the tissue. An increase in xylem tension will thus pull water from surrounding tissues into the transpiration stream. This release of stored water can cause pronounced lags between changes in transpiration in the tree's crown and changes in axial (vertical) water flow through stems (Goldstein *et al.*, 1998; Phillips *et al.*, 2003; Ford *et al.*, 2004; Meinzer *et al.*, 2004a). Trees typically exhibit diel (24-hour) cycles of capacitive discharge of stored water followed by complete recharge (or nearly so) during periods of reduced transpiration later in the day or overnight. Daily reliance on stored water as a percentage of total transpiration varies widely, ranging from about 10 to 50% (Waring *et al.*, 1979; Holbrook and Sinclair, 1992; Loustau *et al.*, 1996; Kobayashi and Tanaka, 2001; Maherali and DeLucia, 2001; Phillips *et al.*, 2003; Meinzer *et al.*, 2004a). There is evidence that relative reliance on stored water increases with tree size in some species (Phillips *et al.*, 2003), but not in others (Meinzer *et al.*, 2004a), and that trees use larger amounts of stored water in drought conditions (Phillips *et al.*, 2003; Ford *et al.*, 2004). In absolute terms, daily utilization of stored water ranges from about 20–50 kg in large, old-growth conifers (Phillips *et al.*, 2003) to 80–100 kg or more in large tropical trees (Meinzer *et al.*, 2004a). During seasonal drought, water withdrawn from storage in the sapwood of large coniferous trees may be sufficient to replace up to 27 mm of transpirational losses before seasonal recharge occurs (Waring and Running, 1978). The behavior of deuterated water (D_2O) injected into trees as a tracer of water movement suggests that maximum sap velocity and water residence time in the tree are strongly dependent on sapwood capacitance among both vessel- and tracheid-bearing trees independent of species. Tracer velocity decreased linearly and tracer residence time increased exponentially with increasing sapwood capacitance among 12 trees representing four tropical angiosperm species and two temperate coniferous species (James *et al.*, 2003; Meinzer *et al.*, 2003, 2006). Tracer velocities for the angiosperm trees were as high as 26 m per day, but generally less than 5 m per day in the conifers, implying that transit times for water taken up by roots to arrive in the upper crown would be at least three weeks in the tallest old-growth conifers. The tracer residence time was 79 days in a 1.43-m-diameter Douglas fir tree, the largest individual injected, and only 4 days in a 0.34-m-diameter tropical tree. These results are consistent with a prominent role for sapwood water storage in determining whole-tree water transport and storage dynamics.

2.2.4 Water Uptake by Roots

In woody plants, resistance to water flow in the root system can equal or even exceed resistance aboveground (Nardini and Tyree, 1999; Sperry *et al.*, 1998; Tyree *et al.*, 1998). Thus, the mechanisms and physical constraints regulating root water uptake are at least as important as aboveground constraints. Due to the difficulties of belowground investigations, however, water transport has not been studied as intensively in roots as in stems and leaves.

The ability of roots to supply water for plant transpiration depends on the hydraulic conductance of the root system (determined by fine-root conductivity and total fine-root surface area), the distribution of roots within the soil profile along with the ability to produce new roots dynamically as soil water is used and replenished, and soil water availability throughout the rooting zone. Hyphae of mycorrhizal fungi can greatly increase the effective hydraulic conductance of roots and therefore can exert considerable influence over transpiration (Hobbie and Colpaert, 2004). The hydraulic conductivity of soil in the rhizosphere is also critically important to root water uptake.

Root conductivity varies because water flows into roots through multiple pathways that are influenced by both osmotic and hydraulic drivers (Steudle, 1994, 2001). Roots generally have very high axial ('lengthwise') conductivity; thus, overall conductance of root systems is generally limited by radial ('crosswise') conductivity as water enters the root from the soil. The low radial conductivity is largely due to a special feature of root anatomy that forces most of the water to cross cell membranes. By forcing water to cross a cell membrane, plants are able 'sieve out' undesirable chemical compounds and favor others, but the flow of water is impeded considerably.

The surface area and demography of fine roots greatly influences the ability of the root system to conduct water. Water uptake primarily occurs in young, unsuberized roots (*suberization* is the development of a waxy, protective layer around roots, and is usually associated with a change in color). The radial hydraulic conductivity of these young roots is 10- to 100-times higher than in older roots. Fine roots continuously emerge, age and die through the favorable growing season and, as with leaves above ground, their physiological characteristics change with age, although at present these developmental changes are poorly characterized (Wells and Eissenstadt, 2003).

The surface area of fine roots is an important parameter in models of plant water transport, particularly in connection with the transpiring surface area of leaves. Increasing root surface area per unit leaf area ($A_r:A_l$) allows water uptake from more soil per transpiring leaf. $A_r:A_l$ is generally greater than 1 and can vary dramatically depending on xylem vulnerability and soil texture properties (Sperry *et al.*, 1998). For example, loblolly pine (*Pinus taeda*) growing in sandy soil had an $A_r:A_l$ ratio of 9.75 compared with 1.68 for the same species in a loam soil (Hacke *et al.*, 2000), whereas five different species of oak (*Quercus* spp.) growing under similar conditions with adequate water but differing in drought tolerance did not differ as much in their $A_r:A_l$ (ranging from 1.45 to 2.37) (Nardini and Tyree, 1999). In coarse soils, soil hydraulic conductance drops more rapidly with decreases in Ψ_{soil} than in finely textured soils (Bristow et al., 1984). Increasing root surface area can help alleviate this conductance limitation in coarse soils.

The axial conductivity of roots can be as much as 40-times greater than that of stems due to the greater diameter of water transport cells, and deeper roots are more conductive than shallow roots, providing a continuum of decreasing conductivity along the xylem pathway (Kavanagh et al., 1999; McElrone et al., 2004). However, roots are also more vulnerable to cavitation than are shoots (Doussan et al., 1998; Kavanagh et al., 1999; Sperry and Ikeda, 1997). Sperry et al. (1998) speculated that surface roots may act as a kind of hydraulic 'fuse'; root xylem failure that is localized in the dry upper soil may allow deeper roots in wetter soil to continue transporting water. Seasonal loss of root conductivity in upper soil has been correlated with decreased stomatal conductance even though deep roots had access to water (Domec et al., 2004). Similarly, the rate of water depletion throughout the soil profile is correlated with the Ψ_{soil} in the upper 20 cm (Warren et al., 2005). These results imply that seasonal declines in Ψ_{soil} and root conductivity in the upper soil may generate signals that induce stomatal closure and limit water uptake even though most water is coming from deeper parts of the soil profile.

Rooting depth and the distribution of roots through the soil profile also has a significant impact on access to water throughout the growing season. Soil resources (nutrients and water) are not evenly distributed within the soil profile and are dynamic seasonally. Generally, the majority of roots are concentrated in the upper soil where nutrient concentrations are high (Jackson et al., 1996; Warren et al., 2005). As a result, this portion of the soil is the first to dry out during periods without rain both from direct evaporation and root water uptake (Brooks et al., 2006; Warren et al., 2005) leaving most of the roots in the driest portion of the soil. However, roots are also located much deeper in the soil and sometimes into rock layers (McElrone et al., 2004; Rose et al., 2003; Zwieniecki and Newton, 1996), especially in areas with low soil moisture. Thus water uptake rates per unit root surface area shift seasonally down the profile. In the early growing season when water is plentiful, the majority of uptake comes from the upper roots, shifting later in the season to the relatively fewer deeper roots (Brooks et al., 2006; Warren et al., 2005; Hacke et al., 2000).

When roots are in contact with soils that vary spatially in moisture content, they may act as conduits for water redistribution through the soil, driven by gradients in soil water potential. The process of water transport from deep to shallow soil layers through roots, termed *hydraulic lift*, has been demonstrated in a large number plant species, including grasses and cacti as well as shrubs and trees (Caldwell and Richards, 1989; Caldwell, 1990; Caldwell et al., 1998; Dawson, 1993; Moreira et al., 2003; Yoder and Nowak, 1999). It occurs horizontally as well as vertically (Brooks et al., 2002; Burgess et al., 1998; Schulze et al., 1998), so the more general term 'hydraulic redistribution' (HR) is now preferred. Downward HR can enhance the rate of recharge of deeper soil layers following rainfall events that are not sufficient to saturate the rooting zone (Burgess et al., 2001; Ryel et al., 2003). HR is most common among deeply rooted species, but can occur even in shallow-rooted species when soil moisture conditions are conducive. Broom snakeweed (*Gutierrezia sorothrae*) rooted to a depth of only 60 cm hydraulically lifts about 15 % of the water it transpires (Wan et al., 1993; Richards and Caldwell, 1987).

The process generally occurs only at night or during periods of heavy cloud cover (Caldwell and Richards, 1989) when stomata are closed, and when there are strong gradients in soil moisture content. Roots then provide a low-resistance pathway for water flow from areas of high to low soil moisture. When the water potential in the shallow roots reaches a certain threshold above the water potential in the surrounding soil (this threshold varies among species), water begins to exude from the roots and into the soil (Baker and van Bavel, 1986; Richards and Caldwell, 1987; Caldwell, 1990; Dawson, 1993). Meinzer et al. (2004b) found that when upper soil layers dry to about −0.4 MPa, they become an effective sink for water from deeper layers.

The amount of water moved by HR is relatively small – less than 0.5 mm m^{-1} soil depth day^{-1} – (Brooks et al., 2006), but may have a significant impact on the rate of soil drying since water uptake from those layers has also slowed considerably at soil water potentials below −0.4 MPa. HR can replace 40–80 % of the daily water used from those upper layers (Brooks et al., 2002; Brooks et al., 2006, Domec et al., 2004; Meinzer et al., 2004b). Brooks et al. (2002) found that HR delayed soils in coniferous forests from reaching water potentials equal to the minimum midday leaf water potential (point at which no water can be obtained from that soil layer) by an additional 16–21 days depending on the system. This delay in soil drying can be critical in decreasing root cavitation and preserving root function in these upper soils (Domec et al., 2004).

In summary, root systems appear to be highly responsive to soil water availability and soil properties such as texture. Rooting depth, total fine root surface area and specific fine root conductivity are dependent on species and site conditions. In addition, root cavitation and hydraulic redistribution also play important roles in regulating root water uptake and plant transpiration.

2.3 Evapotranspiration in Forest Ecosystems

2.3.1 Evaporation and Transpiration

Micrometeorologists and hydrologists often combine evaporation and transpiration into one measurement for a watershed, largely because the methods used to determine evapotranspiration (ET) cannot distinguish between the two fluxes, yet the two processes are quite different. Recently, stable isotopic techniques have become available for helping to separate these two fluxes (Moreira et al., 1997; Wang and Yakir, 2000; Williams et al., 2004; Yepez et al., 2003) because water transpired from leaves is more enriched isotopically than is water evaporated from soil (Yakir et al., 1993). By measuring the isotopic signature of water vapor from leaves and the soil, and measuring the atmospheric water vapor signature over time, it is possible to separate these fluxes using a mixing model approach explained in Moreira et al. (1997). Trees have the potential to greatly increase evaporative losses from an ecosystem because of the increase in evaporative surface and the greater access to soil water through roots. Evaporation from soils is generally restricted to the upper few centimeters; thus, in forests, transpiration generally accounts for most of ET. For example, Moreira et al. (1997) found that in the Amazon forest, transpiration was responsible for nearly all of the loss in water vapor. Wang and Yakir (2000) found that soil evaporation was only 1.5–3.5 % of the evapotranspiration flux from crops in a desert environment. Williams et al. (2004) found that soil evaporation changed from 0 % in an olive orchard prior to irrigation, to 14–31 % for the 5 days following irrigation. Thus, even with wet soils in a system with relatively low canopy cover, transpiration far exceeds soil evaporation.

2.3.2 Transpiration from the Understory

Transpiration can be further divided between understory and overstory components, which can experience very different environmental microclimates. The understory is a relatively sheltered environment with lower radiation and higher relative humidity than the overstory (Blanken and Black, 2004; Scott *et al.*, 2003; Unsworth *et al.*, 2004; Yepez *et al.*, 2003). As a result, transpiration of the understory is generally less than that of the overstory. In a mesic coniferous forest with an LAI of 9.6, understory transpiration was approximately one tenth of the ecosystem vapor flux (Unsworth *et al.*, 2004). However, in a semiarid woodland with LAI of 1.6, understory transpiration was closer to one third to one half of the ecosystem flux during wet periods (Scott *et al.*, 2003). Similarly, in larch (*Larix gmelinii*) and pine (*Pinus sylvestris*) forests in Siberia where 40% of the radiation reaches the understory, understory transpiration can amount to 25–50% of the ecosystem vapor flux (Hamada *et al.*, 2004).

Seasonal variability of understory transpiration is dependent on the seasonal variability of surface soil moisture, R_n and D. In a semiarid woodland, understory transpiration was more variable than overstory transpiration over time because the understory plants had shallow roots in soil layers with highly variable moisture availability, whereas the overstory had deep roots with access to more consistent and reliable water (Scott *et al.*, 2003). However, in more mesic coniferous forests, understory transpiration may be less variable over time as understory radiation and surface soil moisture are less variable over time (Unsworth *et al.*, 2004). A deciduous overstory will also cause more variability in the understory environment that could influence understory transpiration if understory leaves are present when overstory leaves are not, especially in tropical deciduous forests, which have large seasonal variation in rainfall but little variation in temperature.

2.4 Applying Concepts: Changes in Hydrologic Processes through the Life Cycle of Forests

As an example of an application of concepts presented in the preceding sections, we now explore some of the ways that changes in the structure and function of forests through developmental stages impact hydrologic processes. In many parts of the world, one of the most dramatic impacts of forest land use is the alteration of forest age–class structures. The structure and function of forests undergo significant changes through the entire life cycle (Franklin *et al.*, 2002; Bond and Franklin, 2002), and these changes impact evapotranspiration (Harr, 1982; Hicks *et al.*, 1991; Keppeler and Ziemer, 1990; Zimmerman *et al.*, 2000; Law *et al.*, 2001; Moore *et al.*, 2004), fog and rainfall interception and losses (Pypker *et al.*, in press; Zinke, 1967), and streamflow (Harr *et al.*, 1975; Hicks *et al.*, 1991; Jones and Grant, 1996; Thomas and Megahan, 1998). The dramatic impacts of forest harvest and early regeneration on hydrology have been well documented (e.g., Hewlett and Hibbert, 1961; Swank *et al.*, 2001; Jones and Post, 2004). Less well recognized are slow but profound changes that may occur as the composition, structure and function of the new forest continue to develop.

We focus on coniferous forests of the western USA. The details of developmental stages are different in other forest types, but most undergo changes in species composi-

2.4.1 A Summary of Age-related Changes in Forest Composition, Structure, and Function

Many of the compositional, structural and functional changes that occur through forest development can strongly influence hydrology. After a major disturbance, coniferous forests of the Pacific Northwest regenerate quickly, but their density and early growth varies depending on the type of disturbance and legacies from the previous forest (Franklin *et al.*, 2002). Often, fast-growing grasses and broadleaf shrubs and trees, which typically have much higher maximum g_s compared with later successional species, are established in dense stands. Even though the LAI of this pioneer vegetation may be lower than that of the previous forest, vegetation water use may be as high or higher. However, the decoupling coefficient (Ω; Section 2.2.2) between transpiration and D in the pioneer vegetation is likely to be greater than in later developmental stages because of the low stature and comparatively smooth canopy structure.

If conifer regeneration is abundant, the coniferous crown closes and LAI reaches a maximum within a couple of decades – this is less than 5% of the natural life cycle of the forest. In humid regions, the crown at this point is usually dense, and there is little understory vegetation. Subsequently, as trees grow taller, the crown is 'pushed up' in space, but total leaf area does not change much. Trees start to compete with each other: some die, others grow much larger, and small, shade-tolerant, trees (mostly conifers) begin to populate gaps. A thick organic layer forms on the soil surface, which minimizes soil evaporation but also intercepts precipitation. Roots and mycorrhizal hyphae of shade-tolerant trees explore the organic layer and even large woody debris, which can provide a considerable water reservoir that is not utilized by the initial cohort of non-shade-tolerant trees. As the trees grow larger, their biomass also stores increasing amounts of water (Section 2.2.3). This stored water can be important in maintaining transpiration in large trees during seasonal drought, although it is a small component of total site water storage, even in mature forests.

In most forest types, the canopies of older forests become structurally complex both vertically and horizontally, increasing in aerodynamic conductance and interception relative to dense, young forests. In most humid regions, older forests harbor an abundance of bryophytes and lichens that strongly influence the interception of precipitation, serve as a 'spongy' storage of water and also moderate the canopy microclimate. Root systems also become more structurally complex over time, but generalizations are difficult due to limited data.

2.4.2 Impacts of Tree Size on Stomatal Conductance and Whole-tree Water Use

Large trees extract a huge quantity of water from the soil (more than 100 kg day^{-1} for a large conifer) and transport it against gravity and through tiny xylem conduits to foliage that is 50–100 meters above the ground. Increasing height results in a longer hydraulic path length; note that Equation (2.2) predicts an inverse relationship between g_s and L, and therefore between transpiration and L, if all other factors remain constant. The complex, 'tortuous' branching pattern of older trees makes the transport even more difficult, because xylem conductivity (k_s) is especially low at branch junctures. All of these changes tend to reduce hydraulic conductance (K, Section 2.2.3) as trees grow larger (Ryan *et al*., 2000; Phillips *et al*., 2002), although the potential impacts are compensated to some extent by other adjustments (e.g., $A_L:A_S$, Section 2.2.4; McDowell *et al*., 2002a).

As predicted from Equation (2.2), age-related changes in tree hydraulic architecture often affect stomatal conductance at the leaf level. Many studies have shown that g_s is reduced in older trees, although this is not found universally (Bond, 2000). The change in stomatal conductance is indicated most clearly by a change in the composition of carbon isotopes ($\delta^{13}C$) in foliage over a range of tree sizes (Figure 2.5). Although there are several possible explanations for the isotopic change, in most cases the best explanation is a decrease in g_s in older trees (Yoder *et al*., 1994; Koch *et al*., 2004). Direct measurements of water fluxes in conifer stems also show that transpiration per unit leaf area is reduced in older trees, especially in conditions of relatively high water availability in soil and low to moderate D (Ryan *et al*., 2000; Phillips *et al*., 2002; Irvine *et al*., 2004). As a result, transpiration increases exponentially with tree size during the early stages of tree growth, but later reaches an asymptotic maximum (Meinzer *et al*., 2005). However, there are exceptions to these general trends (see, for example, Barnard and Ryan, 2003).

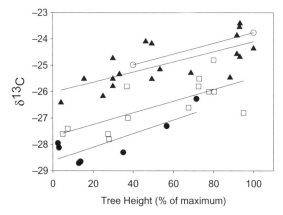

Figure 2.5 Carbon isotope composition of foliage from the tops of nonsuppressed trees of different heights. Each symbol represents a different species: closed circles – Pinus sylvestris (unpublished data courtesy of Maurizio Mencuccini); open squares – Pinus ponderosa (Yoder et al., 1994); closed triangles – Pseudotsuga menziesii, each point is a mean of five samples (unpublished data collected by Bond et al. Wind River, WA and McDowell et al., 2002b); open circles – Quercus oregana, each point is a mean of seven samples (unpublished data collected by McDowell et al., Corvallis OR). Heights are expressed relative to the maximum for the species at the site sampled (heights of Scots pine were estimated from tree ages). Slopes of the relationships between $\delta^{13}C$ and percent maximum height are 0.22, 0.19, 0.257 and 0.20 for P. sylvestris, P. ponderosa, P. menziesii and Q. oregana, respectively

In certain situations (e.g., when resistance of the canopy boundary layer is very low relative to leaf diffusive resistance), it is possible to estimate differences in stomatal conductance using measurements of stable isotope composition of leaf carbon and photosynthetic capacity of foliage from the sunlit tops of tree crowns. The estimation relies on an explicit linear relationship between isotope discrimination and the ratio of carbon dioxide concentrations internal and external to leaves (Farquhar *et al.*, 1982). These carbon dioxide concentrations, in turn, are a function of the ratio between photosynthesis and stomatal conductance, or A/g, also known as 'intrinsic water use efficiency' (Ehleringer *et al.*, 1993). Finally, relative differences in stomatal conductance can be derived using measurements of the response function of photosynthesis to internal carbon dioxide, or 'A/C_i' curves. For the Douglas fir in Figure 2.5, the change in foliage isotope composition indicates a reduction in average g_s of a little over 2% for each 10-percentile change in height, or 20% over the height range of the forest, which in turn suggests a similar reduction in transpiration per unit leaf area as long as the microclimate is similar across the forest age classes.

2.4.3 Age-related Change in Transpiration, Interception and Water Storage on the Forest Stand Level

In general, LAI tends to decrease somewhat in aging forests after they achieve maximum LAI (Ryan *et al.*, 1997). Given that leaf-level transpiration also decreases with age (above), it might be hypothesized that transpiration on the stand level might also decline with age. Alternatively, changes in stand density, species composition or understory characteristics could counterbalance the age-related shifts in LAI and leaf-level transpiration, preserving relative constancy in plant water use.

One problem with this analysis is the assumption that transpiration is closely related to LAI. Although LAI is usually a good predictor of change in transpiration in initial stages of forest development as vegetation cover increases following a disturbance, transpiration does not correspond well with changes in LAI in later developmental stages (Zimmermann *et al.*, 2000). Changes in sapwood basal area, on the other hand, appear to explain much of the variation in transpiration in stands of different ages (Dunn and Connor, 1993; Zimmermann *et al.*, 2000; Moore *et al.*, 2004), consistent with the notion that transpiration is strongly influenced by the supply capacity of the hydraulic system (K; Section 2.2.3). The presence of very large trees in old forests can give a deceptive sense of their water conducting capacity. Moore *et al.* (2004) found that although the total basal area in an old-growth Douglas-fir forest was more than twice that of a 45-year-old forest (about 85 vs 35 m^2 ha^{-1}, respectively), the old forest had lower sapwood basal area (17 vs nearly 22 m^2 ha^{-1} for the old and young forests, respectively). In this study, age-related changes in species composition, tree height, and sapwood basal area all limited seasonal transpiration in the old-growth forest compared with that of the younger stand. Due to their additive influence, the young forest used nearly three times as much water over a growing season compared with the old forest.

In contrast to the findings of Moore *et al.* (2004), a replicated study of three old (about 450 years) and three young (about 25 years) Douglas-fir stands in western Washington, USA, revealed no significant difference in stand level transpiration between the age classes during the growing season (Bond *et al.*, unpublished). However, the young stands had recently been heavily infected by a foliar pathogen (*Phaeocryptopus gaeumannii*) that is known to reduce transpiration (Manter *et al.*, 2003), so the results may say more about the effects of the pathogen than stand age on transpiration. Published data for four conifer species show an average decrease in maximum daily transpiration of about 4% for each ten-percentile increment in forest height (Figure 2.6), although there is a great deal of variability in the data. However, age-related differences in maximum daily transpiration do not necessarily translate into lower total water use on a yearly or seasonal basis. In a water-limited ponderosa pine ecosystem, young forests exhausted soil water reserves earlier in the growing season than did old forests. The young forests endured a longer period of drought than did older forests, and over the growing season water use was similar between the age classes (Irvine *et al.*, 2004).

Interception water losses were the chief cause of reduced water yield following afforestation of pasture or farmed land with eucalypts (Whitehead and Beadle, 2004), and increased interception also affects site water balance in regenerating forests after harvest. As with developmental changes in transpiration, leaf area is an important determinant of change in interception in the early stages of stand development, but it is less important in subsequent stages. A large epiphytic community often develops in older forests, and the mosses and bryophytes are able to intercept and store very large amounts of water; epiphytes double the water storage capacity in canopies of old-growth Douglas-fir forests relative to young forests with similar LAI (Pypker *et al.*, 2005). Pypker *et al.* (2005) found that differences in interception were small during moderate to heavy rain events due to other changes in canopy structure, but interception losses were much greater from old forests during intermittent and small rain events, which can occur frequently in the Pacific

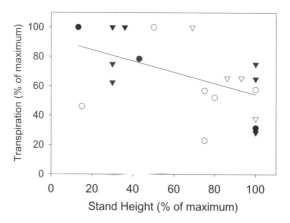

Figure 2.6 Response of maximum stand-level transpiration to increasing stand height. Each symbol represents a different species: closed circles – Pinus ponderosa (Irvine et al., 2004); open circles – Pinus sylvestris (Zimmermann et al., 2000); closed triangles – Pseudotsuga menziesii (Phillips et al., 2002); open triangles – Picea abies (Köstner et al., 2002). Height is expressed as a proportion of estimated maximum for the species and site; transpiration is daily maximum, expressed as a proportion of the maximum in each dataset. The overall r^2 is 0.24

Northwest where this forest type grows. Thus, both interception losses from the canopy as well as a heavier litter layer in the older forests result in less rainfall penetration to the mineral soil of older forests during small rain events (Pypker *et al.*, unpublished data).

In some areas interception and condensation of fog by mature conifer canopies can actually increase the amount of moisture reaching the soil (Dawson, 1998), and extensive harvests in such forests may result in increased precipitation 'downwind' from the harvest (Harr, 1982).

2.4.4 Impacts of Change in Species Composition on Transpiration in Aging Forests

Most forests undergo significant changes in species composition over their normal lifespan. In some cases the changes are dramatic, and often characterized as successional patterns; in others the changes are more subtle. In the conifer-dominated Pacific Northwest of the United States, broadleaf trees and shrubs are often heavy water consumers in young stands. In the case study by Moore *et al.* (2004) outlined above, the sapwood basal area of broadleaf species in a young (45-year-old) Douglas-fir forest was six times greater than in an old-growth forest, and seasonal water use normalized by sapwood basal area was 40% greater for the broadleaf species than for conifer species of the same age. In old-growth forests, late-successional, shade-tolerant conifers with high leaf areas and low transpiration rates constituted most of the sapwood basal area. These shade-tolerant conifers averaged 30% lower water use, again normalized by sapwood area, than the pioneer conifers. In combination, the reduction in broadleaf vegetation and increase in shade tolerant conifers, in addition to lower overall sapwood basal area, resulted in a seasonal water use by the old forest that was only about one-third that of the young forest. In agreement with the preceding patterns, Meinzer *et al.* (2005) found that daily transpiration of individual trees was consistently greater in broadleaf than in coniferous species at a given tree size.

2.4.5 Implications for Predictive Models

The Penman–Monteith equation (Monteith, 1965) is widely accepted as the definitive mechanistic description of relationships between vegetation properties and environmental drivers that influence transpiration, and is arguably one of the most important contributions of its time to vegetation science. However, the difficulty involved in accurate estimation of the conductance terms (G_c and boundary layer) in the Penman–Monteith equation is often overlooked, even though estimates of transpiration for forests are sensitive to these terms due to the strong coupling between stomatal conductance and transpiration (Section 2.2.2). Both canopy stomatal and boundary layer conductance change over stand development, and neither is easily measured or predictable as a function of LAI (although this is commonly done).

However, our current understanding of plant physiological processes suggests using more robust approaches. One approach is to estimate canopy conductance using mechanistic models of environment and plant hydraulic architecture, as described in this chapter. This has been done successfully in at least one model, SPA (Soil–Plant–Atmosphere; Williams *et al.*, 1996). Stable carbon isotopes may offer another, more empirical, approach.

The consistent change in $\delta^{13}C$ in relation to relative maximum tree height (Figure 2.5) suggests that this parameter could be a useful predictor of change in stomatal conductance through stand development. Together with measurements of change in LAI, $\delta^{13}C$ could provide a semi-empirical approach for estimating change in G_c through forest development.

Acknowledgments

This manuscript was greatly improved thanks to careful reading and advice from Georgianne Moore and Chelcy Ford. We also gratefully acknowledge helpful editorial comments from two anonymous reviewers as well as the editor of this volume. The work reported in this chapter was partially supported by the Western Regional Center (WESTGEC) of the National Institute for Global Environmental Change (NIGEC) under cooperative agreement No. DE-FC03-90ER61010, the Department of Forest Science, Oregon State University, US Environmental Protection Agency, and the USDA Forest Service Ecosystem Processes Program. This manuscript has been subjected to the Environmental Protection Agency's peer and administrative review, and it has been approved for publication as an EPA document. Mention of trade names or commercial products does not constitute endorsement or recommendation for use. Opinions, findings and conclusions are those of the authors and do not necessarily reflect the views of the DOE or the EPA.

References

Andrade JL, Meinzer FC, Goldstein G, Holbrook NM, Cavelier J, Jackson P, Silvera K. 1998. Regulation of water flux through trunks, branches and leaves in trees of a lowland tropical forest. *Oecologia*, **115**, 463–471.

Angeles G, Bond B, Boyer B, Brodribb T, Brooks JR, Burns MJ, Cavender-Bares J, Clearwater M, Cochard H, Comstock J, Davis S, Domec J-C, Donovan L, Ewers F, Gartner B, Hacke U, Hinckley T, Holbrook NM, Jones HG, Kavanagh K, Law B, Lopez-Portillo J, Lovisolo C, Martin T, Martinez-Vilalta J, Mayr S, Meinzer FC, Melcher P, Mencuccini M, Mulkey S, Nardini A, Neufeld HS, Passioura J, Pockman WT, Pratt RB, Rambal S, Richter H, Sack L, Salleo S, Schubert A, Schulte P, Sparks JP, Sperry J, Teskey R and Tyree M. 2004. The Cohesion-Tension Theory. *New Phytologist* (Letters), **163**, 451–452.

Anthoni PM, Unsworth MH, Law BE, Irvine J, Baldocchi D, Moore DJ. 2002. Seasonal differences in carbon and water vapor exchange in young and old-growth ponderosa pine ecosystems. *Agricultural and Forest Meteorology*, **111**, 203–222.

Baker JM, van Bavel CHM. 1986. Resistance of plant roots to water loss. *Agronomy Journal*, **78**, 641–644.

Barnard HR, Ryan MG. 2003. A test of the hydraulic limitation hypothesis in fast-growing *Eucalyptus saligna*. *Plant, Cell and Environment*, **26**, 1235–1245.

Benyon RG. 1999. Nighttime water use in an irrigated *Eucalyptus grandis* plantation. *Tree Physiology*, **19**, 853–859.

Blanken PD, Black TA. 2004. The canopy conductance of a boreal aspen forest, Prince Albert National Park, Canada. *Hydrological Processes*, **18**, 1561–1578.

Blanken PD, Black TA, Yang PC, Neumann HHNZ, Staebler R, den Hartog G, Novak MD, Lee X. 1997. Energy balance and canopy conductance of a boreal aspen forest: partitioning overstory and understory components. *Journal of Geophysical Research*, **102**, 28915–28927.

Bond BJ. 2000. Age-related changes in photosynthesis of woody plants. *Trends in Plant Science*, **5**, 349–353.
Bond BJ. 2003. Hydrology and ecology meet – and the meeting is good. *Hydrological Processes*, **17**, 2087–2089.
Bond BJ, Franklin JF. 2002. Aging in Pacific Northwest forests: A selection of recent research. *Tree Physiology*, **22**, 73–76.
Bond BJ, Kavanagh KL. 1999. Stomatal behavior of four woody species in relation to leaf-specific hydraulic conductance and threshold water potential. *Tree Physiology*, **19**, 503–510.
Bristow KL, Campbell GS, Calissendorff C. 1984. The effects of texture on the resistance to water movement within the rhizosphere. *Soil Science Society of America Journal*, **42**, 657–659.
Brooks JR, Meinzer FC, Coulombe R, Gregg JW. 2002. Hydraulic redistribution of soil water during summer drought in two contrasting Pacific Northwest coniferous forests. *Tree Physiology*, **22**, 1107–1117.
Brooks JR, Meinzer FC, Warren JM, Domec JC, Coulombe R. 2006. Hydraulic redistribution in a Douglas-fir forest: lessons from system manipulations. *Plant, Cell and Environment*, **29**, 138–150.
Bucci SJ, Scholz FG, Goldstein G, Meinzer FC, Sternberg LDSL. 2003. Dynamic changes in hydraulic conductivity in petioles of two savanna tree species: factors and mechanisms contributing to the refilling of embolized vessels. *Plant, Cell and Environment*, **26**, 1633–1645.
Bucci SJ, Scholz FG, Goldstein G, Meinzer FC, Hinojosa JA, Hoffmann WA, Franco AC. 2004. Processes preventing nocturnal equilibration between leaf and soil water potential in tropical savanna woody species. *Tree Physiology*, **24**, 1119–1127.
Bucci SJ, Goldstein G, Meinzer FC, Franco AC, Campanello P, Scholz FG. 2005. Mechanisms contributing to seasonal homeostasis of minimum leaf water potential and predawn disequilibrium between soil and plant water potential in Neotropical savanna trees. *Trees*, **19**, 296–304.
Burgess SSO, Adams MA, Turner NC, Ong CK. 1998. The redistribution of soil water by tree root systems. *Oecologia*, **115**, 306–311.
Burgess SSO, Adams MA, Turner NC, White DA, Ong CK. 2001. Tree roots: conduits for deep recharge of soil water. *Oecologia*, **126**, 158–165.
Caldwell MM 1990. Water parasitism stemming from hydraulic lift: a quantitative test in the field. *Israel Journal of Botany*, **39**, 395–402.
Caldwell MM, Richards JH. 1989. Hydraulic lift: water efflux from upper roots improves effectiveness of water uptake by deep roots. *Oecologia*, **79**, 1–5.
Caldwell MM, Dawson TE, Richards JH. 1998. Hydraulic lift: consequences of water efflux from the roots of plants. *Oecologia*, **113**, 151–161.
Canny MJ. 1995. A new theory for the ascent of sap – cohesion supported by tissue pressure. *Annals of Botany*, **75**, 343–357.
Canny MJ. 1998. Transporting water in plants. *American Scientist*, **86**, 152–159.
Cochard H. 1992. Vulnerability of several conifers to air embolism. *Tree Physiology*, **11**, 73–83.
Davies WJ, Zhang J. 1991. Root signals and the regulation of growth and development of plants in drying soil. *Annual Review of Plant Physiology and Molecular Biology*, **42**, 55–76.
Dawson TE. 1993. Hydraulic lift and water use by plants: implications for water balance, performance and plant-plant interactions. *Oecologia*, **95**, 565–574.
Dawson TE. 1998. Fog in the California redwood forest: ecosystem inputs and use by plants. *Oecologia*, **117**, 476–485.
Dixon HH, Joly J. 1894. On the ascent of sap. *Philosophical Transactions of the Royal Society of London B*, **186**, 563–576.
Domec J-C, Gartner BL. 2001. Cavitation and water storage capacity in bole xylem segments of mature and young Douglas-fir trees. *Trees*, **15**, 204–214.
Domec J-C, Gartner BL. 2002. How do water transport and water storage differ in coniferous earlywood and latewood? *Journal of Experimental Botany*, **53**, 2369–2379.
Domec J-C, Warren JM, Meinzer FC, Brooks JR, Coulombe R. 2004. Native root xylem embolism and stomatal closure in stands of Douglas-fir and ponderosa pine: mitigation by hydraulic redistribution. *Oecologia*, **141**, 7–16.

Donovan LA, Grise DJ, West JB, Papport RA, Alder NN, Richards JH. 1999. Predawn disequilibrium between plant and soil water potentials in two cold-desert shrubs. *Oecologia*, **120**, 209–217.

Doussan C, Vercambre G, Pagès L. 1998. Modelling of the hydraulic architecture of root systems: an integrated approach to water absorption–distribution of axial and radial conductances in maize. *Annals of Botany*, **81**, 225–232.

Dunn GM, Connor DJ. 1993. An analysis of sap flow in mountain ash (*Eucalyptus regnans*) forests of different age. *Tree Physiology*, **13**, 321–336.

Ehleringer JR, Hall AE, Farquhar GD. 1993. *Stable isotopes and plant carbon-water relations*. Academic Press: San Diego.

Farquhar GD. 1978. Feedforward responses of stomata to humidity. *Australian Journal of Plant Physiology*, **5**, 787–800.

Farquhar GD, O'Leary MH, Berry JA. 1982. On the relationship between carbon isotope discrimination and the intercellular carbon dioxide concentration in leaves. *Australian Journal of Plant Physiology*, **9**, 121–137.

Feild TS, Holbrook NM. 2000. Xylem sap flow and stem hydraulics of the vesselless angiosperm *Drymis granadensis* (Winteraceae) in a Costa Rican elfin forest. *Plant, Cell and Environment*, **23**, 1067–1077.

Ford CR, Goranson CE, Mitchell RJ, Will RE, Teskey RO. 2004. Diurnal and seasonal variability in the radial distribution of sap flow: predicting total stem flow in *Pinus taeda* trees. *Tree Physiology*, **24**, 951–960.

Franklin JF, Spies TA, Van Pelt R, Carey AB, Thornburgh DA, Berg DR, Lindenmayer DB, Harmon ME, Keeton WS, Shaw DC, Bible K, Chen, J. 2002. Disturbances and structural development of natural forest ecosystems with silvicultural implications, using Douglas-fir forests as an example. *Forest Ecology and Management*, **155**, 399–423.

Franks PJ, Cowan IR, Farquhar GD. 1997. The apparent feedforward response of stomata to air vapour pressure deficit: information revealed by different experimental procedures with two rainforest trees. *Plant, Cell and Environment*, **20**, 142–145.

Goldstein G, Andrade JL, Meinzer FC, Holbrook NM, Cavelier J, Jackson P, Celis A. 1998. Stem water storage and diurnal patterns of water use in tropical forest canopy trees. *Plant, Cell and Environment*, **21**, 397–406.

Gollan T, Passioura JB, Munns R. 1986. Soil water status affects stomatal conductance of fully turgid wheat and sunflower leaves. *Australian Journal of Plant Physiology*, **13**, 459–464.

Granier A, Huc R, Colin F. 1992. Transpiration and stomatal conductance of two rain forest species growing in plantations (*Simarouba amara* and *Goupia glabra*) in French Guyana. *Annals of Forest Science*, **49**, 17–24.

Green SR, McNaughton KG, Clothier BE. 1989. Observations of night-time water use in kiwifruit vines and apple trees. *Agricultural and Forest Meteorology*, **48**, 251–261.

Hacke UG, Sperry GS, Ewers BE, Ellsworth DS, Schafer KVR, Oren R. 2000. Influence of soil porosity on water use in *Pinus taeda*. *Oecologia*, **124**, 495–505.

Hamada S, Ohta T, Hiyama T, Kuwada T, Takahashi A, Maximov TC. 2004. Hydrometeorological behaviour of pine and larch forests in eastern Siberia. *Hydrological Processes*, **18**, 23–39.

Harr RD. 1982. Fog drip in the Bull Run municipal watershed, Oregon. *Water Resources Research*, **18**, 785–789.

Harr RD, Harper, WC, Krygier JT. 1975. Changes in storm hydrographs after road building and clear-cutting in the Oregon coast range. *Water Resources Research*, **11**, 436–444.

Herbst M. 1998. Stomatal behaviour in a beech canopy: an analysis of Bowen ratio measurements compared with porometer data. *Plant, Cell and Environment*, **18**, 1010–1018.

Hewlett JD, Hibbert AR. 1961. Increases in water yield after several types of forest cutting. *International Association Scientific Hydrology Bulletin*, **6**, 5–17.

Hicks BJ, Beschta RL, Harr RD. 1991. Long-term changes in streamflow following logging in western Oregon and associated fisheries implications. *Water Resources Research*, **27**, 217–226.

Hobbie EA, Colpaert JV. 2004. Nitrogen availability and mycorrhizal colonization influence water use efficiency and carbon isotope patterns in *Pinus sylvestris*. *New Phytologist*, **164**, 515–525.

Holbrook NM, Sinclair TR. 1992. Water balance in the arborescent palm, *Sabal palmetto*. II. Transpiration and water storage. *Plant, Cell and Environment*, **15**, 401–409.
Holbrook NM, Burns MN, Field CB. 1995. Negative xylem pressures in plants: a test of the balancing pressure technique. *Science*, **270**, 1193–1194.
Irvine J, Law BE, Kurpius MR, Anthoni PM, Moore D, Schwarz PA. 2004. Age-related changes in ecosystem structure and function and effects on water and carbon exchange in ponderosa pine. *Tree Physiology*, **24**, 753–763.
Jackson RB, Canadell J, Ehleringer JR, Mooney HA. 1996. A global analysis of root distributions for terrestrial biomes. *Oecologia*, **108**, 389–411.
James SA, Meinzer FC, Goldstein G, Woodruff D, Jones T, Restom T, Mejia M, Clearwater M, Campanello P. 2003. Axial and radial water transport and internal water storage in tropical forest canopy trees. *Oecologia*, **134**, 37–45.
Jarvis PG, McNaughton KG. 1986. Stomatal control of transpiration: scaling up from leaf to region. *Advances in Ecological Research*, **15**, 1–49.
Jones JA, Grant GE, 1996. Peak flow responses to clear-cutting and roads in small and large basins, western Cascades, Oregon. *Water Resources Research*, **32**, 959–974.
Jones JA, Post DA. 2004. Seasonal and successional streamflow response to forest cutting and regrowth in the northwest and eastern United States. *Water Resources Research*, **40**: W05203, doi:10.1029/2003WR002952.
Kavanagh KL, Bond BJ, Aitken SN, Gartner BL, Knowe S. 1999. Shoot and root vulnerability to xylem cavitation in four populations of Douglas-fir seedlings. *Tree Physiology*, **19**, 31–37.
Keppeler ET, Ziemer RR. 1990. Logging effects on streamflow: Water yield and summer low flows at Caspar Creek in Northwestern California. *Water Resources Research*, **26**, 1669–1679.
Kobayashi Y, Tanaka T. 2001. Water flow and hydraulic characteristics of Japanese red pine and oak trees. *Hydrological Processes*, **15**, 1731–1750.
Koch, GW, Sillett SC, Jennings GM, Davis SD. 2004. The limits to tree height. *Nature*, **428**, 851–854
Köstner BMM, Falge E, Tenhunen JD. 2002. Age-related effects on leaf area/sapwood area relationships, canopy transpiration and carbon gain of Norway spruce stands (*Picea abies*) in the Fichtelgebirge, Germany. *Tree Physiology*, **22**, 567–574.
Köstner BMM, Schulze E-D, Kelliher FM, Hollinger DY, Byers JN, Hunt JE, McSeveny TM, Meserth R, Weir PL. 1992. Transpiration and canopy conductance in a pristine broad-leaved forest of *Nothofagus*: an analysis of xylem sap flow and eddy correlation measurements. *Oecologia*, **91**, 350–359.
Larcher W. 1975. *Physiological Plant Ecology*. Springer-Verlag: New York.
Law BE, Goldstein AH, Anthoni PM, Unsworth MH, Panek JA, Bauer MR, Fracheboud JM, Hultman N. 2001. Carbon dioxide and water vapor exchange by young and old ponderosa pine ecosystems during a dry summer. *Tree Physiology*, **21**, 299–308.
Loustau D, Berbigier P, Roumagnae P, Arruda-Pacheco C, David JS, Ferreira MI, Pereira JS, Tavares R. 1996. Transpiration of a 64-year-old maritime pine stand in Portugal. 1. Seasonal course of water flux through maritime pine. *Oecologia*, **107**, 33–42.
Maherali H, DeLucia EH. 2001. Influence of climate-driven shifts in biomass allocation on water transport and storage in ponderosa pine. *Oecologia*, **129**, 481–491.
Manter DK, Bond BJ, Kavanagh KL, Stone JK, Filip GM. 2003. Modelling the impacts of the foliar pathogen, *Phaeocryptopus gaeumannii*, on Douglas-fir physiology: net canopy carbon assimilation, needle abscission and growth. *Ecological Modelling*, **164**, 211–226.
McDowell NG, Barnard H, Bond BJ, Hinckley T, Hubbard RH, Ishii KB, Meinzer FC, Marshall JD, Magnani F, Phillips N, Ryan MG, Whitehead D. 2002a. The relationship between tree height and leaf area : sapwood area ratio. *Oecologia*, **132**, 12–20.
McDowell NG, Phillips N, Lunch C, Bond BJ, Ryan MG. 2002b. An investigation of hydraulic limitation and compensation in large, old Douglas-fir trees. *Tree Physiology*, **22**, 763–774.
McElrone AJ, Pockman WT, Martinez-Vilalta J, Jackson RB. 2004. Variation in xylem structure and function in stems and roots of trees to 20 m depth. *New Phytologist*, **163**, 507–517.
Meinzer FC. 2002. Co-ordination of liquid and vapor phase water transport properties in plants. *Plant, Cell and Environment*, **25**, 265–274.

Meinzer FC, Goldstein G, Holbrook NM, Jackson P, Cavelier J. 1993. Stomatal and environmental control of transpiration in a lowland tropical forest tree. *Plant, Cell and Environment*, **16**, 429–436.

Meinzer FC, Goldstein G, Jackson PJ, Holbrook NM, Gutierrez MV. 1995. Environmental and physiological regulation of transpiration in tropical forest gap species: the influence of boundary layer and hydraulic properties. *Oecologia*, **101**, 514–522.

Meinzer FC, Hinckley TM, Ceulemans R. 1997. Apparent responses of stomata to transpiration and humidity in a hybrid poplar canopy. *Plant, Cell and Environment*, **20**, 1301–1308.

Meinzer FC, James SA, Goldstein G, Woodruff D. 2003. Whole-tree water transport scales with sapwood capacitance in tropical forest canopy trees. *Plant, Cell and Environment*, **26**, 1147–1155.

Meinzer FC, James SA, Goldstein G. 2004a. Dynamics of transpiration, sap flow and use of stored water in tropical forest canopy trees. *Tree Physiology*, **24**, 901–909.

Meinzer FC, Brooks JR, Bucci SJ, Goldstein GH, Scholz FG, Warren JM. 2004b. Converging patterns of uptake and hydraulic redistribution of soil water in contrasting woody vegetation types. *Tree Physiology*, **24**, 919–928.

Meinzer FC, Bond BJ, Warren JM, Woodruff DR. 2005. Does water transport scale universally with tree size? *Functional Ecology*, **19**, 558–565.

Meinzer FC, Brooks JR, Domec J-C, Gartner BL, Warren JM, Woodruff DR, Bible K, Shaw DC. 2006. Dynamics of water transport and storage in conifers studied with deuterium and heat tracing techniques. *Plant, Cell and Environment*, **29**, 105–114.

Meinzer FC, Woodruff DR, Shaw DC. 2004c. Integrated responses of hydraulic architecture, water and carbon relation of western hemlock to dwarf mistletoe infection. *Plant, Cell Environment*, **27**, 937–946.

Melcher PJ, Goldstein G, Meinzer FC, Yount DE, Jones T, Holbrook NM, Huang CX. 2001. Water relations of coastal and estuarine *Rhizophora mangle*: xylem pressure potential and dynamics of embolism formation and repair. *Oecologia*, **126**, 182–192.

Mencuccini M, Grace J. 1995. Climate influences the leaf area/sapwood area ratio in Scots pine. *Tree Physiology*, **15**, 1–10.

Monterith, JL, 1965. Evaporation and environment. *Symposium Society Experimental Biology*, **19**, 205–234.

Moore GW, Bond BJ, Jones JA, Phillips N, Meinzer FC. 2004. Structural and compositional controls on transpiration in a 40- and 450-yr-old riparian forest in western Oregon, USA. *Tree Physiology*, **24**, 481–491.

Moreira MZ, Sternberg L da SL, Martinelli LA, Cictoria RL, Barbosa EM, Bonates LCM, Nepstad DC. 1997. Contribution of transpiration to forest ambient vapor based on isotopic measurements. *Global Change Biology*, **3**, 439–450.

Moreira MZ, Scholz FG, Bucci SJ, Sternberg L da SL, Goldstein G, Meinzer FC, Franco AC. 2003. Hydraulic lift in a Netropical Savanna. *Functional Ecology*, **17**, 573–581.

Mott KA, Parkhurst DF. 1991. Stomatal responses to humidity in air and helox. *Plant, Cell and Environment*, **14**, 509–515.

Nardini A, Tyree MT. 1999. Root and shoot hydraulic conductance of seven *Quercus* species. *Annals of Forest Science*, **56**, 371–377.

Oren R, Phillips N, Ewers BE, Pataki DE, Megonigal JP. 1999a. Sap-flux-scaled transpiration responses to light, vapor pressure deficit, and leaf area reduction in a flooded *Taxodium distichum* forest. *Tree Physiology*, **19**, 337–347.

Oren R, Sperry JS, Katul GG, Pataki DE, Ewers BE, Phillips N, Schäfer KVR. 1999b. Survey and synthesis of intra- and interspecific variation in stomatal sensitivity to vapour pressure deficit. *Plant, Cell and Environment*, **22**, 1515–1526.

Pataki D, Oren R, Phillips N. 1998. Responses of sap flux and stomatal conductance of *Pinus taeda* L. trees to stepwise reductions in leaf area. *Journal of Experimental Botany*, **49**, 871–878.

Phillips N, Bond BJ, McDowell NG, Ryan MG. 2002. Canopy and hydraulic conductance in young, mature and old Douglas-fir trees. *Tree Physiology*, **22**, 205–211.

Phillips NG, Ryan MG, Bond BJ, McDowell NG, Hinckley TM, Cermák J. 2003. Reliance on stored water increases with tree size in three species in the Pacific Northwest. *Tree Physiology*, **23**, 237–245.

Pockman WT, Sperry JS, O'Leary JW. 1995. Sustained and significant negative water pressure in xylem. *Nature*, **378**, 715–716.

Pypker TG, Bond BJ, Link TE, Marks D, Unsworth MH. 2005. The importance of canopy structure in controlling the interception loss of rainfall: Examples from a young and old-growth Douglas-fir forests. *Agricultural and Forest Meteorology*, **130**, 113–129.

Pypker TG, Unsworth M, Bond BJ. The role of epiphytes in rainfall interception by forests in the Pacific Northwest: Field measurements at the branch and canopy scale. *Canadian Journal of Forest Research*. In press.

Rayment MB, Loustau D, Jarvis PG. 2000. Measuring and modeling conductances of black spruce at three organizational scales: shoot, branch and canopy. *Tree Physiology*, **20**, 713–723.

Richards JH, Caldwell MM. 1987. Hydraulic lift: substantial nocturnal water transport between soil layers by *Artemisia tridentata* roots. *Oecologia*, **73**, 486–489.

Roberts J, Cabral OMR, De Aguiar LF. 1990. Stomatal and boundary-layer conductances in an Amazonian terra firme rain forest. *Journal of Applied Ecology*, **27**, 336–353.

Rose KL, Graham RC, Parker DR. 2003. Water source utilization by *Pinus jeffreyi* and *Arctostaphylos patula* on thin sois over bedrock. *Oecologia*, **134**, 46–54.

Ryan MG, Binkley D, Fownes JG. 1997. Age-related decline in forest productivity: pattern and process. *Advances in Ecological Research*, **27**, 213–262.

Ryan MG, Bond BJ, Law BE, Hubbard RM, Woodruff D, Cienciala E, Kucera J. 2000. Transpiration and whole-tree conductance in ponderosa pine trees of different heights. *Oecologia*, **124**, 553–560.

Ryel RJ, Caldwell MM, Leffler AJ, Yoder CK. 2003. Rapid soil moisture recharge to depth by roots in a stand of *Artemisia tridentata*. *Ecology*, **84**, 757–764.

Schulze E-D, Hall AE. 1982. Stomatal responses, water loss, and CO_2 assimilation rates of plants in contrasting environments. In *Encyclopedia of Plant Physiology, New Series*, Vol. 12B. Eds O.L. Lange, P.S. Nobel, C.B. Osmond, H. Ziegler. Springer-Verlag: New York, pp. 181–230.

Schulze E-D, Lange OL, Buschbom U, Kappen L, Evenari M. 1972. Stomatal responses to changes in humidity in plants growing in the desert. *Planta*, **108**, 259–270.

Schulze E-D, Caldwell MM, Canadell J, Mooney HA, Jackson RB, Parson D, Scholes R, Sala OE, Trimborn P. 1998. Downward flux of water through roots (i.e. inverse hydraulic lift) in dry Kalahari sands. *Oecologia*, **115**, 460–462.

Scott RL, Watts C, Payan JG, Edwards EA, Goodrich DC, Williams DG, Shuttleworth WJ. 2003. The understory and overstory partitioning of energy and water fluxes in an open canopy, semiarid woodland. *Agricultural and Forest Meteorology*, **114**, 127–139.

Sellin A. 1999. Does pre-dawn water potential reflect conditions of equilibrium in plant and soil water status? *Acta Oecologica*, **20**, 51–69.

Shuttleworth WJ, Gash JHC, LLoyd CR, Moore CJ, Roberts J, Marques ADO, Fisch G, Silva V De Paula, De Nazare Goes Ribeiro M, Molion LCB, De Abreu Sa LD, Nobre JC, Cabral OMR, Patel SR, De Moraes JC. 1984. Eddy correlation measurements of energy partition for Amazonian forest. *Quarterly Journal of the Royal Meteorological Society*, **110**, 1143–1162.

Sparks JP, Black RA. 1999. Regulation of water loss in populations of *Populus trichocarpa*: the role of stomatal control in preventing xylem cavitation. *Tree Physiology*, **19**, 453–459.

Sperry JS, Ikeda T. 1997. Xylem cavitation in roots and stems of Douglas-fir and white fir. *Tree Physiology*, **17**, 275–280.

Sperry JS, Saliendra NZ, Pockman WT, Cochard H, Cruziat P, Davis SD, Ewers FW, Tyree MT. 1996. New evidence for large negative xylem pressures and their measurement by the pressure chamber method. *Plant, Cell and Environment*, **19**, 427–436.

Sperry JS, Adler FR, Campbell GS, Comstock JP. 1998. Limitation of plant water use by rhizosphere and xylem conductance: results form a model. *Plant, Cell and Environment*, **21**, 347–359.

Steudle, E. 1994. Water transport across roots. *Plant and Soil*, **167**, 79–90.

Steudle, E. 2001. The cohesion–tension mechanism and the acquisition of water by plant roots. *Annual Review of Plant Physiology and Molecular Biology*, **52**, 847–875.

Swank WT, Vose JM, Elliot KJ. 2001. Long-term hydrologic and water quality responses following commercial clearcutting of mixed hardwoods on a southern Appalacian catchment. *Forest Ecology and Management*, **143**, 163–178.

Tardieu F, Davies WJ. 1993. Integration of hydraulic and chemical signalling in the control of stomatal conductance and water status of droughted plants. *Plant, Cell and Environment*, **16**, 341–349.

Thomas RB, Megahan WF. 1998. Peak flow responses to clear-cutting and roads in small and large basins, western Cascades, Oregon: A second opinion. *Water Resources Research*, **34**, 3393–3403.

Tognetti R, Michelozzi M, Giovanelli A. 1997. Geographical variation in water relations, hydraulic architecture and terpene composition of Aleppo pine seedlings from Italian provenances. *Tree Physiology*, **17**, 241–250.

Tyree MT, Ewers FW. 1991. Tansley review No. 34. The hydraulic architecture of trees and other woody plants. *New Phytologist*, **119**, 345–360.

Tyree MT, Sperry JS. 1989. Vulnerability of xylem to cavitation and embolism. *Annual Review of Plant Physiology and Plant Molecular Biology*, **40**, 19–36.

Tyree MT, Velez V, Dalling JW. 1998. Growth dynamics of root and shoot hydraulic conductance in seedlings of five neotropical tree species: scaling to show possible adaptation to different light regimes. *Oecologia*, **114**, 293–298.

Tyree MT, Zimmermann MH. 2002. *Xylem Structure and the Ascent of Sap*. Springer: New York.

Tyree MT, Salleo S, Nardini A, Lo Gullo MA, Mosca R. 1999. Refilling of embolized vessels in young stems of laurel. Do we need a new paradigm? *Plant Physiology*, **120**, 11–21.

Unsworth MH, Phillips N, Link T, Bond BJ, Falk M, Harmon ME, Hinckley TM, Marks D, Paw U KT. 2004. Components and controls of water flux in an old-growth Douglas-fir-western hemlock ecosystem. *Ecosystems*, **7**, 468–481.

van den Honert TH. 1948. Water transport in plants as a catenary process. *Discussions of the Faraday Society*, **3**, 146–153.

Vogt UK. 2001. Hydraulic vulnerability, vessel refilling, and seasonal courses of stem water potential of *Sorbus aucuparia* L. and *Sambucus nigra* L. *Journal of Experimental Botany*, **52**, 1527–1536.

Wan CG, Sosebee RE, McMichael BL. 1993. Does hydraulic life exist in a shallow-rooted species? A quantitative examination with a half-shrub *Gutierrezia sarothrae*. *Plant Soil*, **153**, 11–17.

Wang X-F, Yakir D. 2000. Using stable isotopes of water in evapotranspiration studies. *Hydrological Processes*, **14**, 1407–1421.

Waring RH, Running SW. 1978. Sapwood water storage: its contribution to transpiration and effect upon water conductance through the stems of old-growth Douglas-fir. *Plant, Cell and Environment*, **1**, 131–140.

Waring RH, Whitehead D, Jarvis PG. 1979. The contribution of stored water to transpiration in Scots pine. *Plant, Cell and Environment*, **2**, 309–317.

Warren JM, Meinzer FC, Brooks JR, Domec J-C. 2005. Vertical stratification of soil water storage and release dynamics in Pacific Northwest coniferous forests. *Agricultural and Forest Meteorology*, **130**, 39–58.

Wells CE, Eissenstadt DM. 2003. Beyond the roots of young seedlings: the influence of age and order on fine root physiology. *Journal of Plant Growth Regulation*, **21**, 324–334.

Whitehead D. 1998. Regulation of stomatal conductance and transpiration in forest canopies. *Tree Physiology*, **18**, 633–644.

Whitehead D, Beadle CL. 2004. Physiological regulation of productivity and water use in *Eucalyptus*: A review. *Forest Ecology and Management*, **193**, 113–140.

Williams DG, Cable WL, Hultine KR, Hoedjes JCB, Yepez EA, Simonneaux V, Er-Raki S, Boulet G, de Bruin HAR, Chehbouni A, Hartogensis OK, Timouk T. 2004. Evapotranspiration components determined by stable isotope, sap flow and eddy covariance techniques. *Agricultural and Forest Meteorology*, **125**, 241–258.

Williams M, Rastetter EB, Fernandes DN, Goulden ML, Wofsy SC, Shaver GR, Melilli JM, Muhger JW, Fan SM, Nadelhoffer KJ. 1996. Modelling the soil-plant-atmosphere continuum in a Quercus–Acer stand at Harvard Forest: the regulation of stomatal conductance by light, nitrogen and soil/plant hydraulic properties. *Plant, Cell and Environment*, **19**, 911–927.

Yakir D, Berry JA, Giles L, Osmond CB. 1993. The $\delta^{18}O$ of water in the metabolic compartment of transpiring leaves. In Ehleringer JR , Hall AE, Farquhar GD. (eds) *Stable Isotopes and Plant Carbon-Water Relations*. Academic Press: San Diego, pp. 529–540.

Yepez EA, Williams DG, Scott RL, Lin G. 2003. Partitioning overstory and understory evapotranspiration in a semiarid savanna woodland from the isotopic compostion of water vapor. *Agricultural and Forest Meteorology*, **119**, 53–68.

Yoder CK, Nowak RS. 1999. Hydraulic lift among native plant species in the Mojave Desert. *Plant and Soil*, **215**, 93–102.

Yoder BJ, Ryan MG, Waring RH, Schoettle AW, Kaufmann MR. 1994. Evidence of reduced photosynthetic rates in old trees. *Forest Science*, **40**, 513–527.

Zimmerman R, Schulze E-D, Wirth C, Schulze E-D, McDonald KC, Vygodskaya NN, Ziegler W. 2000. Canopy transpiration in a chronosequence of Central Siberian pine forests. *Global Change Biology*, **6**, 25–37.

Zinke PJ. 1967. Forest interception studies in the United States. In Sopper WE, Lull HW (eds), *International Symposium on Forest Hydrology*. Pergamon Press: New York, pp. 137–161.

Zwieniecki MA, Holbrook NM. 1998. Diurnal variation in xylem hydraulic conductivity in white ash (*Fraxinus americana* L.), red maple (*Acer rubrum* L.) and red spruce (*Picea rubens* Sarg.). *Plant, Cell and Environment*, **21**, 1173–1180.

Zwieniecki MA, Newton M. 1996. Seasonal pattern of water depletion from soil-rock profiles in a Mediterranean climate in southwestern Oregon. *Canadian Journal of Forest Research*, **26**, 1346–1352.

3

The Ecohydrology of Invertebrates Associated with Exposed Riverine Sediments

Jonathan P. Sadler and Adam J. Bates

3.1 Introduction

The lateral and longitudinal patterns of river channels and floodplains are ultimately controlled by flow regime and sediment supply, and exhibit a diversity of forms including straight, meandering, wandering and braided (Leopold and Wolman, 1957). Globally, around 90% of all floodplains have been anthropogenically modified as a result of changing land use, water abstraction and impoundment, floodplain development and flood defence engineering (Tockner and Stanford, 2002). Accordingly, the abundance of the more dynamic, wandering and braided channel forms and the degree of connectivity between rivers and their floodplains have drastically declined to the point where they are one of the most denuded and endangered global habitats (Tockner et al., 2006).

In today's highly managed landscapes, the importance of naturally active floodplains as ecosystems that support high levels of alpha (patch), beta (habitat turnover) and gamma (landscape) diversity has been highlighted in recent conceptual papers (Ward, 1998; Ward and Tockner, 2001). This work illustrates that active floodplains are very dynamic systems with many linkages between lotic and riparian habitats and should therefore be viewed as an interconnected 'riverine landscape', or 'riverscape' (Ward et al., 2002). Within these riverscapes, poorly vegetated areas of alluvial sediment, termed exposed riverine sediments (ERS), are key habitats of concern for conservation and restoration as they are associated with a wide range of rare and endangered species (Sadler *et al.*, 2004; Tockner *et al.*, 2006).

Studies of ERS habitats in the UK and elsewhere in Europe (e.g., Andersen and Hanssen, 2005; Eyre *et al.*, 2001a,b; Sadler *et al.*, 2004) have highlighted that species assemblages vary in response to a number of environmental variables, such as sediment diversity, habitat heterogeneity (e.g., profile and structure), vegetation cover, and habitat size (e.g., width and length). Within patches of ERS habitat, species spatio-temporal distributions closely correlate with the underlying distribution of microhabitats that vary in, for example, temperature, humidity, sediment size and vegetation cover (Andersen, 1969, 1983, 1985a; Desender, 1989).

This physical variability is mainly driven by changing flow levels that can subtly alter humidity zones and the position of the aquatic terrestrial interface, or during high flows, can cause extreme changes in the composition, size and position of habitat patches. Flood disturbances in particular are of primary importance in sustaining dynamic riverscapes and structuring the animal and plant communities that rely on them (Plachter and Reich, 1998).

This chapter provides an introduction to the nature and conservation significance of ERS habitats. The important structuring influence of disturbance on ERS invertebrates is evaluated in terms of: (i) the links between the physical variables and ERS invertebrate assemblages; (ii) species adaptations to natural flow regimes, and (iii) the significance of lateral and longitudinal connectivity. The chapter concludes with an assessment of the threats to ERS habitats due to the alteration of the natural hydrological dynamics. Emphasis will be placed throughout on the ecology of invertebrates that almost exclusively rely on ERS for their habitat.

3.2 ERS Habitats

ERS can be defined as 'exposed, within channel, fluvially deposited, sediments (gravels, sands and silts) that lack continuous vegetation cover, whose vertical distribution lies between the levels of bankfull and the typical base flow of the river' (Bates and Sadler, 2005, p. 3). This will include a whole range of complex habitats within the river channel, such as lateral, point, and island bars and their adjacent eroding banks (Figure 3.1). The habitat of ERS species in its widest sense (cf. Dennis *et al.*, 2003) can also include backwater pools, riparian wetlands and woodlands, which can provide important resources at certain times of the year (e.g. overwintering habitat and flow refugia).

3.3 Invertebrate Conservation and ERS Habitats

Fowles (1989) illustrated the significance of shingle ERS as an important habitat for a wide range of rare and nationally scarce species of Coleoptera (beetles). As a result of this work, the UK Biodiversity Steering Group created grouped species action plans for six species of ERS Coleoptera and additional plans exist for two species of Diptera and one species of water beetle (Anon, 1995), that are also seen as ERS specialists. Over the past ten years, the conservation potential of ERS habitats for rare invertebrates in the UK

Figure 3.1 A complex ERS habitat on the River Otter in Devon (UK)

has been systematically assessed in Government Agency funded research projects (e.g., Eyre, 2000; Sadler and Bell, 2002; Sadler *et al.*, 2005; Sadler and Petts, 2000). The results of these studies have been used to classify 131 species of invertebrates as ERS specialists that exhibit high fidelity to the sediments (Bates and Sadler, 2005). Eighty-six (66%) of these currently have some conservation status with 29 classified as Red Data Book (RDB2, RDB3, RDBI, and RDBK) and 57 classified as Nationally Scarce (Na, Nb and N). Survey work on ERS beetles on rivers in England and Wales has served to underline the conservation importance of ERS for these species. For example, Sadler *et al.* (2004) recorded 81 rare beetles in their survey and 42 of these (52%) were considered to be ERS specialists. In a parallel survey of ERS habitats in Scotland and northern England, Eyre *et al.* (2001,b), recorded 115 species with conservation status.

Elsewhere in Europe, active river flood plains are also habitat for a large number of rare invertebrates (e.g., Reich, 1991). The German Red Data Book lists 38 species of beetle associated with ERS (Manderbach and Reich, 1995; Plachter and Reich, 1998). Van Looy (2005) records 21 species of ERS specialist ground beetles from the River Meuse all of which are listed in the Belgium Red Data Book. Andersen and Hanssen (2005) reviewed the situation in Fennoscandia and identified a total of 69 riparian specialist beetles in the region, of which 47 (69%) were listed in the Red Data Books of one the three countries examined (Norway, Sweden and Denmark).

There is a dearth of systematic research on the extent and nature of the ERS conservation resource in Europe and elsewhere in the world and there is a clear need to rectify the situation (e.g., Yoshimura *et al.*, 2005). ERS habitats are clearly associated with a wealth of rare species that are of conservation importance. Effective management of this conservation resource requires a fuller understanding of the relationships between the spatial and temporal dynamics of the physical habitat and the ecology of the organisms.

3.4 Flow Disturbance in ERS Habitats

Disturbance resulting from flooding is the key structuring force for both lotic (Lake, 2000) and riparian biodiversity (Junk, 1999; Naiman and Decamps, 1997; Tockner *et al.*, 2000). It is of pivotal importance for ERS invertebrate ecology (Plachter and Reich, 1998). At a basic level the system is best viewed as a process of patch creation and destruction resulting from the balance between rejuvenating flooding events and ERS stabilisation due to vegetation succession. In natural floodplain systems this balance leads to a shifting mosaic of physical habitats that is heterogeneous and provides extensive littoral margins, variable bar forms, flow refugia and high connectivity, especially at lower flows when a lot of ERS are exposed (Figure 3.2).

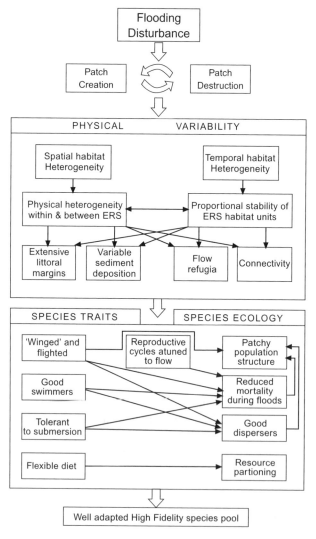

Figure 3.2 The main elements of a hydroecological model for ERS specialist species

For ERS systems, the magnitude, frequency and timing of geomorphologically active disturbance events is central of importance for sustaining flood plain biodiversity (Ward, 1998; Ward and Tockner, 2001). Harris *et al.* (2000) considered the ecological significance of flow magnitude by assessing how it might affect ecological processes. Very high flows have the effect of resetting the channel–floodplain system, whereas low flows strengthen the partitioning of habitat patches. Indeed, it seems possible to relate the magnitude of events with thresholds in the ecological system. Large-scale events may provide a 'resetting or rejuvenating pulse' that causes whole scale reconfigurations of channel morphology. A short term impact of such an event is increased mortality for many aquatic and riparian organisms (Hering *et al.*, 2004). This was observed on the River Isar in Germany, following a very large flood, although numbers of ground beetles recovered within two months, probably as a result of increased habitat availability (Hering *et al.*, 2004). No data exists to establish the longer term impact of such large events on populations of ERS species. Lower magnitude events, or 'flow pulses' (Tockner *et al.*, 2000) may also prove to be essential for preserving ecosystem function, for example by causing spatial and temporal shifts in the food resource availability (Paetzold *et al.*, 2005; Paetzold and Tockner, 2005).

The importance of flood duration for ERS invertebrates is not well documented. In tropical areas, flood plains can be inundated for long periods of time, which has driven the evolution of adaptations to low oxygen conditions and other physiological traits (Adis, 1982; Adis and Junk, 2002; Zerm and Adis, 2003). In temperate areas, however, flow duration may play a lesser role in ecological dynamics, as the flood pulse has shorter residence times. However, increased base flows may accelarate vegetation succession on some riverine systems because of the greater availability of water for plants (Manderbach and Reich, 1995), thereby reducing the available ERS habitat.

The timing of peak flow events is also significant. If there are large magnitude events when particular species are at life stages less able to cope with inundation extensive mortality might result, potentially affecting recruitment in the following year. For most ERS specialist invertebrates and some arachnids this would be in mid to late summer when the larvae/young are active on ERS. Air temperature at the time of inundation might also strongly affect the level of mortality, as invertebrates are less able to fly when the air temperature is low.

3.5 The Importance of Flow Disturbance for ERS Invertebrate Ecology

Bunn and Arthington (2002) posit four basic guiding principles concerning the influence of changing flow regimes (flood disturbance) on aquatic biodiversity: (i) the relationship between biodiversity and the physical nature of the aquatic habitats is driven by large events that affect channel form and shape; (ii) life history patterns have evolved in direct response to the natural flow regimes; (iii) the maintenance of natural patterns of longitudinal and lateral connectivity is essential to the viability of populations of many riverine species, and (iv) the survival and success of introduced species is enhanced by variation in flow regimes. Although designed with aquatic species in mind, such an approach has considerable relevance to ERS invertebrates (Table 3.1). Principle (iv) remains untested for ERS communities but there are supporting data for principles (i), (ii) and (iii).

Table 3.1 ERS invertebrate communities and Bunn and Arthington's (2002) guiding principles

Principle	Description	Biological response	Sources
(i)	Channel form Habitat complexity Patch creation/disturbance	Longitudinal and lateral variation in species assemblages Variations in species diversity and rarity	(Bell et al., 1999; Framenau et al., 2002; Reich, 1991) (Andersen and Hanssen, 2005; Eyre, 2000; Looy et al., 2006; Sadler and Petts, 2000)
(ii)	Seasonal predictability and life cycle synchroneity	Space partitioning (e.g., microhabitat use) Temporal partitioning (e.g., larval activity, overwintering modes) Dispersal and lifecycle cues	(Andersen, 1969; 1978; 1986; Bates et al., 2007) (Andersen, 1984; 1985a,b,c) (Andersen, 1968; Framenau et al., 1996a,b)
(iii)	Lateral and longitudinal connectivity essential to viability of populations	Dispersal dynamics	(Bates et al., 2005; 2006; Stelter et al., 1997).
(iv)	Natural flow regime discourages invasions	No data on ERS invertebrates	Untested

3.5.1 Principle (i): Physical Variability and ERS Invertebrates

Several studies at the patch scale have shown that ERS beetle assemblages are strongly related to patch scale habitat variability (Eyre et al., 2001a,b; Sadler et al., 2004), which exhibits clear longitudinal and lateral patterning. There is an evident gradient of assemblages from upland to lowland ERS patches resulting from a progressive reduction in sediment size, from coarser boulders and cobbles in upland areas to fine shingles and sands further downstream (Eyre et al., 2001b; Framenau et al., 2002; Sadler et al., 2004), although this pattern is interrupted at tributary junctions (Benda et al., 2004; Rice et al., 2001a,b) and other areas where coarse sediment inputs to the river (e.g., unstable hillslopes). Lateral variation in biodiversity across the floodplain is also a key element and this is thought to be a result of inundation frequency in single thread channels. In braided rivers (e.g., Reich, 1991), however, the situation is complicated not only by the contraction and expansion of water across the complex topography of the floodplain (Tockner et al., 2006), but also by variability in food derived from drifting and emerging aquatic insects (Paetzold et al., 2005; 2006).

Although useful in identifying inter-catchment and regional differences in ERS species assemblages and their conservation significance, this comparative work conflated variation acting at a reach, river, or catchment scale with that at the patch scale. More recent work on a 5-km reach of the River Severn in mid-Wales (UK) (Figure 3.3) has focused on distinct ERS patches in the same segment of river, thereby removing this potential source of confounding spatial and temporal variation (Sadler et al., 2006). Statistical analysis of the relationship of beetles with patch-scale environmental data illustrated a number of significant relationships (Figure 3.4).

Species densities and the RDA ordination illustrated that vegetation structure is an important patch variable. Several studies have illustrated the relationship between ERS vegetation and channel form and river hydrology (Gilvear *et al.*, 2000; Parsons and

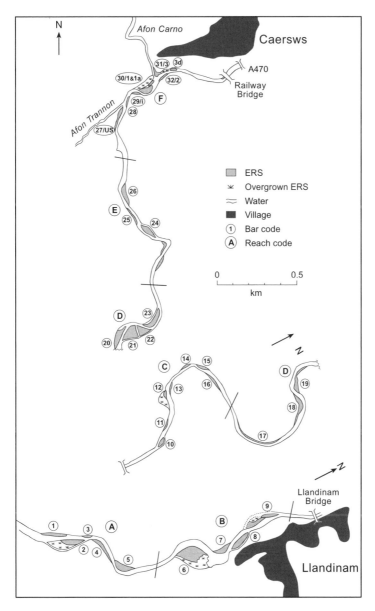

Figure 3.3 *The study reach detailing the approximate position and size of all significant bars in the Upper Severn survey reach. Numbers represent individual bars, letters represent distinct reaches divided by 'sediment transfer zones' with little ERS*

Gilvear, 2002) and invertebrate assemblages (Eyre *et al.*, 2001a,b; Sadler *et al.*, 2004). Increased patch heterogeneity and elements of this diversity, such as variations in ERS profile, were also significant variables affecting the species assemblages (Figure 3.4) illustrating the importance of microhabitat diversity for species richness. ERS size was also related to both overall patch species richness and was a significant variable in the RDA (redundancy analysis) ordination (Figure 3.4). ERS inundation potential was found to be significantly related to all the species metrics (Table 3.2). Although not a quantitative measure of true inundation, the variable relates to the likelihood of flooding of individual patches and hence disturbance.

At smaller spatial scales, variability in the physical habitats within individual ERS patches strongly influences the microspatial distribution of ERS invertebrates. This within-patch variability is mainly related to the channel's fluvial geomorphology and water level (Figure 3.5). There are strong lateral gradients of increasing elevation, falling inundation frequency and duration, sediment fining (e.g., Andersen, 1978, 1983, 1985a, 1988), falling aquatic food subsidy (e.g., Paetzold *et al.*, 2005; Paetzold and Tockner, 2005), increasing temperature, and falling humidity (Hannah, in preparation). This variation, potentially in combination with interspecific competitive interactions, is thought to drive the microspatial distribution of ERS invertebrates on individual patches of ERS (Andersen, 1969, 1983, 1988).

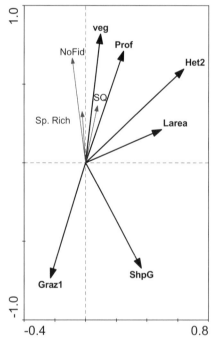

Figure 3.4 RDA biplot with supplementary variables species richness (Sp. Rich), the number of species exhibiting fidelity to the sediments (NoFid) and the species quality score (SQ). Only significant environmental variables are displayed (Sadler et al., 2006). Environmental variables codes as described in Table 3.2

Table 3.2 *Single factor ANOVA and Spearman's rho (Larea only) tests showing the relationship between the species metrics and environmental variables. (Variable codes: IP = inundation potential; Veg1 = vegetation type; graz1, and graz2 = indices of grazing intensity; cattG = cattle grazing intensity; Het2 = an index of ERS heterogeneity; Prof = ERS profile type; Gors = presence of gorse on or adjacent to the ERS; Larea = log of ERS area*

Metric(s)	Variable(s)
Density	IP*, Veg1*, graz2**, graz1*, cattG*, Prof*, Gors*
Species richness (Sp Rich)	IP*, graz1*, Larea*
Number of fidelity species (NoFid)	IP*
Species quality (SQ)	Het2*

*$P \leq 0.05$; **$P \leq 0.01$

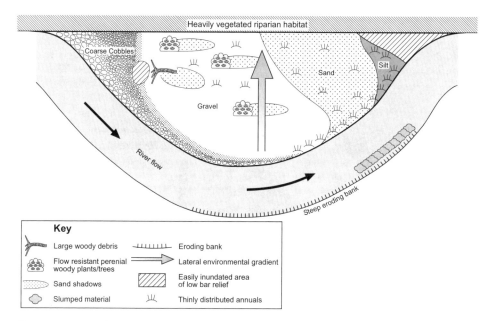

Figure 3.5 *A schematic view of an ERS point bar illustrating a wide range of available microhabitats. The lateral environmental gradient represents increasing elevation, falling inundation frequency and duration, sediment fining, falling aquatic food subsidy, increasing temperature, and falling humidity*

Bates et al. (2007) studied the microscale distribution of a range of ERS species on ERS located at Caersws on the River Severn in Wales (Figure 3.6; Figure 3.3, reach F) using a tightly packed sequence of non-fatal pitfall traps. The results showed strong and significant microspatial zonation for all species that was remarkably temporally stable over the study period. The highest species richness was consistently distributed around the water's edge. Species zonation was mainly associated with elevation and proximity to the water, and several species were consistently spatially associated or disassociated

46 *Hydroecology and Ecohydrology: Past, Present and Future*

Figure 3.6 Downstream organisation of ERS patches on the River Severn at Caersws (mid Wales, UK)

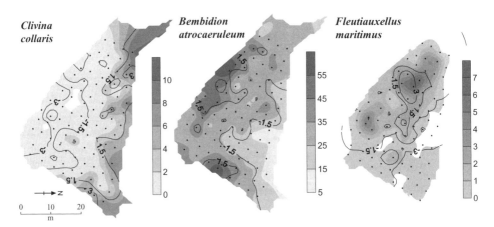

Figure 3.7 Distribution and local clustering of the ERS beetles on an ERS patch on the River Severn at Caersws (mid-Wales, UK). The shading indicates the numbers of individuals, the black dots the location of pitfall traps, and the contours the SADIE statistical indices, where values >1.5 indicate significant clusters and those <1.5 significant gaps

with one another. For example, *Clivina collaris* was situated in the 'upper', more elevated, section of the patch, away from the water's edge, *Bembidion atrocaeruleum* was situated in the 'lower' section of the patch close to the water's edge, and *Fleutiauxellus maritimus* was located in the 'mid' elevation of the bar (Figure 3.7). Although some variability existed across the ERS used in the study, this partitioning in space indicates differential tolerances to variation in microclimate (e.g., climate and humidity) (cf. Antvogel and Bonn, 2001) and different food source and foraging techniques. As Tockner *et al.* (2006) note, repeated wetting and drying resulting from contraction and expansion of the water causes species to move up and down the sediments as individuals track optimal habitat and forage for food.

3.5.2 Principle (ii): Life History Patterns and Function Ecology

Species ecology in aquatic systems is hypothesised to be synchronised with the natural variability of the flow regime (Poff *et al.*, 1997). In ERS systems, flooding leads to variability in the physical habitat and the species themselves have a range of adaptations to deal with the dynamic nature of the environment (Figure 3.2). In most natural flood plains there is a stochastic element in flow regimes, and as a result flood-plain species are characterised by traits that are adapted to sporadic, unpredictable and often high magnitude floods. Andersen (1968; 1985a) illustrated that carabids and probably most other riparian arthropods are all capable swimmers and it seems likely that passive transport downstream with prevailing current is an important form of dispersal, although no data are available to substantiate this. Moreover, laboratory experiments illustrate that individuals of species of Bembidinii (a tribe of ground beetles) can survive immersion for a maximum of to 48 hours (Andersen, 1968; 1985a). In a natural situation such long periods of inundation are quite infrequent.

Desender (1989) records high levels of macroptery in riparian beetle species, a useful trait in a environment where inundation is a sporadic and unpredictable event. Indeed, the proportion of macropterous beetles on unvegetated ERS varies in relation to ERS location in the flood plain. Between 91–99% of the species are capable of flight close to the river edge and this proportion falls to about 76% in areas that are rarely inundated (Plachter, 1986). Bonn (2000) studied the flight activity of carabid beetles in relation to flooding of the River Elbe in northern Germany over a period of 3 years using a sequence of flight interception (window) traps. The results indicate that carabids fly actively after the spring and autumn floods, normally towards the river, presumably in search of newly deposited food resources although species did appear to move away from the river immediately before flood peaks.

Over the annual cycle many ERS species have life cycles that are in tune with the seasonal flood in temperate regions. For example, a large proportion of riparian specialist carabids overwinter as first-year adults in habitats that are less susceptible to winter floods, such as high in trees and/or in grass tussocks higher up on the bank (Lott, 1996), and some will travel considerable distances away from the river (Zulka, 1994). Similarly, other ERS species, such as the large lycosid spider, *Arctosa cinerea*, hibernates in areas less likely to be flooded during winter (Framenau *et al.*, 1996a). There are some notable variants to this general pattern which are related to differences in the timing of peak flows. For example, the carabid *Nebria picicornis,* a large specialist of alpine rivers, usually hibernates as larva and the adults, which can more easily avoid inundation emerge early in summer when floods are more likely (Manderbach and Plachter, 1997). However, some individuals of *N. picicornis* overwinter as adults, providing an exceptional degree of phenological plasticity, which is thought to be a mechanism for dealing with unpredictable environments (Plachter and Reich, 1998).

3.5.3 Principle (iii): Lateral and Longitudinal Connectivity and Population Viability

Data exist to indicate the presence of both longitudinal and lateral patterning in species assemblage (see Principle (i) above), but the critical link to connectivity and its importance is lacking. The high turnover of ERS habitats necessitates the transfer of some individuals between habitat patches, and connectivity may prove to be an important element in enhancing population stability during this process. Few studies have focused on this aspect of floodplain ecology, although Ward and Tockner (2001, p. 814) postulated that for aquatic species diversity is maximised at some intermediate level of 'hydrologic' connectivity. They suggest that at low connectivity, species diversity is reduced because organisms cannot move between water bodies that are not connected. In contrast, excessive connectivity is likely to increase the levels of contact between species, enhancing predation, parasitism and disease transmission and interspecific competition for resources. In situations where such predation and competition are not the main controls of species fitness (e.g., where disturbance or microhabitat availability are of prime importance), high connectivity rather than intermediate connectivity should maximise individual species fitness and hence, diversity.

A few studies of ERS specialist invertebrates have demonstrated some degree of spatial structuring of populations, hinting at the potential significance of connectivity between patches (Manderbach and Reich, 1995; Reich, 1991). Stelter *et al.* (1997) studied a metapopulation of the rare grasshopper *Bryoderma tuberculata* Fabricius on the River Isar in Germany. They noted that females, which are flightless and have limited dispersal abilities, rarely move between bars, and move only when bars become connected during periods of low flow. Moreover, they showed that more stable shingle bars, which are less prone to inundation, were sources from which individuals dispersed after the loss of individuals from their core ERS habitat during flooding. Their model suggested that flooding was necessary to disrupt succession and thereby keep the connectivity of ERS habitat high enough for the species to survive, but that too frequent flooding would not give the species time to repopulate their core habitat. The frequency and intensity of flooding therefore controls dispersal in this species. In contrast, research on the ground beetle *Nebria picicornis* (Fabricius) showed that most individuals often disperse over large areas (Manderbach and Plachter, 1997), suggesting a 'patchy' (cf. Harrison, 1991) population structure not so strongly affected by the level of habitat connectivity.

Bates *et al.* (2006) used mark-resight techniques to investigate the three interdependent stages of dispersal: (a) emigration; (b) inter-patch movement, and (c) immigration of a common patchy population of the ERS specialist beetle, *Bembidion atrocaeruleum* on reach F of the study section of the River Severn in mid-Wales (UK) (Figure 3.3; Figure 3.6). Dispersal was found to be condition-dependent, with a much greater rate (and possibly distance) of dispersal following the complete inundation of its main ERS habitat. Although individuals frequently moved between patches of habitat, the distances moved were small (<20m). It is seems likely that, notwithstanding its ability to fly fairly long distances, this species has evolved behavioural traits that require high flow conditions to trigger long-distance dispersal. This could be an adaptation to the underlying distribution of ERS habitat, which is a sequence of closely linked patches of ERS (Figure 3.6), and ready means of downstream dispersal through hydrochory. As a result, dispersal in this species might also be linked to the frequency and intensity of flooding.

With small and mobile organisms such as those that characterise disturbed ERS habitats in NW Europe, large-scale mark-resight/recapture studies are not possible and alternative methods are needed to understand dispersal characteristics. Population synchrony, which results from the interaction of density dependent and density independent variables and species dispersal, offers a potential means of doing this. Sadler *et al.* (2006) examined the population synchrony of three patchy populations of ERS specialists beetles [*Bembidion atrocaeruleum, Bembidion decorum* (Zenker in Panzer) and *Perileptus areolatus* (Creutzer)] on the River Severn in mid-Wales (UK) (Figure 3.3). As the study patches were all in one section of the same river it meant that any observed differences in population synchrony were unlikely to be due to weather, as these effects are normally on a scale of tens of kilometres (Moran, 1953; Raimondo *et al.*, 2004; Sutcliffe *et al.*, 1996). Any distance population synchrony patterns at this level were likely to be due to smaller scale dispersal effects (Ranta *et al.*, 1995; Sutcliffe *et al.*, 1996).

Perileptus areolatus, which is one of the UK ERS Biodiversity Action Plan (BAP) species, and is considerably rarer than the other two species, was the only species that exhibited any clear spatial patterns. Neither of the species of *Bemdidion* showed any significant population synchrony across the reaches. This is not especially surprising as both species are ERS specialists, capable fliers and able dispersers (Desender, 1989), and there were frequent flood dispersal cues due to the fairly natural hydrological regime of the upper River Severn. In contrast, *P. areolatus* populations showed clear structuring at a local reach scale and the regression plots of inter-annual synchrony against distance between patches suggest that this species very rarely disperses much further than 800 m (Figure 3.8).

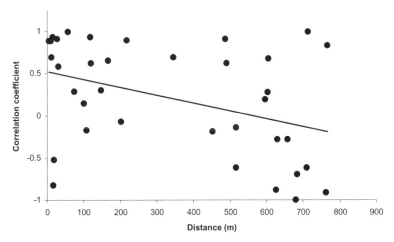

Figure 3.8 Linear regression relationship of the distance between ERS patches and Pearson's correlation coefficient for Perileptus areolatus *for reaches E and F (see Figure 3.3 for the map) ($p \leq 0.001$) (data collected 2002–4). Population synchrony was assessed by calculating the yearly fluctuation in species counts at each individual ERS and correlating these against each other in a pair-wise manner (Site 1 vs Site 2, 1 vs Site 3 and so on). These coefficients were then used as dependent variables and regressed against the independent variable measured by the direct downstream distance between mid-points of each ERS (in metres) (Sutcliffe* et al.*, 1996)*

3.6 How Much Disturbance is Needed to Sustain ERS Diversity?

Connell's (1978) 'Intermediate Disturbance Hypothesis' (IDH) predicts low levels of species diversity in highly disturbed habitats, where fugitive species with rapid colonisation and reproductive cycles dominate. Low species diversity also occurs in stable habitats as superior competitors can monopolize available resources. Intermediate levels of disturbance should maximise diversity as a wider range of organisms can coexist under these conditions. Ward and Tockner (2001) provide evidence to support the presence of IDH patterns in riverine biodiversity on a range of different river systems. However, the interpretation of disturbance diversity patterns in ERS specialist communities is difficult, because at very low levels of disturbance vegetation succession occurs and the habitat is no longer ERS.

The classification of ERS sites undertaken by Sadler *et al.* (2004) provides some corroboratory information supporting an intermediate disturbance diversity response in ERS invertebrates. The classification (Table 3.3) placed sites on a clear disturbance gradient with peaks in the diversity of specialist ERS species (Figure 3.9) and rare species (Figure 3.10) in sites characterised by an intermediate level of extreme disturbances (wandering gravel bed reaches). The diversity of ERS specialist beetles is low in environments with little disturbance (stable meandering and 'straight' reaches), as they require large expanses of bare sediments as habitat, and is also lower in environments with a greater level of disturbance (heavily braided, and mainly bedrock controlled reaches). Unfortunately, viewing disturbance in this manner is an oversimplification as the magnitude, frequency and timing of disturbances can have variable effects on different species. This is a research area that urgently requires further focus.

Table 3.3 Twinspan classification of ERS beetle assemblages from a survey of 69 shingle ERS in England and Wales in 1997–98 (Sadler et al., 2004)

Twinspan end groups	N	Description
1	9	Small upland cobble and boulder ERS in the Upper (Derbyshire) Derwent
2	3	Small upland cobble ERS in the River Wharfe catchment
3	11	Cobble–gravel lowland ERS (rivers Ystwyth, Tywi, Wharfe and Severn). A range of upland and lowland sites of varying size. Largely unvegetated.
4	11	Large heterogeneous sandy ERS (rivers Wye and Usk). Largely unvegetated.
5	12	Medium-large unvegetated lowland ERS from sites in Devon (rivers Teign, Exe, Torridge, Otter and Creedy Yeo).
6	16	Smaller vegetated ERS (rivers Camel, Erme, Thrushel, Frome, Yarty and lower Wharfe)
7	4	Polluted urban ERS on the rivers Calder and Worth
8	3	Small, heavily shaded chert ERS (Highland Water in the New Forest)

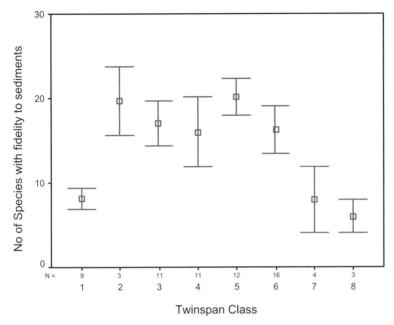

Figure 3.9 Plot of Twinspan end groups and the species richness of ERS specialists (error bars = 2 SE)

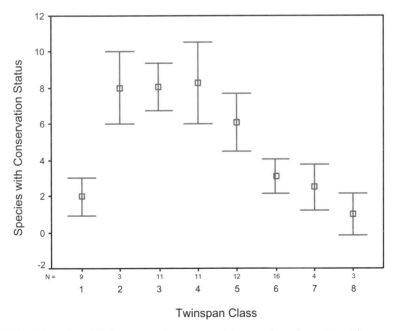

Figure 3.10 Box plot of Twinspan end groups and the number of species with conservation status (error bars = 2 SE)

3.7 Threats to ERS Invertebrate Biodiversity

Individual ERS habitat patches are affected by threats such as localised bank protection to reduce erosion and livestock damage, which normally operate at relatively small spatial and temporal scales. The most significant threats, however, are those that operate over longer timescales and at larger spatial scales (e.g., catchment level). These include anthropogenic activities such as flow regulation via impoundment, channelisation, abstraction, catchment landuse changes and climate change, all of which have the potential to modify natural hydrological disturbance regimes. These are regarded as being the main contributors to the considerable rarity of many ERS specialist invertebrates (Plachter and Reich, 1998; Sadler *et al.*, 2004; Stelter *et al.*, 1997).

The importance of modified hydrological dynamics within river catchments has only been addressed indirectly by studies examining the impacts of river regulation. River regulation can reduce ERS habitat availability and connectivity by increasing the rate of vegetation succession and reducing sediment supply through a reduction in the frequency and magnitude of flood events and the capture of sediment in impoundments (Petts, 1984; 1988; 2000). Sustained minimum flows can also accelerate succession by raising water tables, thus allowing plants to grow in areas that were previously too dry (Plachter and Reich, 1998). Pioneer unvegetated ERS decreased by around 80% between 1925 and 1985 (Plachter and Reich, 1998) after the construction of a dam on the Upper Isar in Germany (Manderbach and Reich, 1995). Significantly, the number of endangered carabid species inhabiting the downstream reach was only two, while 15 were found in reaches above the dam. In the UK, many large-scale changes in ERS habitat abundance appear to be the result of human modification of hydrological regimes through regulation and engineering and/or subtle medium to long-term climatic changes (Brewer *et al.*, 2000). River engineering (e.g., artificial levees, channel straightening) also alters fluvial dynamics and constrains channel migration, often leading to over deepening and the loss of ERS.

The population implications of reductions in the frequency and magnitude of high flow events include (Bates *et al.*, 2006): (i) reduced levels of dispersal both through flight and hydrochory; (ii) increased dispersal costs in an increasingly fragmented landscape to a point where the costs of dispersal outweigh the enhanced fitness gained by that dispersal, and (iii) a reduction in the selective advantage accrued by the investment in flight ability. All of these increases in costs will favour more generalist riparian species, giving them a competitive edge over specialist species. In addition to such competition, if the level of dispersal is reduced to a level that isolates populations, then local extinctions through demographic stochasticity would also be likely to occur. The implications for less common specialists, particularly those with lower dispersal capabilities, will be particularly extreme and local extinctions are likely (cf. Stelter *et al.*, 1997).

3.8 Conclusions

ERS habitats are unique environments globally that exhibit high alpha and beta biodiversity. The dynamic rivers that create them are under continued threat from regulation, catchment modifications, land-use changes and so on, and they are habitat for highly specialised and often rare species, which are threatened on national and international scales.

All the studies discussed in this chapter illustrate that inundation (flooding) is the engine that drives ERS ecological diversity by constantly regenerating old habitats and creating new ones. Population dynamics of the few invertebrates that have been studied intensively suggest that species exhibit metapopulation and patchy population characteristics that, crucially, are strongly affected by disturbance regimes.

Although a reasonable amount of work has been carried out on ERS habitats, there are very few autecological studies on ERS species. There is limited understanding of key ecological processes, such as competition, food-web dynamics, energetics and the relationship between productivity and species diversity (Huston, 1979; Loreau *et al.*, 2001), but the most important avenue of research is probably investigation of the influences of flow variability on ERS populations. This requires two steps: first, it requires the creation of a hydrodynamic model of the flow through various reaches using DEM (digital elevation model) data. This can be used initially to create quantitative outputs of real inundation frequencies for ERS patches (at a range of flow levels), which can be related statistically to variation in invertebrate survey data. Secondly, hydrodynamic models can be coupled with spatially explicit ecological models of species to provide a tool that helps to predict variability in populations in relation to flow variation scenarios (Petts *et al.*, 2006). Lastly, there is also a need to monitor populations over the long term to provide a basis for understanding the linkages between river flow variability, climate variability and ERS population fluctuations.

References

Adis, J. (1982) On the colonization of Central Amazonian inundation-forests (Varzea area) by carabid beetles (Coleoptera). *Archiv für Hydrobiologie*, **95**, 3–15.

Adis, J. and Junk, W.J. (2002) Terrestrial invertebrates inhabiting lowland river floodplains of Central Amazonia and Central Europe: a review. *Freshwater Biology*, **47**, 711–731.

Andersen, J. (1968) The effect of inundation and choice of hibernation sites of Coleoptera living on river banks. *Norsk Entomologisk Tidsskrift*, **15**, 115–133.

Andersen, J. (1969) Habitat choice and life history of *Bembidion* (Col., Carabidae) on river banks in central and northern Norway. *Norsk Entomologisk Tidsskrift*, **17**, 17–65.

Andersen, J. (1978) The influence of substratum on the habitat selection of Bembidiini (Col., Carabidae). *Norwegian Journal of Entomology*, **25**, 119–138.

Andersen, J. (1983) The habitat distribution of species of the tribe Bemdidiini (Coleoptera, Carabidae) on banks and shores in northern Norway. *Notulae Entomologicae*, **63**, 131–142.

Andersen, J. (1984) A re-analysis of the relationship between life cycle patterns and the geographic distribution of Fennoscandian carabid beetles. *Journal of Biogeography*, **11**, 479–489.

Andersen, J. (1985a) Ecomorphological adaptations of riparian Bembidiini species (Coleoptera, Carabidae). *Entomologica Generalis*, **12**, 41–46.

Andersen, J. (1985b) Humidity and responses and water balance of riparian species of Bembidiini (Col., Carabidae). *Ecological Entomology*, **10**, 363–375.

Andersen, J. (1985c) Low thermo-kinesis, a key mechanism in habitat selection by riparian *Bembidion* (Carabidae) species. *Oikos*, **44**, 499–505.

Andersen, J. (1986) Temperature response and heat tolerance of riparian Bembidiini species (Col., Carabidae). *Entomologica Generalis*, **12**, 57–70.

Andersen, J. (1988) Resource partitioning and interspecific interactions among riparian *Bembidion* species (Coleoptera: Carabidae). *Entomologica Generalis*, **13**, 47–60.

Andersen, J. and Hanssen, O. (2005) Riparian beetles, a unique, but vulnerable element of the fauna of Fennoscandia. *Biodiversity and Conservation*, **14**, 3497–3524.

Anon (1995) *Biodiversity: The UK Steering Group Report* DoE., London.

Antvogel, H. and Bonn, A. (2001) Environmental parameters and microspatial distribution of insects: a case study of carabids in an alluvial forest. *Ecography*, **24**, 470–482.

Bates, A.J. and Sadler, J.P. (2005) The ecology and conservation of beetles associated with exposed riverine sediments. Contract Science Report No. 668, CCW, Bangor.

Bates, A.J., Sadler, J.P., and Fowles, A.P. (2006) Condition-dependent dispersal of a patchily distributed riparian ground beetle in response to disturbance. *Oecologia*, **150**, 50–60.

Bates, A.J., Sadler, J.P., and Fowles, A.P. (2007) Microspatial distribution of beetles (Coleoptera) within spatially delimited patches of exposed riverine sediments. *Journal of European Entomology*, **104**, 479–487.

Bates, A., Sadler, J.P., Fowles, A.P., and Butcher, C.R. (2005) Spatial dynamics of beetles living on exposed riverine sediments in the upper River Severn: Method development and preliminary results. *Aquatic Conservation-Marine and Freshwater Ecosystems*, **15**, 159–174.

Bell, D., Petts, G.E., and Sadler, J.P. (1999) The distribution of species in the wooded riparian zone of three rivers in western Europe. *Regulated Rivers: Research and Management*, **15**, 141–158.

Benda, L., Poff, N.L., Miller, D., Dunne, T., Reeves, G., Pess, G., and Pollock, M. (2004) The network dynamics hypothesis: How channel networks structure riverine habitats. *Bioscience*, **54**, 413–427.

Bonn, A. (2000) Flight activity of carabid beetles on a river margin in relation to fluctuating water levels. In *Natural History and Applied Ecology of Carabid Beetles* (eds P. Brandmayr, G.L. Lövei, T.Z. Brandmayr, A. Casale and A.V. Taglianti), Pensoft, Sofia, pp. 147–160.

Brewer, P.A., Maas, G.S., and Macklin, M.G. (2000) A fifty-year history of exposed riverine sediment dynamics on Welsh rivers. In *Water in the Celtic world: Managing Resources for the 21st century*. British Hydrological Society, Occasional Paper No. 11, pp. 245–252.

Bunn, S.E. and Arthington, A.H. (2002) Basic principles and ecological consequences of altered flow regimes for aquatic biodiversity. *Environmental Management*, **30**, 492–507.

Connell, J.H. (1978) Diversity in tropical rain forests and corel reefs. *Science*, **199**, 1302–1310.

Dennis, L.H., Shreeve, T.G., and Van Dyck, H. (2003) Towards a functional resource-based concept for habitat: a butterfly biology viewpoint. *Oikos*, **102**, 417–426.

Desender, K. (1989) Ecomorphological adaptations of riparian carabid beetles. In *Proceeding of the Symposium 'Invertebrates of Belgium'*. L'Institut Royal des Science Naturelles de Belgique, Brussels, pp. 309–314.

Eyre, M.D. (2000) Preliminary assessment of the invertebrate fauna of exposed riverine sediments in Scotland. Scottish Natural Heritage Commissioned Report F97AC306 (unpubl. report).

Eyre, M.D., Lott, D.A., and Luff, M.L. (2001a) The rove beetles (*Coleoptera, Staphylinidae*) of exposed riverine sediments in Scotland and northern England: Habitat classification and conservation aspects. *Journal of Insect Conservation*, **5**, 173–186.

Eyre, M.D., Luff, M.L., and Phillips, D.A. (2001b) The ground beetles (Coleoptera : Carabidae) of exposed riverine sediments in Scotland and northern England. *Biodiversity and Conservation*, **10**, 403–426.

Fowles, A.P. (1989) The Coleoptera of shingle banks on the River Ystwyth, Dyfed. *Entomologist's Record*, **101**, 209–221.

Framenau, V., Dieterich, M., Reich, M., and Plachter, H. (1996a) Life cycle, habitat selection and home ranges of *Arctosa cinerea* (Fabricius, 1771) (Araneae: Lycosidae) in a braided section of the Upper Isar (Germany, Bavaria). *Revue Suisse de Zoologie*, 223–234.

Framenau, V., Reich, M., and Plachter, H. (1996b) Migration and prey of *Arctosa cinerea* (Fabricius, 1771) on a braided section of an Alpine River. *Verhandlungen der Gesellschaft für Ökologie*, **26**, 369–376.

Framenau, V.W., Manderbach, R., and Baehr, M. (2002) Riparian gravel banks of upland and lowland rivers in Victoria (south-east Australia): arthropod community structure and life-history patterns along a longitudinal gradient. *Australian Journal of Zoology*, **50**, 103–123.

Gilvear, D.J., Cecil, J., and Parsons, H. (2000) Channel change and vegetation diversity on a low-angle alluvial fan, River Feshie, Scotland. *Aquatic Conservation-Marine and Freshwater Ecosystems*, **10**, 53–71.

Harris, N.M., Gurnell, A.M., Hannah, D.M., and Petts, G.E. (2000) Classification of river regimes: a context for hydroecology. *Hydrological Processes*, **14**, 2831–2848.

Harrison, S. (1991) Local extinction in a metapopulation context – an empirical evaluation. *Biological Journal of the Linnean Society*, **42**, 73–88.

Hering, D., Gerhard, M., Manderbach, R., and Reich, M. (2004) Impact of a 100-year flood on vegetation, benthic invertebrates, riparian fauna and large woody debris standing stock in an Alpine floodplain. *River Research and Applications*, **20**, 445–457.

Huston, M.A. (1979) A general hypothesis of species diversity. *American Naturalist*, **113**, 81–101.

Junk, W.J. (1999) The flood pulse concept of large rivers: learning from the tropics. *Archiv für Hydrobiologie*, **3**, 261–280.

Lake, P.S. (2000) Disturbance, patchiness, and diversity in streams. *Journal of the North American Benthological Society*, **19**, 573–592.

Leopold, L.B. and Wolman, M.G. (1957) *River channel patterns: Braided, meandering and straight*, Geological Survey Professional Paper 282-B. United States Government Printing Office, Washington.

Looy, K.V., Honnay, O., Pedroli, B., and Muller, S. (2006) Order and disorder in the river continuum: the contribution of continuity and connectivity to floodplain meadow biodiversity doi:10.1111/j.1365–2699.2006.01536.x. *Journal of Biogeography*, **33**, 1615–1627.

Loreau, M., Naeem, S., Inchausti, P., Bengtsson, J., Grime, J.P., Hector, A., Hooper, D.U., Huston, M.A., Raffaelli, D., Schmid, B., Tilman, D., and Wardle, D.A. (2001) Ecology – Biodiversity and ecosystem functioning: Current knowledge and future challenges. *Science*, **294**, 804–808.

Lott, D.A. (1996) Beetles by rivers and the conservation of riparian and floodplain habitats. In *Environmental Monitoring, Surveillance and Conservation using Invertebrates* (ed M.D. Eyre). EMS Publications, Newcastle Upon Tyne, pp. 36–41.

Manderbach, R. and Plachter, H. (1997) Life strategy of the carabid beetle *Nebria picicornis* (Fabr. 1801) (Coleoptera, Carabidae) on river banks. *Beitrage Ökologie*, **3**, 17–27.

Manderbach, R. and Reich, M. (1995) Effects of dams and weirs on the ground beetle communities (Coleoptera, Carabidae) of braided sections of the Isar floodplain. *Archiv für Hydrobiologie Supplement*, **101**, 573–588.

Moran, P.A.P. (1953) The statistical analysis of the Canadian lynx cycle. II. Synchronization and meteorology. *Australian Journal of Zoology*, **1**, 291–298.

Naiman, R.J. and Decamps, H. (1997) The ecology of interfaces: Riparian zones. *Annual Review of Ecology and Systematics*, **28**, 621–658.

Paetzold, A. and Tockner, K. (2005) Effects of riparian arthropod predation on the biomass and abundance of aquatic insect emergence. *Journal of the North American Benthological Society*, **24**, 395–402.

Paetzold, A., Schubert, C.J., and Tockner, K. (2005) Aquatic–terrestrial linkages across a braided river: Riparian arthropods feeding on aquatic insects. *Ecosystems*, **8**, 748–759.

Paetzold, A., Bernet, J.F., and Tockner, K. (2006) Consumer-specific responses to riverine subsidy pulses in a riparian arthropod assemblage. *Freshwater Biology*, **51**, 1103–1115.

Parsons, H. and Gilvear, D. (2002) Valley floor landscape change following almost 100 years of flood embankment abandonment on a wandering gravel-bed river. *River Research and Applications*, **18**, 461–479.

Petts, G.E. (1984) Sedimentation within a regulated river. *Earth Surface Processes and Landforms*, **9**, 125–134.

Petts, G.E. (1988) Accumulation of fine sediments within substrate gravels along two regulated rivers, UK. *Regulated Rivers: Research and Management*, **2**, 141–153.

Petts, G.E. (2000) A perspective on the abiotic processes sustaining the ecological integrity of running waters. *Hydrobiologia*, **422**, 15–27.

Petts, G.E., Morales-Chaves, Y., and Sadler, J.P. (2006) Linking hydrology and biology to assess the water needs of river ecosystems. *Hydrological Processes*, **20**, 2247–2251.

Plachter, H. (1986) Composition of the carabid beetle fauna of natural riverbanks and of man-made secondary habitats. In *Carabid Dynamics: Their Adaptations and Dynamics* (eds P.J. den Boer, M.L. Luff, D. Mossakowski and F. Weber). Gustav Fisher, Stuttgart, pp. 509–535.

Plachter, H. and Reich, M. (1998) The significance of disturbance for population and ecosystems in natural floodplains. In *The International Symposium on River Restoration*. Tokyo, pp. 29–38.

Poff, N.L., Allan, J.D., Bain, M.B., Karr, J.R., Prestegaard, K.L., Richter, B.D., Sparks, R.E., and Stromberg, J.C. (1997) The natural flow regime. *Bioscience*, **47**, 769–784.

Raimondo, S., Liebhold, A.M., Strazanac, J.S., and Butler, L. (2004) Population synchrony within and among Lepidoptera species in relation to weather, phylogeny, and larval phenology. *Ecological Entomology*, **29**, 96–105.

Ranta, E., Kaitala, V., Lindstrom, J., and Linden, H. (1995) Synchrony in population-dynamics. *Proceedings of the Royal Society of London Series B – Biological Sciences*, **262**, 113–118.

Reich, M. (1991) Grasshoppers (Orthoptera, Saltatoria) on alpine and dealpine riverbanks and their use as indicators for natural floodplain dynamics. *Regulated Rivers-Research & Management*, **6**, 333–339.

Rice, S.P., Greenwood, M.T., and Joyce, C.B. (2001a) Macroinvertebrate community changes at coarse sediment recruitment points along two gravel bed rivers. *Water Resources Research*, **37**, 2793–2803.

Rice, S.P., Greenwood, M.T., and Joyce, C.B. (2001b) Tributaries, sediment sources, and the longitudinal organisation of macroinvertebrate fauna along river systems. *Canadian Journal of Fisheries and Aquatic Sciences*, **58**, 824–840.

Sadler, J.P. and Bell, D. (2002) *Invertebrates of Exposed Riverine Sediments – Phase 3 – Baseline communities*, Environment Agency, Bristol.

Sadler, J.P. and Petts, G.E. (2000) *Invertebrates of Exposed Riverine Sediments – Phase 2. Sampling Considerations*, Environment Agency, Bristol.

Sadler, J.P., Bell, D., and Fowles, A.P. (2004) The hydroecological controls and conservation value of beetles on exposed riverine sediments in England and Wales. *Biological Conservation*, **118**, 41–56.

Sadler, J.P., Bell, D., and Hammond, P.M. (2005) *An Assessment of the Distribution of Bembidion testaceum and the Reasons for its Decline*, Environment Agency, Bristol.

Sadler, J.P., Bell, D., and Bates, A.J. (2006) *The abundance and dynamics of Coleoptera on exposed riverine sediments on the River Severn in Mid-Wales*. CCW Contract Science, Report No: 754. CCW, Bangor.

Stelter, C., Reich, M., Grimm, V., and Wissel, C. (1997) Modelling persistence in dynamic landscapes: lessons from a metapopulation of the grasshopper *Bryodema tuberculata*. *Journal of Animal Ecology*, **66**, 508–518.

Sutcliffe, O.L., Thomas, C.D., and Moss, D. (1996) Spatial synchrony and asynchrony in butterfly population dynamics. *Journal of Animal Ecology*, **65**, 85–95.

Tockner, K. and Stanford, J.A. (2002) Riverine flood plains: present state and future trends. *Environmental Conservation*, **29**, 308–330.

Tockner, K., Malard, F., and Ward, J.V. (2000) An extension of the flood pulse concept. *Hydrological Processes*, **14**, 2861–2883.

Tockner, K., Paetzold, A., Karaus, U., Claret, C., and Zettel, J. (2006) Ecology of braided rivers. In *Braided Rivers* (eds G. Sambrook-Smith, J. Best, S. Lane and G.E. Petts). IAS Special Publication, No. 36, Blackwell Publishing, Oxford, pp. 339–359.

Van Looy, K., Vanacker, S., Jochems, H., de Blust, G., and Dufrêne, M. (2005) Ground beetle habitat templets and riverbank integrity. *River Research and Applications*, **21**, 1133–1146.

Ward, J.V. (1998) Riverine landscapes: biodiversity patterns, disturbance regimes, and aquatic conservation. *Biological Conservation*, **83**, 269–278.

Ward, J.V. and Tockner, K. (2001) Biodiversity: towards a unifying theme for river ecology. *Freshwater Biology*, **46**, 807–819.

Ward, J.V., Tockner, K., Arscott, D.B., and Claret, C. (2002) Riverine landscape diversity. *Freshwater Biology*, **47**, 517–539.

Yoshimura, C., Omura, T., Furumai, H., and Tockner, K. (2005) Present state of rivers and streams in Japan. *River Research and Applications*, **21**, 93–112.

Zerm, M. and Adis, J. (2003) Exceptional anoxia resistance in larval tiger beetle, *Phaeoxantha klugii* (Coleoptera : Cicindelidae). *Physiological Entomology*, **28**, 150–153.

Zulka, K.P. (1994) Carabids in a Central European floodplain: species distribution and survival during inundations. In *Carabid Beetles: Ecology and Evolution* (eds K. Desender, M. Dufêne, M. Loreau, M.L. Luff and J.-P. Maelfait). Kluwer, Dordrecht, pp. 399–405.

4

Aquatic–Terrestrial Subsidies Along River Corridors

Achim Paetzold, John L. Sabo, Jonathan P. Sadler, Stuart E.G. Findlay, and Klement Tockner

4.1 Introduction

Rivers are open ecosystems that are characterized by intensive spatial flows of matter and organisms between land and water (Ward, 1989). A number of theoretical concepts have been developed to explain how trans-boundary energy flow varies along a river corridor. The river continuum concept proposes that detritus input from riparian forests fuels aquatic food webs in headwater streams (Vannote *et al.*, 1980). The riverine productivity model and its revised version discuss the relative role of allochthonous input versus autotrophic production in large river food webs (Thorp and Delong, 1994; 2002), and the flood pulse concept stresses the importance of lateral interactions on biogeochemical processes and nutrient transfer rates in large floodplain rivers (Junk *et al.*, 1989; Tockner *et al.*, 2000; Junk and Wantzen, 2004). Common to these models is the focus upon the role of terrestrial inputs into aquatic food webs, while energy exchanges in the opposite direction have received less attention. Recent studies have emphasized the importance of energy flow from aquatic to terrestrial systems (Sabo and Power, 2002a; Power *et al.*, 2004b; Paetzold *et al.*, 2005). Bi-directional aquatic–terrestrial energy flows have, however, not yet been evaluated, either conceptually or empirically, in a landscape context (but see Power *et al.*, 2004 and Baxter *et al.*, 2005).

In this review, we will place the existing knowledge of aquatic–terrestrial interactions along rivers in a conceptual framework. Our main focus is on temperate rivers that have bedrock controlled headwaters, alluvial braided mid-sections and alluvial meandering lower reaches. Finally, we will emphasize the potential effects of human impacts on aquatic–terrestrial linkages and the resulting implications for river management.

4.2 What Controls Aquatic–Terrestrial Flows?

Spatial subsidies describe situations where resources (prey, detritus) from one habitat increase productivity of a recipient consumer in another habitat (Polis *et al.*, 1997). Wherever organic material, prey or consumers move across habitat boundaries, subsidies can occur (Polis and Hurd, 1996a), and the importance of these subsidies is related to the relative productivity disparity between the donor and recipient habitats. Transport pathways across the aquatic–terrestrial boundary are either physical or biotic. In rivers, water transports material and organisms downstream, and flow or flood pulses transfer matter between land and water. Active movements of organisms can vary widely in their direction and spatiotemporal scale.

The productivity gradient, boundary length, and boundary permeability between adjacent systems are primary factors that control the efficiency of spatial subsidies (Polis *et al.*, 1997; Cadenasso *et al.*, 2004; Witman *et al.*, 2004). The flow direction is generally from the more productive to the less productive habitat (Polis *et al.*, 1997). However, the subsidy effect also depends on the food quality of allochthonous resources relative to *in situ* resources. Low food quality of terrestrial detritus inputs compared with algae might explain why in streams with catchments >$10\,km^2$, many aquatic consumers rely on algal carbon, even though most organic matter is terrestrially derived (Finlay, 2001). The rate of flow between habitats is also positively related to boundary length (Polis and Hurd, 1996b). Boundary permeability controls the flux and travel distance of material (organisms and matter) into recipient habitats (Witman *et al.*, 2004). Structural complexity of the boundary will strongly affect the passive flux of material, whereas boundary permeability for actively moving organisms is determined largely by the habitat requirements and movement characteristics of the organisms in question (Puth and Wilson, 2001; Briers *et al.*, 2005).

Inundation characteristics control the passive transport of matter across the aquatic–terrestrial boundary, and these are determined by the flow regime together with local floodplain topography. Inundation magnitude controls the total amount of material in transport while inundation frequency is expected to determine the rate of exchange, and floodplain topography together with the structural complexity of the boundary determines the lateral travel distance. Timing of the flood can also influence the subsidy effect. For instance, a summer flood, when terrestrial organisms could actively escape the water, can result in a net flux from the river to the terrestrial habitat because of stranding of aquatic organisms and organic matter after the flood recedes. However, a winter flood, when most terrestrial organisms are inactive and cannot escape the water, may result in a net input of terrestrial organisms into the aquatic habitat.

Aquatic–terrestrial fluxes can be temporally highly variable. Nakano and Murakami (2001) demonstrated that asynchronous seasonal changes in productivity between a stream and its riparian forest resulted in reciprocal and alternate aquatic–terrestrial subsidies. Mass emergence during short time-periods is common among aquatic insects, and peaks in surface drifting organisms occur during storms through accidental input of terrestrial invertebrates to the stream (Mason and Macdonald, 1982; Sweeney and Vannote, 1982). The subsidy efficiency of such pulsed resources depends on the short-term aggregative response of the recipient consumers (Ostfeld and Keesing, 2000; Paetzold et al., 2006). Power et al. (2004a) found that ground dwelling lycosid spiders efficiently responded to changes in aquatic insect emergence within 24 hours, while the web-building filmy dome spider did not track aquatic insect emergence, probably because filmy dome spiders invest so massively in web production that repositioning is not energy efficient. Paetzold et al. (2006) demonstrated that ground-dwelling riparian arthropods of gravel bars in the Tagliamento River (NE Italy) can respond efficiently to spatially heterogeneous subsidy pulses.

4.2.1 Subsidies from Land to Water

The role of terrestrial detritus as an important subsidy for aquatic invertebrates has been widely recognized (Table 4.1), particularly in temperate forest streams where detritus input exceeds within-stream primary production (Fisher and Likens, 1973; Wallace et al., 1997; 1999). In large lowland floodplains, terrestrial fruits and seeds are important food sources for many fish species (Goulding, 1980; Chick et al., 2003). Terrestrial arthropods that fall into the stream represent another important type of subsidy for fish (Cloe and Garman, 1996; Wipfli, 1997; Nakano et al., 1999; Kawaguchi et al., 2003; Baxter et al., 2005). For example, in a Japanese forest stream terrestrial invertebrates contributed ~50% to the annual prey consumption of salmonid fishes (Nakano et al., 1999b; Kawaguchi and Nakano, 2001). Semi-aquatic animals that feed in terrestrial habitats can transfer large amounts of terrestrial biomass into aquatic habitats (Naiman and Rogers, 1997). For instance, hippopotami feed on terrestrial grasses during night and return to water by day. Thereby, a single hippopotamus transfers approximately 9 tons dry mass of faeces to the aquatic system annually (Heeg and Breen, 1982).

4.2.2 Subsidies from Water to Land

A variety of organisms, including insects, amphibians and reptiles, require both aquatic and terrestrial habitats to complete their life cycle. These shifts in life-history stages can result in significant flows of organisms across the aquatic–terrestrial boundary (Schneider et al., 2002). Emerging aquatic insects represent important subsidies for many terrestrial consumers including arthropods, lizards, birds, and bats (Table 4.1). For instance, some riparian beetles living along braided riverbanks rely almost entirely on emerging aquatic insects (Hering and Plachter, 1997; Paetzold et al., 2005). Given their small size, riparian arthropods can exploit small prey items that are not taken by larger vertebrates, thereby converting aquatic prey into biomass that can then be used by larger riparian predators (Paetzold et al., 2005). Terrestrial predation on emerging aquatic insects can result in a significant transfer of aquatic derived energy into terrestrial food webs (Jackson and Fisher, 1986; Paetzold and Tockner, 2005; Paetzold et al., 2005). For example, in a Sonoran desert stream only 3% of the biomass of emerging insects returned to the stream (Jackson and Fisher, 1986).

Amphibians feed on algae and aquatic invertebrates during their aquatic larval stage and leave the water upon metamorphosis. During their terrestrial stage many amphibians

Table 4.1 Important pathways of aquatic–terrestrial subsidies along rivers

Input pathway	Recipient consumer	References
Land to water		
Detritus	Benthic invertebrates	Fisher and Likens, 1973; Wallace et al., 1997; 1999
Invertebrate	Fish	Mason and Macdonald, 1982; Cloe and Garman, 1996; Wipfli, 1997; Nakano et al., 1999a,b; Nakano and Murakami, 2001; Kawaguchi and Nakano, 2001; Kawaguchi et al., 2003; see review by Baxter et al., 2004
Fruits and seed	Fish	Goulding, 1980; Chick et al., 2003
Water to land		
Insect emergence	Riparian arthropods	Hering and Plachter, 1997; Collier et al., 2002; Kato et al., 2003; Sanzone et al., 2003; Power et al., 2004a; Baxter et al., 2004; Akamatsu et al., 2004; Briers et al., 2005; Paetzold and Tockner, 2005; Paetzold et al., 2005
	Lizards	Sabo and Power, 2002a,b
	Birds	Murakami and Nakano, 2001; Nakano and Murakami, 2001; Iwata et al., 2003
	Bats	Power and Rainey, 2000; Power et al., 2004
Predation on fish	Birds, mammals	Draulans, 1988; Willson and Halupka, 1995; Gende et al., 2002; Willson et al., 2004
Stranding of fish	Insects, vertebrates	Johnson and Ringler, 1979; Willson et al., 2004
Stranding of algae	Riparian grasshoppers	Bastow et al., 2002
Surface drift	Riparian arthropods	Hering and Plachter, 1997; Paetzold, 2005; Paetzold et al., 2006

migrate hundreds of meters away from their breeding habitat, thereby transporting aquatic derived biomass far into surrounding terrestrial habitats (Semlitsch, 1998; Davic and Welsh, 2004). In the north-eastern United States, salamanders can be the dominant vertebrate predators in both headwater streams and in forest ecosystems (Burton and Likens, 1975; Lowe and Bolger, 2002).

Feeding in aquatic habitats is common among several terrestrial vertebrate consumers. Predation by terrestrial vertebrate predators on fish can be an important pathway of aquatic biomass transfer into terrestrial food webs (Draulans, 1988; Willson and Halupka, 1995; Gende et al., 2002). Bears can carry salmon carcasses over 75 m from the stream (Willson et al., 2004) thereby fertilizing the terrestrial riparian system (Ben-David et al., 1998; Hilderbrand et al., 1999; Helfield and Naiman, 2001; Naiman et al., 2002). Direct consumption of aquatic primary production by terrestrial consumers appears to be less common. However, some terrestrial herbivores, such as moose, move regularly into aquatic habitats to feed on aquatic macrophytes (Naiman and Rogers, 1997).

Stranded organisms provide another aquatic–terrestrial subsidy pathway. Stranding and accumulation of senescing and dying fish is an important aquatic subsidy along streams with runs of anadromous semelparous salmon. Most anadromous Pacific salmon species (*Oncorhynchus* spp.) die after spawning and their carcasses are fed upon by a variety of scavengers, such as eagles, corvids, gulls, carnivorous mammals, and flies and beetles soon after deposition on land (Johnson and Ringler, 1979; Willson *et al.*, 2004). Because >95 % of salmon growth is accumulated during the marine phase of their life cycle, migrations of Pacific salmon can transfer significant amounts of marine-derived nutrients to freshwater and adjacent riparian habitats (Helfield and Naiman, 2001; Schindler *et al.*, 2005). Although very little is known about aquatic–terrestrial subsidy dynamics in temporary aquatic habitats, aquatic organisms that are stranded during receding water levels might be a particularly important pathway in temporary streams. Bastow *et al.* (2002) demonstrated that stranded aquatic algae provide an important food source for riparian grasshoppers during summer droughts.

Surface-floating organisms that accumulate along the shoreline or strand during receding flood pulses in terrestrial habitats can be another important subsidy for riparian arthropods, particularly ants (Paetzold *et al.*, 2006). A high percentage of surface floating organisms can be of terrestrial origin (Cloe and Garman, 1996; Langhans, 2000; Tenzer, 2003; Trottmann, 2004; Paetzold *et al.*, 2006). Therefore, this subsidy pathway represents a feedback of upstream-derived terrestrial energy back to downstream terrestrial food webs unless consumed by aquatic organisms and decomposers.

It is clear that there is a variety of subsidies along river margins that can influence aquatic and terrestrial food webs. However, more empirical and experimental research is required to identify factors controlling the magnitude, direction and persistence of these exchanges.

4.3 Aquatic–Terrestrial Flows Along River Corridors

As landscape characteristics and aquatic primary production change along a river corridor (Vannote *et al.*, 1980; Naiman *et al.*, 1987), we can expect concurrent changes in the direction, type and pathways of aquatic–terrestrial flows. Power and Rainey (2000) predicted that aquatic insect emergence and lateral flux distribution change along the river corridor. Here we discuss the bi-directional flow across aquatic–terrestrial boundaries in forested headwater streams and braided gravel-bed rivers to highlight the shifts in importance of various pathways of exchange.

4.3.1 Aquatic–Terrestrial Subsidies in Forested Headwater Streams

In small forested headwater streams, primary production is limited by shade, and inputs of terrestrial detritus provide the bulk of the resource base for primary consumers (Vannote *et al.*, 1980; Wallace *et al.*, 1999). In temperate zones, annual terrestrial litter input varies from ~200 to 900 g m^{-2} ash-free dry mass in small-medium sized streams (Naiman and Décamps, 1997). Terrestrial invertebrates that accidentally fall into the stream can contribute an annual input of ~9–35 g m^{-2} dry weight in forest streams (Mason and Macdonald, 1982; Kawaguchi and Nakano, 2001). This input represents an important subsidy for fish because of their high food quality (i.e., low C:N ratio) and can exceed within-stream production of benthic invertebrates (Mason and Macdonald, 1982).

Detailed investigations in a deciduous forest stream in Japan showed that aquatic–terrestrial fluxes and subsidy effects changed seasonally. Terrestrial insect inputs were highest in summer when aquatic insect biomass in the stream was low and aquatic insect emergence was highest in spring when terrestrial invertebrate biomass in the forest was low. Aquatic insect emergence (~1–13 mg m^{-2} day^{-1}) was about 10-times lower than input of terrestrial invertebrates to the stream (~10–100 mg m^{-2} day^{-1}). However, aquatic insect emergence provided a significant subsidy to riparian birds and web-building spiders, particularly during spring (Nakano et al., 1999a; Kato et al., 2003). Emerged aquatic insects contributed ~26% to the annual resource budget of the forest bird assemblage (Nakano et al., 1999b). The importance of emerged aquatic insects varied among bird taxa with the highest contribution to the annual resource budget being of the great tits (39%) and nuthatches (32%) (Nakano and Murakami, 2001). The proportion of emerged aquatic insects to web-building spider diets also differed among guilds from 53–91% for horizontal orb weavers, 52% for sheet weavers, and <15% for vertical orb weavers between May and July (Kato et al., 2003; 2004). The role of aquatic insect emergence as a food source for ground-dwelling arthropods in riparian forests is, however, unknown.

The structural complexity of riparian zones might also affect food-web subsidies. For example, a high density of hedgerows in an agricultural catchment can reduce lateral movements of emerged aquatic insects into terrestrial habitats (Delettre and Morvan, 2000). Hedgerows might have a filtering effect and provide shelter for adult insects whereas in open landscapes higher wind speeds can passively transport weak flying insects further into the terrestrial habitat (Drake and Farrow, 1998; Delettre and Morvan, 2000). Along forested streams fluxes of adult aquatic insects decline exponentially with distance from the stream, with reductions in abundance by ~50% within 10 m from the streams edge (Petersen et al., 1999; Iwata et al., 2003; Power et al., 2004a).

Structural complexity can also determine the recipient terrestrial consumer assemblage and, therefore, the dominant pathway of aquatic insect emergence to the terrestrial food web. High structural complexity provides habitat for a diverse and abundant web-building spider assemblage that can position their webs close to the stream edge. Birds capture emerged aquatic insects in the air or on the ground (Murakami and Nakano, 2001). However, bat foraging is probably limited in headwaters because prey detection is constrained by interference from turbulent water and dense riparian vegetation, and flight is limited by dense vegetation (von Frenckell and Barclay, 1987; Brigham et al., 1997).

In conclusion, we hypothesize that generally, terrestrial inputs to headwater streams are much higher than are aquatic inputs to the riparian forest. However, in some headwater streams, predation on fish can result in a significant energy transfer into riparian habitats, particularly those with salmon runs. In forested headwater streams, a significant proportion of aquatic insect production is fuelled by terrestrial litter input. During emergence, aquatic insects transfer some of this energy back into the terrestrial system, transformed into high-quality food for riparian predators (Paetzold et al., 2005). Floods should result in an increased terrestrial input to the streams because they can wash litter from the forest floor and terrestrial invertebrates into the stream (Power et al., 2004a).

4.3.2 Aquatic–Terrestrial Subsidies in a Braided River Reach

Aquatic habitats in the active corridor of braided rivers tend to be spatially separated from riparian vegetation by extensive gravel areas (Figure 4.1). Consequently, they are likely to be characterized by low levels of shading and lateral inputs of terrestrial detritus and invertebrates. In a Sonoran desert stream, where the riparian forest was 1–30 m from the stream edge, Jackson and Fisher (1986) found very low input rates of terrestrial insects. Thus, in braided rivers aquatic primary production should be high and lateral terrestrial matter inputs very low compared with headwaters.

In the island-braided reach of the Tagliamento River (Tockner *et al.*, 2003; Figure 4.1), NE Italy, average aquatic insect emergence between April and October was 30 mg m^{-2} day^{-1} along the wetted channel edge (Paetzold and Tockner, 2005; Paetzold *et al.*, 2005). Additional emergence of aquatic insects (~10 mg m^{-2} day^{-1}) occurred on the terrestrial part of the shore and was dominated by stoneflies, which often crawl on land for emergence (Paetzold and Tockner, 2005). In this river reach, the average area of aquatic habitat and shoreline length was ~130 m^2 river m^{-1} and ~14 m river m^{-1}, respectively (Arscott *et al.*, 2000; van der Nat *et al.*, 2002). Aquatic insect emergence of a river reach appears to be positively correlated with channel length (Iwata *et al.*, 2003), and accordingly, high aquatic habitat area and shoreline length in braided rivers should result in high emergence of aquatic insects.

The riparian arthropod community along braided rivers is dominated by ground-dwelling carnivorous ground beetles, rove beetles, spiders, and ants (Plachter, 1986; Lude *et al.*, 1999; Sadler *et al.*, 2004; Paetzold *et al.*, 2005), many of which rely substantially

Figure 4.1 Island-braided reach along the Tagliamento River (NE Italy) at base flow (width of the active corridor: 800m). Note the high shoreline length, variety of aquatic habitats (secondary channels, backwaters, ponds) in the active corridor, and the separation of aquatic habitats from the riparian forest by open gravel surfaces

on aquatic subsidies (Paetzold *et al.*, 2005). Along the Tagliamento River, the dominant ground beetles feed entirely on aquatic insects, predominantly stoneflies that leave the water before emergence (Figure 4.2). The diets of the dominant rove beetle, *Paederidus rubrothoracicus*, and lycosid spiders consist of 80% and ~50% aquatic insects respectively. However, the contribution of emerging aquatic insects to the diets of riparian arthropods can be also determined by the availability of alternative terrestrial prey. Along gravel banks of small streams, where the influence of the adjacent forests is higher, the contribution of aquatic insects to ground beetle diets can be as low as 34% as a result of a greater availability of alternative terrestrial food sources relative to the level of aquatic insect emergence (Hering and Plachter, 1997).

The aquatic–terrestrial boundary in braided rivers is very permeable to both aquatic insects and riparian arthropods. For instance, stoneflies crawl on land as larvae before emergence and spiders can prey on emerging aquatic insects directly from the water surface (Figure 4.2) (Paetzold and Tockner, 2005). Field exclosure experiments showed

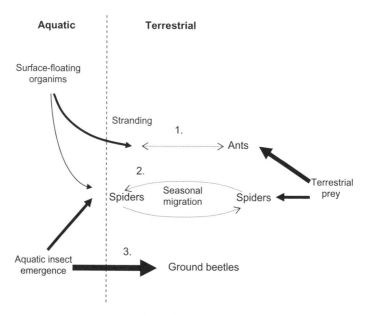

Figure 4.2 *Aquatic–terrestrial subsidy pathways along a gravel bank in the Tagliamento River (NE Italy). 1. Ants feed on surface-floating organisms that strand during receding flows. Ants further transport riverine subsidies into the terrestrial habitat by their movements between foraging sites and nests. 2. Lycosid spiders feed on emerging aquatic insects and surface floating organisms directly from the water surface. Spiders can transfer aquatic derived energy into terrestrial habitats through seasonal migrations away from the stream edge. 3. Ground beetles feed entirely on aquatic insects, particularly on those that crawl on land for emergence, such as many stoneflies. Black arrows show feeding linkages and their sizes indicate the relative strength of the feeding linkages. Dotted lines indicate the movements of riparian arthropods*

that riparian arthropods consume up to 45% of aquatic insects emerging along the river's edge during certain periods of the year, with even stronger effects on certain taxa, such as caddis flies (Paetzold and Tockner, 2005). Arthropod predation, therefore, is an important pathway for the transfer of aquatic secondary production into the terrestrial system.

Surface-floating organisms that accumulate along the stream edge represent another important subsidy along braided rivers (Figure 4.2). Biomass of surface floating organisms close to the stream edge varied from ~60–600 mg m^{-2} day^{-1} along the Tagliamento River (Paetzold et al., 2006). Terrestrial arthropods contributed a high proportion (58–67%) to the biomass of floating organisms, indicating inputs from further upstream where riparian vegetation bordered the river channel. Spiders may use this resource by direct predation from the water surface. It is likely that a high proportion of surface-floating organisms accumulate in the porous gravel matrix along the stream edge on the receding limb of flow pulses (see Tockner et al., 2000). Because storms result in an increased input of terrestrial invertebrates, most of the stranded organisms should be of terrestrial origin (Mason and Macdonald, 1982; Langhans, 2000). Ants, for example, opportunistically feed on stranded organisms, particularly after floods (Paetzold et al., 2006). The ability of ants to form attractive trails between the nest and the food source allows them to exploit pulsed resources very effectively and to transport this riverine subsidy further into the terrestrial habitat. For this type of subsidy the river functions more like a conveyer that transports terrestrial invertebrate biomass produced further upstream to downstream riparian consumers.

Dominant subsidy pathways can change seasonally because of changes in both the composition of the riparian consumer community and the timing of aquatic insect emergence. Along the Tagliamento River, ground beetles were abundant on the river's edge in spring and autumn when they fed predominantly on aquatic insects that crawl onto land for emergence (Paetzold and Tockner, 2005). Spiders moved to the river's edge in summer and were able to feed on emerging aquatic insects directly from the water surface. Seasonal separation and resource partitioning by riparian arthropods resulted in aquatic fluxes to the terrestrial system throughout the active period of terrestrial arthropods (March–October). Thus, along braided rivers riparian consumers are subsidized by aquatic inputs during most of the year, whereas in headwater streams significant aquatic fluxes into the terrestrial system occur predominantly in spring.

Low structural complexity of gravel banks allows the transport of surface-floating material far into the active flood plain during floods. With receding water levels, organisms and matter accumulate in discrete drift lines on the gravel banks. Depressions in the flood plain, which are connected with the main channel during floods, can trap aquatic invertebrates and fish when the water level decreases. In the island-braided reach of the Tagliamento River 21 out of 39 ponds in the active floodplain were temporary (Karaus et al., 2005). When these habitats dry out stranded aquatic organisms are easy prey for terrestrial scavengers and predators. Inputs of terrestrial organisms to the river during floods should be low because riparian arthropods of braided rivers are highly mobile or can survive floods being submerged (Andersen, 1968; Plachter and Reich, 1998; Tockner et al., 2005). Consequently, floods in braided rivers are likely to increase the flow of aquatic resources into terrestrial habitats. Only exceptionally large floods that erode vegetative islands and disturb refuges of riparian arthropods may result in a significant input of terrestrial matter and organisms into the aquatic system (cf. Hering *et al.*, 2004).

The predaceous arthropod community along braided rivers is dominated by ground-dwelling taxa because low structural complexity together with a high inundation frequency limits the abundance of web building spiders along braided river banks. Lizards, birds and bats might be other important predators that rely directly or indirectly on aquatic subsidies. For instance, exposed riverine sediments provide favorable thermal environments for lizards, resulting in high lizard densities along river margins in some areas (Sabo, 2003; Power *et al.*, 2004a). Lizards can feed directly on emerged aquatic insects and during times of low aquatic insect emergence they might also feed on ground-dwelling arthropods that have derived most of their energy from aquatic habitats (Sabo and Power, 2002a,b). Riparian ground beetles and spiders are also important food sources for bats, shrews, and birds (Kolb, 1958; Thiele, 1977; Forster and Forster, 1999). Birds and bats are likely to be important predators on emerged aquatic insects in braided rivers because the open space over aquatic habitats can facilitate aerial predation (Iwata *et al.*, 2003). Standing water bodies and slowly flowing waters, which are abundant in active braided river floodplains, are preferred foraging habitats for bats (von Frenckell and Barclay, 1987; Power and Rainey, 2000). Bats can transfer significant amounts of aquatic derived nutrients further into the terrestrial habitat by guano deposition at their feeding roosts (Power *et al.*, 2004a). Many riparian arthropods might transfer aquatic derived energy further into terrestrial habitats by seasonal habitat movements, such as the lycosid spider *Arctosa cinerea* and some riparian ground beetles that move away from the stream edge for winter diapause or hibernation, respectively (Andersen, 1968; Framenau *et al.*, 1996). The different aquatic habitats of braided river corridors provide breeding sites for a diverse and abundant amphibian fauna (Klaus *et al.*, 2001). Their migration, after metamorphosis, into surrounding forests might represent another important vector of aquatic energy into terrestrial habitats.

In conclusion, the high productivity gradient between aquatic habitats and adjacent gravel banks, the high boundary length and permeability along braided rivers are all prerequisites for an intensive transfer of aquatic-derived energy into surrounding terrestrial habitats. Surface-floating matter and organisms derived from upstream terrestrial inputs accumulate along braided rivers and get returned into downstream terrestrial food webs.

4.3.3 Aquatic–Terrestrial Subsidies in Temperate Lowland Rivers

For lowland rivers, almost no data are available on aquatic–terrestrial organism exchanges. Riparian zones of lowland rivers tend to be vegetated but the importance of terrestrial inputs for aquatic biota is likely to be lower compared with headwater streams because of the higher instream primary productivity and a larger volume of water relative to stream edge. Aquatic inputs to riparian habitats can be high because open channel and low flowing waters are ideal foraging sites for birds and bats. Henschel *et al.* (2001) showed that aquatic insect emergence made up ~50% of the diet of riparian spiders along a lowland reach of the Main River, Germany. The diet of riparian web-building spiders consisted of 54%, on average, of aquatic insect in the middle reach of the Chikuma River, Japan (Akamatsu *et al.*, 2004). High reciprocal aquatic–terrestrial fluxes can be expected in oxbows and flood-plain ponds because such habitats receive high terrestrial inputs from the surrounding forest (K. Tockner, unpublished data), are breeding sites for amphibians, and aquatic insect emergence is likely to be higher than off-channel habitats (Power and Rainey, 2000). During floods, large amounts of organic matter from the floodplain floor together with organisms might get transported into the aquatic system. Further investigation of entire consumer assemblages and their relative reliance on allochthonous food sources are required the better to understand reciprocal trophic fluxes along a river corridor.

4.4 Influence of Human Impacts on Aquatic–Terrestrial Subsidies

Human impacts, such as flow regulation, channelization, and riparian land-use changes, can impair the reciprocal energy exchange between the river and its riparian zone (Nakano and Murakami, 2001). Effects on trophic linkages are likely to be caused by alterations of the aquatic and terrestrial biota and the aquatic–terrestrial boundary characteristics. While riparian deforestation directly affects the input of terrestrial detritus to the stream, other impacts can function more indirectly, for instance through invasions of exotic species (Kawaguchi and Nakano, 2001; Baxter et al., 2004) or alteration in thermal regime. Little is known about the effects of human impact on the bidirectional flux across the aquatic–terrestrial boundary. However, we can hypothesize some potential effects based on general mechanisms that control spatial subsidies between systems.

4.4.1 Riparian Deforestation

Riparian deforestation generally results in reduced inputs of terrestrial detritus and invertebrates to the stream and higher in-stream primary production because of reduced shading (Bilby and Bisson, 1992; Kawaguchi and Nakano, 2001). Reduced canopy density generally increases aquatic primary production, which can also result in increased aquatic insect emergence (Newbold et al., 1980; Hawkins et al., 1982; Bilby and Bisson, 1992; Tait et al., 1994). However, riparian deforestation can also cause stream narrowing, resulting in a loss of benthic habitat per unit of channel length and macroinvertebrate abundance (Sweeney et al., 2004). Deforestation reduces boundary complexity and this may result in an increased travel distance of emerging insects into the terrestrial system (Delettre and Morvan, 2000). However, Petersen et al. (2004) found that most adult aquatic insects stayed close to the stream channel independent of difference in landuse of the catchment (coniferous plantation forest, cleared forest, moorland). Reduced complexity of the riparian habitat may allow an increased abundance of recipient consumers that depend on open habitats for foraging, such as many bats, while abundance of consumers utilizing forested habitats will decrease (Brigham et al., 1997). Thus, deforestation is likely to change aquatic–terrestrial interactions but further research is needed to understand how deforestation affects energy flux from streams to surrounding terrestrial habitats.

4.4.2 River Channelization and Regulation

River channelization can result in a drastic reduction in shoreline length, particularly in braided river reaches where shoreline length is naturally very high (Ward et al., 2002).

Consequently, trophic linkages among aquatic–terrestrial systems will be reduced because most of the aquatic–terrestrial interactions occur directly along the shoreline (Polis and Hurd, 1996b; Iwata et al., 2003). Channelization of braided rivers can alter inundation dynamics of remaining gravel banks resulting in a decrease of riparian arthropod abundance and diversity (Paetzold, 2005). Regulation of the flow regime, such as hydropeaking, can also cause reductions in abundance and diversity of aquatic and riparian invertebrates (Petts, 1984; Moog, 1993; Paetzold, 2005). Consequently, river channelization and regulation are likely to reduce the flow of aquatic organisms into the terrestrial system. Flow regulation in braided rivers with reduced floods can cause losses of open gravel habitats by enhanced vegetation succession (Brewer et al., 2000; Gilvear et al., 2000). With increasing vegetation development the productivity gradient between aquatic and riparian systems will decrease and as a consequence the flux from aquatic to terrestrial habitats might be reduced.

4.5 Conclusions

Aquatic and terrestrial food webs are tightly linked by trophic fluxes across aquatic–terrestrial boundaries through multiple pathways. The prominent direction of aquatic–terrestrial fluxes appears to shift from forested headwater streams, where terrestrial input dominates, to open braided rivers, which exhibit a net transfer of aquatic-derived energy in the form of organisms into riparian food webs. Riparian land use and river regulation can affect the bidirectional flow of matter and organisms along river corridors. However, further research is needed to understand what are the effects of human alterations of riverine landscapes on aquatic–terrestrial food web interactions and their consequences on ecosystem functioning. Developing a better understanding of aquatic–terrestrial feeding exchanges is essential for the management of both habitats, as changes in one will inextricably affect the other. These exchanges further highlight the considerable importance of hydroecology as a distinct research field.

4.6 Future Research

To understand better the complex linkages between aquatic and terrestrial systems we propose some directions for future research:

(i) How does the diversity of aquatic and terrestrial consumers affect the transfer between the two systems?
(ii) How does the availability and diversity of spatial subsidies affect the populations, communities, and ecosystem processes in recipient habitats?
(iii) Can terrestrial predation affect aquatic insect population dynamics?
(iv) How does the spatial extent of aquatic subsidies into terrestrial habitats change along a river corridor?
(v) What are the human impacts on aquatic–terrestrial subsidies?

References

Akamatsu, F., H. Toda, and T. Okino. 2004. Food source of riparian spiders analyzed by using stable isotope ratios. *Ecological Research*, **19**:655–662.

Andersen, J. 1968. The effect of inundation and choice of hibernation sites of Coleoptera living on river banks. *Norsk Entomologisk Tidsskrift*, **15**:115–133.

Arscott, D. B., K. Tockner, and J. V. Ward. 2000. Aquatic habitat diversity along the corridor of an Alpine floodplain river (Fiume Tagliamento, Italy). *Archiv für Hydrobiologie*, **49**:679–704.

Bastow, J. L., J. L. Sabo, J. C. Finlay, and M. E. Power. 2002. A basal aquatic–terrestrial trophic link in rivers: algal subsidies via shore-dwelling grasshoppers. *Oecologia*, **131**:261–268.

Baxter, C. V., K. D. Fausch, M. Murakami, and P. L. Chapman. 2004. Fish invasion restructures stream and forest food webs by interrupting reciprocal prey subsidies. *Ecology*, **85**:2656–2663.

Baxter, C. V., K. D. Fausch, and W. C. Saunders. 2005. Tangled webs: reciprocal flows of invertebrate prey link streams and riparian zones. *Freshwater Biology*, **50**:201–220.

Ben-David, M., T. A. Hanley, and D. M. Schell. 1998. Fertilization of terrestrial vegetation by spawning salmon: the role of flooding and predator activity. *Oikos*, **83**:47–55.

Bilby, R. E., and P. A. Bisson. 1992. Allochthonous versus autochthonous organic matter contributions to the trophic support of fish populations in clear-cut and old growth forested streams. *Canadian Journal of Fisheries and Aquatic Sciences*, **49**:540–551.

Brewer, P. A., G. S. Maas, and M. G. Macklin. 2000. A fifty-year history of exposed riverine sediment dynamics on Welsh rivers. *BHS Occasional Paper*, **11**:245–252.

Briers, R. A., H. M. Cariss, R. Geoghegan, and J. H. R. Gee. 2005. The lateral extent of the subsidy from an upland stream to riparian lycosid spiders. *Ecography*, **28**:165–170.

Brigham, R. M., S. D. Grindal, M. C. Firman, and J. L. Morissette. 1997. The influence of structural clutter on activity patterns of insectivorous bats. *Canadian Journal of Zoology*, **75**:131–136.

Burton, T., and G. E. Likens. 1975. Energy flow and nutrient cycling in salamander populations in Hubbard Brook experimental forest. *Ecology*, **56**:1068–1080.

Cadenasso, M. L., S. T. A. Pickett, and K. C. Weathers. 2004. Effect of landscape boundaries on the flux of nutrients, detritus, and organisms. In G. A. Polis, M. E. Power, and G. R. Huxel (eds). *Food Webs at the Landscape Level*. Chicago Press, Chicago, pp. 154–168.

Chick, J. H., R. J. Cosgriff, and L. S. Gittinger. 2003. Fish as potential dispersal agents for floodplain plants: first evidence in North America. *Canadian Journal of Fisheries and Aquatic Sciences*, **60**:1473–1439.

Cloe III, W. W., and G. C. Garman. 1996. The energetic importance of terrestrial arthropod inputs to three warm-water streams. *Freshwater Biology*, **36**:105–114.

Collier, K. J., S. Bury, and M. Gibbs. 2002. A stable isotope study of linkages between stream and terrestrial food webs through spider predation. *Freshwater Biology*, **47**:1651–1659.

Davic, R. D., and H. H. J. Welsh. 2004. On the ecological role of salamanders. *Annual Review of Ecology and Systematics*, **35**:405–434.

Delettre, Y. R., and N. Morvan. 2000. Dispersal of adult Chironomidae (Diptera) in agricultural landscapes. *Freshwater Biology*, **44**:399–411.

Drake, V. A., and R. A. Farrow. 1998. The influence of atmospheric structure and motions on insect migration. *Annual Review of Entomology*, **33**:183–210.

Draulans, D. 1988. Effects of fish-eating birds on freshwater fish stocks: an evaluation. *Biological Conservation*, **44**:251–263.

Finlay, J. C. 2001. Stable-carbon-isotope ratios of river biota: implications for energy flow in lotic food webs. *Ecology*, **82**:1052–1064.

Fisher, S. G., and G. E. Likens. 1973. Energy flow in Bear Brook, New Hampshire: An integrative approach to stream ecosystem metabolism. *Ecological Monographs*, **43**:421–439.

Forster, R., and L. Forster. 1999. *Spiders of New Zealand and their Worldwide Kin*. University of Otago Press, Dunedin.

Framenau, V., M. Dietrich, M. Reich, and H. Plachter. 1996. Life cycle, habitat selection and home ranges of *Arctosa cinerea* (Fabricius, 1777) (Araneae: Lycosidae) in a braided section of the upper Isar (Germany, Bavaria). *Revue Suisse de Zoologie*, **vol. hors serie 1**:223–234.

Gende, S. M., R. T. Edwards, M. F. Willson, and M. S. Wipfli. 2002. Pacific salmon in aquatic and terrestrial ecosystems. *BioScience*, **52**:917–928.

Gilvear, D. J., J. Cecil, and H. Parsons. 2000. Channel change and vegetation diversity on a low-angle alluvial fan, River Feshie, Scotland. *Aquatic Conservation: Marine and Freshwater Systems*, **10**:53–71.

Goulding, M. 1980. *The Fishes and the Forest: Explorations in Amazonian Life History*. University of California, Berkely.

Hawkins, C. P., M. L. Murphy, and N. H. Anderson. 1982. Effects of canopy, substrate composition, and gradient on the structure of macroinvertebrate communities in the Cascade Range streams of Oregon. *Ecology*, **63**:1840–1856.

Heeg, J., and C. Breen. 1982. *Man and the Pongolo Floodplain*. Council for Scientific and Industrial Research, Pretoria (South Africa).

Helfield, J. M., and R. J. Naiman. 2001. Effects of salmon-derived nitrogen on riparian forest growth and implications for stream productivity. *Ecology*, **82**:2403–2409.

Henschel, J. R., D. Mahsberg, and H. Stumpf. 2001. Allochthonous aquatic insects increase predation and decrease herbivory in river shore food webs. *Oikos*, **93**:429–438.

Hering, D., and H. Plachter. 1997. Riparian ground beetles (Coeloptera, Carabidae) preying on aquatic invertebrates: a feeding strategy in alpine floodplains. *Oecologia*, **111**:261–270.

Hering, D., M. Gerhard, R. Manderbach, and M. Reich. 2004. Impact of a 100-year flood on vegetation, benthic invertebrates, riparian fauna and large woody debris standing stock in an Alpine floodplain. *River Research and Applications*, **20**:445–457.

Hilderbrand, G. V., T. A. Hanley, C. T. Robbins, and C. C. Schwartz. 1999. Role of brown bears (*Ursus arctos*) in the flow of marine nitrogen into a terrestrial ecosystem. *Oecologia*, **121**:546–550.

Iwata, T., S. Nakano, and M. Murakami. 2003. Stream meanders increase insectivorous bird abundance in riparian deciduous forests. *Ecography*, **26**:325–337.

Jackson, J. K., and S. G. Fisher. 1986. Secondary production, emergence, and export of aquatic insects of a Sonoran desert stream. *Ecology*, **67**:629–638.

Johnson, J. H., and N. H. Ringler. 1979. The occurrence of blow fly larvae (Diptera: Calliphoridae) on salmon carcasses and their utilization as food by juvenile salmon and trout. *Great Lakes Entomologist*, **12**:137–140.

Junk, W. J., and K. M. Wantzen. 2004. The flood pulse concept: new aspects, approaches, and applications – an update. In R. Welcomme and C. Barow (eds). *The Second Large River Symposium* (LARS). Pnom Penh, Cambodia, pp. 117–149.

Junk, W. J., P. B. Bayley, and R. E. Sparks. 1989. The flood pulse concept in river-floodplain systems. *Canadian Special Publication in Fisheries and Aquatic Sciences*, **106**:110–127.

Karaus, U., L. Alder, and K. Tockner. 2005. Concave islands. Habitat heterogeneity of parafluvial ponds in a gravel-bed river. *Wetlands*, **25**:26–37.

Kato, C., T. Iwata, S. Nakano, and D. Kishi. 2003. Dynamics of aquatic insect flux affects distribution of riparian web-building spiders. *Oikos*, **103**:113–120.

Kato, C., T. Iwata, and E. Wada. 2004. Prey-use by web-building spiders: stable isotope analyses of trophic flow at a forest-stream ecotone. *Ecological Research*, **19**:633–643.

Kawaguchi, Y., and S. Nakano. 2001. Contribution of terrestrial invertebrates to the annual resource budget for salmonids in forest and grassland reaches of a headwater stream. *Freshwater Biology*, **46**:303–316.

Kawaguchi, Y., M. Taniguchi, and S. Nakano. 2003. Terrestrial invertebrate inputs determine the local abundance of stream fishes in a forest stream. *Ecology*, **84**:701–708.

Klaus, I., C. Baumgartner, and K. Tockner. 2001. Die Wildflusslandschaft des Tagliamento (Italien, Friaul) als Lebensraum einer artenreichen Amphibiengesellschaft. *Zeitschrift für Feldherpetologie*, **8**:21–30.

Kolb, A. 1958. Nahrung und Nahrungsaufnahme bei Fledermäusen. *Zeitschrift für Säugetierkunde*, **23**:84–95.

Langhans, S. D. 2000. *Schwemmgut: Indikator der ökologischen Integrität einer Flussaue*. ETH. Zürich.

Lowe, W. H., and D. T. Bolger. 2002. Local and landscape-scale predictors of salamander abundance in New Hampshire headwater streams. *Conservation Biology*, **16**:183–193.

Lude, A., M. Reich, and H. Plachter. 1999. Life strategies of ants in unpredictable floodplain habitats of Alpine rivers (Hymenoptera: Formicidae). *Entomologica Generalis*, **24**:75–91.

Mason, C. F., and S. M. Macdonald. 1982. The input of terrestrial invertebrates from tree canopies to a stream. *Freshwater Biology*, **12**:305–311.

Moog, O. 1993. Quantification of daily peak hydropower effects on aquatic fauna and management to minimize environmental impacts. *Regulated Rivers: Research and Management*, **8**:5–14.

Murakami, M., and S. Nakano. 2001. Species-specific foraging behavior of birds in a riparian forest. *Ecological Research*, **16**:913–923.

Naiman, R. J., and H. Décamps. 1997. The ecology of interfaces: riparian zones. *Annual Review of Ecology and Systematics*, **28**:621–658.

Naiman, R. J., and K. H. Rogers. 1997. Large animals and system-level characteristics in river corridors. *BioScience*, **47**:521–529.

Naiman, R. J., J. M. Melillo, M. A. Lock, T. E. Ford, and S. R. Reice. 1987. Longitudinal patterns of ecosystem processes and community structure in a subarctic river continuum. *Ecology*, **68**:1139–1156.

Naiman, R. J., R. E. Bilby, D. E. Schindler, and J. M. Helfield. 2002. Pacific salmon, nutrients, and the dynamics of freshwater and riparian ecosystems. *Ecosystems*, **5**:399–417.

Nakano, S., and M. Murakami. 2001. Reciprocal subsidies: Dynamic interdependence between terrestrial and aquatic food webs. *Proceedings of the National Academy of Sciences of the United States of America*, **98**:166–170.

Nakano, S., Y. Kawaguchi, M. Taniguchi, H. Miyasaka, Y. Shibata, H. Urabe, and N. Kuhara. 1999a. Selective foraging on terrestrial invertebrates by rainbow trout in a forested headwater stream in northern Japan. *Ecological Research*, **14**:351–360.

Nakano, S., H. Miyasaka, and N. Kuhara. 1999b. Terrestrial–aquatic linkages: riparian arthropod inputs alter trophic cascades in a stream food web. *Ecology*, **80**:2435–2441.

Newbold, J. D., D.-C. Erman, and K. B. Roby. 1980. Effect of logging on macroinvertebrates in streams with and without buffer strips. *Canadian Journal of Fisheries and Aquatic Sciences*, **37**:1076–1085.

Ostfeld, R. S., and F. Keesing. 2000. Pulsed resources and community dynamics of consumers in terrestrial ecosystems. *Trends in Ecology and Evolution*, **15**:232–237.

Paetzold, A. 2005. Life at the edge – aquatic–terrestrial interactions along rivers. PhD thesis, ETH, Zurich.

Paetzold, A., and K. Tockner. 2005. Effects of riparian arthropod predation on the biomass and abundance of aquatic insect emergence. *Journal of the North American Benthological Society*, **24**:395–402.

Paetzold, A., C. J. Schubert, and K. Tockner. 2005. Aquatic–terrestrial linkages along a braided river: Riparian arthropods feeding on aquatic insects. *Ecosystems*, **8**:748–759.

Paetzold, A., J. Bernet, and K. Tockner. 2006. Consumer-specific responses to riverine subsidy pulses in a riparian arthropod assemblage. *Freshwater Biology*, **51**:1103–1115.

Petersen, I., J. H. Winterbottom, S. Orton, N. Friberg, A. G. Hildrew, D. C. Spiers, and W. S. C. Gurney. 1999. Emergence and lateral dispersal of adult Plecoptera and Trichoptera from Broadstone Stream, UK. *Freshwater Biology*, **42**:401–416.

Petersen, I., Z. Masters, A. G. Hildrew, and S. J. Ormerod. 2004. Dispersal of adult aquatic insects in catchments of differing land use. *Journal of Applied Ecology*, **41**:934–950.

Petts, G. E. 1984. *Impounded Rivers*. John Wiley & Sons, Ltd, Chichester.

Plachter H. 1986. Composition of the carabid beetle fauna of natural riverbanks and man-made secondary habitats. In P. J. den Boer, M. L. Luff, D. Mossakowski, and F. Weber (eds). *Carabid Beetles: Their Adaptations and Dynamics*. G. Fischer Verlag, Stuttgar, pp. 509–538.

Plachter, H., and M. Reich. 1998. The significance of disturbance for populations and ecosystems in natural floodplains. *Proceedings of the International Symposium on River Restoration*, Tokyo, Japan, pp. 29–38.

Polis, G. A., and S. D. Hurd. 1996a. Allochthonous input across habitats, subsidized consumers, and apparent trophic cascades: Examples from the ocean-land interface. In G. A. Polis, and K. O. Winemiller (eds). *Food Webs: Integration of Patterns and Dynamics.* Chapman & Hall, New York, pp. 275–285.

Polis, G. A., and S. D. Hurd. 1996b. Linking marine and terrestrial food webs: Allochthonous input from the ocean supports high secondary productivity on small islands and coastal land communities. *The American Naturalist,* **147**:396–423.

Polis, G. A., W. B. Anderson, and R. D. Holt. 1997. Toward an integration of landscape and food web ecology: the dynamics of spatially subsidized food webs. *Annual Review of Ecology and Systematics,* **28**:289–316.

Power, M. E., and W. E. Rainey. 2000. Food webs and resource sheds: towards spatially delimiting trophic interactions. In M. J. Hutchings, E. A. John, and A. J. A. Stewart (eds). *The Ecological Consequences of Environmental Heterogeneity.* Blackwell Science, Cambridge.

Power, M. E., W. E. Rainey, M. S. Parker, J. L. Sabo, A. Smyth, S. Khandwala, J. C. Finlay, F. C. McNeely, K. Marsee, and C. Anderson. 2004a. River-to-watershed subsidies in an old-growth conifer forest. In G. A. Polis, M. E. Power, and G. A. Huxel (eds). *Food Webs at the Landscape Level.* The University of Chicago Press, Chicago and London, pp. 217–240.

Power, M. E., M. J. Vanni, P. Stapp, and G. A. Polis. 2004b. Subsidy effects on managed ecosystems: implications for sustainable harvest, conservation, and control. In G. A. Polis, M. E. Power, and G. R. Huxel (eds). *Food Webs at the Landscape Level.* University of Chicago Press, Chicago, pp. 387–409.

Puth, L. M., and K. A. Wilson. 2001. Boundaries and corridors as a continuum of ecological flow control: lessons from rivers and streams. *Conservation Biology,* **15**:21–30.

Sabo, J. L. 2003. Hot rocks or no rocks: overnight retreat availability and selection by a diurnal lizard. *Oecologia,* **136**:329–335.

Sabo, J. L., and M. E. Power. 2002a. Numerical responses of lizards to aquatic insects and short-term consequences for terrestrial prey. *Ecology,* **83**:3023–3036.

Sabo, J. L., and M. E. Power. 2002b. River-watershed exchange: Effects of riverine subsidies on riparian lizards and their terrestrial prey. *Ecology,* **83**:1860–1869.

Sadler, J. P., D. Bell, and A. Fowles. 2004. The hydroecological controls and conservation value of beetles on exposed riverine sediments in England and Wales. *Biological Conservation,* **118**:41–56.

Sanzone, D. M., J. L. Meyer, E. Marti, E. P. Gardiner, J. L. Tank, and N. B. Grimm. 2003. Carbon and nitrogen transfer from a desert stream to riparian predators. *Oecologia,* **134**:238–250.

Schindler, D. E., P. R. Leavitt, C. S. Brock, S. P. Johnson, and P. D. Quay. 2005. Marine derived-nutrient, commercial fisheries, and production of salmon and lake algae in Alaska. *Ecology,* **86**:3225–3231.

Schneider, R. L., E. L. Mills, and D. C. Josephson. 2002. Aquatic–terrestrial linkages and implications for landscape management. In J. Liu, and W. W. Taylor (eds). *Integrating Landscape Ecology into Natural Resource Management.* Cambridge University Press, Cambridge, pp. 241–262.

Semlitsch, R. D. 1998. Biological definition of terrestrial buffer zones for pond-breeding salamanders. *Conservation Biology,* **12**:1113–1119.

Sweeney, B. W., and R. L. Vannote. 1982. Population synchrony in mayflies: A predator satiation hypothesis. *Evolution,* **36**:810–821.

Sweeney, B. W., T. L. Bott, J. K. Jackson, L. A. Kaplan, J. D. Newbold, L. J. Standley, W. C. Hession, and R. J. Horwitz. 2004. Riparian deforestation, stream narrowing, and loss of stream ecosystem services. *Proceedings of the National Academy of Sciences of the United States of America,* **101**:14132–14137.

Tait, C. K., J. L. Li, G. A. Lamberti, T. N. Pearsons, and H. W. Li. 1994. Relationships between riparian cover and the community structure of high desert streams. *Journal of the North American Benthological Society,* **13**:45–56.

Tenzer, C. 2003. Ausbreitung terrestrischer Wirbelloser durch Fliessgewässer. PhD thesis. Philipps Universität, Marburg.

Thiele, H. U. 1977. *Carabid Beetles in their Environments.* Springer, Berlin.

Thorp, J. H., and M. D. Delong. 1994. The riverine productivity model: an heuristic view of carbon sources and organic processing in large river ecosystems. *Oikos*, **70**:305–308.

Thorp, J. H., and M. D. Delong. 2002. Dominance of autochthonous autotrophic carbon in food webs of heterotrophic rivers. *Oikos*, **96**:543–550.

Tockner, K., F. Malard, and J. V. Ward. 2000. An extension of the flood pulse concept. *Hydrological Processes*, **14**:2861–2883.

Tockner, K., J. V. Ward, D. B. Arscott, P. J. Edwards, J. Kollmann, A. M. Gurnell, G. E. Petts, and B. Maiolini. 2003. The Tagliamento River: A model ecosystem of European importance. *Aquatic Sciences*, **65**:239–253.

Tockner, K., A. Paetzold, U. Karaus, C. Claret, and J. Zettel. 2005. Ecology of braided rivers. In G. H. Sambroock Smith, J. L. Best, C. S. Bristow, and G. Petts (eds). *Braided Rivers – IAS Special Publication*. Blackwell, Oxford.

Trottmann, N. 2004. Schwemmgut – Ausbreitungsmedium terrestrischer Invertebraten in Gewässerkorridoren. Master thesis. ETH Zurich, Dübendorf.

van der Nat, D., A. P. Schmidt, K. Tockner, P. J. Edwards, and J. V. Ward. 2002. Inundation dynamics in braided floodplains: Tagliamento River, Northeast Italy. *Ecosystems*, **5**:636–647.

Vannote, R. L., G. W. Minshall, K. W. Cummins, J. R. Sedell, and C. E. Cushing. 1980. The river continuum concept. *Canadian Journal of Fisheries and Aquatic Sciences*, **37**:130–137.

von Frenckell, B., and R. M. R. Barclay. 1987. Bat activity over calm and turbulent waters. *Canadian Jornal of Zoology*, **65**:219–222.

Wallace, J. B., S. L. Eggert, J. L. Meyer, and J. R. Webster. 1997. Multiple trophic levels of a forest stream linked to terrestrial litter inputs. *Science*, **277**:102–104.

Wallace, J. B., S. L. Eggert, J. L. Meyer, and J. R. Webster. 1999. Effects of resource limitation on a detrital based ecosystem. *Ecological Monographs*, **69**:409–442.

Ward, J. V. 1989. The four-dimensional nature of lotic ecosystems. *Journal of the North American Benthological Society*, **8**:2–8.

Ward, J. V., K. Tockner, D. B. Arscott, and C. Claret. 2002. Riverine landscape diversity. *Freshwater Biology*, **47**:517–539.

Willson, M. F., and K. C. Halupka. 1995. Anadromous fish as keystone species in vertebrate communities. *Conservation Biology*, **9**:489–497.

Willson, M. F., S. M. Gende, and P. A. Bisson. 2004. Anadramous fishes as ecological links between ocean, freshwater, and land. In G. A. Polis, M. E. Power, and G. A. Huxel (eds). *Food Webs at the Landscape Level*. University of Chicago Press, Chicago and London.

Wipfli, M. S. 1997. Terrestrial invertebrates as salmonid prey and nitrogen sources in streams: contrasting old-growth and young-growth riparian forests in southern Alaska, USA. *Canadian Journal of Fisheries and Aquatic Sciences*, **54**:1259–1269.

Witman, J. D., J. C. Ellis, and W. B. Anderson. 2004. The influence of physical processes, organisms, and permeability on cross-ecosystem fluxes. In G. A. Polis, M. E. Power, and G. R. Huxel (eds). *Food Webs at the Landscape Level*. University of Chicago Press, Chicago, pp. 335–358.

5
Flow-generated Disturbances and Ecological Responses: Floods and Droughts

Philip S. Lake

5.1 Introduction

In recent years hydrologists and ecologists have come to realize the importance of flow-generated disturbances in streams and rivers. Such disturbances (floods and droughts) can exert a strong, if not dominant, influence on the ecology of flowing waters. Knowledge of floods and droughts, their different modes of action and effects, is vital in elucidating the links between the hydrology and ecology of running waters. Thus, this chapter aims to document and compare the nature of floods and droughts as disturbances and the consequential effects of these disturbances on the ecology of flowing waters. The chapter covers floods and droughts in both constrained and floodplain rivers and primarily concentrates on fauna (invertebrates and fish). Our understanding of the ecology of floods is better known than that of droughts and this imbalance currently hinders detailed comparisons.

5.2 Definition of Disturbance

Disturbance may be defined as occurring 'when potentially damaging forces are applied to habitat space occupied by a population, community, or ecosystem' (Lake, 2000, p. 574). The forces themselves may be physical (e.g., high shear stress), chemical (e.g., deoxygenation) or biological (a new pathogen) or combinations of these. Disturbances usually change the nature of habitats and deplete organisms and their resources. These effects have been used to measure the strength of disturbances, but such characterization of disturbance prevents objective comparison between disturbance events. As forces, disturbances should be evaluated by their abiotic properties. Thus, floods could be characterized by their peak flow, their intensity (shape of hydrograph), average recurrence interval (Gordon *et al.*, 2004), spatial extent, duration and predictability (Smith and Ward, 1998). Droughts could be characterized by the duration of extreme low flow, average recurrence interval of extreme low flows from low flow frequency curves (McMahon and Finlayson, 2003), spatial extent and predictability.

It is important in disturbance ecology to distinguish between the characteristics of disturbance and the subsequent biotic responses. Following the definition by Bender *et al.* (1984), and modified by Glasby and Underwood (1996), perturbation encompasses both the disturbance and the consequential responses of the biota. The response contains two different elements: resistance – the capacity to withstand the disturbance; and resilience – the capacity to recover from the disturbance (Webster *et al.*, 1983; Fisher and Grimm, 1991).

5.3 Disturbances and Responses

Initially two types of disturbance, pulses and presses, were recognized (Bender *et al.*, 1984), to which Lake (2000) added ramp disturbances as a third (Figure 5.1). Pulses are sharply defined events exemplified by floods. Pulses may scale from being rapid events operating at a small scale to large-scale events of much longer duration. Press disturbances increase in intensity to reach a constant pressure. Although natural presses in rivers are rare, human activities, such as agriculture on catchments causing stream sedimentation, can produce persistent press disturbances. Lake (2000) defined ramp disturbances as those that steadily increase in strength with time without a clear endpoint. Droughts are best considered as ramps (Lake, 2003b; Boulton, 2003) because they steadily build as water availability declines and may steadily subside as precipitation increases.

Clearly the disturbance and response need to be separated. Responses may be either pulses, presses or ramps (Glasby and Underwood, 1996; Lake, 2000) (Figure 5.2). Thus, a flood may be a pulse, the response of the invertebrates may be a pulse, whereas the response of the riparian vegetation may be a ramp. For different floods, such as seasonal flood pulses, invertebrates may show a pulse response, but with aseasonal or very large floods the responses may be press (e.g., Giller, 1996).

Floods and droughts may act alone, but in many cases they interact with other disturbances, such as sedimentation and exotic species. To understand the manifold effects of flow-generated disturbances these interactions need to be recognized. Disturbances in riverine ecosystems generate habitat patchiness. The spatial extent and severity of impact may vary within the ecosystem, and patchiness may vary between disturbance events. By depleting pre-existing patches and creating new patches of varying sizes and spatial distribution, disturbances act as catalysts for subsequent succession (Fisher, 1983; Townsend, 1989; Fisher and Grimm, 1991). Disturbance can act as a filter to set the stage for subsequent colonization, succession and community assembly (Poff, 1997; Matthews,

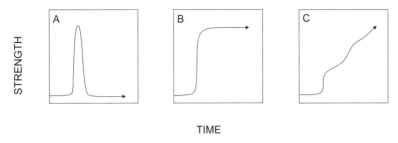

Figure 5.1 The three forms of disturbance (A) pulse, (B) press and (C) ramp

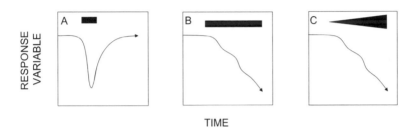

Figure 5.2 Three types of disturbance–response interaction. In A a pulse disturbance induces a pulse response; in B a press disturbance induces a ramp response, and in C a ramp disturbance induces a ramp response

1998). The prevailing disturbance regime in a river may act as a strong selection force, moulding traits (morphological, life history, or behavioural) so that species can persist (Poff, 1997; Lytle, 2001; Lytle and Poff, 2004). Such adaptations may be expressed both in terms of resistance to flow events and in the capacity to recover (resilience). Townsend *et al.* (1997) found that traits related to resistance ('small size, high adult mobility, habitat generalist') and to resilience ('clinger, streamlined, two or more life stages outside the stream') were significantly related to the intensity of disturbance by floods. Distinctive traits have evolved to allow biota to contend with droughts, but they have not been systematically described. Traits for resistance may include desiccation-resistant life stages and high temperature and low oxygen tolerance, whereas traits for resilience include refugia-seeking behaviour and high adult mobility.

5.4 Disturbance and Refugia

Biota in flowing waters can resist the impacts of disturbance by the use of refugia. Sedell *et al.* (1990) defined refugia as 'habitats or environmental factors that convey spatial and temporal resistance and/or resilience to biotic communities impacted by biophysical disturbances' (p. 711). Lancaster and Belyea (1997) expanded the definition to include attributes of organisms' life histories. Refugia can increase the resistance of biota to disturbance and bolster resilience after disturbance by acting as patches from which recovery can occur by providing colonists.

Organisms may use refugia that operate between generations and between habitats. An example of intergenerational refugia is the egg banks of microinvertebrates on dry floodplains that hatch with flooding. Interhabitat, refugial behaviour is shown by belostomatid bugs in desert streams that use rainfall as a signal to leave streams and avoid flash floods (Lytle and Smith, 2004).

The two other forms of refugia operate at smaller spatial scales and each involves changes in habitat of the biota or biota changing position within habitats. The movement by surficial benthic invertebrates into the hyporheic zone during floods exemplifies habitat change (Dole-Olivier *et al.*, 1997), whereas the persistence of invertebrates and algae (Matthaei *et al.*, 2004) on stable stones during and after floods exemplifies refugia use within a habitat. For most fauna, whether the use of refugia is an active or passive process remains poorly understood.

5.5 Floods

5.5.1 The Disturbance

A flood 'is a body of water which rises to overflow land which is not normally submerged' (Ward, 1978, p. 5). Such a definition accommodates overbank flow and floodplain inundation. In upland constrained streams, only rarely do floods produce overbank flow, while there are more frequent high-flow events or spates that constitute disturbances (Gordon *et al.*, 2004). Similarly, lowland floodplain rivers can have high flow events that stay in the channel. Puckridge *et al.* (1998) called such events 'flow pulses' to distinguish them from the flood pulse (Junk *et al.*, 1989).

Basically, two types of natural flood are recognized, river floods and estuarine–coastal floods (Smith and Ward, 1998). The latter may exacerbate river floods in coastal lowlands through high tides and storm surges. Most river floods arise from high rainfall events and vary from flash floods, extreme in the case of hurricanes, to the long durational, seasonal flood pulses.

Floods in constrained streams with a rapid increase in flow velocity and shear stress may sweep away biota – algae, macrophytes, benthos, riparian plants – and move sediments, benthic organic matter and coarse wood (e.g., Fisher *et al.*, 1982; Molles, 1985; Lytle, 2000), and create major changes in channel morphology (Gordon *et al.*, 2004). They are marked by a great increase in suspended solids with larger particles being transported at higher flow velocities and forces. The entrainment and transport of sediments result in patches of the streambed, especially in upland streams, being scoured and other patches being filled by sediment (Matthaei *et al.*, 2004).

Floods and spates create substratum patches by moving and burying stones (Downes *et al.*, 1998), scouring and filling the streambed (Matthaei *et al.*, 2004), and abrading solid surfaces (Bond, 2004) in gravel-bed upland streams. In sandy streams, floods move large amounts of sand, creating large patches of scour and fill and depleting algae and invertebrates (Fisher *et al.*, 1982; O'Connor and Lake, 1994; Grimm and Fisher, 1989). The dynamic mosaics of patches created become platforms for the development of biotic assemblages.

Lowland rivers consist of a channel and an extensive floodplain. With major floods, water moves out from the channel inundating the floodplain filling wetlands, anabranches and flood runners, generating the flood pulse (Junk *et al.*, 1989; Tockner *et al.*, 2000). Such floods may be highly predictable as seasonal events, such as with large tropical and temperate rivers in mesic regions (Puckridge *et al.*, 1998; Tockner *et al.*, 2000) or unpredictable as in arid and semi-arid regions (Puckridge *et al.*, 2000).

With a major rainfall event, small-order streams of a river's upper catchment are subjected to a fast moving flood – a pulse disturbance. In the constrained valley section the floods from all the small streams may fuse to build a powerful flood, a large pulse disturbance, that results in overbank flow for a short period of time. In the lowland section the flood flows over the banks and inundates the floodplain with a slow moving flood, which recedes slowly leaving a legacy of replenished wetlands. The flood pulse is transformed from being a damaging disturbance in small-order streams to be a major regenerator of biodiversity and production in lowland floodplains.

5.6 Droughts

5.6.1 The Disturbance

There are many definitions of drought. It can be defined in meteorological or hydrological terms and in terms of its effects on human activities, especially agriculture. Wilhite (2000) delineated three forms of drought: meteorological, agricultural and hydrological, with the latter form being defined in terms of effects on water availability arising from shortfalls in precipitation. Droughts constitute an unusual type of natural disturbance in that they are caused by a deficiency (a deficiency of water).

As for floods, it would help our understanding of droughts if they were characterized in hydrological terms. Such hydrological measures include low-flow frequency curves and low-flow spell duration (Gordon *et al.*, 2004). When examining low-flow frequency curves, as the average recurrence interval increases there comes a point where there is a break in slope of the curve indicating very infrequent flows and the onset of drought conditions (Gordon *et al.*, 2004). For streams in the UK, it appears that this break point occurs at a low-flow exceedance probability of ~65% (Institute of Hydrology, 1980), whereas for streams in Victoria, Australia, this break point occurs at around 80% (Nathan and McMahon, 1990). Other characteristics of droughts include spatial extent, timing and abruptness of onset.

Ecologically two types of drought are recognized. First, there are predictable *seasonal droughts* that may vary in duration and severity in different years, e.g. Mediterranean systems zone (e.g., Gasith and Resh, 1999; Magalhães *et al.*, 2002) and the wet-dry tropics (Douglas *et al.*, 2003). More damaging and unpredictable are the aseasonal, or *supra-seasonal droughts* that are marked by a decline in precipitation across seasons. Seasonal droughts act more like a press disturbance, whilst supra-seasonal droughts are ramp disturbances (Lake, 2003b). In many streams supra-seasonal drought may be imposed on seasonal droughts. Both types of drought result in decreasing stream flow, and cessation of flow, and stream fragmentation, giving rise to a 'stepped' response by the biota to critical environmental events, such as when connectivity is broken (Boulton, 2003).

Our understanding of the ecology of drought is much less than that for floods (Lake, 2003b). In terms of the biota, the research that has been undertaken has concentrated on invertebrates and fish. The ecological effects of drought are better known for seasonal compared to supra-seasonal droughts.

5.7 The Responses to Floods

5.7.1 Constrained Streams

As a generalization, the biota of flowing waters have a low resistance to floods. This is is better documented for the biota of constrained rivers than for the biota of unconstrained floodplain rivers. Most of our knowledge on the ecological effects of floods has come from studies in perennial systems.

Floods in constrained streams deplete and generate habitat space, sweep away fauna, algae and macrophytes, remove attached biota (e.g., biofilms), damage, if not kill biota, by sediment agitation, scour and deposition, remove debris dams, deposit sediments and deplete riparian vegetation (Resh *et al.*, 1988; Lake, 2000). Fish may use floods to move around barriers (see David and Closs, 2002). Catastrophic drift of invertebrates may increase as a result of flood conditions. However, Brooks (1997) by sampling drift before, during and after a major flood, found that drift decreased during the flood and then increased after it. The decrease in drift he ascribed to the benthos using refugia during the flood.

The amount of damage that a flood does is related to its magnitude (Molles, 1985; Grimm and Fisher, 1989) and to its timing, with aseasonal floods being more damaging than normal seasonal ones (see Giller, 1996; Lytle, 2003). Large floods and flash floods in constrained streams deplete refugia by such means as lowering the stability of the streambed and scouring the hyporheic zone (Matthaei *et al.*, 2004). However, refugia appear to be numerous. They include moving behind stable stones (Matthaei *et al.*, 2004), behind wood and leaf dams, into areas where shear stress is low (Lancaster and Hildrew, 1993), into pools (Brooks, 1997; Fausch *et al.*, 2002), along lateral margins of streams, into crevices and the hyporheic zone (see Dole-Olivier *et al.*, 1997) and simply leaving the stream (Lytle and Smith, 2004). The refugia allow survival during the floods, but also serve as bases for recolonization of streambed patches. Floods can exert varying effects on different biota at the same location.

Resistance by the biota to floods in constrained systems is regarded as being low and resilience is deemed to be high, with some exceptions. Recovery following floods is usually a case of rather rapid secondary succession driven by colonization. Colonization of the streambed by micro-algae, notably diatoms, occurs rapidly after floods (Fisher *et al.*, 1982; Steinman and McIntire, 1990). Refugia, such as crevices and stable rocks and boulders (Matthaei *et al.*, 2003) may serve as sources for colonizers. In Sycamore Creek, Arizona (USA), post-flood succession in patches can be rapid, leading to filamentous algae. With time, the supply of dissolved nitrogen lessens and grazer-resistant, nitrogen-fixing, cyanobacteria become dominant (Fisher *et al.*, 1982; Grimm and Fisher, 1989).

Floods may create patches of scour and fill and stable patches may be unaffected (Matthaei *et al.*, 2004). For some time after a flood these patches may differ in the composition and abundance of invertebrates, strongly suggesting that they are shaped by their local disturbance history. Whereas such a legacy effect may be expected for sedentary organisms such as algae, it is surprising that such an effect occurs with highly mobile invertebrates.

The high mobility of most stream animals (invertebrates and fish) allows them to use refugia during floods and to colonize the recently disturbed channel (Mackay, 1992; Lake, 2000). Such recolonization occurs via a number of means. Rapid surface movement, swimming and drift from refugia is a major form of recolonization in perennial streams (Mackay, 1992), whereas in temporary streams aerial colonization may be important (Gray and Fisher, 1981). Filter feeders, such as simuliid larvae, appear to be early colonizers (Downes and Lake, 1991) followed by highly mobile grazers-scrapers, such as mayfly nymphs. Case-dwelling grazers (e.g., glossosomatid caddis larvae), net-spinning caddis and shredders are intermediate colonizers (Mackay, 1992). Predators take their time to colonize, possibly due to their need for worthwhile prey densities. Even though streams are highly dynamic environments inhabited by highly mobile organisms, the trajectory of recolonization of disturbed patches is relatively smooth to reach a clear endpoint (Minshall and Peterson, 1985) suggesting some form of regulation of assemblage structure.

Floods can disrupt metabolic processes. In Sycamore Creek, Arizona, floods greatly reduced gross primary production and ecosystem respiration (Fisher *et al.*, 1982). However, soon after flooding occurred, gross primary production and respiration increased indicating short-term recovery. The system rapidly became autotrophic and stabilized with a P/R ratio averaging 1.46 (Fisher *et al.*, 1982). Floods in a rocky stream reduced gross primary production by 37 and 53% at two sites, whilst ecosystem respiration declined by only 14 and 24% (Uehlinger, 2000). P/R ratios were autotrophic (>1) before the floods, whereas after the floods they were heterotrophic (<1), a situation that persisted for months. Resistance to floods was greater for respiration than for primary production, whilst respiration recovery was less rapid than that of primary production.

Floods may alter interspecific interactions. Stream invertebrates can compete for space and the outcomes of competition may be altered by floods. In a Californian stream, winter floods favour space occupation by simuliids, but lack of floods favours hydropsychid caddis (Hemphill and Cooper, 1983). Predation can be disrupted by floods with predators and prey being differentially affected. Thomson (2002) found that with floods, prey were more resilient than their predators.

Resistance to the invasion of exotic species may be strengthened by the prevailing disturbance regime. In desert streams of southwest USA, the non-native mosquito fish (*Gambusia affinis*) displaces, by predation, the native Sonoran topminnow (*Poeciliopsis occidentalis*) in streams with few floods. In more variable streams, mosquito fish populations are decimated by flash floods whereas the topminnow, adapted to such events, survives them with little loss (Meffe, 1984).

Trophic subsidies occur between streams and their riparian zones and catchments. A well-known subsidy is the input of allochthonous organic matter from the catchment into streams. This subsidy can be changed by floods and disturbances on the catchment. Furthermore, there are prey subsidies whereby terrestrial biota, such as riparian insects, fall into the stream and are eaten by fish, and where emerging stream insects are consumed by a host of predators (Baxter *et al.*, 2005). Feasibly, such subsidies in streams are disrupted by floods reducing emerging stream insect densities or greatly reducing the density of riparian predators such as spiders.

5.7.2 Floodplain Rivers

With flooding in lowland unconstrained rivers, water moves out onto the floodplain generating a great diversity of connective pathways between the channel and various topographical components of the floodplain (Ward, 1989; Ward *et al.*, 2002; Amoros and Bornette, 2002). The floods may be regular seasonal events – flood pulses (Junk *et al.*, 1989; Tockner *et al.*, 2000) or rather unpredictable events as in dryland rivers (Puckridge *et al.*, 1998, 2000). Flow pulses (Puckridge *et al.*, 1998) may run down the channel moving sediments and organic matter and stimulating primary production.

Floods in floodplain rivers are powerful regenerative events for aquatic biota and ecological processes on the floodplain, yet are disturbances for floodplain terrestrial biota as many organisms may be killed, seek refugia or leave. Floods may favour some terrestrial invertebrates. On southern Australian floodplains, floods favoured predatory hydrophilic arthropods, such as carabid beetles and spiders, but reduced others, such as ants (Ballinger *et al.*, 2005).

Floods move nutrients, sediments, organic matter and biota onto the floodplain. On the floodplain, nutrients and dissolved organic matter (DOM) are released from sediments into the water column (O'Connell *et al.*, 2000; Valett *et al.*, 2005) stimulating bacterial and phytoplankton production and with time, macrophyte production. With inundation floodplain respiration is stimulated, but deoxygenation may occur in stagnant parts of floodplain wetlands. Particulate detritus, DOM and nutrients from sediments fuel production in forested floodplains (Ward, 1989; Junk *et al.*, 1989; Tockner *et al.*, 2000). In dryland rivers, with scanty vegetation and detritus on the floodplains, primary production by algae may be the major food source feeding secondary production (Bunn *et al.*, 2003).

Floods on floodplains are expansion–contraction ecosystem events (Stanley *et al.*, 1997) that stimulate and regulate biodiversity and production. Indeed, in the period of expansion with rapid succession and growth of populations, floodplains are one of the most productive environments on earth. Inundated floodplains of dryland rivers become hotspots of production and for waterbirds. Such floods are not predictable and can cover very large geographical areas where they occur. Waterbirds move over great distances, from flooded wetlands to wetlands, taking advantage of high levels of production (Roshier *et al.*, 2002).

The recession of the floodwaters brings with it the movement of mobile fauna, such as fish and invertebrates, nutrients and both particulate and dissolved organic matter back into the river channel (Amoros and Bornette, 2002). In contrast to the expansion part of a flood event, the ecological significance of the contraction/recession has not been substantially investigated.

5.8 Responses to Drought

5.8.1 Impacts

There are three longitudinal patterns of drying that occur in streams (Lake, 2003b). Many streams originate in headwaters as springs that persist in drought (and act as refugia) whilst drying and fragmentation occur downstream (e.g. Erman and Erman, 1995).

Second, and most commonly, with drought the headwaters and upstream sections dry and the downstream reaches persist as pools or low flow streams (e.g., Larimore *et al.*, 1959; Covich *et al.*, 2003; Dodds *et al.*, 2004). Finally, both headwaters and downstream reaches persist in drought and drying occurs in the mid reaches (Bond and Lake, 2005a).

The early effects are similar for both seasonal and supra-seasonal droughts, but the effects of supra-seasonal droughts are more severe, less predictable, and lasting in terms of lag effects. Water levels in the channel drop and surface water recedes from littoral margins and from the riparian zone (Stanley *et al.*, 1997): a weakening of lateral connectivity. With receding water depths, longitudinal fragmentation ultimately occurs, at first in shallow areas of the stream such as riffles and runs (Stanley *et al.*, 1997; Boulton, 2003). The cessation of flow, or very low flow, disrupts the longitudinal transport of nutrients, fine sediments and organic matter and may lower nutrient and DOM levels in pools (Dahm *et al.*, 2003). The creation of lentic conditions can occur in pools as algae may bloom (Dahm *et al.*, 2003) producing nocturnal deoxygenation. Cessation of flow eliminates fauna that rely on currents, such as filterers like simuliids and hydropsychids (e.g., Harrison, 1966; Boulton and Lake, 1992). In pools, high temperatures may lead to elevated conductivity (e.g., Caruso 2002) and in pools thermal stratification may occur. Both habitat space and quality decline. Lateral connectivity is weakened and prey subsidies may decline due to aquatic habitat contraction from riparian margins (Baxter *et al.*, 2005). As water levels fall, cyanobacteria and microalgae on solid surfaces may be exposed to the air and become desiccated, but not necessarily dead (Stanley *et al.*, 1997; Robson and Matthews, 2004). The development of resistance depends on the rate of drying, with rapid drying preventing resistant mechanisms from developing (Stanley *et al.*, 2004). With time, macrophytes and filamentous algae die with lasting effects on assemblage structure when the drought breaks (Ladle and Bass, 1981; Holmes, 1999). Major changes in aquatic plant assemblage structure with drought undoubtedly affect invertebrate and fish assemblages (Boulton, 2003).

Drought, as a disturbance of deficit and of long duration, probably places greater pressure on biota to evolve effective adaptations to survive than do floods. However, such adaptations to drought have not received much attention. Between generations, in common with the biota of temporary waters, drought refugia include desiccation-resistant propagules (e.g., Boulton, 1989). As drought builds, fish may migrate to pools and with the breaking of the drought, juveniles from surviving adults colonize newly inundated stream sections (Labbe and Fausch, 2000; Magoulick and Kobza, 2003). The persistence of fish in dryland rivers during drought appears to depend on the maintenance of waterholes at a large spatial extent (Arthington *et al.*, 2005). Refugia within generations can involve changes in habitat or within-habitat tolerance. Changes in habitat may be exemplified by the movement of surface-dwelling invertebrates into the hyporheic zone as drying occurs (e.g., Clinton *et al.*, 1996). Refugium use within habitats is shown by insects, especially beetles, and by crayfish that remain under stones and logs without free water, but with high humidities (Boulton, 1989).

As streams dry, especially when this occurs rapidly, fauna from insects to fish become stranded or entrapped in shallow pools that subsequently disappear (Matthews and Marsh-Matthews, 2003). Riffle-dwelling mussels, being sedentary, can suffer high mortality

(Gagnon et al., 2004). Hyporheic zones may contract away from the channel surface, a contraction that can give rise to increased faunal abundance and diversity in the deeper parts of the zone (Boulton and Stanley, 1995). Mobile invertebrates (e.g., Stanley et al., 1994; Covich et al., 2003) and fish may move to pools (see Magalhães et al., 2002) and refugia (Magoulick and Kobza, 2003). In pools, invertebrates (e.g., Stanley et al., 1994; Miller and Golladay, 1996) and fish (Magoulick and Kobza, 2003) may be concentrated, leading to an increase in the intensity of interactions such as predation and competition.

Pools with accumulated detritus, warm temperatures, high levels of DOM and low oxygen levels create conditions that may kill fish and invertebrates (Lake, 2003b). In south-eastern Australia, with a summer leaf fall from eucalypts, during drought this leaf input produces large amounts of dissolved organic carbon, predominantly polyphenols (O'Connell et al., 2000) and low levels of oxygen. Such harsh conditions can severely stress fish and yet some, such as western carp gudgeon (*Hyseleotris klunzingeri*) and pygmy perch (*Nannoperca australis*), are very tolerant of high DOM and low oxygen levels (D. McMaster, personal communication, 2005). This lends support to the idea that drought acting as a strong selection force may produce a distinct assemblage of highly tolerant biota (Magoulick and Kobza, 2003; Matthews and Marsh-Matthews, 2003).

With pool formation, the density and diversity of predators may increase (Boulton, 2003; Lake, 2003b). These predators may be fully aquatic species (e.g., odonatan nymphs, fish), mobile semi-aquatic species (e.g., dytiscid beetles and hemipterans), and fully terrestrial animals (e.g., birds, mammals). Lentic predators (e.g., beetles and hemipterans) invade the pools in great numbers and disappear when flow returns. Predation by fish may be a major factor in deciding population survival until the drought breaks (Magoulick and Kobza, 2003). The fauna trapped in drying pools may be prey to a host of terrestrial predators. Competition may be expected to strengthen as habitat space and resources decrease. Covich et al. (2003), in a drought-affected Puerto Rican stream, found that limited habitat space and depleted resources resulted in strong competition between shrimp and reduced reproductive outputs. Drought can alter invasion. Poor water quality in pools during drought in an Australian stream killed brown trout, an invading species, and a predator of native galaxiid fish. Galaxiids tolerated the poor water conditions and with the drought breaking, they colonized the trout-free sections of the stream (Closs and Lake, 1995). Overall drought, especially supra-seasonal drought, greatly reduces invertebrate and fish populations and substantially alters the structure of faunal assemblages (Lake, 2003b; Boulton, 2003; Matthews and Marsh-Matthews, 2003).

The effects of droughts on ecosystem processes have been neglected. In Sycamore Creek, Arizona, nitrogen is a limiting element and its availability is altered by drying, with some pools receiving high levels of nitrogen from fixation and hyporheic upwelling while others do not (Dent et al., 2001). Given the localized supply of nitrogen to various pools, a mosaic of pools at the landscape scale is generated with quite different communities and production (Stanley et al., 1997, 2004).

Besides greatly affecting rivers, droughts can have a strong impact on estuaries. With a reduction in freshwater input to estuaries, poor water quality may develop and small estuaries may become disconnected from the sea (Gasith and Resh, 1999). With the reduction of freshwater inputs there can be reductions in both freshwater and marine biota (Attrill and Power, 2000).

5.8.2 Recovery from Drought

Whereas, for floods biotic resistance is low and resilience is high, in streams subjected to seasonal drought, both resistance and resilience appear to be relatively high. For supra-seasonal droughts, resistance may steadily decline with time and resilience is highly variable, with in some cases long lags in recovery.

With rewetting of the stream channel, pools form before the arrival of flow (Stanley et al., 2004). Such pools in small tropical streams may have high temperatures and conductivities and harbour a non-insect fauna, consisting of crustaceans and oligochaetes (Douglas et al., 2003). At this time there is the hazard of 'false starts' (Boulton, 2003) when pools briefly fill and then dry out. Droughts may break with floods, a severe transition with the removal of organic matter in the channel and localized scouring and/or infilling of the streambed. With the return of flow, algal and cyanobacterial colonization of surfaces may proceed rapidly either from resuscitation of dormant cells or from colonization by upstream propagules (Ledger and Hildrew, 2001; Robson and Matthews, 2004; Stanley et al., 2004). Algal primary production may rapidly recover to provide food for early invertebrate colonizers.

Drought reduces macrophyte diversity and density in streams. With the return of flow after a 'severe' drought (1973–1974), Ladle and Bass (1981) observed a rapid recovery of macrophyte cover and a major change in the aquatic plant community structure. In another 'severe' drought (1989–1992), Holmes (1999) found, in one stream, a complete lack of aquatic plants, and in another stream the dry channel was invaded by nonaquatic plants. In both streams the recovery period was no more than two years.

Colonization of stream sections by invertebrates is variable and appears to depend on the severity of the drought (e.g., Boulton and Lake, 1992; Wood and Armitage, 2004), whether the stream is normally perennial or intermittent (Stanley et al., 1994; Miller and Golladay, 1996), refugium availability (abundance, types and distance between indivigual refugia) (Larimore et al., 1959; Wood and Petts, 1999; Ledger and Hildrew, 2001; Boulton, 1989, 2003; Fritz and Dodds, 2004), and whether there are lag effects due to previous droughts (Boulton, 2003). Means of colonization vary from system to system with downstream movement from refugia being important (e.g., Fritz and Dodds, 2004; Boulton, 2003) along with emergence from desiccation-resistant eggs (see Douglas et al., 2003). In some systems insect colonization occurs through flying adults laying eggs (see Harrison, 1966). In pools persisting through drought a standing water biota may develop, which in recovery is eliminated and replaced by a stream biota.

The recovery of invertebrates in intermittent streams from seasonal drought follows a predictable pattern of assemblage development, provided that there are no lag effects from past droughts (Boulton and Lake, 1992; Boulton, 2003). Early colonists are groups with short life cycles, such as chironomids and simuliids (Harrison, 1966; Fritz and Dodds, 2004). Species richness steadily increases as longer-lived collectors and grazers build up with predators arriving as species richness levels off (Boulton and Lake, 1992; Boulton, 2003). There is a strong indication of adaptations, such as in life histories, to seasonal drought, but this area is largely unstudied.

In perennial British streams (e.g., Wood and Petts, 1999; Wright et al., 2004), streams of mid-western USA (Larimore et al., 1959; Fritz and Dodds, 2004) and of New Zealand

(Caruso, 2002), recovery from supra-seasonal drought was found to be relatively complete with only a few uncommon taxa not returning. However, recovery from supra-seasonal drought, even in streams subjected to seasonal drought, can be somewhat unpredictable due to the deletion of species and the arrival of new or uncommon species. Ladle and Bass (1981) reported great reductions in the abundant amphipod *Gammarus pulex* and a very slow recovery following drought. Ledger and Hildrew (2001) found after a severe summer drought that dried up a perennial stream, that recovery of invertebrate abundance was rapid, but the community structure was quite different from the predrought one and was initially dominated by a previously uncommon chironomid. It was only with the return of high flows that the community came to resemble that existing prior to the drought. It is clear that supra-seasonal droughts can leave lingering signals in community structure. By drastically reducing the abundances of species and creating new habitat, drought may create the conditions to favour new or uncommon species.

Drought may create a mosaic of pools, in which fish populations persist (see Labbe and Fausch, 2000; Magoulick and Kobza, 2003). Pools differ in their species composition and population structure (see Magalhaes *et al.*, 2002) and with the breaking of seasonal drought, remixing occurs and fish recovery is relatively rapid (Matthews and Marsh-Matthews, 2003). Recovery from severe supra-seasonal droughts is variable. In some instances recovery was rapid – a matter of months (Larimore *et al.*, 1959), but in other cases recovery was slow and incomplete with long lag effects (Matthews and Marsh-Matthews, 2003). We have a poor understanding of how drought affects the ecology of stream fish. To understand the dynamics of fish with drought, it is necessary to work at spatial scales at which the fish react to the drought and to commit to long-term studies. By acting as a harsh filter (Poff, 1997; Matthews, 1998), drought may, especially in regions with marked seasonality or with semi-arid conditions, determine the composition of the fish fauna.

Drought is a natural phenomenon and biota have evolved many different adaptations to contend with it. Like floods, droughts may serve as an environmental filter to eliminate or reduce some biota and create opportunities for others. In this way as resetting or scene changing events, they may be vital for the long-term persistence of lotic ecosystems and their biota (Everard, 1996).

5.9 Summary

We have much to learn about hydrological disturbances, especially as regards droughts. As indicated in Table 5.1, it is possible to derive some general characteristics of floods and droughts. In the case of floods there are marked differences between constrained and floodplain rivers in the impacts of floods: either damaging or replenishing. The resistance and resilience of stream ecosystems to drought depend on the duration; whether seasonal or supra-seasonal. Not surprisingly, there are few parallels between the effects of floods and droughts, with the latter not being the symmetrical opposite of floods. The major differences between the two disturbance types are largely due to the differences in the acuity of onset and termination and the duration: a rapid event or a creeping calamity.

Table 5.1 Characteristic attributes of floods and droughts in flowing waters

Attribute of disturbance	**Floods**	**Droughts**
Type of disturbance	Pulse	Ramp
Cause	Excess of water	Extended deficit of water
Major abiotic effects	Reconfiguration of streambed in constrained streams Habitat regeneration with ecosystem expansion in floodplain rivers	Contraction of ecosystem, drying of streambed, pool formation and low water quality
Type of response	Pulse	Ramp
Resistance	Low in constrained streams Variable in floodplain rivers	Medium to high for seasonal drought Varies with duration for supraseasonal drought
Refugia	Wide variety on both constrained and floodplain systems	Wide variety for seasonal drought Variety declines with duration in supraseasonal drought
Resilience	Medium to high in constrained streams Pulse of growth and production in floodplain systems.	High for seasonal drought Varies with duration for supraseasonal drought, may have lingering lags.

5.10 Hydrological Disturbances and Future Challenges

Landscape ecology seeks to generate an understanding of the links between spatial complexity in landscapes and ecological processes. The need to apply the landscape ecology approach to streams, and between streams and land, has been stressed by Wiens (2002). Concepts such as riverscape (Fausch *et al.*, 2002) and riverine landscape (Ward *et al.*, 2002), to describe spatial complexity and processes within rivers and to areas that 'are directly influenced by the river' have been proposed. It may be simpler to adopt a scape ecology approach (Lake, 2003a), applying principles developed in landscape ecology within and across habitats and domains (rivers and land).

Floods and droughts are generated by changes in water availability in catchments. Headwater streams cover 70–80% of catchments and contribute most of the water. However, we have a poor understanding of basic headwater stream processes and their role in generating disturbances. Floods may be localized, being formed in particular headwater streams in a catchment, whereas droughts tend to be large-scale phenomena and can affect many headwater streams. Within different river systems, dependent on their morphology, there will be different impacts of floods down the rivers and different positions where floods change from being damaging forces to floodplain replenishers. As floods and droughts arise, build and subside in different ways within the same catchment and exert different effects, it is important that an overall scape ecology approach be taken from headwaters to floodplain. The scape ecology of floods and droughts may be developed more readily by dealing with intermittent systems, as scape changes can be readily assessed by changes in the extent of channel inundated or dry (Dodds *et al.*, 2004).

For floods we have some understanding of their scape ecology – the land–water links and processes and within-channel processes, but for droughts our understanding is rudimentary (Fisher *et al.*, 2001, 2004; Stanley *et al.*, 2004). A major problem with the scape ecology of disturbance is that hitherto most research has been carried out at the site level – an approach that is inadequate for understanding the spatial heterogeneity and ecological processes at the required spatial extent (Fausch *et al.*, 2002). The challenge in the disturbance ecology of flowing waters is to carry out research at the appropriate spatial and temporal scales, which will be invariably much larger and longer than the site or reach level.

Human activities both in streams (e.g., flow regulation) and on stream catchments (e.g., urbanization creating impervious surfaces) can exacerbate both droughts and floods. Global climate change is projected to increase the frequency and severity of extreme events – floods and droughts (Poff *et al.*, 2002; Pittock, 2003). Thus, we need to have a comprehensive understanding of floods and droughts. Furthermore, we will need to develop ways by which we can restore 'natural' disturbance regimes where needed, and to develop and implement effective restoration measures to cope with the impacts of global climate change.

Acknowledgements

I wish to thank Nick Bond and Paul Reich of the School of Biological Sciences, Monash University, for advice and criticism, and Land and Water Australia for funding.

References

Amoros C, Bornette G. 2002. Connectivity and biocomplexity in waterbodies of riverine floodplains. *Freshwater Biology*, **47**: 761–776.

Arthington AH, Balcombe SR, Wilson GA, Thoms MC, Marshall J. 2005. Spatial and temporal variation in fish-assemblage structure in isolated waterholes during the 2001 dry season of an arid-zone floodplain river, Cooper Creek, Australia. *Marine and Freshwater Research*, **56**: 25–36.

Attrill MJ, Power M. 2000. Effects on invertebrate populations of drought-induced changes in estuarine water quality. *Marine Ecology Progress Series*, **203**: 133–143.

Ballinger A, Mac Nally R, Lake PS. 2005. Immediate and longer-term effects of managed flooding on floodplain invertebrate assemblages in south-eastern Australia: generation and maintenance of a mosaic landscape. *Freshwater Biology*, **50**: 1190–1205.

Baxter CV, Fausch KD, Saunders WC. 2005. Tangled webs: reciprocal flows of invertebrate prey link streams and riparian zones. *Freshwater Biology*, **50**: 201–220.

Bender EA, Case TJ, Gilpin ME. 1984. Perturbation experiments in community ecology; theory and practice. *Ecology*, **65**: 1–13.

Bond NR. 2004. Spatial variation in fine sediment transport in small upland streams: the effects of flow regulation and catchment geology. *River Research and Applications*, **20**: 705–717.

Bond N, Lake PS. 2005. Ecological restoration and large-scale ecological disturbances: effects of drought on a stream habitat restoration experiment. *Restoration Ecology*, **13**: 39–48.

Boulton AJ. 1989. Over-summering refuges of aquatic macroinvertebrates in two intermittent streams in central Victoria. *Transactions of the Royal Society of South Australia*, **113**: 23–34.
Boulton AJ. 2003. Parallels and contrasts in the effects of drought on stream macroinvertebrate assemblages. *Freshwater Biology*, **48**: 1173–1185.
Boulton AJ, Lake PS. 1992. The ecology of two intermittent streams in Victoria, Australia. III. Temporal changes in faunal composition. *Freshwater Biology*, **27**: 123–138.
Boulton AJ, Stanley EH. 1995. Hyporheic processes during flooding and drying in a Sonoran Desert Stream. II. Faunal dynamics. *Archiv für Hydrobiologie*, **134**: 27–52.
Brooks S. 1997. Impacts of flood disturbance on the macroinvertebrate assemblage of an upland stream. PhD thesis, Monash University, Australia.
Bunn SE, Davies PM, Winning M. 2003. Sources of organic carbon supporting the food web of an arid zone floodplain river. *Freshwater Biology*, **48**: 619–635.
Caruso BS. 2002. Temporal and spatial patterns of extreme low flows and effects on stream ecosystems in Otago, New Zealand. *Journal of Hydrology*, **257**: 115–133.
Clinton SM, Grimm NB, Fisher SG. 1996. Response of a hyporheic invertebrate assemblage to drying disturbance in a desert stream. *Journal of the North American Benthological Society*, **15**: 700–712.
Closs GP, Lake PS. 1995. Drought, differential mortality and the coexistence of native and an introduced fish species in a south-east Australian intermittent stream. *Environmental Biology of Fishes*, **47**: 17–26.
Covich AP, Crowl TA, Scatena FN. 2003. Effects of extreme low flows on freshwater shrimps in a perennial tropical stream. *Freshwater Biology*, **48**: 1199–1206.
Dahm CN, Baker MA, Moore DI, Thibault JR. 2003. Coupled biogeochemical and hydrological responses of streams and rivers to drought. *Freshwater Biology*, **48**: 1219–1231.
David BO, Closs GP. 2002. Behavior of a stream-dwelling fish before, during and after high-discharge events *Transactions of the American Fisheries Society*, **131**: 762–771.
Dent CL, Grimm NB, Fisher SG. 2001. Multiscale effects of surface–subsurface exchange on stream water nutrient concentrations. *Journal of the North American Benthological Society*, **20**: 162–181.
Dodds WK, Gido K, Whiles MR, Fritz KM, Matthews WJ. 2004. Life on the edge: the ecology of Great Plains prairie streams. *BioScience*, **54**: 205–216.
Dole-Olivier M-J, Marmonier P, Beffy J-L. 1997. Response of invertebrates to lotic disturbance: is the hyporheic zone a patchy refugium? *Freshwater Biology*, **37**: 257–276.
Douglas MM, Townsend SA, Lake PS. 2003. Stream dynamics. In *Fire in Tropical Savannas: An Australian Study* (Eds Andersen A, Cook GD, Williams RJ). Springer-Verlag, New York, pp. 59–78.
Downes BJ, Lake PS. 1991. Different colonization patterns of two closely related stream insects (*Austrosimulium* spp.) following disturbance. *Freshwater Biology*, **26**: 295–306.
Downes BJ, Lake PS, Glaister A, Webb JA. 1998. Scales and frequencies of disturbance: rock size, bed packing and variation among upland streams. *Freshwater Biology*, **40**: 625–639.
Erman NA, Erman DC. 1995. Spring permanence, Trichoptera species richness and the role of drought. *Journal of the Kansas Entomolgical Society*, **68** (Suppl.): 50–64.
Everard M. 1996. The importance of periodic droughts for maintaining diversity in the freshwater environment. *Freshwater Forum*, **7**: 33–50.
Fausch KD, Torgersen CE, Baxter CV, Li HW. 2002. Landscapes to riverscapes: Bridging the gap between research and conservation of stream fishes. *BioScience*, **52**: 483–498.
Fisher SG 1983. Succession in streams. In *Stream Ecology: Applications and Testing of General Ecological Theory* (Eds Barnes JR, Minshall GW). Plenum Press, New York, pp. 7–27.
Fisher SG, Grimm NB. 1991. Streams and disturbance: Are cross-ecosystem comparisons useful? In *Comparative Analyses of Ecosystems: Patterns, Mechanisms and Theories* (Eds Cole JJ, Lovett GM, Findlay SEG). Springer-Verlag, New York, pp. 196–221.
Fisher SG, Gray LJ, Grimm NB, Busch DE. 1982. Temporal succession in a desert stream following flooding. *Ecological Monographs*, **52**: 93–110.
Fisher SG, Welter J, Schade J, Henry J. 2001. Landscape challenges to ecosystem thinking: Creative flood and drought in the American Southwest. *Scientia Marina*, **65** (Suppl.2): 181–192.

Fisher, SG, Sponseller RA, Heffenan JB. 2004. Horizons in stream biogeochemistry: flowpaths to progress. *Ecology*, **85**: 2369–2379.

Fritz KM, Dodds WK. 2004. Harshness: Characterization of intermittent stream habitat over space and time. *Marine and Freshwater Research*, **56**: 13–23.

Gagnon PM, Golladay SW, Michener WK, Freeman MC. 2004. Drought responses of freshwater mussels (Unionidae) in coastal plain tributaries of the Flint River basin, Georgia. *Journal of Freshwater Ecology*, **19**: 667–679.

Gasith A, Resh VH. 1999. Streams in Mediterranean climate regions: Abiotic influences and biotic responses to predictable seasonal events. *Annual Review of Ecology and Systematics*, **30**: 51–81.

Giller PS. 1996. Floods and droughts: the effects of variations in water flow on streams and rivers. In *Disturbance and Recovery of Ecological Systems* (Eds Giller PS, Myers AA). Royal Irish Academy, Dublin, Eire, pp. 1–19.

Glasby TM, Underwood AJ. 1996. Sampling to differentiate between pulse and press perturbations. *Environmental Monitoring and Assessment*, **42**: 241–252.

Gordon ND, McMahon TA, Finlayson BL, Gippel CJ, Nathan RJ. 2004. *Stream Hydrology. An Introduction for Ecologists*. Second Edition. John Wiley & Sons, Ltd, Chichester, UK.

Gray LJ, Fisher SG. 1981. Postflood recolonization pathways of macroinvertebrates in a lowland Sonoran Desert stream. *American Midland Naturalist*, **106**: 249–257.

Grimm NB, Fisher SG. 1989. Stability of periphyton and macroinvertebrates to disturbance by flash floods in a desert stream. *Journal of the North American Benthological Society*, **8**: 293–307.

Harrison AD. 1966. Recolonisation of a Rhodesian stream after drought. *Archiv für Hydrobiologie*, **62**: 405–421.

Hemphill N, Cooper SD. 1983. The effect of physical disturbance on the relative abundances of two filter-feeding insects in a small stream. *Oecologia*, **58**: 376–382.

Holmes NTH. 1999. Recovery of headwater stream flora following the 1989–1992 groundwater drought. *Hydrological Processes*, **13**: 341–354.

Institute of Hydrology. 1980. *Low Flow Studies*. Volumes 1–4. Institute of Hydrology, Wallingford, England, UK.

Junk WJ, Bayley PB, Sparks RE. 1989. The flood pulse concept in river-floodplain systems. In *Proceedings of the International Large River Symposium*. Canadian Special Publication of Fisheries and Aquatic Sciences 106: 110–127.

Labbe TR, Fausch KD. 2000. Dynamics of intermittent stream habitat regulate persistence of a threatened fish at multiple scales. *Ecological Applications*, **10**: 1774–1791.

Ladle M, Bass JAB. 1981. The ecology of a small chalk stream and its responses to drying during drought conditions. *Archiv für Hydrobiologie*, **90**: 448–466.

Lake PS. 2000. Disturbance, patchiness and diversity in streams. *Journal of the North American Benthological Society*, **19**: 573–592.

Lake PS. 2003a. Scape ecology: the study of patchiness. *Watershed*, May: 4–5.

Lake PS. 2003b. Ecological effects of perturbation by drought in flowing waters. *Freshwater Biology*, **48**: 1161–1172.

Lancaster J, Belyea LR. 1997. Nested hierarchies and scale-dependence of mechanisms of flow refugium use. *Journal of the North American Benthological Society*, **16**: 221–238.

Lancaster J, Hildrew AG. 1993. Characterizing in-stream flow refugia. *Canadian Journal of Fisheries and Aquatic Sciences*, **50**: 1663–1675.

Larimore RW, Childers WF, Heckrotte C. 1959. Destruction and re-establishment of stream fish and invertebrates affected by drought. *Transactions of the American Fisheries Society*, **88**: 261–285.

Ledger ME, Hildrew AG. 2001. Recolonization by the benthos of an acid stream following a drought. *Archiv für Hydrobiologie*, **152**: 1–17.

Lytle DA. 2000. Biotic and abiotic effects of flash flooding in a montane desert stream. *Archiv für Hydrobiologie*, **150**: 85–100.

Lytle, DA. 2001. Disturbance regimes and life-history evolution. *American Naturalist*, **157**: 525–536.

Lytle DA. 2003. Reconstructing long-term flood regimes with rainfall data: Effects of flood timing on caddisfly populations. *Southwestern Naturalist*, **48**: 36–42.

Lytle DA, Poff NL. 2004. Adaptation to natural flow regimes. *Trends in Ecology and Evolution*, **19**: 94–100.

Lytle DA, Smith RL. 2004. Exaptation and flash flood escape in the giant water bugs. *Journal of Insect Behavior*, **17**: 169–178.

Mackay RJ. 1992. Colonization by lotic macroinvertebrates: a review of processes and patterns. *Canadian Journal of Fisheries and Aquatic Sciences*, **49**: 617–628.

Magalhães MF, Beja P, Canas C, Collares-Pereira MJ. 2002. Functional heterogeneity of dry-season fish refugia across a Mediterranean catchment: The role of habitat and predation. *Freshwater Biology*, **47**: 1919–1934.

Magoulick DD, Kobza, RM. 2003. The role of refugia during drought: A review and synthesis. *Freshwater Biology*, **48**: 1186–1198.

Matthaei CD, Guggelberger C, Huber H. 2003. Effects of local disturbance history on benthic river algae. *Freshwater Biology*, **48**: 1514–1526.

Matthaei CD, Townsend CR, Arbuckle CJ, Peacock KA, Guggelburger C, Küster CE, Huber H. 2004. Disturbance, assembly rules, and benthic communities in running waters: A review and some implications for restoration projects. In *Assembly Rules and Restoration Ecology. Bridging the Gap between Theory and Practice* (Eds Temperton VM, Hobbs RJ, Nuttle T, Halle S). Island Press, Washington, USA, pp. 367–388.

Matthews WJ. 1998. *Patterns in Freshwater Fish Ecology*. Chapman and Hall, New York, USA.

Matthews WJ, Marsh-Matthews E. 2003. Effects of drought on fish across axes of space, time and ecological complexity. *Freshwater Biology*, **48**: 1232–1253.

McMahon TA, Finlayson BL. 2003. Droughts and anti-droughts: the low flow hydrology of Australian rivers. *Freshwater Biology*, **48**: 1147–1160.

Meffe GK. 1984. Effects of abiotic disturbance on coexistence of predator and prey fish species. *Ecology*, **65**: 1525–1534.

Miller AM, Golladay SW. 1996. Effects of spates and drying on macroinvertebrate assemblages of an intermittent and a perennial prairie stream. *Journal of the North American Benthological Society*, **15**: 670–689.

Minshall GW, Petersen RC. 1985. Towards a theory of macroinvertebrate community structure in stream ecosystems. *Archiv für Hydrobiologie*, **104**: 49–76.

Molles MC. 1985. Recovery of a stream invertebrate community from a flash flood in Tesuque Creek, New Mexico. *The Southwestern Naturalist*, **30**: 279–287.

Nathan RJ, McMahon TA. 1990. Practical aspects of low flow frequency analysis. *Water Resources Research*, **26**: 2135–2141.

O'Connell M, Baldwin DS, Robertson AI, Rees D. 2000. Release and bioavailability of dissolved organic matter from floodplain litter: influence of origin and oxygen levels. *Freshwater Biology*, **45**: 333–342.

O'Connor NA, Lake PS. 1994. Long-term and seasonal large-scale disturbances of a small lowland stream. *Australian Journal of Marine and Freshwater Research*, **45**: 243–255.

Pittock B. 2003. *Climate Change: An Australian Guide to the Science and Potential Impacts*. Australian Greenhouse Office, Canberra, Australia.

Poff NL. 1997. Landscape filters and species traits: towards mechanistic understanding and prediction in stream ecology. *Journal of the North American Benthological Society*, **16**: 391–411.

Poff NL, Brinson MM, Day JW. 2002. *Aquatic Ecosystems and Global Climate Change. Potential Impacts on Inland Freshwater and Coastal Wetland Ecosystems in the United States*. Pew Center on Global Climate Change, Arlington, USA.

Puckridge JT, Sheldon F, Walker KF, Boulton AJ. 1998. Flow variability and the ecology of large rivers. *Marine and Freshwater Research*, **49**: 55–72.

Puckridge JT, Walker KF, Costelloe JF. 2000. Hydrological persistence and the ecology of dryland rivers. *Regulated Rivers: Research and Management*, **16**: 385–402.

Resh VH, Brown AV, Covich AP, Gurtz ME, Li HW, Minshall GW, Reice SR, Sheldon AL, Wallace JB, Wissmar R. 1988. The role of disturbance in stream ecology. *Journal of the North American Benthological Society*, **7**: 433–455.

Robson BJ, Matthews TG. 2004. Drought refuges affect algal recolonization in intermittent streams. *River Research and Applications*, **20**: 753–763.

Roshier, DA, Robertson AI, Kingsford RT. 2002. Responses of waterbirds to flooding in an arid region of Australia and implications for conservation. *Biological Conservation*, **106**: 399–411.

Sedell JR, Reeves GH, Hauer FR, Stanford JA, Hawkins CP. 1990. Role of refugia in recovery from disturbances: modern fragmented and disconnected river systems. *Environmental Management*, **14**: 711–724.

Smith K, Ward R. 1998. *Floods. Physical Processes and Human Impacts.* John Wiley & Sons, Chichester, UK.

Stanley EH, Buschman DL, Boulton AJ, Grimm NB, Fisher SG. 1994. Invertebrate resistance and resilience to intermittency in a desert stream. *American Midland Naturalist*, **131**: 288–300.

Stanley EH, Fisher SG, Grimm NB. 1997. Ecosystem expansion and contraction in streams. *BioScience*, **47**: 427–435.

Stanley EH, Fisher SG, Jones JB. 2004. Effects of water loss on primary production: a landscape-scale model. *Aquatic Sciences*, **66**: 130–138.

Steinman AD, McIntire CD. 1990. Recovery of lotic periphyton communities after disturbance. *Environmental Management*, **14**: 589–604.

Thomson JR. 2002. The effect of hydrological disturbance on the densities of macroinvertebrate predators and their prey in a coastal stream. *Freshwater Biology*, **47**: 1333–1351.

Tockner K, Malard F, Ward JV. 2000 An extension of the flood pulse concept. *Hydrological Processes*, **14**: 2861–2883.

Townsend CR. 1989. The patch dynamics concept of stream community ecology. *Journal of the North American Benthological Society*, **8**: 36–50.

Townsend CR, Dolédec S, Scarsbrook MR. 1997. Species traits in relation to temporal and spatial heterogeneity in streams: A test of habitat templet theory. *Freshwater Biology*, **37**: 367–387.

Uehlinger U. 2000. Resistance and resilience of ecosystem metabolism in a flood-prone river system. *Freshwater Biology*, **45**: 319–332.

Valett HM, Baker MA, Morrice JA, Crawford CS, Molles MC, Dahm CN, Moyer DL, Thibault JR, Ellis LM. 2005. Biogeochemical and metabolic responses to the flood pulse in a semiarid floodplain. *Ecology*, **86**: 220–234.

Ward JV. 1989. Riverine-wetland interactions. *Freshwater Wetlands and Wildlife, 1989. Conf. – 8603101, DOE Symposium Series No. 61* (Eds Sharitz RR, Gibbons JW) USDOE Office of Scientific and Technical Information, Oak Ridge, Tennessee, USA, pp. 385–400.

Ward JV, Tockner K, Arscott DB, Claret C. 2002. Riverine landscape diversity *Freshwater Biology*, **47**: 517–539.

Ward R. 1978. *Floods. A Geographical Perspective.* Macmillan, London, UK.

Webster JR, Gurtz ME, Hains JJ, Meyer JL, Swank WT, Waide JB, Wallace JB. 1983. Stability of stream ecosystems. In *Stream Ecology: Applications and Testing of General Ecological Theory* (Eds Barnes JR, Minshall GW). Plenum Press, New York, USA, pp. 355–395.

Wiens JA. 2002. Riverine landscapes: taking landscape ecology into the water. *Freshwater Biology*, **47**: 501–515.

Wilhite DA. 2000. Drought as a natural hazard. Concepts and definitions. In *Drought. Volume 1. A Global Assessment* (Ed Wilhite DA). Routledge London, UK, pp. 3–18.

Wood PJ, Petts GE. 1999. The influence of drought on chalk stream invertebrates. *Hydrological Processes*, **13**: 387–399.

Wood PJ, Armitage PD. 2004. The response of the macroinvertebrate community to low-flow variability and supra-seasonal drought within a groundwater dominated stream. *Archiv für Hydrobiologie*, **161**: 1–20.

Wright JF, Clarke RT, Gunn RJM, Kneebone NT, Davy-Bowker J. 2004. Impact of major changes in flow regime on the macroinvertebrate assemblages of four chalk stream sites, 1997–2001. *River Research and Applications*, **20**: 775–794.

6

Surface Water–Groundwater Exchange Processes and Fluvial Ecosystem Function: An Analysis of Temporal and Spatial Scale Dependency

Pascal Breil, Nancy B. Grimm and Philippe Vervier

6.1 Introduction

There is general agreement that river flow variability is a major driving force in shaping fluvial hydrosystems and in determining ecological strategies and species richness in water courses (Southwood, 1977; Hildrew and Townsend, 1987; Townsend, 1989; Poff, 1997). Flow variability creates conditions for water fluxes and linked material exchanges between surface and groundwater (SGW), resulting in an active interface zone with varying boundaries in time and space. This interface or ecotone is termed the hyporheic zone, where flow direction and strength of changes over time and space varies, providing dynamic conditions for metabolic and biogeochemical processes. Hydrologic and geomorphic contexts (including SWG exchange) establish a template upon which biotic processes operate (Poff, 1997). Thus, effectively to address natural trends as well as human-modified changes in fluvial ecosystems, ecological and hydrologic perspectives must be merged. For management, the priority is to identify the timing, frequency, magnitude, and duration in flow variability that is needed to maintain processes as near to natural conditions in space and time as possible (Petts *et al.*, 2005). Hence, information

94 *Hydroecology and Ecohydrology: Past, Present and Future*

on flow variability is key to judging ecosystem resilience in response to natural or human-induced dynamic change (Connel and Wayne, 1983; Bunn *et al.*, 2006; Thoms, 2006).

The objective of this chapter is to explore how flow variability affects SGW interactions and thereby functions of the fluvial ecosystem along the 'hyporheic corridor' from head waters to river mouth (Stanford and Ward, 1993). The chapter begins by defining fluvial ecosystems in terms of two aspects of their structure and function – biogeochemical processes and biotic structure – that are expected to respond to flow variability and SGW exchange. Data from a field experiment serve to illustrate dimensions and variability of SGW exchange at one location in space and time. The range of space and time scales over which SGW interactions occur is then considered in a case study. This analysis takes the form of a cross-continental comparison of river basins in humid (Europe) and arid/semi-arid (southwestern USA) settings, generalizing the dynamics of SGW exchange in contrasting environments as recommended by Poff *et al.* (2006). Finally, the relationship between patterns of flow variability and biotic processes is explored, considering important biological and anthropogenic influences on SGW exchange at different spatio-temporal scales within fluvial ecosystems.

6.2 Fluvial Ecosystems: The Hydrogeomorphic Template and Ecosystem Function

Fluvial ecosystems are defined herein as comprising the annually flooded channel (i.e., surface water, adjacent parafluvial zone with saturated sediment deposits, and underlying hyporheic zone) plus the laterally connected riparian zone. The hydrogeomorphic template is defined as the geomorphic and hydrologic features of the fluvial ecosystem – and their spatial and temporal variability – that control the intensity and direction of water movement, and thus establish conditions experienced by biotic communities. Vertical, lateral and longitudinal dimensions of the river–riparian–upland ecotone are expected to vary from headwaters to river mouth in predictable ways (Gregory *et al.*, 1991; Grimm and Fisher, 1992; Stanford and Ward, 1993; Dahm *et al.*, 1998). Depending upon location along the water course, connectivity between the river and its landscape should be enhanced or limited, shaping the limits and fluctuations of the hyporheic system. Heterogeneity in the fluvial ecosystem, including paleochannels in riverbed sediments or riparian floodplains, creates preferential flows. The magnitude and direction of these flows influences nutrient transport and biotic activity in the hyporheic environment, creating hot spots of biodiversity and production (Stanford and Gonser, 1998; McClain *et al.*, 2003).

6.2.1 Fluvial Ecosystem Function: Biogeochemical Dynamics

Biogeochemical dynamics in fluvial ecosystems can be described by conceptual models that recognize the opposing processes of hydrologic transport and retention, for example nutrient spiralling (Elwood *et al.*, 1983; Mulholland *et al.*, 1990). Nutrient spiralling and nutrient transformations that contribute to whole-ecosystem nutrient retention are influenced by flow variability and SGW interactions through: (i) effects on biota that are responsible for nutrient processing; (ii) development of microhabitat conditions, such as redox conditions, that favour certain processes over others, and (iii) delivery of essential materials (macronutrients, organic carbon) from uplands to fluvial ecosystems or among different compartments of fluvial ecosystems. Finally, the biogeochemistry of fluvial ecosystems is a function of the biogeochemical processing in each component subsystem, linked by the fluxes of water among them (Fisher *et al.*, 1998).

Effects of SGW interactions on biogeochemical dynamics will vary depending on basin size. Small streams are subject to short-duration but severe floods. Fluxes from the bank to the clearly delimited riverbed are direct, with consequent extensive interactions between soil and water. The contributions of material and water fluxes are usually unidirectional (basin to river). Small streams in their valleys are best visualized as the visible fraction of underground flows (Winter, 2001), rather than as disconnected gutters draining catchments. By contrast, large floodplain rivers have low bank length:volume ratios, and the river both deposits and entrains sediments and nutrients from parafluvial zones during floods. Major alluvial aquifers and extended floodplains are important storage reservoirs for water and nutrients that are mobilized during periodic floods. The contributions of groundwater from uplands are proportionately lower than for small streams, with a vast hyporheic zone interacting with the river and creating a fluvial 'riparian corridor' or 'hyporheic corridor' (e.g., Stanford and Ward, 1988, 1993). Thus, the magnitude and direction of connections between the catchment and its river vary downstream. Intermediate-sized rivers exhibit riffle–pool morphology featuring downwelling zones, where surface water enters the subsurface, alternating with upwelling zones, where hyporheic water is discharged to the surface. Matter fluxes are lateral (bidirectional exchanges with parafluvial and riparian groundwater) and vertical (bidirectional exchanges with the hyporheic zone). These patterns of SGW exchange strongly influence river biogeochemistry because of differential processing of nutrients and organic matter in surface and subsurface environments (e.g., Dent *et al.*, 2001), and because groundwater inputs often are chemically distinct from the river.

6.2.2 Fluvial Ecosystem Structure: Biotic Communities

Superimposed upon the hydrogeomorphic template of fluvial ecosystems are its habitats, which determine the structure and dynamics of biotic communities and their component populations. Extensive research and theory connects this template (especially flow dynamics) to communities of fish and macroinvertebrates (e.g., Southwood, 1977; Connell, 1978; Townsend, 1989). More recently, the importance of SGW exchanges and the different zones or 'biotopes' they create has been recognized (e.g., Boulton *et al.*, 1992; Lafont *et al.*, 2006). In sandy-bed rivers, surface benthos is often absent owing to scour, whereas deep hyporheos is protected from export and may thrive due to high rates of SGW exchange, which delivers organic material through the sandy matrix (Boulton, 1993; Rouch *et al.*, 1997; Fellows *et al.*, 2001). Of course, biotic communities include organisms within functional groups ranging from denitrifiers and nitrifiers (bacteria) to active predators that transform large biotic particles to small particles and dissolved nutrients (fish), and so these communities are essential components of biogeochemical dynamics.

6.3 Flow Variability and SGW Water Movements

Water flow drives nutrient fluxes in fluvial ecosystems. Flow variability induces fluctuations in water depth of river channel cross-sections (laterally) and along the long profile (downstream). Just as variation in surface water hydraulic gradient (a measure of the difference in water total energy between two points in space) drives water movement between points on a river, water movement between free-flowing water and connected unconfined porous media can be approximated using the hydraulic gradient (HG; see Chapter 9). Local conditions can create or enhance HG along a water course, generating local flowpaths in riverbed sediments and banks at angles from the main flow direction. The extent of SGW exchange will depend upon both HG and saturated hydraulic conductivity, according to Darcy's Law (e.g., Maidment, 1993). Variations in HG and hydraulic conductivity in space and time yield variations in water flow and SGW exchanges. Hydraulic conductivity values of the channel-lining layer range from 10^{-7} to $10^{3}\,\mathrm{m\,sec^{-1}}$ (Calver, 2001). As a general rule, hydraulic conductivity of riverbed substrates is greater than that of bank sediments by a factor of 100 to 1000, such that water flux will be greater in bed sediments than in the banks at any given HG (e.g., Wroblicky *et al.*, 1998).

6.3.1 In Space

SGW exchange can occur in several directions. At small scales, exchange between surface and hyporheic waters is dictated by variation in form of the bed (Savant *et al.*, 1987; Thibodeaux and Boyle, 1987; Harvey and Bencala, 1993) or the influence of structural features – including biotic features such as macrophyte beds or fish nests (e.g., White, 1990; Dent *et al.*, 2000). These features cause local variation in vertical HG (VHG) or hydraulic conductivity, and hence SWG exchange. Heads of riffles are infiltration zones (downwelling) of surface water in the hyporheic zone, and discharge zones (upwelling) occur at the end of hyporheic flow paths, often at the tails of riffles. Similar variations occur at larger scales, such as variation in VHG near riffle-pool transitions (e.g., Dent *et al.*, 2001) and variation in hydraulic conductivity associated with abandoned channels in oxbows of meandering rivers. Such lateral water movements may also be linked to an existing HG between a river and shallow groundwater in the adjacent floodplain. Along the river corridor, hydraulic conductivity tends to decrease from headwaters, with their coarse sediments, to lowlands, where fine sediments dominate. This generalization is expressed in the hyporheic corridor concept (Stanford and Ward, 1993) as an increase in the ratio of annual overland flows to interstitial flows on floodplains from headwaters to river mouth.

Longitudinal components of SGW fluxes correspond to shortcuts in meanders, again as a result of HG and bank hydraulic conductivity. In general, bottom gradients are naturally steeper upstream within a river network, with more riffle-pool sequences per km and more pronounced steps, creating greater opportunity for SGW exchanges regardless of river flow regime. This idea was investigated using a three-dimensional mechanistic hydraulic model under steady-flow conditions (Cardenas *et al.*, 2004), parameterized with field hydraulic conductivity data for a channel bend. Heterogeneity of hydraulic conductivity, bed-form profile, and channel curvature had significant influences on hyporheic zone geometry, SGW flux magnitude, and residence time. The major cause of SGW exchange was found to be the spatial density and magnitude of change in water-surface topography. Although these results are revealing, the steady-flow condition is a restricted case. In a natural river flow regime, a succession of transient HG occurs and particular combinations of HG duration and magnitude shape the three-dimensional boundaries of SGW exchange. To investigate this fuller suite of conditions, results from a field experiment are presented (Ruysschaert and Breil, 2004). Figures 6.1 and 6.2 illustrate how the

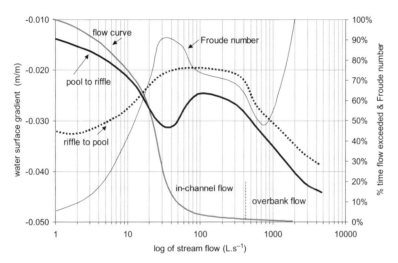

Figure 6.1 Observed mean water surface gradients as a function of flow rate in a sequence of pool–riffle and riffle–pool geomorphic units for a small stream near Lyon, France (curves were manually smoothed from empirical data.)

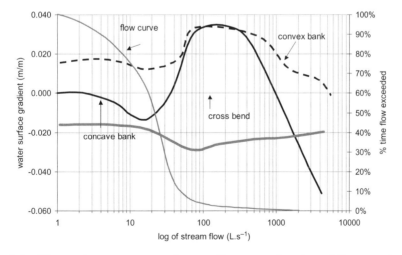

Figure 6.2 Observed mean water–surface gradients as a function of flow rate in lateral direction between channel and convex and concave banks and longitudinal direction crossing a meander (cross bend) for a small stream near Lyon, France

combined effect of vertical and planar geomorphic features – pool-riffle sequences and associated meanders – can influence the local free surface-water gradient at varying discharge. Local variations of these gradients create conditions for streambed VHG to change. Gradient–discharge curves are based upon mean observed values for hourly discharge class intervals over a period of 1 year, when discharge varied from no flow to a large flood. Negative gradient values indicate potential water movement from the free surface water in a lateral or downstream direction.

Surface-water hydraulic gradient in the pool–riffle sequence decreased with flow rate, exhibiting a minimum for an extended range of flows from 50 to $400 \, \text{ls}^{-1}$ (Figure 6.1). At discharge values above this threshold, only overbank flows can lead to larger gradients.

The pool-to-riffle gradient curve exhibited a more complex pattern, with an in-channel maximum for intermediate discharge. Comparison of gradient values indicated that the water surface becomes steeper in the pool–riffle sequence than in the riffle–pool at intermediate flows. A maximum hydraulic gradient at intermediate discharge makes sense when compared to the corresponding Froude hydraulic number (e.g., Maidment, 1993), which also exhibits a maximum (left scale in Figure 6.1) for the same flow range. The Froude number expresses the combined effect of stream channel roughness and overall bend curvature (which slow down water velocity) plus the driving gravitational force that results from stream gradients. The pool-riffle sequence locally increases channel roughness, inducing steeper HG. The opposite patterns of HG magnitude–discharge curves for the pool-riffle and riffle-pool units indicate local and temporarily unbalanced SGW exchange.

Lateral bank flows in both convex and concave directions are similar to the combined pattern of the pool-riffle and riffle-pool curves (Figure 6.2), showing that pool–riffle–pool sequence influence lateral SGW exchanges in a similar way. Only the longitudinal bank or cross bend flow seems insensitive to this effect but the gradient remains high regardless of flow rate, increasing slightly at intermediate flows. Bank gradient in the convex bank always flows toward the river (positive gradient) with an increase for larger flows. This pattern was not expected but it was confirmed with other measures. It reveals a persistent cross-bend flow near the channel. In the concave bank, the water surface gradient indicates flow from bank to river for a low-to-intermediate stream flow range and from river to the bank for larger stream flows.

Considering water surface gradient, corresponding flow rates and some range of values for hydraulic conductivity in the stream bed sediments and banks, it is possible roughly to compare rates of exchange between the longitudinal, lateral and vertical components. Results indicate that lateral flow in the convex bank and vertical flows in the pool–riffle and riffle–pool sequences account for almost 30% each of the yearly water exchange, but longitudinal bank and lateral concave bank flows account for only 10%. This specific case study example is not intended as a generalization but rather as an illustration of how complexity of fluvial geomorphology can produce spatially heterogeneous SGW exchanges. For larger water courses, the effect of bottom gradient should be less important than hydraulic conductivity of the flood plain (especially preferential flow-paths and paleochannels) in determining SGW exchange rates (Stanford and Ward, 1993).

6.3.2 In Time

Fluxes across the SGW interface are known to vary seasonally (e.g. Evans and Petts, 1997; Malcolm *et al.*, 2004). Based upon the case study presented above, larger HGs support downwelling in the channel at medium to high flows. These flow conditions occur ~60% of the year for this case study, compared with just 10% of the time (at low flow) when upwelling dominates. Three-dimensional SGW exchanges can persist with low, almost steady, flows except for the lateral concave bank. Gradient magnitude is emphasized with intermediate flows, which corresponds to unsteady flow conditions. Of course, these results may not apply in other contexts, such as streams or rivers that are strongly groundwater fed, where near-channel head gradients should be modified. During drier periods in more humid climates, groundwater feeds the river as baseflow. During wet weather, alternating rise and fall of river stage generates exchange fluxes from surface water to hyporheic and floodplain waters, and the reverse also occurs (Evans and Petts, 1997). In arid regions, VHG strengthens during the dry season, and the direction of SGW exchange may even shift from discharge to recharge (Stanley and Boulton, 1995). In human-controlled, steady low-flow conditions, vertical flow should predominate as a result of reduction in lateral head gradient between SW and shallow GW surface, but as shown in Figure 6.2, lateral and longitudinal flux can persist in river banks as a result of planar and vertical geomorphic features. Flow variations in time can modify SGW boundaries, producing greater changes than expected given the geomorphic template (Figures 6.1 and 6.2). Rapid change in flow emphasizes vertical hydraulic gradients between the channel and surrounding shallow floodplain groundwater, reinforcing lateral exchange rates. The latter case was shown to occur in a matter of hours during flooding of a mid-sized river in Arizona, USA (Marti *et al.*, 2000). The temporal distribution of flow variations and antecedent events are critical variables in determining SGW exchange boundaries, because VHG and even hydraulic conductivity can change rapidly. Thus, the total annual flux across the SGW interface is not likely to be the same for small VHGs over long durations as for large VHGs over short periods of time. Measurement of flow duration across a range of discharges is therefore needed to characterize the influence of flow variability on SGW exchanges rates.

6.3.3 An Analysis of Flow Variability Dependency with Basin Area

Both the duration and intensity of flows are expected to control SGW exchange rate as well as the spatial extent of the hyporheic zone. Comparison of flow variability for range of basin areas can provide scaling factors to judge how the strength and direction of SGW exchanges changes with system size.

In this analysis, time series of daily river flows from the Colorado (Arizona, USA) and the Loire (western France) basins were used to: (i) evaluate how SGW exchange scales with basin area in two contrasting climatic regimes, and (ii) compare the effect of arid and humid temperate climates on flow variability. River gauging stations were chosen to span a range of basin areas from 100 to 100 000 km^2, selecting as far as possible a common monitoring period in each basin, but giving priority to natural flow conditions. The duration of each time-series used was 10 years.

The Colorado River basin in southwestern USA is a ~489 000 km² catchment with headwaters in the Rocky Mountains. The Colorado drains mountainous and lowland arid and semi-arid terrain, and has been heavily modified to meet water-supply demands for agriculture, mining, and urban growth in the region. Discharge from the mainstem Colorado to the Gulf of California has dropped to a trickle as a result of this water use. Annual precipitation ranges nearly 1000 mm in the mountains, which falls mainly as snow, to under 200 mm in the deserts. Temperatures exhibit a similarly wide range (<0–>45 °C), with cold winters in the mountains and hot, dry summers in the deserts. The basins selected for analysis ranged in area from 100 to just under 300 000 km², and are drained by unimpounded, free-flowing rivers.

The Loire River in western France drains 110 000 km² of gentle terrain in the west and highly eroded mountainous lands in the east that account for a quarter of the total area. The basin is covered with a patchwork of land uses from forested uplands to agricultural and urban areas on flat lands. Climate in this region is humid, with annual precipitation ranging from 700 mm in the plains to 1000 mm in the mountains, and is characterized by cool, rainy winters and mild summers. River flow alterations include hydropeaking in winter and abstractions for agricultural water supply in the summer; however, the latter impacts mainly the Loire mainstem. Basin selection covers a range from 100 to 110 000 km² in area, with nearly natural flows for <5000 km² catchments. The specific discharge for small to large Colorado and Loire basins, respectively, ranges from 0 to 4 and from 5 to 15 $l s^{-1} km^{-2}$.

Flow variability has two dimensions: the first reflects deviation from a mean value, and the second concerns the timing of given flow conditions. For the first dimension, the coefficient of variation (CV) was calculated from the discharge time series. This metric is a dimensionless measure of the dispersion around the mean discharge; hence, it can be used for comparison between basins and across spatial scales. Higher values of the CV for natural flow regimes often reflect a high ratio of peak-to-mean flows. The temporal distribution of flow fluctuations was assessed using mean event duration (D-value) and mean number of events per year (N-value) versus basin area. Events were defined as flows exceeding a threshold, with duration of exceedance computed for each event. Median discharge was used as the threshold because, by definition, it gives the flow value that is exceeded (or not reached, in the case of low-flow events) half of the time. The median reference is calculated only for the period of flow for intermittent streams to focus on water-movement indicators. To illustrate the dynamics of flow, the relationship between mean time of rise to peak flow and mean event-flow duration at each station was considered. To judge the magnitude with duration, the rate of change was calculated as the peak event flow divided by the median flow to give an M-value.

The comparative analysis of discharge time series revealed differences in flow variability for the arid/ semi-arid Colorado basin compared to the humid Loire basin. Overall, CV was greater in the arid basins than in the humid basins. CV decreased from small to large basins above a threshold of ~1000 km² in both climatic situations, although the pattern is more marked for the Colorado (Figure 6.3). Below this threshold, CV was positively correlated with basin size for the Loire basin. In contrast, the Colorado exhibited no clear relationship between CV and basin size for small basin areas. It may be that local basin features, like topography and geology, more strongly influence flow variability of small rivers in arid climates, but a regional study should be developed to confirm this

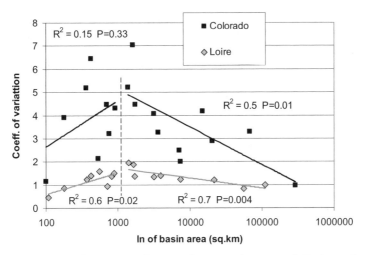

Figure 6.3 Coefficient of variation of stream flow as a function of discharge for variously sized catchments in the Colorado River basin, southwestern USA, and the Loire basin, France (coefficient of variation calculated from 10 years' data in each system.)

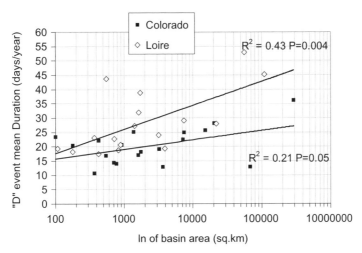

Figure 6.4 Mean event duration per year versus log of basin size, for variously sized catchments in the Colorado River basin, southwestern USA, and the Loire basin, France

assertion. These results suggest that a trend toward reduced flow variability with basin size is a general phenomenon. Considering SGW exchanges, these results lead to the expectation that in arid climates temporal gradients in water level will be greater and should induce more intense exchanges.

102 Hydroecology and Ecohydrology: Past, Present and Future

As discussed earlier, the frequency and duration of flow fluctuations are expected to influence HGs and thus SGW exchanges. Mean event number per year decreases as mean event duration increases according to the mathematical relationships that link these variables (Figure 6.5), except for arid intermittent streams with no flow periods (out points from curve). Means of event duration and event number both significantly differed between climates; the probability that both belong to the same statistical population is <1% and <3%, respectively. Overall, the Colorado basin exhibits flow events with a mean duration of 20 days repeated on average nine times a year, whereas event duration for the Loire basin is 28 days and frequency is seven times per year.

The Colorado at its outlet (~300 000 km^2 basin area) shows a major change in mean event duration D-value, from 36 to 8 days, as a result of dam operation (white triangle). Such a value corresponds to basins <100 km^2 in area (Figure 6.5). D-Value also correlates positively with log basin area, with the strongest relationship in the humid climate (Figure 6.4). We conclude from this analysis, which is limited to a relatively small dataset but encompasses a wide range of basin areas, that event durations increase and number of events decrease as basin area increases.

Assuming that a large number of events favours increased SGW exchange (owing to abrupt changes in hydraulic gradient), event duration should determine total SGW exchange per unit distance. This hypothesis was tested by regressing the relative magnitude of events, calculated as the ratio for each event of the peak flow to the median flow, the event duration, and the time of rise to peak on basin area (Figure 6.6). Both Colorado and Loire data exhibit a similar pattern for the two indices, with almost the same range of values.

Relative flow magnitude gives an indirect measure of water-level increase in the river channel, since river cross sections are sized to easily transfer the median flow. Because log

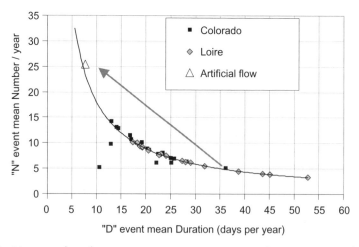

Figure 6.5 Mean number of events per year versus event mean duration per year, for variously sized catchments in the Colorado River basin, southwestern USA, and the Loire basin, France

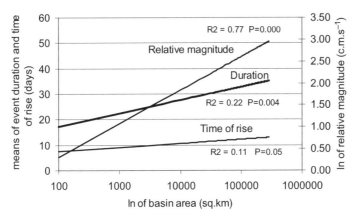

Figure 6.6 *Magnitude, duration, and time of rise to peak flow versus basin scale; regressions presented for Loire and Colorado Rivers combined*

relative magnitude is strongly correlated with log basin area ($R^2 = 0.77$; Figure 6.6), larger basins should exhibit larger water-level elevations during events as a percentage of median flow, and vertical hydraulic gradients between SW and GW should be correspondingly greater. Also, since event duration is positively related to basin area, event duration and relative magnitude are correlated (for the Loire and Colorado, $R^2 = 0.44$ and 0.22, respectively). Such a relation between magnitude and duration is clearly confirmed for maximum values and forms the basis for flood magnitude–duration–frequency analysis to calculate recurrence intervals of major events (Javelle *et al.*, 2002). Time of rise to peak flow is only weakly correlated with basin area (Figure 6.6) but the trend does imply that water level rises faster in small than in large basins. Again, rapid changes in water elevation might lead to strong, short-term vertical hydraulic gradients favouring SGW exchanges.

6.3.4 Linkage Between SGW and Flow Dynamics

Given the patterns of change in flow variability with the spatial scale that have been pointed out, it follows that rivers draining small- to medium-sized basins should provide nutrients to the hyporheic zone more efficiently than do rivers of larger basins. From our results it is not possible to get an idea on the 'quantity' of water exchanges, but clearly the temporal and spatial dimensions of concern are not the same for small to medium and large basin sizes. This has clear implications for the validation of ecological concepts at appropriate scales (Minshall, 1988).

Hydrologic dynamics described in the previous section interact with the fluvial landscape to create variability in the strength of SGW exchanges, with consequences for the structure and function of fluvial ecosystems. However, few studies are conducted for a range of scales and flow rates. The analysis presented here of how flow variability scales with basin area may provide a starting point for hypotheses about major controls of SGW exchanges for a range of fluvial ecosystem scales, considering duration (D), Frequency (N) and Intensity (M) of events over a frequent discharge like the median discharge in this example (Table 6.1).

Table 6.1 Hydrologic factors likely to control SGW exchanges at a range of basin areas, based on scaling analysis of changes in flow variability with basin area. For each basin area, the relevant spatial scale within the fluvial ecosystem, and the important biological and anthropogenic influences on SGW exchange, are listed

Basin area (km^2)	Hydrologic factors time scale	Indicative fluvial ecosystem dimension of concern	Biologic influences	Sensitive anthropogenic influences
>10 000	D > 25–45 days N < 4 days M > 2	Sector (hundredths of meters)	Humans	Large dams Flow diversion
1000 to 10 000	D = 22–35 days N < 6 days M > 1	Reach (tenths of meters)	Long-lived trees (evapo-transpiration) Beavers	Straightening Bridge construction
100 to 1000	D = 20–25 days N < 8 days M > 0.5	Channel unit (meters)	Riparian vegetation (coarse woody debris)	Bank incision Bank stabilization Canalization Large trash
<100	D = 15–20 days N = 10 days M < 0.5	Sub-unit (meter)	Fish bioturbation Algal mats	Introduction/removal of species Channel modifications
<10	D^{**} = hours $N^{**} \gg$ 10 days $M^{**} \ll$ 0.5	Particle (decimetre)	Invertebrate bioturbation	Pollution (clogging pores, killing inverts and microbiota) Sedimentation

(*) D: event mean duration per year, N: event mean number per year, M: event mean relative magnitude per year (log measure). ** *expected time scale, not tested*

6.4 Implications of Flow Variability for SGW Exchange and Fluvial Ecosystem Structure and Function

6.4.1 Material Delivery to and within Fluvial Ecosystems

The catchment is the ultimate source of nutrients and organic matter entering the river. As water flows from uplands to fluvial ecosystems, then downstream along riverine flowpaths, it transports chemical elements from one compartment or functional unit to another. Abiotic and biotic reactions within the catchment buffer material concentrations in the river via pore water; hence, concentrations of dissolved organic carbon (DOC) and non-reactive minerals (chemical signatures) of rivers are relatively stable and insensitive to catchment disturbances (Bormann and Likens, 1979; Neal *et al.*, 1992). More reactive substances are altered during upland–river transport. Differential export of organic carbon compounds from the soil to the river, for example, results from an interaction between degradation reactions and adsorption, a concept reflected in the 'regional chromatography model' advanced by Hedges *et al.* (1986). Hedges *et al.* (1994) further suggest that greater attention should paid to nitrogen (N) content as a factor modulating organic matter exports.

These material fluxes, both to and within fluvial ecosystems, are dependent upon seasonal distributions of hydrologic events (i.e., high flows), as well as vegetation activity and soil temperature. The mobile products of leaching, decomposition, and other terrestrial processes are transported in subsurface flow to the river. Thus, the hydrologic cycle strongly influences quality and quantity of nutrients and organic matter fluxes during water movement from catchment via subsoil waters to the river (Cirmo and McDonnell, 1997; Salmon *et al.*, 2001; McGlynn and McDonnell, 2003). This action is exerted on various scales within the catchment and fluvial ecosystem, and at some scales or in some rivers, the direction of fluxes may be reversed (i.e., from river to riparian zone; Marti *et al.*, 1997). Studies of nutrient dynamics require analyses of spatial and seasonal variations of transport through, and activity in, different permeable media, such as upland soil, the riparian zone, and the hyporheic zone. These dynamics are driven by the physical movement of water in the fluvial ecosystem as a whole. Each functional unit may work as a filter (mechanical, chemical and biochemical filters). Stronger active zones (hot spots) are places where filters are particularly effective.

6.4.2 Modulation of Nutrient and Organic Matter Delivery by the Riparian Interface Zone

The river is coupled with its catchment via the riparian zone or floodplain, whose nature and width modulate transfers of organic matter and nutrients (Gold *et al.*, 2001; Burt and Pinay, 2005). Floodplain width and development can vary along river courses; thus, there is variation in the capacity to modulate sediment or nutrient delivery to the river from the catchment during floods. Within the fluvial landscape, heterogeneity in both biotic activity and hydrologic properties strongly affects biogeochemical function. For instance, the riparian zone usually exhibits a high potential for denitrification, owing to high organic matter content and fine sediments favorable to denitrifying bacteria (Pinay *et al.*, 1990, 1995). However, at the ecosystem scale, zones of high denitrification potential can represent zones of low hydraulic conductivity. Therefore, the total capacity for N transformation within these active zones can be low (Figure 6.7).

Permeable media within the fluvial ecosystem act as biogeochemical reactors whose function varies with hydrologic conditions. For example, transport of organic matter from riparian soil to the river can be by lateral seepage or by episodic runoff (McDowell and Likens, 1988). In the former case, the organic matter is in prolonged contact with the mineral phase in soil interstices, leading to low concentrations. During storms, the system is short-circuited, and large quantities of organic matter leached from surface litter or vegetation are moved rapidly to the river without passing through soil horizons, where microbial uptake or fixation on mineral surfaces might have taken place. Under these conditions, the organic matter reaching the river is less degraded due to its short residence time in the soil. Such organic matter pulses are transitory and quickly absorbed, but in the stream rather than the soil. The organic matter residence time is a few days, or even a few hours.

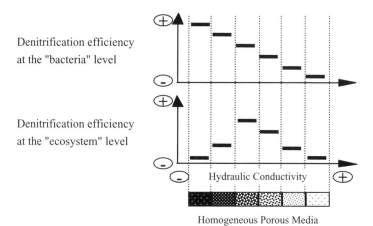

Figure 6.7 *The whole-system capacity for denitrification (i.e., efficiency of the nitrogen 'filter') is a function of both bacterial efficiency (top) and hydrologic properties (bottom), since both high biological activity and substantial hydrologic flux are required for an effective nutrient filter*

In the model just described, near-stream riparian zones are of considerable importance as an organic matter source to the river. High DOC concentrations associated with runoff from storms or snowmelt often come mainly from the near-stream riparian zones (Boyer *et al.*, 1997), whereas infiltration water has a more distant origin and tends to be lower in DOC concentration and more refractory. These suppositions are reasonable for soils of the Hubbard Brook Valley (New Hampshire, USA) with its impermeable horizon that functions like a podzol. Here, more river water comes from superficial horizons and deep infiltration is significant. The model requires testing in more porous soils.

6.4.3 In-stream Biogeochemical Function and Flow Variability

The hyporheic zone lies at the interface of the epigean and hypogean river subsystems, with its size dynamically determined by porosity and the volume of water interacting with the surface (Triska *et al.*, 1989). Dimensions of the hyporheic zone will change in response to flow variability, as shown above for our case study. Like SGW exchange rate, biogeochemical activity in the hyporheic zone changes in response to surface water flow and its variations (Stanley and Boulton, 1995; Valett *et al.*, 1996). The hyporheic zone can serve as a mechanical, physical and biochemical filter. Hyporheic sediments store and process organic matter and nutrients mostly received from exchange with surface water (Grimm and Fisher, 1984; Jones *et al.*, 1995a; Boulton *et al.*, 1998; Grimaldi and Chaplot, 2000). Riffle zones are important sites of SGW exchange (Speaker *et al.*, 1984; Pusch, 1996); downwelling zones supply organic matter that fuels heterotrophic microbial metabolism (Pusch and Schwoerbel, 1994; Jones *et al.*, 1995b; Mulholland *et al.*, 1997), which in turn supports a diverse and unique hyporheic community (Danielopol, 1989; Boulton *et al.*, 1992; Findlay *et al.*, 2003). Thus, temporal dynamics of hyporheic-zone biogeochemical processes are highly dependent upon SGW exchanges that affect delivery of these materials.

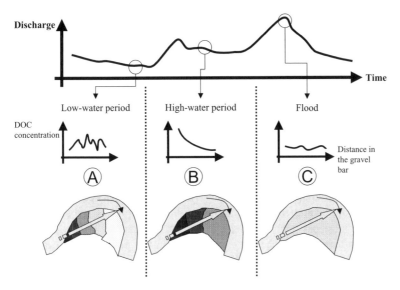

Figure 6.8 Changes in patterns of DOC within gravel-bar flowpaths under different flow conditions. Top: schematic of hydrograph. Middle: variation in DOC concentration with distance along flowpath. Bottom: Gravel-bar flowpath traversing (A) heterogeneous patchwork of habitats under low flow; (B) less heterogeneous gravel bar at high flow; and (C) less heterogeneous gravel bar during flood

Variations in flow can also influence biogeochemical function via effects on the strength or direction of SGW exchanges. An example is provided by DOC dynamics in a river gravel bar (Figure 6.8). Biogeochemical processes within the gravel bar result in strong retention of DOC, variable DOC patterns, or no change, depending on the magnitude of the hydrologic event. During the low flow period, DOC concentrations within the hyporheic zone are controlled by very local parameters such as grain size or organic matter richness of sediments. During floods, the hydraulic conductivity of the sediments can be increased by removal of very fine interstitial particles, and retention does not occur.

Finally, the flow regime controls biological nutrient use via its influence on plant and algal biomass in rivers. Large rivers with constant, slow flows can support water-column phytoplankton that directly use dissolved nutrients, whereas more variable flows restrict algal growth to fixed substrata. Biomass of attached algae or biofilms varies substantially owing to scour and export during high flows, followed by low-flow periods when biomass accumulates (Biggs and Close, 1989; Grimm and Fisher, 1989; Battin *et al.*, 2003a). It is this net growth that can account for a high rate of nutrient uptake from the water column (e.g., Grimm, 1987). However, the relationship between nutrient content of the water column and biofilm growth is not always direct (Ameziane *et al.*, 2002; Sauvage *et al.*, 2003). In turn, biofilms can have an effect on the rate of SGW exchange and other hydrodynamic variables like transient storage (Battin *et al.*, 2003b).

6.5 Conclusion

Dynamics of SGW exchange processes and associated ecosystem functions in river corridors clearly relate to hydrologic variability. At the local scale of a channel reach geomorphic unit, direction and magnitude of exchange were demonstrated to vary with discharge range and flow dynamics. Investigation of these dynamics for medium to large water courses was performed by generating hydrologic indices of flow variability for two contrasting regions, and examining how they changed with increase in basin area. The strength of scaling relationships depended to some extent upon climatic setting (arid or temperate). Biogeochemical process dynamics are likely to covary with these hydrologic gradients, based upon a literature survey of SGW exchange and ecosystem functions.

An ecohydrologic approach that marries hydrologic approaches to flow variability in space and time and ecological understanding of biotic responses can improve our understanding of fluvial ecosystems and lead to better management under increasing human manipulation and indirect impacts. Ecohydrology in this sense refers to a collaborative enterprise, with ecologists bringing understanding of ecosystem structure and function, and hydrologists bringing knowledge of energy and water fluxes, flow variability, and SGW exchange. Flow variability can affect fluvial ecosystems by changing the dynamics of SGW interactions that are key to biotic community structure and biogeochemical function at a wide range of scales. Concepts recognizing spatial heterogeneity and multiscaled phenomena that influence SGW interactions will be most useful to advancing understanding of fluvial ecosystems.

Acknowledgments

This chapter is a result of exciting and fruitful discussions we had during a four days seminar with E. Larson (Arizona State University), M. Lafont, J.F. Perrin and Ph. Namour (all from Cemagref). The River Loire experimental data set was provided by the OTHU field observatory (http://www.graie.org/othu/) and Cemagref technical staff.

References

Ameziane T, Garabetian F, Dalger D, Sauvage S, Dauta A, Capblancq J. 2002. Epilithic biomass in a large gravel-bed river (the Garonne, France): A manifestation of eutrophication? *River Research and Applications*, **18**: 343–354.

Battin TJ, Kaplan LA, Newbold JD, Cheng XH, Hansen C. 2003a. Effects of current velocity on the nascent architecture of stream microbial biofilms. *Applied and Environmental Microbiology*, **69**: 5443–5452.

Battin TJ, Kaplan LA, Newbold JD, Hansen CME. 2003b. Contributions of microbial biofilms to ecosystem processes in stream mesocosms. *Nature*, **426**: 439–442.

Biggs BJF, Close ME. 1989. Periphyton biomass dynamics in gravel bed rivers – the relative effects of flows and nutrients. *Freshwater Biology*, **22**: 209–231.

Bormann FH, Likens GE. 1979. Catastrophic disturbance and the steady-state in Northern Hardwood Forests. *American Scientist*, **67**: 660–669.

Boulton, AJ, 1993. Stream ecology and surface hyporhic hydrologic exchange – Implications, techniques and limitations. *Australian Journal of Marine and Freshwater Research*, **44**:553–564.

Boulton AJ, Valett HM, Fisher SG. 1992. Spatial-distribution and taxonomic composition of the hyporheos of several Sonoran Desert streams. *Archiv für Hydrobiologie*, **125**: 37–61.

Boulton AJ, Findlay S, Marmonier P, Stanley EH, Valett HM. 1998. The functional significance of the hyporheic zone in streams and rivers. *Annual Review of Ecology and Systematics*, **29**: 59–81.

Boyer EW, Hornberger GM, Bencala KE, McKnight DM. 1997. Response characteristics of DOC flushing in an alpine catchment. *Hydrological Processes*, **11**: 1635–1647.

Bunn SE, Thoms MC, Hamilton SK, Capon SJ. 2006. Flow variability in dryland rivers: Boom, bust and the bits in between. *River Research and Applications*, **22**: 179–186.

Burt TP, Pinay G. 2005. Linking hydrology and biogeochemistry in complex landscapes. *Progress in Physical Geography*, **29**: 297–316.

Calver A. 2001. Riverbed permeabilities: Information from pooled data. *Ground Water*, **39**: 546–553.

Cardenas MB, Wilson JL, Zlotnik VA. 2004. Impact of heterogeneity, bed forms, and stream curvature on subchannel hyporheic exchange. *Water Resources Research*, **40**: Art. No. W08307.

Cirmo CP, McDonnell JJ. 1997. Linking the hydrologic and biogeochemical controls of nitrogen transport in near-stream zones of temperate-forested catchments: a review. *Journal of Hydrology*, **199**: 88–120.

Connell JH, Wayne PS. 1983. On the evidence needed to judge ecological stability or persistence. *The American Naturalist*, **121**: 798–824.

Connell JH. 1978. Diversity in tropical rain forests and coral reefs. *Science*, **199**: 1302–1310.

Dahm CN, Grimm NB, Marmonier P, Valett HM, Vervier P. 1998. Nutrient dynamics at the interface between surface waters and groundwaters. *Freshwater Biology*, **40**: 427–451.

Danielopol DL. 1989. Groundwater fauna associated with riverine aquifers. *Journal of the North American Benthological Society*, **8**: 18–35.

Dent CL, Schade JD, Grimm NB, Fisher SG. 2000. Subsurface influences on surface biology. In J.B. Jones, Jr. and P.J. Mulholland (Editors), *Streams and Ground Water*. Academic Press, San Diego, USA.

Dent CL, Grimm NB, Fisher SG. 2001. Multiscale effects of surface–subsurface exchange on stream water nutrient concentrations. *Journal of the North American Benthological Society*, **20**: 162–181.

Elwood JW, Newbold JD, O'Neill RV, Van Winkle W. 1983. Resource spiraling: an operational paradigm for analyzing lotic ecosystems. In T.D. Fontaine and S.M. Bartell (Editors), *Dynamics of Lotic Ecosystems*. Ann Arbor Science, AnnArbor, Michigan, USA, pp. 3–28.

Evans EC, Petts GE. 1997. Hyporheic temperature patterns within riffles. *Hydrological Sciences Journal–Journal Des Sciences Hydrologiques*, **42**: 199–213.

Fellows CS, Valett HM, Dahm CN. 2001. Whole-stream metabolism in two montane streams: Contribution of the hyporheic zone. *Limnology and Oceanography*, **46**: 523–531.

Findlay SEG, Sinsabaugh RL, Sobczak WV, Hoostal M. 2003. Metabolic and structural response of hyporheic microbial communities to variations in supply of dissolved organic matter. *Limnology and Oceanography*, **48**: 1608–1617.

Fisher SG, Grimm NB, Marti E, Holmes RM, Jones JB. 1998. Material spiraling in stream corridors: A telescoping ecosystem model. *Ecosystems*, **1**: 19–34.

Gold AJ, Groffman PM, Addy K, Kellogg DQ, Stolt M, Rosenblatt AE. 2001. Landscape attributes as controls on ground water nitrate removal capacity of riparian zones. *Journal of the American Water Resources Association*, **37**: 1457–1464.

Gregory SV, Swanson FJ, McKee WA, Cummins KW. 1991. An ecosystem perspective of riparian zones. *Bioscience*, **41**: 540–551.

Grimaldi C, Chaplot V. 2000. Nitrate depletion during within-stream transport: Effects of exchange processes between streamwater, the hyporheic and riparian zones. *Water Air and Soil Pollution*, **124**: 95–112.

Grimm NB, Fisher SG. 1984. Exchange between interstitial and surface-water: implications for stream metabolism and nutrient cycling. *Hydrobiologia*, **111**: 219–228.

Grimm NB. 1987. Nitrogen dynamics during succession in a desert stream. *Ecology*, **68**: 1157–1170.

Grimm NB, Fisher SG. 1989. Stability of periphyton and macroinvertebrates to disturbance by flash floods in a desert stream. *Journal of the North American Benthological Society*, **8**: 293–307.

Harvey JW, Bencala, KE. 1993. The effect of streambed topography on surface–subsurface water exchange in mountain catchments. *Water Resources Research*, **29**: 89–98.

Hedges JI, Clark WA, Quay PD, Richey JE, Devol AH, Santos UD. 1986. Compositions and fluxes of particulate organic material in the Amazon River. *Limnology And Oceanography*, **31**:717–738.

Hedges JI, Cowie GC, Richey JR, Quay PD. 1994. Origins and processing of carbohydrates and amino-acids in the Amazon River. Abstracts of papers of *The American Chemical Society*, **207**: 63-GEOC.

Hildrew AG, Townsend CR. 1987. Organization in freshwater benthic communities. In J.H.R. Gee and P.S. Giller (Editors), *Organization of Communities, Past and Present*. 27th Symposium of the British Ecological Society. Blackwell Scientific Publications, Oxford, England, pp. 347–371.

Javelle P, Ouarda TBMJ, Lang M, Bobee B, Galea G, Gresillon JM. 2002. Development of regional flood-duration–frequency curves based on the index-flood method. *Journal of Hydrology*, **258**: 249–259.

Jones JB, Fisher SG, Grimm NB. 1995a. Nitrification in the hyporheic zone of a desert stream ecosystem. *Journal of the North American Benthological Society*, **14**: 249–258.

Jones JB, Fisher SG, Grimm NB. 1995b. Vertical hydrologic exchange and ecosystem metabolism in a Sonoran Desert stream. *Ecology*, **76**: 942–952.

Lafont M, Vivier A, Nogueira S, Namour Ph, Breil P. 2006. Surface and hyporheic oligochaete assemblages in a French suburban stream. *Hydrobiologia*, **564**: 183–193.

Maidment DR. 1993. Chapter V in *Handbook of Hydrology*, McGraw-Hill, New York, pp. 5.1–5.51.

Malcolm IA, Soulsby C, Youngson AF, Hannah DM, McLaren IS, Thorne A. 2004. Hydrological influences on hyporheic water quality: Implications for salmon egg survival. *Hydrological Processes*, **18**: 1543–1560.

Marti E, Grimm NB, Fisher SG. 1997. Pre- and post-flood retention efficiency of nitrogen in a Sonoran Desert stream. *Journal of the North American Benthological Society*, **16**: 805–819.

Marti E, Fisher SG, Schade JD, Grimm NB. 2000. Flood frequency and stream-riparian linkages in arid lands. In J.B. Jones, Jr. and P.J. Mulholland (Editors), *Streams and Ground Waters*. San Diego, Academic Press.

McClain ME, Boyer EW, Dent CL, Gergel SE, Grimm NB, Groffman PM, Hart SC, Harvey JW, Johnston CA, Mayorga E, McDowell WH, Pinay G. 2003. Biogeochemical hot spots and hot moments at the interface of terrestrial and aquatic ecosystems. *Ecosystems*, **6**: 301–312.

McDowell WH, Likens GE. 1988. Origin, composition, and flux of dissolved organic carbon in the Hubbard Brook valley. *Ecological Monographs*, **28**: 177–195.

McGlynn BL, McDonnell JJ. 2003. Role of discrete landscape units in controlling catchment dissolved organic carbon dynamics. *Water Resources Research*, **39(11)**:1310.SCW 2-1-SCW 2-20.

Minshall GW. 1988. Stream ecosystem theory: A global perspective. *Journal of the North American Benthological Society*, **7**: 263–288.

Mulholland PJ, Steinman AD, Elwood JW. 1990. Measurement of phosphorus uptake length in streams: comparison of radiotracer and stable PO4 releases. *Canadian Journal of Fisheries and Aquatic Sciences*, **47**: 2351–2357.

Mulholland PJ, Marzolf ER, Webster JR, Hart DR, Hendricks SP. 1997. Evidence that hyporheic zones increase heterotrophic metabolism and phosphorus uptake in forest streams. *Limnology and Oceanography*, **42**: 443–451.

Neal C, Neal M, Warrington A, Ávila A, Piñol J, Rodà F. 1992. Stable hydrogen and oxygen studies of rainfall and streamwaters for two contrasting holm oak areas of Catalonia, northeastern Spain. *Journal of Hydrology*, **140**: 163–178.

Pinay G, Decamps H, Chauvet E, Fustec E. 1990. Functions of ecotones in fluvial systems. In R. J. Naiman and H. Decamps (Editors), *Ecology and Management of aquatic–terrestrial ecotones*. UNESCO and Parthenon Publishing Group, Paris, pp. 141–169.

Pinay G, Ruffinoni C, Fabre A. 1995. Nitrogen cycling in two riparian forest soils under different geomorphic conditions. *Biogeochemistry*, **30**: 9–29.

Petts G, Nienhuis PH, Tockner K, Brittain J, Breil P, Soulsby C, Wood JP, Ledger ME, Baker A, Capra H, Sadler J, Tabacchi E, Baptist MJ. 2005. Emerging concepts for integrating human and environmental water needs in river basin management. *US Army European Research Office*. WQTS final report. 14 sections. http://el.erdc.usace.army.mil/elpubs/pdf/trel05-13.pdf.

Poff NL. 1997. Landscape filters and species traits: Towards mechanistic understanding and prediction in stream ecology. *Journal of the North American Benthological Society*, **16**: 391–409.

Poff NL, Olden JD, Pepin DM, Bledsoe BP. 2006. Placing global stream flow variability in geographic and geomorphic contexts. *River Research and Applications*, **22**: 149–166.

Pusch M. 1996. The metabolism of organic matter in the hyporheic zone of a mountain stream, and its spatial distribution. *Hydrobiologia*, **323**: 107–118.

Pusch M, Schwoerbel J. 1994. Community respiration in hyporheic sediments of a mountain stream (Steina, Black Forest). *Archiv für Hydrobiologie*, **130**: 35–52.

Rouch R, Mangin A, Bakalowicz M, Dhulst D. 1997. The hyporheic zone: Hydrogeological and geochemical study of a stream in the Pyrenees mountains *Internationale revue der gesamten Hydrobiology*, **82**: 357–378.

Ruysschaert F, Breil P. 2004. Assessment of the hyporheic fluxes in a headwater stream exposed to combined sewer overflows, *Proceeding of the Fifth International Symposium on Ecohydraulics* 12–17 September 2004, Madrid, pp. 1148–1153.

Salmon CD, Walter MT, Hedin LO, Brown MG. 2001. Hydrological controls on chemical export from an undisturbed old-growth Chilean forest. *Journal of Hydrology*, **253**: 69–80.

Sauvage S, Teissier S, Vervier P, Ameziane T, Garabetian F, Delmas F, Caussade B. 2003. A numerical tool to integrate biophysical diversity of a large regulated river: Hydrobiogeochemical bases. The case of the Garonne River (France). *River Research and Applications*, **19**: 181–198.

Savant SA, Reible DD, Thibodeaux LJ. 1987. Convective-transport within stable river sediments. *Water Resources Research*, **23**: 1763–1768.

Southwood TRE. 1977. Habitat, templet for ecological strategies – presidential-address to the British Ecological Society, 5 January 1977. *Journal of Animal Ecology*, **46**: 337–365.

Speaker RW, Moore K, Gregory SV. 1984. Analysis of the process of retention of organic matter instream ecosystems. *Verhandlungen Internationale Vereinigung Limnologie*, **22**: 1835–1841.

Stanford JA, Gonser T. 1998. Rivers in the landscape: Riparian and groundwater ecology – Preface. *Special Issue of Freshwater Biology*, **40**: 401–401.

Stanford J, Ward JV. 1988. The hyporheic habitat of river ecosystems. *Nature*, **335**: 64–66.

Stanford JA, Ward JV. 1993. An ecosystem perspective of alluvial rivers: Connectivity and hyporheic corridor. *Journal of the North American Benthological Society*, **12**: 48–60.

Stanley EH, Boulton AJ. 1995. Hyporheic processes during flooding and drying in a Sonoran Desert stream. 1. Hydrologic and chemical-dynamics. *Archiv für Hydrobiologie*, **134**: 1–26.

Thibodeaux LJ, Boyle JD. 1987. Bedform-generated convective transport in bottom sediment. *Nature*, **325**: 341–343.

Thoms MC. 2006. Variability in riverine ecosystems. *River Research and Applications*, **22**: 115–121.

Townsend CR. 1989. The patch dynamics concept of stream community ecology. *Journal of the North American Benthological Society*, **8**: 36–50.

Triska FJ, Kennedy VC, Avanzino RJ, Zellweger GW, Bencala KE. 1989. Retention and transport of nutrients in a 3rd-order stream in northwestern California: hyporheic processes. *Ecology*, **70**: 1893–1905.

Valett HM, Morrice JA, Dahm CN, Campana ME. 1996. Parent lithology, surface–groundwater exchange, and nitrate retention in headwater streams. *Limnology and Oceanography*, **41**: 333–345.

White DS. 1990. Biological relationships to convective flow patterns within stream beds. *Hydrobiologia*, **196**: 149–158.

Winter TC. 2001. The concept of hydrologic landscapes. *Journal of the American Water Resources Association*, **37**: 335–349.

Wroblicky GJ, Campana ME, Valett HM, Dahm CN. 1998. Seasonal variation in surface–subsurface water exchange and lateral hyporheic area of two stream–aquifer systems. *Water Resources Research*, **34**: 317–328.

7
Ecohydrology and Climate Change

Wendy S. Gordon and Travis E. Huxman

7.1 Introduction

A focus on climate change research over the last twenty years has clearly demonstrated the interrelated nature of the hydrologic cycle and the terrestrial biosphere (i.e., vegetation) (Kabat *et al.*, 2004). As global change research began in the 1980s, it was widely believed that vegetation was a passive component of the hydrologic and energy cycles, or at least its effect was small at large scales. It was thought that the Earth system was driven by ocean–atmosphere dynamics and vegetation did not ameliorate these in any significant way even at the land–atmosphere boundary. The first-generation land-surface modeling schemes treated the vegetation canopy and soil as a single 'bucket' (Sellers *et al.*, 1997). The second generation of these models differentiated between these two components, though the land surface was still represented by static or prescribed vegetation schemes with limited differentiation among plants with different functional attributes, such as rooting depth and leaf demography (e.g., BIOME (Prentice *et al.*, 1992) and MAPSS (Neilson, 1995)). It is now widely accepted that the land surface plays a key role in the partitioning of available water between evaporative fluxes and runoff. Intercomparison projects such as PILPS have demonstrated that differences in land-surface schemes translate into significantly different responses to a given set of atmospheric forcings (Henderson-Sellers *et al.*, 1996). The premise of this book – the interconnectedness of hydrological and ecological processes – has now emerged as an important concept for understanding the effects of global climate change. One cannot realistically discuss one component without its companion. It is clear that the interrelatedness of these processes implies that they will both be affected by increasing temperatures or atmospheric gases and will give rise to feedback schemes that are only now being incorporated into dynamic global vegetation models (DGVMs).

Hydroecology and Ecohydrology: Past, Present and Future, Edited by Paul J. Wood, David M. Hannah and Jonathan P. Sadler
© 2007 John Wiley & Sons, Ltd.

This chapter focuses on the potential effects of climate change on the land-based water cycle using ecohydrology as the underlying paradigm. This chapter aims to: (i) explain the processes of interest and their ecohydrological linkages in the context of different climate forcers; (ii) evaluate the modeling frameworks that have been produced for understanding hydrological and ecological processes at large scales; (iii) review the experimentally derived linkages between ecological and hydrologic cycles from a number of large-scale manipulations of climate in natural ecosystems, and (iv) provide some insight into areas of disconnection that are constraining science's ability to move forward in developing a mechanistic understanding of soil–plant–atmosphere coupling.

Climate is changing as evidenced by increases in atmospheric carbon dioxide (CO_2) concentrations, rising surface temperatures (Houghton *et al.*, 2001b), and more extreme precipitation events (Karl and Knight, 1998; Kiely, 1999). Increases in streamflow have been documented across many regions of the United States and Europe during the Twentieth century (e.g., Lettenmaier *et al.*, 1994; Kiely, 1999; Lins and Slack, 1999). While these trends are expected to increase, it is not possible to predict with certainty the magnitude or extent of continued change. In part, this uncertainly stems from the different processes that dominate the water cycle in different regions (e.g., snowmelt-fed streamflow in mountainous regions versus groundwater domination in karst environments). Uncertainty also arises because the relationship among available water, evaporative demand, and compensatory growth responses of vegetation varies greatly between locations. The availability of water strongly structures our environment through its influences on net primary productivity (e.g., Stephenson, 1990; Nemani *et al.*, 2002). Changes to the timing and availability of water are likely to reshape our landscape profoundly. Hence, the interactions of climate, plants, and water amount to a tangled web dominated by both direct and indirect feedbacks that may vary in importance by location.

7.2 Ecohydrological Controls on Streamflow

The accurate partitioning of precipitation between the components of evapotranspiration (separating transpiration and bare-soil evaporation) and runoff (including surface runoff and root zone drainage) is critical for predicting effects of future climate change (Loik *et al.*, 2004; Huxman *et al.*, 2005). In addition to catchment characteristics such as soil type, geology, and topography, controls on streamflow generation are known to vary according to climate and vegetation (Figure 7.1). In an examination of the water balance of five experimental catchments representing a precipitation gradient (both in amount and timing) and a range of forest types, Post *et al.* (2000) were able to use the comparative method to understand the dynamics of these watersheds and deduce the driving variables of streamflow. For example, Hubbard Brook (New Hampshire, USA) and Casper Creek (California, USA) receive similar amounts of annual rainfall, yet discharge at Hubbard Brook is higher, presumably because of the interaction of climate and vegetation (i.e., warm winter temperatures at Casper Creek lead to high rates of evapotranspiration whereas cold winter temperatures, leafless conditions, and snow fall at Hubbard Creek lead to high rates of soil recharge). Differences in the water balance at the sites depicted in Figure 7.1 illustrate the complexities associated with predicting ecohydrological responses to rising atmospheric CO_2 concentrations and temperature. Streamflow patterns are highly depen-

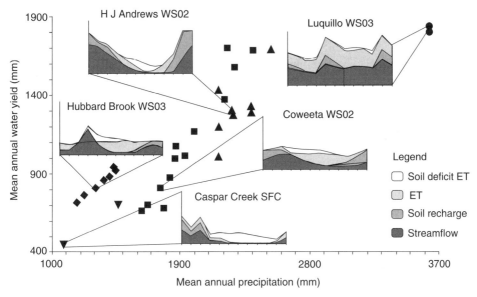

Figure 7.1 *Mean annual streamflow of five forested catchments located along a precipitation gradient. Each individual graph depicts the distribution of precipitation, streamflow, evapotranspiration, and soil recharge/deficit throughout the year (Jan.–Dec.) for one representative catchment at each of the Long-Term Ecological Research sites. The scale of all five plots is identical, with the maximum y-axis value being 450 mm. The topmost line represents monthly precipitation. From Post et al., 2000, reproduced with permission from American Water Resources Association.*

dent upon seasonal factors and isolating the driving variable at a specific location requires detailed knowledge of the relative strengths of climate, vegetation, and landscape controls on streamflow. The prospect of plant physiological mechanisms compensating for or ameliorating climate change effects, such as species-specific acclimation of leaf material and energy exchanges to rising CO_2 (e.g., Curtis and Wang, 1998), further complicates the task of predicting ecohydrological responses.

How precipitation rates and intensity may change in the future is just as uncertain as how those changes might be amplified (or not) into runoff. Nonlinearities in the response of streamflow to precipitation are common. One example of nonlinearity comes from the limestone (karst) regions of central Texas. The predominant woody species on the Edwards Plateau is Ashe juniper (*Juniperus ashei* Buchholz). This evergreen species is estimated to intercept as much as 70% of incoming rainfall (Thurow and Hester, 1997), the majority of which is directly evaporated back into the atmosphere under normal rainfall conditions (storms of <30 mm precipitation) (Wilcox *et al.*, 2005). However, under intense storm conditions, runoff averages 25% of precipitation (with a maximum of 40%) versus less than 4% for small storms. If precipitation events increase in intensity, as has been predicted by Karl and Knight (1998), to the point that the majority of rainfall events exceed the 30-mm threshold in this system, significant increases in runoff relative to precipitation change would occur.

7.3 Simulation Studies of Ecohydrological Effects of Climate Change

The interactions among climate, plants, and water have posed serious challenges to the climate change modeling community and only recently has sufficient understanding of some of the feedbacks been translated into model development (e.g., Krinner *et al.*, 2005). Experimentation has demonstrated that plants can respond to elevated atmospheric CO_2 concentration(s) through increased water-use efficiency and increased rates of photosynthesis at the leaf scale (Carlson and Bazzaz, 1980; Bazzaz, 1990). Whether these short-term dynamics of plant response to CO_2 persist beyond the limited periods of experimentation is unknown (Nowak *et al.*, 2004a). In any event, there is great potential for vegetation change and feedbacks as both temperature and atmospheric CO_2 concentration continue to rise. As part of this greater recognition of the role of vegetation in the climate system, general circulation models (GCMs) have gradually moved from static representations of prescribed vegetation distributions insensitive to changing atmospheric CO_2 concentrations (e.g., Terrestrial Ecosystem Model (TEM) (McGuire *et al.*, 1992; Tian *et al.*, 2000)) to dynamic models in which photosynthesis and transpiration are linked (e.g., Lund–Postsdam–Jena Dynamic Global Vegetation Model, or LPJ (Sitch, 2003)). The latter provide a more physiologically based treatment of plant processes influenced by environmental variables (e.g., temperature, available moisture, CO_2, radiation). However, beyond dynamic vegetation modeling, a critical link in climate change modeling is the feedback from vegetation models to the GCMs. Climate change models have generally excluded this feedback, but an early example of the importance of such feedback comes from the modeling work of Cox *et al.* (2000) in which a fully coupled climate/carbon-cycle simulation resulted in much higher atmospheric CO_2 concentrations by 2100 than an uncoupled simulation. This was primarily due to reduced carbon storage resulting from higher temperatures.

Some simulations conducted under conditions of twice current CO_2 concentrations have yielded increases in annual mean runoff resulting from stomatal closure (i.e., enhanced water-use efficiency) (Cox *et al.*, 1999). However, vegetation is known to be opportunistic (Scanlon *et al.*, 2005), and given additional water net primary productivity is as likely to increase as runoff (Betts *et al.*, 1997). This may be particularly true in arid to semi-arid environments where plants are normally water limited. These types of tradeoff and response are not sufficiently understood in field settings to allow us to model the types of feedback that may well exist. Hence, hindsight modeling with model validation is an essential component of climate change studies (Gordon *et al.*, 2004a).

In a modeling exercise involving the Vegetation Ecosystem Modeling and Assessment Program (VEMAP), terrestrial ecosystem models run with output from two GCMs simulating changes in Twenty-first century climate, projected changes in runoff were often of a different magnitude compared with the underlying changes in precipitation. Projected changes in runoff were most pronounced in arid regions (Gordon and Famiglietti, 2004). The direction of runoff change was not always predictable from precipitation changes. In some instances, runoff decreased despite increases in precipitation. The nonlinear nature of runoff response was consistent with results from other studies (e.g., Karl and Riebsame, 1989; Arora and Boer, 2001; Sankarasubramanian and Vogel, 2003). While there are multiple, plausible explanations for such behavior, one possibility is that the timing of precipitation (i.e., during the warm or cool season) might have influenced the sensitivity of runoff to changes in precipitation by affecting the magnitude of evaporative losses. In the Karl and Riebsame (1989) study of watersheds relatively unimpacted by human activity, greater amplification of runoff occurred when more precipitation fell in the warm season.

One major challenge that remains for modelers is to understand and incorporate important spatial variability or spatially explicit processes required for up-scaling ecohydrological linkages. For example, it is widely acknowledged that the spatial distribution and variability in rainfall interacts with soil properties to translate into complex patterns of soil moisture that may vary on the scale of less than a meter. This patterning of soil moisture helps to explain a large proportion of the variability in vegetation patterns observed in nature (such as banded vegetation and the importance of re-distribution in semi-arid zones – e.g., Ludwig *et al.*, 2005). Hence, it may be difficult to predict with any certainty what will happen to ecological communities at a scale of 30 m or less without an understanding of ecohydrological processes, such as runoff-redistribution interactions with vegetation (Loik *et al.*, 2004). However, changes to the broad-scale climatic, geologic, and hydrologic forces acting to shape ecosystems at the landscape scale can probably be predicted with a higher degree of certainty. Such spatial constraints in prediction are unfortunate, and reflect the lack of a unified soil-vegetation-atmosphere dynamics theory (Kerkoff *et al.*, 2004).

7.4 Experimental Studies of Ecohydrological Effects of Climate Change

Scientific understanding of the ecohydrological response of terrestrial systems to climate change relies heavily on information derived from small-scale experimentation. Many studies utilize individual plants exposed to multiple factors in highly controlled environment settings, or fairly large-scale vegetation complexes experiencing natural variation in many climate drivers, but exposed to single-factor treatments (Shaw *et al.*, 2005). For example, there are hundreds of studies of leaf-level responses of photosynthesis and transpiration to elevated concentrations of atmospheric CO_2 using growth chambers and glasshouses, but fewer than 15 experimental manipulations relying on soil–vegetation complexes of natural ecosystems in undisturbed locations with plot sizes greater than $200\,m^2$ (Nowak *et al.*, 2004a). Considering the scale-dependent nature of hydrologic processes such as runoff, the limited size of experimental units and the lack of data collected in an integrated ecohydrological context are key constraints to broadening our understanding of ecohydrological interactions. As a result, we have only recently been able to evaluate experimentally the important scale-dependent feedbacks between ecological and hydrological cycles.

One common response of terrestrial plants to changes in atmospheric CO_2 concentration is a reduction in stomatal conductance (Morgan *et al.*, 2004). This response derives from changes in the critical balance plants maintain in optimizing water-use efficiency – the relationship between carbon gain and water loss (Cowan and Farquhar, 1977). Decreases in stomatal conductance can result in reduced leaf transpiration, which has been observed for both C_3 and C_4 herbaceous plant species (Morison, 1985), and has also been observed for vegetation from grasslands, Mediterranean shrublands, evergreen and deciduous forests and dry, hot deserts (Jackson *et al.*, 1994; Oechel *et al.*, 1995; Bremer *et al.*, 1996; Nijs *et al.*, 1997; Huxman and Smith, 2001). Stomatal conductance in the dominant C_4 species of the tall-grass prairie in North America decreased substantially in both wet and dry years (Knapp *et al.*, 1993), and in a Mediterranean grassland in California, USA, elevated atmospheric CO_2 concentrations led to a 45% decrease in conductance in mid-season (Shaw *et al.*,). Short-lived, herbaceous, species often show the largest decreases in stomatal conductance and changes in water-use efficiency, while longer-lived, woody, species tend to display less marked responses (Curtis and Wang, 1998; Ellsworth, 1999). Changes in stomatal conductance vegetation in arid and semi-arid ecosystems tend to occur during periods of high water availability when plant growth and leaf gas exchange is greatest (Huxman *et al.*, 1998; Pataki *et al.*, 2000; Morgan *et al.*, 2004).

These changes in leaf-level water use have two implications for the coupling of hydrological and ecological processes. First, under conditions where elevated atmospheric CO_2 concentrations decreases both leaf-level transpiration and canopy water use, increased storage of water in soils is expected (Field *et al.*, 1995). A number of studies have shown that in dry years, elevated CO_2 decreases ecosystem evapotranspiration leading to an increase in soil water storage within the time scale of a growing season (Fredeen and Field, 1995; Ham *et al.*, 1995; Field *et al.*, 1997; Grunzweig and Korner, 2001). Increases in soil water storage may lead to greater deep-soil drainage and/or streamflow (Figure 7.2).

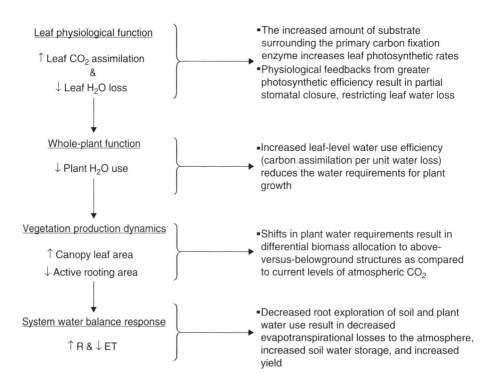

Figure 7.2 Expected effects of rising atmospheric CO_2 concentrations on soil water as a function of different leaf, plant, and canopy responses. In this scenario the response is dominated by the simple scaling of leaf-level savings of water use to the ecosystem.

Alternatively, changes in transpiration and soil water storage can lead to a longer growing season, with plants extending their period of activity into seasonally dry portions of the year (Morgan *et al.*, 2004; Shaw *et al.*, 2005). This pattern results from a shift in plant stress tolerance at elevated atmospheric CO_2 concentrations whereby changes in water-use efficiency promote plant growth during periods of water deficit that typically act to restrict or end the growing season (Tyree and Alexander, 1993; Shaw *et al.*, 2005). Increases in leaf level water-use efficiency under elevated atmospheric CO_2 concentrations are partially a result of photosynthetic stimulation, which leads to greater carbon gain per unit water loss (Huxman *et al.*, 1998). The direct and indirect effects of changes in plant water use induced by atmospheric CO_2 suggests alternative ecological and hydrological process responses at the ecosystem scale. In one scenario, rising atmospheric CO_2 concentrations are expected to increase ecosystem water yield to nonbiological sinks (runoff, deep drainage), while in a second, greater water consumption by vegetation may be expected due to enhanced plant productivity or extended periods of water use by vegetation (Figure 7.3).

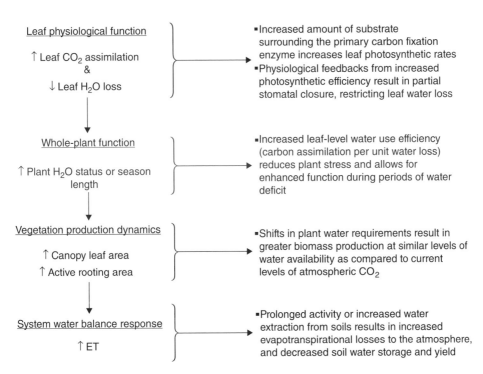

Figure 7.3 Expected effects of rising atmospheric CO_2 concentrations on soil water as a function of different leaf, plant, and canopy responses. In this scenario the response is dominated by changes in plant water stress that influence soil water extraction or total growing season length.

Table 7.1 The effects of elevated CO_2 on soil moisture from a large-scale manipulations carried out in natural conditions for different ecosystem types. Data are presented for the primary growing season and seasonal periods of plant inactivity. The percentage change from ambient to elevated CO_2 conditions is presented where results are significant and 'ns' refers to non-significant responses between treatments.

System-type	CO_2 treatment	Soil H_2O response		Citation
		Growing Season	Non-Growing Season	
Un-grazed grassland (Gissen, Germany)	1.2 × AMB	ns	ns	Kammann et al. (2005)
Grazed, calcareous grassland (NW Switzerland)	2 × AMB	+22%	+6%	Niklaus et al. (1998)
Annual grassland (California, USA)				Field et al. (1997)
(1) Sandstone soil	2 × AMB	+9%	ns	
(2) Serpentine soil	2 × AMB	+5%	ns	
Desert Shrubland (Nevada, USA)	1.5 × AMB	ns	ns	Nowak et al. (2004b)
Managed forest (North Carolina, USA)	1.5 × AMB	ns	ns	Ellsworth (1999), Schafer et al. (2002)
Shortgrass Steppe (Colorado, USA)	2 × AMB	+16%	–	Nelson et al. (2004)

Soil water storage at high atmospheric CO_2 concentrations has been demonstrated in the tallgrass prairie (Owensby *et al.*, 1999), the shortgrass steppe (Morgan *et al.*, 2001), and calcareous grasslands (Niklaus *et al.*, 1998) (see Table 7.1 for additional examples). Increases in atmospheric CO_2 concentrations result in increases of as much as 22% in seasonal soil water content depending upon annual rainfall, despite changes in species composition or productivity (Morgan *et al.*, 2004). In the most xeric ecosystem studied (the Mojave Desert of North America), no significant soil water storage has been detected (Nowak *et al.*, 2004b), most likely due to compensatory changes in leaf-level water use and whole plant growth (Pataki *et al.*, 2000). This suggests that the hydroclimatic region may be an important factor influencing the coupling of hydrological and ecological processes to constrain responses to rising atmospheric CO_2 concentrations. The consequences of changes in soil moisture on runoff, deep drainage, and nutrient leaching have been mostly limited to modeling analyses (Aston, 1984; Idso and Brazel, 1984; Hatton *et al.*, 1992; Jackson *et al.*, 1998). Modeling assessments that have been constrained by experimental data have shown that deep drainage may increase up to 9% in grassland ecosystems at twice ambient atmospheric CO_2 concentrations (Casella *et al.*, 1996; Lund, 2001). Increases in drainage will likely be greatest when substantial differences in evapotranspiration at elevated CO_2 concentrations, compared to ambient conditions, coincide with heavy precipitation to create a large pool of stored soil water that is not immediately lost to the atmosphere (Huxman *et al.*, 2005). Likewise, increases in plant use of soil water and loss to the atmosphere at elevated atmospheric CO_2 concentrations may occur where interactions with other limiting factors (such as nitrogen availability) result in greater canopy development (Hungate et al., 1999). Experimental assessment of these issues is in its infancy.

It appears that the consequences of rising atmospheric CO_2 concentrations on ecohydrological processes can also result from changes in ecosystem water balance as a function of variation in precipitation alone (Morgan et al., 2004). Anthropogenic forcing is also expected to influence global and regional precipitation regimes (Houghton et al., 2001a). Alterations are likely to be ecologically important, occurring through shifts in the seasonality of precipitation (winter versus summer) and in total annual rainfall (Weltzin et al., 2003). Changes in total precipitation in many ecosystems potentially far exceed the impact of rising atmospheric CO_2 concentrations and temperature for specific processes (Weltzin et al., 2003). As a result, a number of studies have been initiated to evaluate the importance of changes in different aspects of precipitation on ecohydrological processes (for example, see Knapp et al., 2002). Changes in the volume and timing of precipitation events at local scales and resultant impacts on soil water content, such as infiltration depth, are expected to have dramatic implications for terrestrial vegetation in both natural and managed ecosystems. Throughout the western USA, plant growth, community composition, microbial diversity and dynamics, and ecosystem processes are primarily limited by the duration and degree of water deficit (Neilson, 1995; Smith et al., 1997; Weltzin et al., 2003; Zak, 2005).

Plants have evolved to use water when it is available, but are limited by morphological (i.e., rooting zone depth), physiological (i.e., root, leaf, and canopy surface area limitations), and phylogenetic (i.e., evolutionary history) constraints. These constraints of plant form and function, and their relationship to climate, have been used to describe the differences in vegetation types in the deserts of North America (Smith et al., 1997). For example, the presence of a large number of perennial C_4 species in the Sonoran and Chihuahuan Deserts has been attributed to the prevalence of high summer precipitation (Ehleringer, 1985). Likewise, winter precipitation fuels woody plant growth by recharging deep soil water while summer precipitation supports the growth of grasses and other more shallowly rooted species (e.g., Scott et al., 2000). Despite these broad regional patterns, the underlying mechanisms linking depth distributions of soil water availability and the establishment and survival of different plant functional types is poorly understood for any given site. Such an understanding is critical in terms of predicting how climate may alter water availability in an ecosystem and affect species diversity, plant functional type composition, and the subsequent interactions between ecological and hydrological processes.

7.5 Differing Perspectives of Hydrologists and Ecologists

The spatial scale at which ecologists and hydrologists conduct research has traditionally differed. Plant ecologists have focused on plot-scale, single-species, assessments of water use, while hydrologists have generally worked at the catchment scale when calculating water budgets, subsuming species' differences in water use in their calculations. Relatively new and integrative methodologies now permit ecologists to determine mixed-species' water use over larger areas. For example, eddy covariance systems allow scientists to measure fluxes of carbon dioxide and water vapor over a footprint whose size can vary from 200–800 m (Baldocchi *et al.*, 2001). Eddy covariance measurements can provide insights into processes controlling ecosystem productivity and evapotranspiration at a scale that is more compatible with catchment modeling than traditional individual-based sampling. While the hydrologic approach of modeling vegetation as an undifferentiated unit may be justified in regional or larger modeling efforts (e.g., Gordon *et al.*, 2004b), such an approach does not easily facilitate an understanding of what will happen to plant communities and the water budget of particular areas under climate change scenarios.

Difficulties in modeling future states may be compounded when interactions between vegetation characteristics and site water balance that will govern the response to climate change are considered. Even where overlap of experimental scale may exist, as in hillslope studies, site-specific conditions such as soil type, topography, or meteorological attributes present a challenge to the extrapolation of findings from one study site to another. It is unusual to be able to instrument a catchment to measure accurately all the variables comprising the water balance and to tease apart the role and response of individual species. In some instances, this type of work has been carried out under the guise of paired catchment experiments in order to test specific hypotheses about land cover influences on runoff or recharge (e.g., Dugas and Hicks, 1998; Jones, 2000). Additional studies of this type, testing hypotheses about the importance of individual plant species or plant communities in controlling the water balance across ecosystems, would be one means of advancing the ecological–hydrological research agenda. Yet, a challenge for scientists studying climate change is having the opportunity to track watersheds over long time periods to discern the fingerprint of changing climate. Funding for long-term studies can be difficult to acquire; the Long-Term Ecological Research stations in the USA (funded by the National Science Foundation) are widely viewed as one success story (Kaiser, 2001). At the same time, streamflow gauges maintained by the United States Geological Survey and others continue to be abandoned at a rate that will hamper our ability to analyze streamflow trends for climate change impacts in the future.

7.6 Future Research Needs

The role that plants play in linking ecological and hydrological processes is well established. However, a mechanistic incorporation of vegetation ecophysiology and composition dynamics is lacking in many models. For example, many physically-based hydrologic models do not account for biological control of evapotranspiration, relying instead on simplistic estimates of land-surface resistance as a function of either light and/or vapor pressure deficit (Garcia-Quijano and Barros, 2005). However, plants regulate the dynamics of leaf-level water loss by opening or closing stomata in order to either (i) maximize CO_2 uptake (Cowan, 1977), (ii) optimize water loss as a function of CO_2 uptake (Cowan, 1977; Smith *et al.*, 1997), or (iii) avoid catastrophic failure of the plant hydraulic pathway (Sperry *et al.*, 1998; see Bond *et al.*, 2007, Chapter 2 in this volume). These ultimate factors produce a number of functional stomatal responses that make it important to consider feedbacks among soil moisture, photosynthesis, and transpiration carefully, in order properly to predict relationships between soil moisture, evapotranspiration, and primary productivity at the ecosystem scale.

Even models that use estimates of water-use efficiency to scale leaf-level function to canopy fluxes in order to relate transpiration and production, often rely on simplistic representations of the diversity of plant functional types and, therefore, miss important compensatory dynamics in plant communities. For example, despite differences in water-use efficiencies of foliage of dominant species across a precipitation gradient from deserts to tropical forests, ecosystem rain-use efficiencies are quite similar during periods of historic drought (Huxman et al., 2004). This suggests important interactions between transpirational dynamics of dominant vegetation, seasonality of production, and bare soil evaporation so as to negate differences in growth potential of vegetation with respect to water availability from divergent biomes.

Similar shortcomings have been identified in terrestrial biogeochemistry models that rely on the photosynthetic model of Farquhar et al. (1980). Water stress cannot limit photosynthetic rates in these models because water availability is only a factor when a stomatal component is included that is linked to plant water status. Thus, it is quite possible that biogeochemistry models relying solely on Farquhar's model may overestimate carbon assimilation when potential photosynthetic rates may be limited by soil moisture availability. Recent research suggests that water stress at a sub-daily time scale is a limiting factor of photosynthesis and does constrain carbon assimilation (Garcia-Quijano and Barros, 2005). However, at annual time scales, the idea has also been put forth that, at least in water-limited systems, vegetation dynamics may have little effect on transpiration and production (Williams and Albertson, 2005). In their model, rainfall explains more than 85% of the temporal variation in annual water and carbon fluxes.

In order to assess climate-induced changes in watersheds and their ecosystems, scientists must continue to make improvements to vegetation and hydrologic models by realistically accounting for the feedbacks between land surface and the atmosphere. Furthermore, simulations must be compared to experimental data for validation (e.g., Gordon et al., 2004a). However, there is a need for experimental manipulations of landscapes beyond the correlative paired-watershed design. Large-scale manipulations of climate features, such as rainfall additions and exclusions, have the means to link hydrological and ecological processes experimentally in real settings.

7.7 Postscript

The focus on and concern over potential climate change impacts appears to be justified given the far-reaching scale of the perturbation (global) and the extent to which impacts have already been recorded (Parmesan and Galbraith, 2004). However, some in the scientific community have argued that land-use changes (past, present, and future) resulting from activities such as deforestation, agriculture, intensive grazing, and urbanization may affect regional climate, ecosystems, and water resources to a degree that is similar in magnitude to, if not greater than, changes driven by greenhouse gases (Pielke et al., 2002). For example, one study in Colorado, USA, documents recent regional cooling thought to have been initiated by agricultural replacement of natural vegetation, which in turn triggered a cascade of mesoscale climate processes leading to cooling (Stohlgren et al., 1998).

Streamflow increased, possibly resulting from reduced rates of transpiration across the landscape and also from changes in snowmelt dynamics that occurred. As a result, it is important to acknowledge that while climate change remains a threat to ecosystems, their processes, and their services, other anthropogenically induced changes warrant research and consideration as the scientific community seeks to unravel the interconnections between ecology and hydrology.

References

Arora VK, Boer GJ. 2001. Effects of simulated climate change on the hydrology of major river basins. *Journal of Geophysical Research*, **106**: 3335–3348.

Aston AR. 1984. The effect of doubling atmospheric CO_2 on streamflow: A simulation. *Journal of Hydrology*, **67**: 273–280.

Baldocchi D, Falge E, Gu L, Olson R, Hollinger D, Running S, Anthoni P, Bernhofer Ch, Davis K, Evans R, Fuentes J, Goldstein A, Katul G, Law B, Lee X, Malhi Y, Meyers T, Munger W, Oechel W, Paw KT, Pilegaard K, Schmid HP, Valentini R, Verma S, Vesala T, Wilson K, Wofsy S. 2001. FLUXNET: A new tool to study the temporal and spatial variability of ecosystem-scale carbon dioxide, water vapor, and energy flux densities. *Bulletin of the American Meteorological Society*, **82**: 2415–2434.

Bazzaz FA. 1990. The response of natural ecosystems to the rising global CO_2 levels. *Annual Review of Ecology and Systematics*, **21**: 167–196.

Betts RA, Cox PM, Lee SE, Woodward FI. 1997. Contrasting physiological and structural vegetation feedbacks in climate change simulations. *Nature*, **387**: 796–799.

Bremer DJ, Ham JM, Owensby CE. 1996. Effects of elevated atmospheric carbon dioxide and open top chambers on transpiration in a tallgrass prairie. *Journal of Environmental Quality*, **25**: 691–701.

Carlson RW, Bazzaz FA. 1980. The effects of elevated CO_2 concentrations on growth, photosynthesis, transpiration, and water use efficiency of plants. In J. Singh and A. Deepak (eds) *Environmental and Climatic Impact of Coal Utilization*. Academic Press: New York, pp. 609–623.

Casella E, Soussana JF, Loiseau P. 1996. Long-term effects of CO_2 enrichment and temperature increase on a temperate grass sward. 1. Productivity and water use. *Plant and Soil*, **182**: 83–99.

Cowan IR. 1977. Stomatal behavior and environment. *Advances in Botanical Research*, **4**: 117–228.

Cowan IR, Farquhar GD. 1977. Stomatal function in relation to leaf metabolism and environment. *Symposium for the Society for Experimental Biology*, **31**: 471–505.

Cox PM, Betts RA, Bunton CB, Essery RLH, Rowntree PR, Smith J. 1999. The impact of new land-surface physics on the GCM simulation of climate and climate sensitivity. *Climate Dynamics*, **15**: 183–203.

Cox PM, Betts RA, Jones CD, Spall SA, Totterdell IJ. 2000. Acceleration of global warming due to carbon-cycle feedbacks in a coupled climate model. *Nature*, **408**: 184–187.

Curtis PA, Wang X. 1998. A meta-analysis of elevated CO_2 effects on woody plant mass, form, and physiology. *Oecologia*, **113**: 299–313.

Dugas WA, Hicks RA. 1998. Effect of removal of *Juniperus ashei* on evaporatranspiration and runoff in the Seco Creek watershed. *Water Resources Research*, **34**: 1499–1506.

Ehleringer J. 1985. Annuals and perennials of warm deserts. In Chabot BF, Mooney HA (eds) *Physiological Ecology of North American Plant Communities*, Chapman and Hall: New York, pp. 162–180.

Ellsworth DS. 1999. CO_2 enrichment in a maturing pine forest: are CO_2 exchange and water status in the canopy affected? *Plant Cell and Environment*, **22**: 461–472.

Farquhar GD, von Caemmerer S, Berry JA. 1980. A biochemical model of photosynthetic CO_2 assimilation in leaves of C_3 species. *Planta*, **149**: 78–90.

Field CB, Jackson RB, Mooney HA. 1995. Stomatal responses to increased CO_2: implications form the plant to the global scale. *Plant, Cell and Environment*, **18**: 1214–1226.

Field CB, Lund CP, Chiariello NR, Mortimer BE. 1997. CO_2 effects on the water budget of grassland microcosm communities. *Global Change Biology*, **3**: 197–206.

Fredeen AL, Field CB. 1995. Contrasting leaf and 'ecosystem' CO_2 and H_2O exchange in *Avena fatua* monoculture: growth at ambient and elevated CO_2. *Photosynthesis Research*, **43**: 263–271.

Garcia-Quijano JF, Barros AP. 2005. Incorporating canopy physiology into a hydrological model: photosynthesis, dynamic respiration, and stomatal sensitivity. *Ecological Modeling*, **185**: 29–49.

Gordon WS, Famiglietti JS. 2004. Response of the water balance to climate change in the United States over the 20th and 21st centuries: results from the VEMAP phase 2 model intercomparisons. *Global Biogeochemical Cycles*, **18**: GB1030, doi 10.1029/2003GB002098.

Gordon WS, Famiglietti JS, Fowler NL, Kittel TGF, Hibbard KA. 2004a. Validation of simulated runoff from six terrestrial ecosystem models: results from VEMAP. *Ecological Applications*, **14**: 527–545.

Gordon WS, Crews-Meyer KA, Famiglietti JS. 2004b. Assessing land cover change in hydroclimatic data network watersheds using NALC imagery. *GIScience and Remote Sensing*, **41**: 322–346.

Grunzweig JM, Korner C. 2001. Growth, water and nitrogen relations in grassland model ecosystems of the semi-arid Negev of Israel exposed to elevated CO_2. *Oecologia*, **128**: 251–262.

Ham JM, Owensby CE, Coyne PI, Bremer DJ. 1995. Fluxes of CO_2 and water vapor from a prairie ecosystem exposed to ambient and elevated atmospheric CO_2. *Agricultural and Forest Meteorology*, **77**: 73–93.

Hatton TJ, Walker J, Dawes WR, Dunin FX. 1992. Simulations of hydroecological responses to elevated CO_2 at the catchment scale. *Australian Journal of Botany*, **40**: 679–696.

Henderson-Sellers A, McGuffe K, Pitman A. 1996. The project for the intercomparison of land-surface parameterization schemes (PILPS): 1992–1995. *Climate Dynamics*, **12**: 849–859.

Houghton JT, Ding Y, Griggs DJ, Noguer M, van der Linden PJ, Xiaosu D (eds). 2001a. *IPCC, 2001: The Scientific Basis.* Contribution of Working Group I to the Third Assessment Report of the Intergovernmental Panel on Climate Change. Cambridge University Press: UK.

Houghton JT, Ding Y, Griggs DJ, Noguer M, van der Linden PJ, Xiaosu D, Maskell K, Johnson CA. 2001b. *Climate Change, 2001: The Science of Climate Change.* The Third Assessment Report of the Intergovernmental Panel on Climate Change. Cambridge University Press: UK.

Hungate BA, Dijkstra P, Johnson DW, Hinkle CR, Drake BG. 1999. Elevated CO_2 increases nitrogen fixation and decreases soil nitrogen mineralization in Florida scrub oak. *Global Change Biology*, **5**: 781–89.

Huxman TE, Smith SD. 2001. Photosynthesis in an invasive grass and native forb at elevated CO_2 during an El Nino year in the Mojave Desert. *Oecologia*, **128**: 193–201.

Huxman TE, Hamerlynck EP, Moore BD, Smith SD, Jordan DN, Zitzer SF, Nowak RS, Coleman JS, Seemann JR. 1998. Photosynthetic down-regulation in *Larrea tridentata* exposed to elevated atmospheric CO_2: Interaction with drought under glasshouse and field (FACE) exposure. *Plant, Cell and Environment*, **21**: 1153–1161.

Huxman TE, Smith, MD, Fay PA, Knapp AK, Shaw MR, Loik ME, Smith SD, Tissue DT, Zak JC, Weltzin JF, Pockman WT, Sala OE, Haddad BM, Harte J, Koch GW, Schwinning S, Small EE, Williams DG. 2004. Convergence across biomes to a common rain-use efficiency. *Nature*, **429**: 651–654.

Huxman TE, Wilcox BP, Scott RL, Snyder K, Hultine K, Small E, Breshears D, Pockman W, Jackson RB. 2005. Ecohydrological implications of woody plant encroachment. *Ecology*, **86**: 308–319.

Idso SB, Brazel AJ. 1984. Rising atmospheric carbon dioxide may increase streamflow. *Nature*, **312**: 51–53.

Jackson RB, Sala OE, Field CB, Mooney HA. 1994. CO_2 alters water use, carbon gain, and yield in a natural grassland. *Oecologia*, **98**: 257–262.

Jackson R, Sala O, Paruelo J, Mooney H. 1998. Ecosystem water fluxes for two grasslands in elevated CO_2: a modeling analysis. *Oecologia*, **113**: 537–46.

Jones JA. 2000. Hydrologic processes and peak discharge response to forest removal, regrowth, and roads in 10 small experimental basins, western Cascades, Oregon. *Water Resources Research*, **36**: 2621–2642.

Kabat P, Claussen M, Dirmeyer PA, Gash JHC, Bravo de Guenni L, Meybeck M, Pielke RS, Vörösmarty CJ, Hutjes RWA, Lütkemeier S (eds). 2004. *Vegetation, Water, Humans and the Climate*. Springer: New York.

Kaiser J. 2001. An experiment for all seasons. *Science*, **293**: 624–627.

Kammann C, Grunhage L, Gruters U, Janze S, Jager HJ. 2005. Response of aboveground grassland biomass and soil moisture to moderate long-term CO2 enrichment. *Basic and Applied Ecology*, **6**: 351–365.

Karl TR, Knight RW. 1998. Secular trends of precipitation amount, frequency, and intensity in the United States. *Bulletin of the American Meteorological Society*, **79**: 231–241.

Karl TR, Riebsame, WE. 1989. The impact of decadal fluctuations in mean precipitation and temperature on runoff: a sensitivity study over the United States. *Climatic Change*, **15**: 423–447.

Kerkhoff AJ, Martens SN, Milne BT. 2004. An ecological evaluation of Eagleson's optimality hypotheses. *Functional Ecology*, **18**: 404–413.

Kiely G. 1999. Climate change in Ireland from precipitation and streamflow observations. *Advances in Water Research*, **23**: 141–151.

Knapp AK, Fay PA, Blair JM, Collins SL, Smith MD, Carlisle JD, Harper CW, Danner BT, Lett MS, McCarron JK. 2002. Rainfall variability, carbon cycling, and plant species diversity in a mesic grassland. *Science*, **298**: 2202–2205.

Knapp AK, Hamerlynck EP, Owensby CE. 1993. Photosynthetic and water relations responses to elevated CO_2 in the C_4 grass *Andropogoneradii*. *International Journal of Plant Science*, **154**: 459–466.

Krinner G, Viovy N, de Noblet-Ducoudré N, Ogée J, Polcher J, Friedlingstein P, Ciais P, Sitch S, Prentice IC. 2005. A dynamic global vegetation model for studies of the coupled atmosphere-biosphere system. *Global Biogeochemical Cycles*, **19**, GB1015, doi:10.1029/2003GB002199.

Lettenmaier DP, Wood EF, Wallis JR. 1994. Hydro-climatological trends in the continental United States, 1948–88. *Journal of Climate*, **7**: 586–607.

Lins HF, Slack, JR. 1999. Streamflow trends in the United States. *Geophysical Research Letters*, **26**: 227–230.

Loik ME, Breshears DD, Lauenroth WK, Belnap J. 2004. A multi-scale perspective of water pulses in dryland ecosystems: climatology and ecohydrology of the western USA. *Oecologia*, **141**: 269–281.

Ludwig JA, Wilcox BP, Breshears DD, Tongway DJ, Imeson AC. 2005. Vegetation patches and runoff-erosion as interacting ecohydrological processes in semiarid landscapes. *Ecology*, **86**: 288–297.

Lund CP. 2001. *Ecosystem carbon and water budgets under elevated atmospheric carbon dioxide concentration in two California grasslands*. PhD Dissertation. Stanford: Stanford University.

McGuire AD, Melillo JM, Joyce LA, Kicklighter DW, Grace AL, Moore 2I B, Vörösmarty CJ. 1992. Interactions between carbon and nitrogen dynamics in estimating net primary productivity for potential vegetation in North America. *Global Biogeochemical Cycles*, **6**: 101–124.

Morison JIL. 1985. Sensitivity of stomatal and water use efficiency to high CO_2. *Plant, Cell and Environment*, **8**: 467–474.

Morgan JA, LeCain DR, Mosier AR, Milchunas DG. 2001. Elevated CO_2 enhances water relations and productivity and affects gas-exchange in C3 and C4 grasses of the Colorado shortgrass steppe. *Global Change Biology*, **7**: 451–466.

Morgan JA, Pataki DE, Dorner C, Clark H, Del Grosso SJ, Grunzweig JM, Knapp AK, Mosier AR, Newton PCD, Niklaus PA, Nippert JB, Nowak RS, Parton WJ, Polley HW, Shaw MR. 2004. Water relations in grassland and desert ecosystems exposed to elevated atmospheric CO_2. *Oecologia*, **140**: 11–25.

Neilson RP, King GA, Koerper G. 1992. Toward a rule-based biome model. *Landscape Ecology*, **7**: 27–43.

Neilson RP. 1995. A model for predicting continental-scale vegetation distribution and water balance. *Ecological Applications*, **5**: 362–385.

Nelson JA, Morgan JA, LeCain DR, Mosier AR, Milchunas DG, Parton BA. 2004. Elevated CO_2 increases soil moisture and enhances plant water relations in a long-term field study in semi-arid shortgrass steppe of Colorado. *Plant and Soil*, **259**: 169–179.

Nemani R, White M, Thornton P, Nishida K, Reddy S, Jenkins J, Running S. 2002. Recent trends in hydrologic balance have enhanced the terrestrial carbon sink in the United States. *Geophysical Research Letters*, **29**: 1468, doi:10.1029/2002GL014867.

Nijs I, Ferris R, Blum H, Hendry G, Impens I. 1997. Stomatal regulation in a changing climate: a field study using Free Air Temperature Increase (FATI) and Free Air CO_2 Enrichment (FACE). *Plant, Cell and Environment*, **20**: 1041–1050.

Niklaus PA, Spinnler D, Korner C. 1998. Soil moisture dynamics of calcareous grassland under elevated CO_2. *Oecologia*, **117**: 201–208.

Nowak RS, Ellsworth DS, Smith SD. 2004a. Functional responses of plants to elevated atmospheric CO_2 – do photosynthetic and productivity data from FACE experiments support early predictions? *New Phytologist*, **162**: 253–280.

Nowak RS, Zitzer SE, Babcock D, Smith-Longozo V, Charlet TN, Coleman JS, Seemann JR, Smith SD. 2004b. Elevated atmospheric CO_2 does not conserve soil water in the Mojave Desert. *Ecology*, **85**: 93–99.

Oechel WC, Hastings SJ, Vourlitis GL, Jenkins MA, Hinkson CL. 1995. Direct effects of elevated CO_2 in chaparral and Mediterranean-type ecosystems. In Moreno JL, Oechel WC (eds). *Global Change and Mediterranean-Type Ecosystems*. Springer: New York, pp. 58–75.

Owensby CE, Ham JM, Knapp AK, Auen LM. 1999. Biomass production and species composition change in a tallgrass prairie ecosystem after long-term exposure to elevated atmospheric CO_2. *Global Change Biology*, **5**: 497–506.

Parmesan C, Galbraith H. 2004. *Observed Impacts of Global Climate Change in the US*. Pew Center for Global Climate Change, Washington, DC. 56 pp.

Prentice IC, Cramer W, Harrison SP, Leemans R, Monserud RA, Solomon AM. 1992. A global biome model based on plant physiology and dominance, soil properties and climate. *Journal of Biogeography*, **19**: 117–134.

Pataki DE, Huxman TE, Jordan DN, Zitzer SF, Coleman JS, Smith SD, Nowak RS, Seemann JR. 2000. Water use of Mojave Desert shrubs under elevated CO_2. *Global Change Biology*, **6**: 889–890.

Pielke Sr RA, Marland G, Betts RA, Chase TN, Eastman JL, Niles JO, Niyogi DS, Runnin SW. 2002. The influence of land-use change and landscape dynamics on the climate system: relevance to climate-change policy beyond the radiative effect of greenhouse gases. *Philosophical Transactions of the Royal Society of London* A, **360**: 1705–1719.

Post DA, Jones JA, Grant GE. 2000. Datasets from long-term ecological research (LTER) sites and their use in ecological hydrology. *Water Resources IMPACT*, **2**: 37–40.

Sankarasubramanian A, Vogel RM. 2003. Hydroclimatology of the continental United States. *Geophysical Research Letters*, **30**: 1363, doi:10.1029/2002GLO15937.

Scanlon BR, Levitt DG, Reedy RC, Keese KE, Sully MJ. 2005. Ecological controls on water-cycle response to climate variability. *Proceedings of the National Academy of Sciences*, **102**: 6033–6038.

Schafer KVR, Oren R, Lai CT, Katul GG. 2002. Hydrologic balance in an intact temperature forest ecosystem under ambient and elevated atmospheric CO_2 concentration. *Global Change Biology*, **8**: 895–911.

Scott RL, Shuttleworth WJ, Goodrich DC. 2000. The water use of two dominant vegetation communities in a semiarid riparian ecosystem. *Agricultural and Forest Meteorology*, **105**: 241–256.

Sellers PJ, Dickinson RE, Randall, DA, Betts AK, Hall FG, Berry JA, Collatz GJ, Denning AS, Mooney HA, Nobre CA, Sato N, Field CB, Henderson-Sellers A. 1997. Modeling the exchanges of energy, water and carbon between continents and the atmosphere. *Science*, **275**: 502–509.

Shaw MR, Huxman TE, Lund CP. 2005. Modern and future semi-arid and arid ecosystems. In Ehleringer J, Cerling T, Dearling D (eds). *The Impact of CO_2 on Plants, Animals and Ecosystems*. Springer: New York, pp. 415–440.

Sitch S, Smith B, Prentice IC, Arneth A, Bondeau A, Cramer W, Kaplan JO, Levis S, Lucht W, Sykes MT, Thonicke K, Venevsky S. 2003. Evaluation of ecosystem dynamics, plant geography and terrestrial carbon cycling in the LPJ dynamic global vegetation model. *Global Change Biology*, **9**: 161–185.

Smith SD, Monson RK, Anderson JE. 1997. *Physiological Ecology of North American Desert Plants*. Springer: Berlin. 286 pp.

Sperry JS, Alder FR, Campbell GS, Comstock JP. 1998. Limitation of plant water use by rhizosphere and xylem conductance: results from a model. *Plant Cell and Environment*, **21**: 347–359.

Stephenson NL. 1990. Climate control of vegetation distribution: the role of the water balance. *American Naturalist*, **135**: 649–670.

Stohlgren TJ, Chase TN, Pielke, Sr RA, Kittel TGF, Baron JS. 1998. Evidence that local land use practices influence regional climate, vegetation, and stream flow patterns in adjacent natural areas. *Global Change Biology*, **4**: 495–504.

Thurow TL, Hester JW. 1997. How an increase or a reduction in juniper cover alters rangeland hydrology. *Juniper Symposium Proceedings*, Texas A&M University, San Angelo, Texas, pp. 9–22.

Tian H, Melillo JM, Kicklighter DW, McGuire AD, Helfrich J, Moore 2I B, Vörösmarty CJ. 2000. Climatic and biotic controls on annual carbon storage in Amazonian ecosystems. *Global Ecology and Biogeography*, **9**: 315–327.

Tyree M, Alexander J. 1993. Plant water relations and the effects of elevated CO_2 – A review and suggestion for future research. *Vegetatio*, **104**: 47–62.

Weltzin JF, Loik ME, Schwinning S, Williams DG, Fay P, Haddad B, Harte J, Huxman TE, Knapp AK, Lin G, Pockman WT, Shaw MR, Small E, Smith MD, Smith SD, Tissue DT, Zak JC. 2003. Assessing the response of terrestrial ecosystems to potential changes in precipitation. *BioScience*, **3**: 41–952.

Williams CA, Alberston JD. 2005. Contrasting short- and long-timescale effects of vegetation dynamics on water and carbon fluxes in water-limited ecosystems. *Water Resources Research*, **41**: W06005, doi: 10.1029/2004WR003750.

Wilcox BP, Owens MK, Knight RW, Lyons R. 2005. Do woody plants affects streamflow on semiarid karst rangelands? *Ecological Applications*, **15**: 127–136.

Zak JC. 2005. Fungal communities in desert ecosystems: links to climate change. In Dighton J, White J, Oudemans P. (eds). *The Fungal Community: Its Organization and Role in the Ecosystem*, third Edition. Marcel Dekker and Sons: New York.

8

The Value of Long-term (Palaeo) Records in Hydroecology and Ecohydrology

Tony G. Brown

This chapter examines: (i) the value of palaeoenvironmental data for studies of ecological dynamics in aquatic systems (hydroecology), (ii) the use of palaeoecological data in the analysis of hydrological and geomorphological dynamics in contemporary aquatic systems (ecohydrology), and (iii) the assessment of medium term hydrological change. In many ways these are merely an extension of the hydrological explanations and inferences made by palaeoecologists when they interpret palaeoenvironmental data derived from cores or sediments from lakes and alluvial floodplains. It also forms one of the methodologies of the hydrological subdiscipline of palaeohydrology which seeks to estimate past discharge regimes and peak flows (Brown, 2000). The chapter first examines the limitations of monitoring river–floodplain–lake systems, before evaluating key concepts, followed by a review of the use of palaeoecological methods in palaeohydrology with the aid of two case studies, one from small floodplains and the other from lake studies.

8.1 River–Floodplain–Lake Systems and the Limits of Monitoring

The rivers, lakes and wetlands we see today are, in most parts of the world, heavily altered by human activity. Indeed many floodplains can be regarded as artifacts of human activity (Brown, 1997) as well as archives of climate change (Macklin and Lewin, 2003). Approximately 60% of the world's river flow is regulated (McAllister *et al.*, 2001) and it has been suggested that there are no unaltered floodplains in North West Europe and few in most other inhabited parts of the globe (Tockner and Stanford, 2002). Interactions between biology and geomorphology are dependent upon the ecological state of the system, and as a result of millennia of change and modification the vast majority of hydroecological systems exist in an altered state. Contemporary hydroecological studies, therefore, can only work within this framework as they are constrained by past regional and local human-driven ecological change. Even without human intervention hydroecological systems are naturally dynamic and dependant upon climate change and the resultant geomorphological response (Macklin *et al.*, 2005). Today we generally make geomorphological distinctions between rivers, lakes and wetlands. This is to a large extent due to the reduction in hydrological connectivity (Brown, 2002; Amoros and Bornette, 2002) caused by human actions such as channel engineering, drainage, and reclamation. Even in the recent past, low lying areas of Europe and North America were a complex interconnected patchwork of river channels, shallow lakes and mires. Remnants of these systems can still be seen in the Everglades, Mississippi delta, central Ireland, Baltic states, Danube delta and Lower Volga. Most, however, have been completely modified such as the Dutch delta, English fenlands and the Camargue in France.

Whilst process studies can analyse interactions between hydrology and ecology over annual to decadal timescales, it is not possible to observe directly interactions over the centennial or millennial scales. This problem is compounded by the probabilistic nature of both ecological and geomorphological processes and their potential interaction such as seed dispersal/germination and the occurrence of destructive floods (Richards *et al.*, 2002). This makes attempts at modelling, such as those undertaken based upon empirical studies of bar colonization and succession by Hughes *et al.* (2000), difficult to upscale both in time and space.

In the absence of longer term data (>200–1000 years), which lies outside of the range of current monitoring schemes, the major source of knowledge of the natural variability of most hydrological systems derives mostly from palaeoecology. This is the basis of the argument that palaeoecology provides a baseline or reference point for the analysis and restoration of modern systems (Brown, 2002), as well as providing important information regarding longer term system dynamics and cycles of change (Birks, 1997).

It is pertinent at this point to consider the nature of the restricted temporal scale of observation of many hydroecological concepts and the clear need for an improved theoretical framework that considers the long-term variability of systems as an important element in understanding their function.

8.2 Key Concepts

Several key concepts in ecology have particular relevance to the analysis of long-term records in ecohydrology. In some cases the theories are temporally contingent but others require validation over long timescales.

The river continuum concept (Vannote *et al.*, 1980) describes a downstream continuum of biotic adjustments and loading, transport, utilization and storage of organic matter due to the gradient of physical conditions. This involves synchronized species replacements and replacement functions with downstream communities utilizing upstream processing inefficiencies. Whilst difficult to test using palaeodata, the theory does emphasise that the downstream integration of ecological processes, such as the primary productivity to respiration ratio (P/R) and pattern of organisms is clearly dependent upon the downstream distribution of physical variables such as stream size (or stream order), velocity and temperature. As this a predictable relationship then it follows that palaeo studies showing changes in river systems over time can be used to imply changing ecological processes and species distributions. However, as Amoros and Van Urk (1989) have noted, it is precisely because the river continuum concept is virtually impossible to test, due to human intervention, that palaeoecological studies are invaluable.

The flood pulse concept (Junk, 1982) was first developed to describe the relationships between the seasonal pulse in water levels in the Amazon floodplains to functional dynamics and species diversity. For example, the relationship of the functional ecology of fish to variations in inundation extent and patterns has been examined by Gutreuter et al. (1999). The concept is clearly most applicable to large, relatively pristine, systems and just the existence of the flood pulse, or its restoration, may not be enough to restore floodplain functionality (Middleton, 2002). Its importance in palaeo-studies is twofold: firstly, the taphonomy of preserved organisms, such as Coleoptera, is almost certainly sensitive to variations in the flood pulse (Davis et al., in preparation). Secondly, the extent and magnitude of the flood pulse should overtime alter the distribution of floodplain species and succession.

Flood plains are highly dynamic environments that are in a state of constant flux with repeated erosional and sedimentational processes resulting from inundation events (Junk et al., 1989). The habitat is thus strongly influenced by channel kinetics determined in part by the frequency and magnitude of flood events (Ward et al., 1999; Tockner et al., 2000), and this (sometimes seasonal) disturbance is thought to maximise biological processes (Naiman and Decamps, 1997) and both in-stream and riparian biodiversity (Naiman and Decamps, 1997; Naiman et al., 1993; Ward, 1998). If there were no disturbance in the long-term floodplains, biodiversity would decrease as successional processes stabilized habitats by, for example, the infilling of cutoff channels (Amoros and Wade, 1986) and enhanced vegetation growth on gravel bars and shoals (Sadler et al., 2004). Disturbance caused by scouring reduces competition between floodplain organisms and favours species with appropriate traits (Adis and Junk, 2002) such as those that disperse via hydrochory (Johansson and Nilsson, 1993; Goodson et al., 2003; Janssen et al., 2005). However, if disturbance was almost constant and of high magnitude, species survival would be much reduced as would colonization success. It has been argued for woodlands that maximal biodiversity and system homeostasis occurs at intermediate disturbance levels (the intermediate disturbance hypothesis), as long as the disturbance is within normal ecological tolerances (Brown et al., 2001). This theory has obvious application to ecohydrological systems, especially in forested floodplains (Brown, 1998; Renöfält et al., 2005; Dunn et al., 2006). The question remains, however, as to what is an intermediate disturbance regime? The most obvious answer is to define it as a 'natural' regime and there is abundant evidence that rivers that are subject to less reduced disturbance, such as those with modified flood regimes downstream of dams have lower overall diversity (McAllister et al., 2001) and disrupted biogeochemical cycles (Friedl and Wuest, 2002). Indeed, the vast majority of human intervention in hydrological systems tends to reduce disturbance, although there are notable exceptions (e.g., military training areas) and the constant changes in water level in water supply reservoirs which result in the almost complete absence of marginal wetland communities due to high (and constant) disturbance.

Groundwater, floodplain and channel connectivity are all believed to be related to both α and β biodiversity (Amoros, 2001) and can be seen as part of the four-dimensional view of floodplains (Ward, 1989, 1998). The hydrogeomorphological processes that influence biodiversity operate at three scales of river sectors or reaches, floodplain water bodies and mesohabitats within each water body (Amoros, 2001). Combining the intensity of the driving forces, such as flow velocities and flood disturbance, connectivity and groundwater supply, can increase habitat diversity between water bodies. This approach has been used in the restoration of side arms of the Rhine River in France (Amoros, 2001). In large systems such as alluvio-lacustrine deltaic systems, a gradient of connectivity is assumed to be one of the key determinants of high biodiversity. This has been illustrated by studies of the the connectivity gradient of lakes in the Danube delta in Romania (RIZA, 2000) and one of the potential applications of palaeolimnological studies can be to document such gradients (Oosterberg *et al.*, 2000).

8.3 Palaeoecology and Palaeohydrology: Proxies and Transfer Functions

In recent years palaeoecology has increasingly taken a wider and quantitative approach to past prediction (or retrodiction). The aim has been to quantify the relationship between preserved biological parameters, such as tree ring width, or an assemblage of a particular organism such as Coleoptera, and climatic variables. In theory almost any biological parameter could be used but some are more closely related to climate and have higher preservation potential. Many organisms can also be used as indicators of the ecological status of hydrosystems and some of these are commonly found preserved in sediments. The principal proxies of relevance here are dendrohydrology, and the analyses of Coleoptera, chironomids, Cladocera, diatoms and pollen.

8.3.1 Dendrohydrology

Tree ring widths are related to growing conditions and therefore to both local site factors and overriding climatic factors, such as temperature and rainfall. This has given rise to the reconstruction of climate from tree rings (dendroclimatology) and the reconstruction of stream flow series (dendrohydrology). They have been used since the 1940s in the USA (Schulman, 1945) and the 1980s in the UK (Hughes *et al.*, 1982) for the extension of the limited instrumental flow record through calibration with instrumental records (Garfin, 2005). As with any biological proxy there are problems of multiple factors and noise. The methodology involves selecting suitable sites near stream gauges, deriving a tree-ring chronology (de-trending, matching, etc.), and calibration of a regression model based on correlation between annual runoff and ring-width indices, verification, recalibration and finally reconstruction using the best available model. The best sites and species are those that will show maximum variation between dry and wet years. In the USA suitable species include western juniper (*Juniperus occidentalis*) in California (Garfin, 2005), Douglas fir in north-central Wyoming (Gray *et al.*, 2004) and oak in the UK (Jones et al., 1984). Tree-ring width variation can account for 50–80% of flow variance as both growth and river flow respond to changes in soil moisture. The classic studies by Stockton (1975), Stockton and Jacoby (1976), and Stockton and Bogess (1983) used the methodology to provide long stream flow series for the Upper Colorado basin, which was used to determine the long term value of the instrumental record and set expectations of future extremes. It emerged that the instrumental period had the longest series of high flows in the last 450 years, and that droughts between 1564 and 1600 and between 1868 and 1892 were of longer duration and greater magnitude than any in the instrumental record. They also suggested by analogy with the past that anthropogenic warming (greenhouse effect) could reduce stream flow by about 35%.

These studies have important implications for future management of streams and floodplains as it may be just as important to design for future increases in low flow magnitude and frequency, with all its associated biological problems (e.g. algal blooms) as for increased flooding (Arnell, 1996).

8.3.2 Coleoptera (Beetles)

Originally studied and used for climatic reconstruction using the mutual climatic range technique (MCR) (Atkinson *et al.*, 1987; Elias, 2001), Coleoptera (beetles) have recently been employed as an indicator of the ecological status of rivers and floodplains. Work by Brayshay and Dinnin (1998) at Bole Ings on the floodplain of the River Trent in Nottinghamshire showed how coleopteran data could provide a spectrum of information on floodplain vegetation and management, channel substrate and flow conditions. Studies of alluvial sediments have also suggested that Coleoptera can indicate changes in river bed conditions and hence fine sediment loading in the past (Osborne, 1988). Smith and Howard (2004) have taken this approach a stage further and shown that from suitable floodplain deposits, Coleoptera can indicate varying discharge and changing patterns of floodplain evolution. Their methodology can be used to distinguish between local high-energy and low-energy fluvial deposition, where other evidence such as coarse sediments are lacking. They do, however, point out that more palaeoentomological data are required from low gradient alluvial contexts and more information from modern ecological analogues. This approach is discussed further in the case study presented later in this chapter.

8.3.3 Chironomids (Non-biting Midges)

The head capsules of chironomids (Insecta: Diptera) are well preserved in sediments and can be recognised to the generic and occasionally species level. Work on lake sediments has shown that chironomids can be accurate indicators of water and indirectly air temperature (Brooks and Birks, 2001; Langdon *et al.*, 2004), dissolved oxygen (Quinlan and Smol, 2002), and nutrient status (particularly total phosphorus and chlorophyll a) (Saether, 1979; Ruse and Wilson, 1984), as well as heavy metal pollution (Wilson, 1988; Janssens de Bisthoven *et al.*, 1998; Ilyashuk *et al.*, 2003). Chironomids, along with diatoms, have also been used to infer phosphorus loading in small lakes (Brooks *et al.*, 2001). Chironomid reconstructions also provide an additional indicator of human activity with the potential to provide quantitative data on the character of aquatic environments associated with archaeological sites. Ruiz *et al.* (2006) have highlighted their value in a recent project where human activity has been inferred from archaeological sediments from three small water bodies in the UK; namely (i) a lake core at the edge of a lake settlement (crannog);

(ii) a palaeochannel infill adjacent to a multiperiod settlement site, and (iii) a Roman well deposit from a floodplain environment. The chironomid assemblages varied significantly both between and within the sites and reflected the immediate environment and the adjacent area. The lake sediment assemblage reflected the construction of the crannog through elevated levels of organic detritus, wood and woody debris. The palaeochannel assemblage revealed changing natural conditions and nutrient enrichment, probably associated with settlement during the Saxon period. The well assemblage was taxonomically restricted, but indicative of organic debris, dead plant material, animal dung and possibly human effluent deposited after abandonment of the well. Chironomids have also recently been used to assess the aquaculture-related eutrophication of the LaCloche channel, Lake Huron by establishing pre-aquaculture baseline conditions (Clerk *et al.*, 2003). The chironomid assemblage revealed decreased hypolimnetic oxygen levels from 1989 onwards whereas diatoms indicated an improvement in water quality after the cessation of cage farming in 1998. Similar studies have been undertaken in small shallow lakes such as the Norfolk Broads in eastern England (Ayres *et al.*, 2006) and in Danish lakes (see later discussion and Brodersen and Lindegaard, 1999). These studies clearly illustrate the potential of palaeolimnological studies using chironomids for the investigation of hydroecological problems.

8.3.4 Cladocera (Water Fleas)

Cladocera are small (generally <1 mm long) aquatic crustaceans most commonly found in freshwater habitats from small puddles to large lakes, and they can be identified by their head shield, which preserves very well in fluvial sediments (Frey, 1976). The order includes herbivores and carnivores, and some species are planktonic while others are benthic and some are epiphytic. Cladocera have been used to infer lake history and the impact of climate change from Langsee, Austria by Schmidt *et al.* (1998). They showed that the lake was of low trophic status during the Late Glacial but that the enrichment of the lake by organic matter in the early Holocene caused pseudeutrophic lake stratification prior to the onset of human impact. Forest clearance in the Bronze Age (about 3000 b.p.) caused a change in sediment type and temporary nutrient enrichment. Cladocera have also been used to evaluate the effects of reservoir construction on natural water bodies (Krzanowski, 1971) and stagnant waters such as oxbow lakes, floodplain pools and backwaters (Amoros and Van Urk, 1989). In a study of a former anastomosed sidearm of the upper Rhône in France (Amoros and Van Urk, 1989) Cladocera indicated that the water body had gone through a series of phases associated with macrophyte succession and terrestrialisation. They also indicated that the water body had remained eutrophic whereas the cutoff channel of a braided arm exhibited oligotrophic to mesotrophic stages, probably due to the more open and disturbed environment. The use of Cladocera in lakes and fluvial deposits undoubtedly has the potential to contribute valuable data on ecological conditions, particularly macrophytes and water quality.

8.3.5 Diatoms

Diatoms are single-cell algae with resistant silica cell walls (frustules). They are found in a remarkably wide range of environments but are very common in lakes, ponds and rivers. Diatoms have been used to examine human impacts on lakes over a variety of timescales from the Pleistocene (Olschesky and Laws, 2002), to the Holocene (Selby and Brown, 2007) to the last few decades (Cremer *et al.*, 2004). They are well known to be sensitive to water chemistry and in particular salinity, pH (Flower, 1986), trophic status (Stoermer *et al.*, 1996; Battarbee *et al.*, 1999) and flow environment (velocity, vegetation and bed conditions). High precision transfer functions now exist for pH (Birks *et al.*, 1990; Cameron *et al.*, 1999) and total P (Bennion *et al.*, 1995), and this allows the quantitative reconstruction of lake history that can be used in lake restoration (Battarbee, 1999). Werner and Smol (2005) have shown that in southwestern Ontario, Canada, the stongest transfer function for TP comes from polymictic lakes, probably due to low macrophyte cover and leading to strong benthic–pelagic coupling. This study illustrates that different transfer functions should be used for polymictic and dimictic lakes. Less work has been done on flow environment, largely due to a lack of modern studies on the ecology and taphonomy on modern riverine diatoms. Diatoms were probably the first palaeoecological tool specifically used in attempts to determine the cause of recent changes to hydroecological systems. The first was the use of diatoms to try and determine whether lake acidification was caused by acid rain or land-use change. Although land-use effects could be seen in some lakes, the long-term estimates provided by the diatoms from lake sediments strongly implicated acid rain in changing lake water chemistry (Battarbee, 1984).

8.3.6 Pollen and Spores

Pollen and spores (palynomorphs) are the most common source of palaeoecological data and have been used for climatic reconstruction for nearly a hundred years. It is not pertinent here to review the various methodologies for climatic reconstruction; however, two methods, the transfer function method (Norton and Wigley, 1986) and the biome method (Prentice *et al.*, 1996) are relevant because they have been used to predict past rainfall patterns. Of more significance here is the use of pollen to describe past ecologies of floodplains and wetlands. There are two principal problems with using palymorphs in this way: first, the influence of the catchment area of sites, and second, the effect of differential productivity and preservation of pollen and spores from different plants. Most alluvial sites are very small and so their pollen catchment area is predominantly local and extra local (Jacobsen and Bradshaw, 1981); however, several studies have shown that lakes can integrate pollen from the entire fluvial catchment (Brown, 1985; Brown *et al.*, submitted). The differential productivity and preservation problem is arguably a more significant issue. It is often addressed by analysing wood and plant macrofossils from the same deposits but they are neither always present nor representative. An example of the problem is provided by the genus *Taxus* (yew). The only native yew in the UK is *Taxus baccata*, which is now associated with dry land and in particular churchyards and burial grounds. Its pollen is sparse and rarely encountered and so its ecological past has remained little known. However, recent archaeological excavations in the Thames floodplain have uncovered a preserved yew woodland and pollen from cores through the sediments reveal significant quantities of yew pollen (15 % land pollen, Batchelor, personal communication). This indicates that our knowledge of the importance of yew in the woodlands of the UK has been adversely affected by the poor dispersal and recovery of its pollen. In fact there are historical references to yew growing on floodplains, including in the

Gearagh on the River Lee in Ireland, noted by Braun Blanquet and Praeger (Brown *et al.*, 1995). Other potential floodplain trees that have been underestimated by pollen analysis include beech (*Fagus*), ash (*Fraxinus*) and poplar (*Populus*). The situation with willows (*Salix*) is complicated and it appears to be well represented in the early Holocene and after floodplain deforestation in the late Holocene (Brown, 1988). This is due to successional status, the 'swamping effect' of other trees that produce high quantities of pollen, particularly alder (*Alnus*), and the presence of more than one species of willow, hybridization and dioceocianism. It is also difficult to determine the structure of past vegetation from palynology. Post deforestation European floodplains were characterised by a variety of ecologies from patchy remnant woodlands surrounded by pasture, through single specimen trees surrounded by pasture (parklands) to hedges and plantations surrounded by pasture.

Pollen analysis can also be used to reveal the age of a diverse and valued ecological system such as the Gearagh which is an alluvial woodland on the River Lee in South West Ireland (Brown *et al.*, 1995). What remained of the Gearagh alluvial forest after 1954 when most of it was flooded by a reservoir, was regarded to be one of the closest analogies to natural alluvial woodland surviving in the UK and Ireland. However, coring and pollen analysis through the islands revealed that at least the central belt only dates from the Twelfth to Fourteenth centuries AD, and that in all probability it was created by the confining of the anastomosing river between flood walls and a reduction in grazing pressure. Whilst its floristic diversity is moderately high, its high invertebrate diversity reflects the fully functioning nature of the system with intensely coupled biotic and hydrological processes (Harwood and Brown, 1993). Palaeo-studies of highly valued ecological sites have the potential to reveal how they were created and this provides valuable case studies for modern habitat restoration.

Many studies of lakes and floodplains refer to changes in the type and abundance of aquatic pollen but they have rarely been used in a systematic manner. An inconvenience ignored by most palynologists is that the majority of pollen entering most lakes is derived not directly from the atmosphere but from river flow from the catchment (Brown, 1985; Brown and Carpenter, in preparation). By using the flood pollen spectra from catchments with different vegetation cover types, it is possible to produce synthetic pollen diagrams that simulate a change in catchment vegetation (Figure 8.1). These diagrams suggest that relatively small and subtle changes in lake pollen can result from significant changes in catchment land use. In many lake cores, spikes or fluctuations in pollen concentration are probably due to changing hydrological conditions (Carpenter, 2006; Brown and Carpenter, in preparation). Whilst this has yet to be exploited it may be that pollen concentration from long cores could be used as a surrogate for river discharge.

Despite all these problems palynology remains the most applicable data source for most hydrological systems at least for the estimation of the major components of the wetland ecology and understanding how they have varied over longer timescales.

8.4 Palaeoecohydrology, Restoration and Enhancement

The preceding sections have shown how palaeotechniques can be used in hydroecology and ecohydrology. How, and precisely which methods are used depends upon the

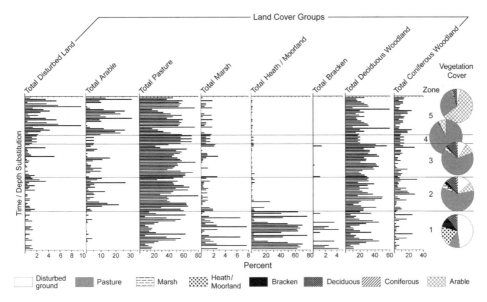

Figure 8.1 *Synthetically generated pollen spectra from sediments in the River Exe, Devon (UK). The results relate to Scenario 1: pasture/deciduous (zone 5), pasture (zone 4), arable (zone 3), increased arable (zone 2), arable/pasture (zone 1)*

objective of the research study. If the objective of hydroecology is to investigate the effect of hydrological variables on plant or animal autecology and communities, then the retrodiction of past ecological conditions is probably the key element in the research design, whereas in ecohydrology the study of the same interactions is focused at the hydrological system; this is particularly appropriate for applied studies such as hydrorestoration. In fact the most obvious application of palaeotechniques in ecohydrology is to produce one or more estimates of a baseline, target state or restoration goal for river, floodplain or lake restoration.

8.5 Case study I. The River Culm in South-west England

A project on a small Fourth order stream in the SW of England, the Culm, attempted to exploit all the palaeomethodologies available for producing estimates of past ecological state and detail changes in floodplain ecology. The Culm displays elements of a multi-channel (anastomosing) planform and has been relatively unimpacted by urbanization and arable agriculture. Geomorphological mapping revealed the true extent of the large palaeochannel network, parts of which are still hydrologically connected to the present channels. Pollen analyses from five sites showed that the floodplain was deforested around 3500 years b.p. although some woodland persisted far later in some areas of the floodplain. The dominant floodplain tree was alder (*Alnus glutinosa*) with oaks (*Quercus*) on the higher areas of floodplain. The pollen and plant macrofossils combined suggest that an important persisting element in the floodplain ecology was very small patches of alder, a fringe of alder along channels and palaeochannels and a variety of poorly connected groundwater fed small wetlands, as well as highly connected traps of organic debris. Coleoptera were used to develop this picture of habitat variability (Figure 8.2) further by using principal components analysis (PCA) to characterize and place different beetle samples into like groups. Radiocarbon dating was employed to constrain the samples by age and although there was a temporal trend, spatial habitat diversity was also a strong driver of community change, at least after the initial phase of deforestation (Davis *et al.*, in preparation). These habitats can also be ranked in terms of connectivity. An unexpected finding was that Elmid (riffle) beetles had persisted in the river until the present day, suggesting that the prevailing view in the literature that they largely disappeared due to an increase in fine sediment in lowland streams during the Bronze–Roman period (Osborne, 1988; Smith, 2000) is region or river specific. Diatoms and chironomids also showed a variety of habitats that varied in their connectivity (Davis *et al.*, in preparation). Three diatom samples from two upstream reaches (Five Fords and Smithincott) show a broadly similar trend of nutrient-enriched, circumneutral waters whereas the samples analysed from the most downstream reach (Columbjohn) appear to represent waters that were considerably more oxygenated, probably with additional environmental niches developing on the peripheries of the river (Selby, personal communication).

In 2000, about 20 ha of the floodplain was afforested by The Woodland Trust at Hunkin Wood (Figure 8.3). This used native species and the planting was designed to avoid the palaeochannels. Hydrological interaction began almost immediately with the formation of trash mounds and scour holes during floods, and the erosion of saplings into the channel despite a 10-m buffer strip. However, an ecological problem arose as the plantation was infested by an attack of the leaf beetle *Phratora vitellinae*. This beetle was present in the palaeosamples but at low levels. It is likely that the conditions of an even age stand with even spacing was ideal for a population explosion. Whilst this may not have been avoidable, it is instructive as it illustrates the problems of trying to induce artificially a major change in floodplain ecology in a very short period of time.

Overall the study suggests that restoration or enhancement should not aim at any particular past state, as the palaeoecology has shown the system to be extremely variable. Instead it should seek to maximize the variety of habitats present and allow a condition characterised by intermediate levels of disturbance and restore connectivity. This work has illustrated the importance of maintaining a dynamic interaction between the channels and floodplain vegetation in at least selected areas of the flood plain. One management option available to do this, while maintaining flood evacuation capacity, is to allow controlled avulsion into palaeochannels and a consequent re-initiation of elements of an anastomosing channel system.

8.6 Case Study II. The Changing Status of Danish Lakes

Danish lake monitoring through the Danish Nationwide Aquatic Monitoring Programme commenced in 1989 (Kronvang *et al.*, 1996) due to nutrient over-enrichment and cultural eutrophication, which led to long-term reductions in the depth distribution of macrophyte communities, changes in the species composition of macrophyte communities, and increases in reports of harmful algal blooms. As it was necessary to identify pre-industrial

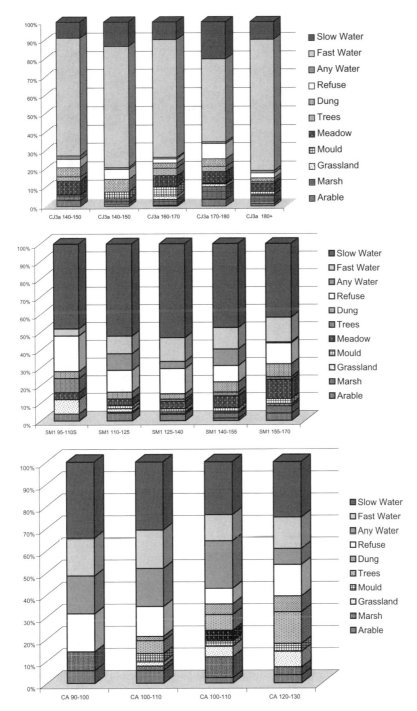

Figure 8.2 Variation in subfossil beetle assemblage composition of three site types on the River Culm (Devon, UK). Top Columjohn 3, middle: Smithincott 1 and bottom: Cutton Alders

140 *Hydroecology and Ecohydrology: Past, Present and Future*

Figure 8.3 *Photographs of Hunkin Wood on the floodplain of the river Culm in Devon, UK: (top) 2001 one year after planting and after a flood; (bottom) the same wood in 2005 during the infestation by* Phratora vitellinae

and pre-agricultural baseline conditions, palaeolimnological techniques were employed. This involved the collation of an archive of palaeolimnological data by GEUS (Geological Survey of Denmark and Greenland) as a crucial resource relevant to the implementation of the European Union's Water Framework Directive (2000). This work initially employed pollen analysis, which revealed the scale and timing of land use change around the lakes (Anderson and Odgaard, 1994). The archive of palaeolimnological data collated by the Geological Survey over many years, was utilized in order to keep the costs low but in addition, new analyses of diatoms and cladocera were undertaken for 21 lakes for four time-slices: 1850, 1900, 1950 and 2000. The project has used the diatom records to infer

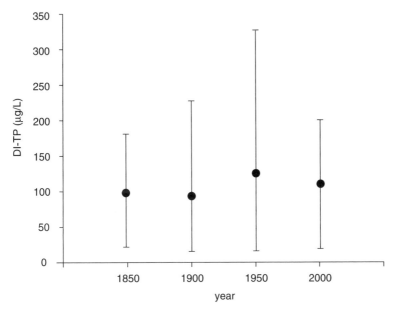

Figure 8.4 *Mean diatom-inferred total phosphorus values (DI-TP) for the additional Danish lake sites studied for the Water Framework Directive. Maximum and minimum values are shown as 'error' bars. Reproduced by permission of the Geological Survey of Denmark and Greenland from Bradshaw and Rasmussen (2003)*

past nutrient status (Bennion *et al.*, 1996; Bradshaw, 2001; Bradshaw *et al.*, 2002). Many other sources of palaeodata have been used including chironomids for trophic status (Brodersen and Lindegaard, 1999). Diatoms have been used to estimate total phosphorus since 1850 in 17 lakes (Figure 8.4). The study has indicated that contrary to popular belief the pollution of Danish lakes that the pollution of Danish lakes is not a relatively modern phenomenon, associated with post-industrial population growth and agricultural development but that although many of the lakes underwent eutrophication during the past 150 years, nutrient concentrations were already high in 1850. As the report points out, this is in fact a conclusion supported by a number of previous studies of longer lake sediment sequences such as; Langesø (Anderson and Odgaard, 1994), Dallund Sø (Bradshaw, 2001) and Gudme Sø and Sarup Sø (unpublished data, P. Rasmussen and E.G. Bradshaw), which demonstrate that some lakes have been impacted by human activities over a scale of centuries and millennia (Bradshaw and Rasmussen, 2003).

8.7 Conclusions

In a world transformed by human activity the state of water bodies is the result of a combination of natural physicochemical conditions and human loadings of sediment, organic matter, nutrients and pollutants. The definition of 'good ecological status' as used in the EU Water Framework Directive is, in many cases, not self evident nor can it be determined by monitoring if the lake has been significantly impacted by human activity. One way of defining 'good ecological status' is to determine pre-impact ecological conditions, and here there is a crucial role for palaeostudies. Whilst most work focuses on lakes, fluvio-lacustrine and fluvial systems also have considerable potential as palaeoenvironmental archives. In most cases, lakes are a part of perennial fluvial networks and changes in fluvial conditions can change lake status. Additionally, in northern hemisphere, temperate, low-gradient environments, rivers have often evolved from fluvio-lacustrine systems and retain many of the physical and biological characteristics of shallow flow-through lakes. An understanding of the coupling of lakes with alluvial river systems is important in understanding and modeling catchment scale ecological conditions. Here the key concepts identified earlier are important conceptual tools that can inform the analysis of palaeodata. Such data is not only applicable to management options such as river restoration and enhancement but also to the determination of stable yet dynamic ecological conditions in the face of global climate change.

Acknowledgements

The author thanks C. Carpenter, S. Davis, K. Selby and Z. Ruiz for advice and data used in this chapter. The work described in the first case study was funded by the Natural Environmental Research Council (NER/A/S/2000/00395).

References

Adis, J. and Junk, W.J. 2002. Terrestrial invertebrates inhabiting lowland river floodplains of Central Amazonia and Central Europe: a review. *Freshwater Biology*, **47**, 711–731.

Amoros, C. and Wade, P. M. 1996. Ecological successions In: G.E. Petts and C. Amoros (Eds) *Fluvial Hydrosystems*, Chapman & Hall, London, 211–241.

Amoros, C. 2001. The concept of habitat diversity between and within ecosystems applied to river side-arm restoration. *Environmental Management*, **28**, 805–817.

Amoros, C. and Bornette, G. 2002. Connectivity and biocomplexity in waterbodies of riverine floodplains. *Freshwater Biology*, **47**, 761–776.

Amoros, C. and Van Urk, G. 1989. Palaeoecological analysis of large rivers: some principles and methods. In G.E. Petts (Ed.) *Historical Change of Large Alluvial Rivers*, John Wiley & Sons, Ltd, Chichester, 143–165.

Anderson, N.J. and Odgaard, B.V. 1994. Recent palaeolimnology of three shallow Danish lakes. *Hydrobiologia*, **256**, 411–422.

Arnell, N. 1996. *Global Warming, River Flows and Water Resources*. John Wiley & Sons, Ltd, Chichester.

Atkinson, T.C., Briffa, K.R. and Coope, G.R. 1987. Seasonal temperatures in Britain during the past 22 000 years, reconstructed using beetle remains. *Nature*, **352**, 587–592.

Ayres, K.R., Sayer, C.D., Skeate, E.R. and Perrow, M.R. 2006. The contribution of palaeolimnology to shallow lake management: an example from Upton Great Broad, Norfolk, UK. *Biodiversity and Conservation*, In press.

Battarbee, R.W. 1984. Diatom analysis and the acidification of lakes. *Philosophical Transactions of the Royal Society*, **305**, 451–477.

Battarbee, R.W. 1999. The importance of palaeolimnology to lake restoration. *Hydrobiologia*, **395**, 149–159.

Battarbee, R.W., Charles, D.F., Sushil, S.D. and Renberg, I. 1999. Diatoms as indicators of lake eutrophication. In F. Stoermer and J.P. Smol (Eds) *The Diatoms: Applications to Environmental and Earth Sciences*, Cambridge University Press, Cambridge.

Bennion, H., Wunsam, S. and Schmidt, R. 1995. The validation of diatom-phosphorus transfer functions: an example from Mondsee, Austria. *Freshwater Biology*, **34**, 271–283.

Bennion H., Juggins S. and Anderson, N.J. 1996. Predicting epilimnetic phosphorus concentrations using an improved diatom-based transfer function and its application to lake eutrophication management. *Environmental Science and Technology*, **30**, 2004–2007.

Birks, H.J.B. 1997. Environmental change in Britain: a long-term palaeoecological perspective. In A.W. Mackay and J. Murlis (Eds) *Britain's Natural Environment: A State of the Nation Review*. Proceedings of a one-day symposium held by the Environmental Change Research Centre, 20 September, 1996. Ensis Ltd, London, pp. 23–28.

Birks, H.J.B., Line, M., Juggins, S., Stevenson, A.C. and ter Braak, C.J.F. 1990. Diatoms and pH reconstruction. *Philosophical Transactions of the Royal Society of London, Series B Biological Sciences*, **327**, 263–278.

Bradshaw, E.G. 2001. *Linking land and lake. The response of lake nutrient regimes and diatoms to long-term land-use change in Denmark*. Unpublished PhD thesis, University of Copenhagen, Denmark. 118 pp.

Bradshaw, E.G. and Rasmussen, P. 2003. Using the geological record to assess the changing status of Danish lakes. *Geological Survey of Denmark and Greenland, Nr. 4, Review of Survey Activities* 2003, 37–40.

Bradshaw, E.G., Anderson, N.J., Jensen, J.P. and Jeppesen, E. 2002. Phosphorus dynamics in Danish lakes and the implications for diatom ecology and palaeoecology. *Freshwater Biology*, **47**, 1963–1975.

Brayshay, B.A. and Dinnin, M.H. 1998. Integrated palaeoecological evidence for biodiversity at the floodplain forest margin. *Journal of Biogeography*, **26**, 115–131.

Brodersen, K.P. and Lindegaard C. 1999. Classification, assessment and trophic reconstruction of Danish lakes using chironoimids, *Freshwater Biology*, **42**, 143–157.

Brooks, S.J. and Birks, H.J.B. 2001. Chironomid-inferred air temperatures from late-glacial and Holocene sites in north-west Europe: progress and problems. *Quaternary Science Reviews*, **20**, 1723–1741.

Brooks, S.J., Bennion, H. and Birks, H.J.B. 2001. Chironomid- and diatom-phosphorus inference models and their application to a sediment core from Betton Pool, Shropshire, England. *Freshwater Biology*, **46**, 511–532.

Brown, A.G. 1985. The potential of pollen in the identification of suspended sediment sources. *Earth Surface Processes and Landforms*, **10**, 27–32.

Brown, A.G. 1988. The palaeoecology of *Alnus* (alder) and the postglacial history of floodplain vegetation: pollen percentage and influx data from the West Midlands, UK. *New Phytologist*, **110**, 425–436.

Brown, A.G. 1997. *Alluvial Environments: Geoarchaeology and Environmental Change*. Cambridge University Press, Cambridge.

Brown, A.G. 1998. The maintenance of biodiversity in multiple-channel floodplains. In R.G. Bailey, P.V. José and B.R. Sherwood (Eds) *United Kingdom Floodplains*. Linnean Society, Westbury Press, pp. 83–92.

Brown, A.G. 2000. Palaeohydrology. In P.L. Hancock and B.J. Skinner (Eds) *The Oxford Companion to The Earth*, Oxford University Press, pp. 780–781.

Brown, A.G. 2002. Learning from the Past: Palaeohydrology and Palaeoecology. *Freshwater Biology*, **47**, 817–830.

Brown, A.G., Stone, P. and Harwood, K. 1995. *The Biogeomorphology of a Wooded Anastomosing River: The Gearagh on the River Lee in County Cork, Ireland*. Occasional Papers in Geography, No. 32, University of Leicester.

Brown, J.H., Witham, T.G., Ernest, S.K.M. and Gehring, C.A. 2001. Complex interactions and the dynamics of ecological systems: long-term experiments. *Science*, **293**, 643–650.

Brown, A.G., Carpenter, R. and Walling, D.E. in press. Monitoring Fluvial Pollen Transport, its Relationship to Catchment Vegetation and Implications for Palaeoenvironmental Studies. *Palaeo3*.

Cameron, N.G., Birks, H.J.B., Jones, V.J. *et al*. 1999. Surface sediment and epilithic diatom pH calibration sets for remote European mountain lakes (AL:PE project) and their comparison with the surface waters acidification programme (SWAP) calibration set. *Journal of Palaelimnology*, **22**, 291–317.

Carpenter, R. 2006. *Fluvial pollen and its relationship to catchment vegetation: implications for suspended sediment sopurce tracing and palaeoenvironmental investigations.* Unpublished PhD Thesis, University of Exeter.

Clerk, S., Selbie, D.T. and Smol, J.P. 2003. Cage aquaculture and water-quality changes in the LaCloche Channel, Lake Huron, Canada: a palaeolimnological assessment. *Canadian Journal of Fisheries and Aquatic Science*, **61**, 1691–1701.

Cremer, H., Buijse, A.D., Lotter, A.F., Oosterberg, W. and Staras, M. 2004. The palaeolimnological potential of diatom assemblages in floodplain lakes of the Danube Delta, Romania: a pilot study. *Hydrobiologia*, **513**, 7–26

Dunn, R.M., Colwell, R.K. and Nilsson, C. 2006. The river domain: why are there more species halfway up the river? *Ecography*, **29**, 251–259.

Elias, S.A. 2001. Mutual Climatic range reconstructions of seasonal temperatures based on late Pleistocene fossil beetle assemblages in Eastern Beringia. *Quaternary Science Reviews*, **20**, 77–91.

EU Water Framework Directive 2000. Directive 2000/60/EC of the European Parliament and of the Council establishing a framework for the Community action in the field of water policy. *Official Journal* (OJ L 327).

Flower, R.J. 1986. The relationship between surface sediment diatom assemblages and pH in 33 Galloway lakes: some regression models for reconstructing pH and their application to sediment cores. *Hydrobiologia*, **143**, 93–103.

Frey, D.G. 1976. Interpretation of Quaternary palaeoecology from cladocera and midges and prognosis regarding usability of other organisms. *Canadian Journal of Zoology*, **54**, 2208–2226.

Friedl, G. and Wuest, A. 2002. Disrupting biogeochemical cycles – consequences of damming. *Aquatic Sciences*, **64**, 55–65.

Garfin, G.M. 2005. *Climate, Drought and Water Management.* Arizona Water Summit. Flagstaff, Arizona.

Goodson, J.M., Gurnell, A.M., Angold, P.G. and Morrissey, I.P. 2003. Evidence for hydrochory and the deposition of viable seeds within winter flow-deposited sediments: the River Dove, Derbyshire, UK. *River Research and Applications*, **19**, 317–334.

Gutreuter, S., Bartels, A.D., Irons, K. and Sandheinrich, M.B. 1999. Evaluation of the flood pulse concept based on statistical models of growth of selected fishes of the upper Mississippi River system. *Canadian Journal of Fisheries and Aquatic Science*, **56**, 2282–2291.

Harwood, K. and Brown, A.G. 1993. Changing in-channel and overbank flood velocity distributions and the morphology of forested multiple channel (anastomosing) systems. *Earth Surface Processes and Landforms*, **18**, 741–748.

Hughes, M. K., Kelly, P.M., Pilcher, J.R. and LaMarche, V.C. 1982. *Climate from Tree Rings.* Cambridge University Press, Cambridge.

Hughes, F.M., Barsoum, N., Richards, K.S. Winfield, M. and Hayes, A. 2000. The response of male and female black poplar (*Populus nigra* L. subsp. *betulifolia* (Pursh) W. Wettst.) cuttings to different water table depths and sediment types: implications for flow management and river corridor biodiversity. *Hydrological Processes*, **14**, 3075–3098.

Ilyashuk, B., Ilyashuk, E., and Dauvalter, V. 2003. Chironomid responses to long-term metal contamination: a palcolimnological study in two bays of lake Imandra, Kola Peninsula, northern Russia. *Journal of Paleolimnology*, **30**, 217–230.

Jacobsen, G.L. and Bradshaw, R.H.W. 1981. The selection of sites for palaeoecological analysis. *Quaternary Research*, **16**, 80–96.

Janssens de Bisthoven, L., Nuyts, P., Goddeeris, B. and Ollevier, F. 1998. Sublethal parameters in morphologically deformed *Chironomus* larvae: clues to understanding their bioindicator value. *Freshwater Biology*, **39**, 179–191.

Jansson, R., Zinko, U., Merritt, D.M. and Nilsson. C. 2005. Hydrochory increases riparian plant species richness: a comparison between a free-flowing and a regulated river. *Journal of Ecology*, **93**, 1094–1103.

Johansson, M.E. and Nilsson C. 1993. Hydrochory, population dynamics and distribution of the clonal aquatic plant *Ranunculus lingua*. *Journal of Ecology*, **81**, 81–91.

Jones, P.D., Briffa, K.R. and Pilcher, J.R. 1984. Riverflow reconstruction from tree rings in southern Britain. *Journal of Climatology*, **4**, 461–472.

Junk, W.J. 1982. Amazonian floodplains: Their ecology, present and potential use. In: *Proceedings of the International Scientific Workshop on Ecosystem Dynamics in Freshwater Wetlands and Shallow Water Bodies*, pp. 98–126. Scientific Committee on Problems of the Environment (SCOPE), United Nations Environment Program (UNEP), New York.

Junk, W.J., Bayley, P.B. and Sparks, R.E. 1989. The flood pulse concept in river-floodplain systems. In: D.P. Dodge (Ed.) *Proceedings of the International Large River Symposium. Canadian Speial Publication Fisheries and Aquatic Sciences*, **106**, 23–34.

Kronvang, B., Svendsen, L.M., Larsen, S.E. and Jensen, J.P. 1996. Monitoring and modelling of nutrient loads in Danish streams and lakes. In: Riverine input to coastal areas notes from a workshop on methodology, *Tema Nord*, **529**, 5360.

Krzanowski, W. 1971. Zooplankton of the dam reservoir on the Sola at Tresna in the first years after its construction. *Acta Hydrobiologia*, **13**, 323–333.

Langdon, P.G., Barber, K.E. and Lomas-Clarke, S.H. 2004. Reconstructing climate and environmental change in Northern England through chironomid and pollen analyses: evidence from Talkin Tarn, Cumbria. *Journal of Paleolimnology*, **32**, 197–213.

Macklin, M.G. and Lewin, J. 2003. River sediments, great floods and centennial-scale Holocene climate change. *Journal of Quaternary Science*, **18**, 102–105.

Macklin, M.G., Johnstone, E. and Lewin, J. 2005. Pervasive and long-term forcing of Holocene river instability and flooding by centennial-scale climate change. *The Holocene*, **15**, 937–943.

McAllister, D.E., Craig, J.F., Davidson, N., Delany, S. and Seddon, M. 2001. *Biodiversity Impacts of Large Dams*. Background Paper No. 1. Prepared for IUCN/UNEP/WCD. International Union for Conservation of Nature and Natural Resources and the United Nations Environmental Programme.

Middleton, B.A. 2002. The flood pulse concept in wetland restoration. In B.A. Middleton (Ed.) *Flood Pulsing in Wetlands: Restoring the Natural Hydrological Balance*, John Wiley & Sons, Ltd, Chichester, pp. 1–7.

Naiman, R.J., Decamps, H. and Pollock, M. 1993. The role of riparian corridors in maintaining regional biodiversity, *Ecological Applications*, **3**, 209–212.

Naiman, R.J. and Decamps, H. 1997. The ecology of interfaces: riparian zones. *The Annual Review of Ecology and Systematics*, **28**, 621–658.

Norton, D.A. and Wigley, T.M.L. 1986. Analyses of modern pollen-climate relationships in New Zealand indigenous forests. *New Zealand Journal of Botany*, **24**, 331–342.

Olschesky, K.S. and Laws, R.A. 2002. Data report: Pliocene-late Pleistocene diatom biostratigraphic data from ODP Leg 185, Hole 1149A. In J.N. Ludden, T. Plank and C. Escutia (Eds) *Proc. ODP, Sci. Results*, 185 [Online]. Available from World Wide Web: <http://www-odp.tamu.edu/publications/185_SR/007/007.htm>

Oosterberg, W., Buijse, A.D., Coops, H., Ibelings, B.W., Menting, G.A.M., Stara, M., Bogdan, L., Constantinescu, A., Hanganau, J., Nvodaru, T. and Rorok, L. 2000. *Ecological Gradients in the Danube Delta lakes*. RIZA Report 2000.015, The Netherlands.

Osborne, P.J. 1988. A late Bronze Age insect fauna from the River Avon, Warwickshire, England: its implications for the terrestrial and fluvial environment and for climate. *Journal of Archaeological Science*, **15**, 715–727.

Prentice, I.C., Guiot, J., Huntley, B., Jolly, D. and Cheddadi, R. 1996. Reconstructing biomes from palaeoecological data: a general method and its application to European pollen data at 0 and 6 ka. *Climatic Dynamics*, **12**, 185–194.

Quinlan, R. and Smol, J.P. 2002. Chironomid-based inference models for estimating end-of summer hypolimnetic oxygen from south-central Ontario lakes. *Freshwater Biology*, **46**, 1529–1551.

Renöfält, B.M., Nilsson, C. and Jansson, R. 2005. Spatial and temporal patterns of species richness in a riparian landscape. *Journal of Biogeography*, **32**, 2025–2037.

Richards, K., Brassington, J. and Hughes, F. 2002. Geomorphic dynamics of floodplains: ecological implications and a potential modeling strategy. *Freshwater Biology*, **47**, 559–580.

RIZA. 2000. *Ecological gradients in the Danube Delta lakes: Present state and man-induced changes*. RIZA Report 2000.015, Netherlands.

Ruiz, Z., Brown, A.G. and Langdon, P.G. 2006. The potential of chironomid (Insecta: Diptera) larvae in archaeological investigations of floodplain and lake settlements. *Journal of Archaeological Science*, **33**, 14–33.

Ruse, L.P. and Wilson, R.S. 1995. Long-term assessment of water and sediment quality of the River Thames using Chironomid pupal skins. In: P. Cranson (Ed.) *Chironomids: From genes to ecosystems*. CSIRO Australia, 482p.

Sadler, J.P., Bell, D. and Fowles, A.P. 2004. The hydroecological controls and conservation value of beetles on exposed riverine sediments in England and Wales. *Biological Conservation*, **118**, 41–56.

Saether, O.A. 1979. Chironomid communities as water quality indicators. *Holartic Ecology*, **2**, 65–74.

Schmidt, R., Wunsam, S., Brosch, U., Fott, J., Lami, A., Löffler, H., Marchetto, A., Müller, H.W., Prazakov, M. and Schwaighofer, B. 1998. Late and post-glacial history of meromictic Längsee (Austria), in respect to climate change and anthropogenic impact. *Aquatic Sciences*, **60**, 56–88.

Schulman, E. 1945. Root growth-rings and chronology. *Tree-Ring Bulletin*, **12**, 2–5.

Selby, K.A. and Brown, A.G. 2007. The Holocene development, spatial variations and anthropogenic record of a shallow lake system in central Ireland as recorded by diatom stratigraphy. *Journal of Palaeolimnology*.

Smith, D.N. 2000. Disappearance of elmid 'riffle beetles' from lowland river systems – the impact of alluviation. In T. O'Connor and R. Nicholson (Eds) *People as an Agent of Environmental Change*, Oxbow Books, Oxford, pp. 75–80.

Smith, D.N. and Howard, A.J. 2004. Identifying changing fluvial conditions in low gradient alluvial archaeological landscapes: can coleopteran provide insights into changing discharge rates and floodplain evolution? *Journal of Archaeological Science*, **31**, 109–120.

Stockton, C.W. 1975. Long-term streamflow records reconstructed from tree rings. *Papers of the Laboratory of Tree-Ring Research 5*. The University of Arizona Press, Tucson. 111 pp.

Stockton, C.W. and Jacoby, G.C. 1976. Long-term surface-water supply and streamflow trends in the Upper Colorado River basin based on tree-ring analyses. *Lake Powell Research Project Bulletin*, 18, 1–70.

Stockton, C.W. and Bogess, W.R. 1983. Tree-ring data – valuable tool for reconstructing annual and seasonal streamflow and determining long-term trends. *Transportation Research Record*, **922**, 10–17.

Stoermer, E.F., Emmert, G., Julius, M.L. and Schelske, C.L. 1996. Palaolimnologic evidence of rapid change in Lake Erie's trophic status. *Canadian Journal of Fisheries and Aquatic Sciences*, **53**, 1451–1458.

Tockner K., Malard F. and Ward J.V. 2000. An extension of the flood pulse concept. *Hydrological Processes*, **14**, 2861–2883.

Tockner, K. and Stanford, J.A. 2002. Riverine flood plains: present state and future trends. *Environmental Conservation*, **29**, 308–330.

Vannote, R.L., Minshall, G.W., Cummins, K.W., Sedell, J.R. and Cushing, C.E. 1980. The river continuum concept. *Canadian Journal of Fisheries and Aquatic Science*, **37**, 130–137.

Ward, J.V. 1989. The four-dimensional nature of lotic ecosystems. *Journal of the North American Benthological Society*, **8**, 2–8.

Ward, J.V. 1998. Riverine landscapes: biodiversity patterns, disturbance regimes, and aquatic conservation. *Biological Conservation*, **83**, 269–278.

Ward J.V., Malard F., Tockner K. and Uehlinger U. 1999. Influence of ground water on surface water conditions in a glacial flood plain of the Swiss Alps. *Hydrological Processes*, **13**, 277–293.

Werner, P. and Smol, J.P. 2005. Diatom-environment relationships and nutrient transfer functions from contrasting shallow and deep limestone lakes in Ontario, Canada. *Hydrobiologia*, **533**, 145–173.

Wilson, R.S. 1988. A survey of the zinc-polluted River Nent (Cumbria) and the East and West Allen (Northumberland), England, using chirnomid pupal exuvia. *Spixiana Supplement*, **14**, 167–174.

9
Field Methods for Monitoring Surface/Groundwater Hydroecological Interactions in Aquatic Ecosystems

Andrew J. Boulton

9.1 Introduction

Much of the hydroecological and ecohydrological literature refers to 'overt' interactions in surface ecosystems such as the response of fish to hydraulics in a river (Nestler *et al.*, Chapter 12 in this volume) or the hydroecological variability evident in vegetation distribution in wetlands (Burt *et al.*, Chapter 14 in this volume). It is often easy to see how ecological features such as fish behaviour or water plant distribution are associated with hydrological variables. Less obvious are the 'covert' interactions where groundwater hydraulics play a role in the ecology of surface ecosystems, either through the direct supply of water for so-called groundwater dependent ecosystems (GDEs, Boulton and Hancock, 2005) or by acting as a vector for critical nutrients (Jones *et al.*, 1995) or thermal energy (Power *et al.*, 1999) required by surface biota and processes. However, as many lakes and rivers are largely fed by groundwater (Winter *et al.*, 1998; Wood *et al.*, 2001), the ecohydrological significance of this covert interaction is greater than often appreciated, and this has important ramifications for management of these aquatic ecosystems (Woessner, 2000).

The relationships between groundwaters and surface water in many aquatic ecosystems have been recognised by hydrologists for many years (e.g., Hubbert, 1940). However, their ecological significance has only fully been appreciated relatively recently as aquatic ecologists have started to use hydrological techniques for measuring and monitoring the direction and strength of groundwater exchanges in lakes and streams. Salmonid fisheries biologists had long been aware of the importance of groundwater inputs for spawning and egg development (Benson, 1953) but it was Hynes's (1983) influential paper that drew most stream ecologists' attention to the ecological significance of groundwater. Other seminal work with Williams (Williams and Hynes, 1974) corroborated the studies in Europe that revealed diverse invertebrate assemblages in saturated sediments below many rivers, and indicated how limnology needed to extend into the groundwater realm (historical reviews in Danielopol, 1989 and Hancock et al., 2005).

A decade after Hynes' paper, a special issue of the *Journal of the North American Benthological Society* hailed a productive merger of hydrologists and ecologists in studies of groundwater–surface water interactions in the hyporheic zone, the saturated sediments below streams (Valett et al., 1993), although availability of suitable techniques was identified as one limitation in this field (Palmer, 1993). By 1996, hydrological techniques involving mini-piezometers and dye tracers to track groundwater pathways were being presented in textbooks on methods (e.g., Dahm and Valett, 1996) and there are now detailed technical manuals on groundwater sampling for ecohydrologists and hydroecologists (Malard et al., 2001). An equivalent convergence of hydrological and ecological approaches to sampling groundwater–surface water interactions in lakes occurred earlier (e.g., Lee and Cherry, 1978; Lock and John, 1978) and entered the textbook literature (Fetter, 1988).

This chapter reviews field techniques commonly used for monitoring surface/groundwater (SGW) hydroecological interactions in aquatic ecosystems. Although these methods are discussed in the context of rivers and lakes, many of them are equally applicable to SGW interactions in estuaries, marine habitats, and littoral or riparian zones. Typical research contexts for the use of these methods are initially identified, as these dictate the best techniques for a particular objective. 'Direct' and 'indirect' hydrological field methods for assessing and monitoring SGW hydroecological interactions are then summarised, using relevant case studies to illustrate their application. This review is not intended as an exhaustive list of techniques and literature but more an introduction for ecohydrologists to the main approaches for studying SGW interactions.

9.2 Research Contexts: Questions, Scales, Accuracy and Precision

Before reviewing methods for assessing and monitoring SGW hydroecological interactions, it helps to specify the research context in which these methods are to be employed. This includes identification of potential mismatches of spatial and temporal scale, accuracy of field measurements, and precision of scientific predictions that may arise from the different 'knowledge structures' when hydrologists and ecologists share common goals (Benda et al., 2002). In general, ecohydrologists exploring the ecological significance of SGW interactions are seeking the effects of surface–groundwater exchanges on either surface or subsurface biochemistry, ecosystem processes, and biota. For example, benthic algal abundance was found to differ between two New Mexican streams varying in number of retentive structures and hyporheic (subsurface) processing of organic matter

(Coleman and Dahm, 1990). Although benthic algae in both streams were nitrogen limited, the higher algal abundance in one of the streams was attributed to the greater inorganic nitrogen availability at the sediment–water interface where hyporheic ammonium concentrations were 25-times greater than in the other stream. Upwelling subsurface water supplied the limiting nutrient to the benthic algae, providing an example of surface biota and processes (photosynthesis) being governed by SGW interactions.

As an example of the converse process, Rouch (1988) demonstrated strong associations between the occurrence of various species of crustaceans at 60 cm depth in the hyporheic zone of a 75-m^2 section of the Lachein stream in the French Pyrenees, and the direction of hydrological exchange with the surface stream. Where surface water downwelled into areas of preferential flow and interstitial oxygen concentrations were high, the hyporheic assemblage was dominated by numerous harpacticoid copepods, whereas obligate groundwater invertebrates such as syncarids were found in areas of reduced permeability, lower dissolved oxygen, and groundwater upwelling.

As most of these studies have been undertaken at spatial scales of 1–100 m (e.g., lake littoral zones, stream reaches), the ecological significance of SGW interactions in many aquatic ecosystems is best known at this level (Boulton *et al.*, 1998). However, biogeochemical processes, especially those mediated by microbes, frequently occur at much finer spatial scales in the interstitial zone or at the sediment–water interface (Dahm *et al.*, 1998; Duff and Triska, 2000). This contrasts with the spatial scale of groundwater mapping in many hydrogeological studies, which can extend out to entire catchments (Winter, 1999), and it is here that the first potential hydroecological mismatch is evident. Is there a single technique that will provide useful data on hydrological exchanges between surface waters and groundwater at fine, intermediate, and broad scales or do we need to integrate results from multiple methods?

Accuracy and precision of field measurements are dictated by the research question and the spatial scale of the study. Fine-scale studies attempting to assess ecohydraulics of biochemical reactions at the sediment–water interface will probably require more precise and accurate measurements of, for example, vertical hydraulic gradient (see below) than a reach-scale study. However, this needs to be tempered by the recognition of the immense spatial 'patchiness' of sediments, their permeability, and their interstitial flow paths at a variety of scales (Malard *et al.*, 2002). Furthermore, these spatial patterns are likely to vary dramatically with changes in surface water levels and groundwater tables. Even in small lakes, seasonal changes in lake level and water table can alter the direction and magnitude of shallow, local groundwater flow, and may even control the permanence of surface water (Figure 9.1). Therefore, when assessing suitable methods for monitoring SGW hydroecological interactions, the spatial and temporal heterogeneity of the hydrological *and* ecological processes must be considered. Subtle changes in water table or groundwater flowpaths may be critical to some ecological processes but remain undetected by relatively coarse-scale hydrological techniques, thus compromising conclusions about ecohydrological mechanisms.

Many research objectives often include some predictive phase, typically presented as verbal or mathematical models (Sanford, 2002). It is here that one of the main mismatches occurs between ecologists and hydrologists because the predictive power and sophistication of mathematical hydrological models typically far exceeds the largely qualitative models extracted from ecological data (Benda *et al.*, 2002). A fruitful avenue of research

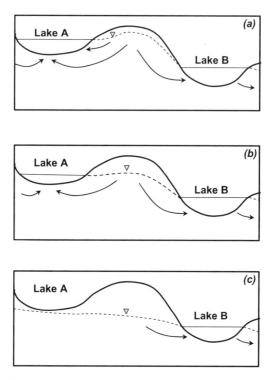

Figure 9.1 Seasonal changes in SGW and groundwater flowpaths in two hypothetical lakes. At the end of the wet season, the groundwater table (marked with a triangle) is high and both lakes receive water (a). With the onset of drier weather (b), lake levels fall until Lake A dries (c) because it no longer receives groundwater input. Such seasonal changes in groundwater have major ecological repercussions for both lakes

in hydroecology is the development of mutual models of prediction of hydrological and ecological responses, and the need to test these models empirically means that the principles of the field methods and their practical limitations must be fully appreciated. Again, issues of scale, accuracy, and precision are paramount here, and the practitioner must explicitly identify the research context of their objectives and how the data are to be used before embarking on any study.

9.3 Direct Hydrological Methods for Assessing SGW Interactions

For convenience, the field methods for assessing SGW interactions are divided into 'direct' methods that entail measuring or estimating hydrological exchanges directly based on hydraulics versus those where SGW interactions are inferred from 'indirect' evidence such as differences in thermal features or chemical composition of surface waters and groundwater, and use of tracers. Generally, it appears that a combination of direct and indirect methods is most popular in many studies of SGW interactions. These field methods for assessing SGW interactions in lakes and rivers are listed in Table 9.1 where each method's principle is described briefly with an ecohydrological example of its application and relevant references.

9.3.1 Seepage Meters

Many lakes gain and lose surface water to the groundwater (Figure 9.1), and this flux can be measured using a seepage meter (Lee, 1977; Lee and Hynes, 1977; Lock and John, 1978). An open-bottomed container with a tube leading into a plastic bag is forced into the sediments (Figure 9.2), and the rate of change of water volume in the bag is a direct measure of seepage flux. If the meter is placed in a downwelling zone where surface water drains to the groundwater, it is necessary to add a known volume of water to the bag. Although Lee and Cherry (1978) presented a formula for calculating the specific discharge, recognition of anomalies in bag filling (Shaw and Prepas, 1990) led to the following equation for corrected seepage rate (q):

$$q = (V_2 - V_1)/A(t_2 - t_1)$$

where V_1 and V_2 are volume changes during short and long exposure times (t_1 and t_2 respectively), and A is the area of the seepage meter.

Seepage meters are typically used in standing waters but have proved less successful in streams (Lee and Hynes, 1977) where coarse sediments hamper placement while erosion around the meter may alter local hydrological exchange patterns (Boulton, 1993). The effect of stream currents on the unprotected seepage bag also contributes to variance in readings (Isiorho and Meyer, 1999) although electromagnetic flowmeters can be used to replace the problematic bag (Rosenberry and Morin, 2004). This has the advantage of also being able to provide continuous measurements to assess temporal changes in seepage rates. Depending on the basal area of the seepage meter, its spatial scale of measurement ranges up to a metre. Meters can be linked to integrate measurements across multiple meters, reducing labour costs and increasing the representative area of measurement (Rosenberry, 2005).

Lesack (1995) measured fluxes of water entering an Amazon tropical lake using seepage meters made from commercially available, plastic water jugs (20cm diameter). As the jugs were transparent, it was possible to see whether appreciable volumes of gas had accumulated in the device and whether installation caused undue disturbance. In this example, seepage rates (up to $13.95 \, lm^{-2}h^{-1}$) were high enough for suitable volumes of water to be collected without needing a larger meter. Devices were installed 60–80cm underwater, and by pooling daily data from ten replicates, the standard deviation of the measurements ranged from 30–130% of the average daily seepage rate. These data were used to construct a water budget of this tropical lake, and were considered relevant to assessing controls on biogeochemical cycling of nutrients and the ecology of primary producers on the floodplain (Lesack, 1995).

9.3.2 Mini-piezometers and Groundwater Mapping

Piezometers are routinely used by hydrogeologists to measure hydraulic head in saturated groundwater sediments, and comprise pipes with slotted tips that are embedded in the

Table 9.1 *Direct and indirect field methods for assessing SGW interactions in lakes and rivers. Each method's principle is described with an ecohydrological example. The spatial scale of the measurements spans the application in the literature, a subset of which is listed in the final column*

Method	Principle	Example of ecohydrological application	Spatial scale (m)	Relevant references
Seepage meters	Directly measures seepage across the SGW interface (usually of standing water) by covering an area of sediment with an open-bottomed container and measuring the time and change in water volume in a plastic bag connected to the container.	Water budget of an Amazon floodplain lake, with implications for biogeochemistry and primary production of the floodplain (Lesack, 1995).	0.1 to 1 (or greater if linked meters used).	Lee, 1977; Lee and Hynes, 1977; Shaw and Prepas, 1990; Rosenberry and Morin, 2004; Rosenberry, 2005
Mini-piezometers	Directly measures depth to water table or relative differences in water level between the surface water and water within a pipe inserted below the water table or into the saturated sediments of a lake or river.	Association of subsurface water chemistry and invertebrates with upwelling/downwelling zones in a desert stream (Stanley and Boulton, 1995; Boulton and Stanley, 1995).	0.01 (single) to >100 (when used as a net of wells).	Lee and Cherry, 1978; Boulton, 1993; Dahm and Valett, 1996; Winter 1999; Constantz et al., 2003
Synoptic measures of surface discharge or lake levels	Direct measurements of lake water levels or stream discharges collected at multiple locations within a short time period to reveal locations and amounts of groundwater input or loss in a lake or river.	Applied to several reaches along the Middle Fork of the John Day River, northeast Oregon, USA, to identify whether they were gaining or losing groundwater (Wright et al., 2005).	10 to >1000 (applied along river sections or whole lakes).	Buchanan and Somers, 1969; Riggs, 1972; Kaleris, 1998; Zacharias et al., 2003; Becker et al., 2004

Water temperature and thermal patterns	Indirect indications of SGW interactions are provided by differences in water temperature between surface and subsurface samples, and empirical relationships can be derived with direct measures of flux (Conant, 2004).	A 95-cm long steel temperature probe and thermistor was used to measure temperatures at the bed surface and various depths in the beds of three riffles in the East Branch of the Maple River, USA, to reveal subsurface flow paths (White, 1990).	0.001 (single measure) to >1000 (when used along a lake shore or stream bed).	Silliman et al., 1995; Evans and Petts, 1997; Constantz et al., 2003; Malcolm et al., 2004
Water chemistry and chemical signatures	Natural chemical composition of water samples from surface and subsurface habitats provides indirect indications of SGW interactions, often reflecting prevailing weathering patterns or redox conditions.	Ratios of Na^+, SiO_2, and Mg^{2+} were used to indicate the contribution of groundwater to surface waters in a glacial stream in the Swiss Alps (Malard et al., 2003).	0.01 (single measure) to >1000 (along a lake shore or stream bed).	Rutherford and Hynes, 1987; Brunke and Gonser, 1997; Kobayashi et al., 2000; Gurrieri and Furniss, 2004
Added dyes and tracers	Dyes can be used to visualise SGW exchanges whereas changes in concentrations of added tracers reveal pathways and retention. Co-injected conservative and non-conservative tracers demonstrate uptake and transformation of nutrients.	Dyes used to supplement data from thermal patterns (White, 1990); co-injected chloride (conservative tracer) and nitrate illustrated nitrogen dynamics in Little Lost Man Creek, California, USA (Triska et al., 1989a, b).	0.01 (single measure) to >1000 (along a lake shore or stream bed).	Triska et al., 1990; Harvey and Wagner, 2000; Constantz et al., 2003

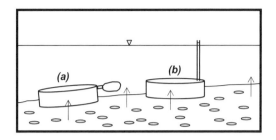

Figure 9.2 Seepage meter (a) as designed by Lee (1977). A plastic bag attached to the tube at the right collects seepage water. The meter can also be used for determining head relative to lake level (b) marked with a triangle. Arrows indicate groundwater flowpaths

ground, sometimes to depths of many metres (Freeze and Cherry, 1979). They operate on the principle that the water level in the slotted wells is the same as the surrounding water table. Mini-piezometers are similar but smaller (Lee and Cherry, 1978) and can often be inserted manually (Boulton, 1993) or with power augers and drilling equipment (Dahm and Valett, 1996). The typical mini-piezometer is made of PVC pipe (e.g., 20 mm electrical conduit pipe) and can either have the end of the pipe open or sealed but with a band of slots or tiny holes near the base of the tube (Figure 3). A mini-piezometer can be inserted into the bed of a lake or stream by threading the PVC pipe onto a steel T-bar whose diameter is large enough to provide a snug fit within the tube. A small sledge hammer can be used to drive the T-bar and PVC pipe into the sediments to the desired depth. While holding the mini-piezometer firmly, the T-bar is slowly drawn from the PVC tube. In fine sediments such as sand, this withdrawal must be done carefully to avoid generating a 'syringe' effect that might suck sediments into the mini-piezometer. The water within the mini-piezometer can then be pumped out with either a syringe or a small bilge-pump (e.g., Boulton and Stanley, 1995), checking that the mini-piezometer refills after bailing to ensure that it is not clogged.

Mini-piezometers may be connected to a hydraulic potentiomanometer to measure differences in hydraulic head relative to surface water (Winter *et al.*, 1988; Boulton, 1993). This overcomes the problems of trying to see differences in head that are slightly above or below the water surface (Lee and Cherry, 1978). The manometer is an air-tight, inverted Y-tube with one lower arm connected to the piezometer and the other to a stilling well in the surface water (Figure 9.3). The upper arm is connected to a hand-pump or syringe that creates the pressure vacuum. Water is sucked up into the manometer, and care taken to expel all air bubbles within the apparatus (this can be done by vigorously tapping the tubes). Pressure is then released until the tops of the two parallel columns of water in the lower arms of the inverted Y are visible against the graduated scale (Figure 9.3), revealing the difference in vertical hydraulic head.

Darcy's Law (Freeze and Cherry, 1979) describes the flux of groundwater (volume per unit time) Q as:

$$Q = -KA(dh/dl)$$

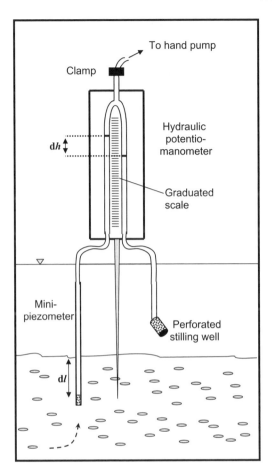

Figure 9.3 Mini-piezometer and hydraulic potentiomanometer in an upwelling zone (broken arrow). The vertical hydraulic gradient is given by dh/dl where dh is the head difference and dl is the depth of the mini-piezometer below the sediment–water interface. Surface water level is marked with a triangle

where K is the hydraulic conductivity (i.e., ease with which water flows through the sediments), A is the area through which the flow occurs, and dh/dl is the vertical hydraulic gradient (VHG) expressed as the ratio of the head difference (dh) and the depth of the mini-piezometer below the sediment–water interface (dl, Figure 9.3). If VHG is positive, this implies that water is upwelling from the groundwater into the surface system; if negative, the mini-piezometer lies in a downwelling zone. However, unlike the seepage meter where actual seepage is collected, VHG is only a measure of *potential* exchange and does not always reveal the true movement of water (Boulton, 1993; Woessner, 2000). Layers of sediments with low hydraulic conductivity may lie between the base of the mini-piezometer and the surface water, confounding assessments of hydrological exchange. Nonetheless, mini-piezometers provide a rapid indication of VHG and have been used for ecohydrological studies of the relationships between SGW exchanges, interstitial water chemistry, and the distribution of benthic and interstitial invertebrates (e.g., Cooling and Boulton, 1993; Winter, 1999; Kang *et al.*, 2005).

An example of the use of mini-piezometers for assessing ecohydrological associations is a study of temporal changes in VHG, water chemistry, and hyporheic fauna at three sites along a Sonoran Desert stream, Arizona, USA (Stanley and Boulton, 1995; Boulton and Stanley, 1995). Mini-piezometers inserted along or perpendicular to the stream channel and to different depths revealed changes in the strength and direction of vertical hydrological exchange between the surface stream and its hyporheic zone in response to surface discharge. Concomitant changes in water chemistry (Stanley and Boulton, 1995) and hyporheic invertebrate assemblage composition (Boulton and Stanley, 1995) were associated with changes in VHG during flooding and drying. In this desert stream where sediments are relatively homogenous and hydraulic conductivity is high, VHG provided a reliable guide to actual hydrological exchange and hydrology explained much of the variance in water chemistry and hyporheic faunal composition. However, where sediments have variable pore size and interstitial velocity, other environmental factors, such as availability of dissolved oxygen or organic matter, may contribute to patchiness in subsurface conditions that are not explained by hydrological variables alone (Malard *et al.*, 2002).

Vertical, lateral, and longitudinal variation in SGW interactions can be addressed by judicious placement of nests of mini-piezometers (Wondzell and Swanson, 1996; Woessner, 2000), and their small size and cost relative to seepage meters provides opportunities for better replication. Their versatility as sampling ports into the subsurface zones of lakes and streams is another advantage, and they would now appear to be the most popular tool for assessing SGW interactions at fine to intermediate scales with their spatial extent dictated by the diameter, length, and arrangement of the tubes. It is plausible that 'micro'-piezometers could be used to supplement pore-water samplers (e.g., Cleven and Meyer, 2003) for very fine-scale assessments of biochemical activity and its association with hydrological variables at the SGW interface.

At the broader scale of lake shores, stream reaches, and catchments, mini-piezometers and piezometers have been used to map groundwater flowpaths and flow fields, assuming that differences in the height of water in the wells indicate the direction of flow (Fetter, 1988). Various mathematical approaches (reviews in Winter, 1999; Sanford, 2002) can use these data to generate multidimensional models to show local, intermediate, and regional flowpaths into and out of lakes and river channels. For example, Crystal Lake, northern Wisconsin, USA, and its catchment lie on sandy glacial deposits over crystalline bedrock. As precipitation in this area exceeds evaporation by 180 mm, much of the time, Crystal Lake has seepage outflow to groundwater across most of its bed. Using nested piezometers inserted on four sides of the lake, Anderson and Cheng (1993) collected data over a decade spanning 5 years of relatively wet conditions followed by several drier years. Subsurface flow patterns as deep as 10 m into the groundwater system differed between the wet and dry periods. Piezometric mapping has also been used to assess seasonal and storm-induced changes in SGW exchange in streams and to calibrate groundwater flow models (e.g., MODFLOW, Wondzell and Swanson, 1996). With increasing recognition of the ecological significance of groundwater and its potential to transport contaminants, nests of piezometers down to different depths and across flow lines have been fundamental to mapping equipotential lines of shared hydraulic heads.

9.3.3 Synoptic Surveys of Surface Discharge or Lake Levels

Synoptic surveys, where discharge at multiple locations is measured over a short time period, can reveal potential inputs from or losses to groundwater. This is most commonly done in streams where gauging might be done at a number of cross sections, preferably simultaneously, and discharge measurements are compared using standard methods (Buchanan and Somers, 1969; Riggs, 1972). Ideally, the comparisons are done during a period of steady flow. Gauging locations are situated in stream sections where surface diversions and tributary inputs can be accounted for. At several reaches along the Middle Fork of the John Day River, northeast Oregon, USA, Wright *et al.* (2005) measured VHG and stream discharge at two to four locations in each reach, repeating the measurements three times at each location to maximize precision and accuracy. At each reach, the measurements were treated as simultaneous (typically taken within 2 h), and with the assumption of steady flow, any differences in discharge were attributed to error in measurement, groundwater exchange, or losses due to evapotranspiration along the reach. Synoptic data confirmed results from VHG measurements, and even in reaches that were predicted to be gaining water from the groundwater, there was no evidence of input within the margin of error (4%) (Wright *et al.*, 2005). The combination of the broad reach-scale data from the synoptic measurements with the more site-specific data from the mini-piezometers is a powerful hydroecological approach and provided strong support for the observations in this study.

Synoptic gauging methods are often used in conjunction with other techniques as a cross reference and to supplement finer-scale data with results collected from whole rivers, reaches or stream segments (Becker *et al.*, 2004). In many streams, the reliability of the gauging data is influenced by the length of reach and the complexity of hydraulic resistance of the bed (Kaleris, 1998). For deep lakes, geographic information systems (GIS) software has been used with digital terrain models to integrate water level measurements to provide estimates of groundwater recharge (Zacharias *et al.*, 2003).

9.4 Indirect Hydrological Methods for Assessing SGW Interactions

9.4.1 Water Temperature and Thermal Patterns

One of the most noticeable effects of groundwater inputs into lakes and streams can be localized pockets of cooler or warmer water relative to the surrounding surface water. The changes or buffering in water temperature caused by upwelling groundwater has many ecohydrological influences (Power *et al.*, 1999) so thermal signatures have been an obvious tool for indirect assessments of groundwater contribution to lakes and streams (Silliman *et al.* 1995; Evans and Petts 1997; Constantz *et al.* 2003) as well as an indicator of surface water inputs to groundwater ecosystems (Cooling and Boulton, 1993).

Although subsurface water temperatures can be measured on freshly collected samples from mini-piezometers or other sampling devices, the use of probes and thermistors is more common. For example, White (1990) used a 95-cm long stainless steel temperature probe and thermistor to measure subsurface water temperatures at the bed surface and at depths of 5, 10, 15, 20, 30, 40 and 50 cm into the beds of three riffles in the East Branch of the Maple River, Michigan, USA, to generate maps of subsurface flow and SGW exchange. Data loggers can also be used to collect temporal data. Claret and Boulton (2003) buried sensors at the surface, 10 cm, and 30 cm in three habitats in an Australian sand-bed stream to identify diel variation in interstitial water temperatures at different depths. Water temperature fluctuations were dampened with sediment depth in these downwelling zones, and the lateral translation of thermal peaks indicated the slow movement of water through this sand-bed stream.

Using water temperature as an indicator of groundwater movement into and out of lakes and streams has many advantages. Not only does water temperature have direct ecological significance for aquatic biota and many ecosystem processes (Power et al., 1999; Jones and Mulholland, 2000), measurements are cheap, instantaneous, and frequently collected at the same time as conductivity and dissolved oxygen when using field meters. In streams, temperature and flow surveys can be combined using a simple heat balance to obtain better estimates of streamflow than those obtained with current meters alone, so the approach has potential as a cost-effective method of quantifying groundwater dynamics where inflow is highly heterogeneous (Becker et al., 2004). Water temperature data are also ideal for tracking fine-scale differences in SGW gradients in streams (Malcolm et al., 2004) at a scale relevant to many ecological processes. Simple empirical relationships can be drawn between water temperatures and hydrologic flux derived from mini-piezometer data so that exchanges may be calculated at locations where only water temperature data are available (Conant, 2004). Water temperature and seepage data have been used concurrently to illustrate how some seepage in streams may be derived from surface water that has only recently down-welled rather than being true groundwater discharge (Alexander and Caissie, 2003).

9.4.2 Water Chemistry and Chemical Signatures

Exchanging surface waters and groundwaters act as vectors for dissolved gases and chemicals, many of which are vital for biota and biochemical processes in surface and subsurface habitats (Brunke and Gonser, 1997; Jones and Mulholland, 2000; Hancock et al., 2005). Pollutants and contaminants are also transported, and can be used to identify water sources to lakes and streams (Landmeyer et al., 2003). Typically, downwelling surface water is rich in dissolved oxygen and oxidized forms of nitrogen and phosphorus, whereas upwelling groundwater has far less oxygen and higher concentrations of reduced forms of nutrients such as ammonium (Duff and Triska, 2000). Concentrations of dissolved organic carbon also differ and can be used as an indicator of groundwater influence (Rutherford and Hynes, 1987). Regional groundwater often has a distinctive chemical 'signature' of ionic composition, and the direction and extent of SGW interactions can be gleaned from comparisons of the chemical signatures of water collected from the surface and different depths (Gurrieri and Furniss, 2004).

Malard et al. (2003) explored the association between hyporheic invertebrate composition and zones of upwelling in the Roseg River, a glacial stream in the Swiss Alps. To characterize the sources of water, ionic composition was used because weathering of rocks in the catchment yielded higher concentrations of Na^+, SiO_2, and Mg^{2+} in hillslope groundwater than in glacial surface water or shallow alluvial groundwater fed by the stream. Samples were collected from habitats on the upper and lower floodplains, and by plotting hyporheic versus surface concentrations of the ions and comparing the distribution of the points along an equivalence line, the extent of groundwater influence on the surface water could be ascertained. Natural isotopes also serve as useful indicators of groundwater flow. For example, Kobayashi *et al.* (2000) used concentrations of the stable isotopes of hydrogen (δ^2H) and oxygen (δ^{18}O) and the radioactive isotopes tritium and radon (^{222}Rn) to estimate groundwater discharge in Lake Biwa, Japan. These isotopes were effective in showing the origin of groundwater from the mountains as well as indicating the residence time of shallow groundwater to be about 2–3 years while that of deep groundwater was approximately 7 years.

9.4.3 Dyes and Added Tracers

One of the earliest methods used to complement hydrological and thermal techniques (White, 1990) and indicate flow paths was the use of dyes. These have the advantage of providing visual confirmation of predicted zones of input and output of groundwater into lakes and streams, and can also be used to illustrate plumes of dispersion in interstitial sediments. One elegant adaptation of the dye method entailed injecting red ink into a sand bed stream, leaving the needle in the sediments to mark the point of injection and then, 2–6 h later, taking a sediment core at the point of injection and freezing it (Mutz and Rohde, 2003). By breaking the core in the plane of the injection flow path, the maximum distance traveled by the ink and its dispersion could readily be seen.

Tracers also serve the same purpose to illustrate flow paths and SGW exchange, and have the further benefit that they can be detected chemically at extremely low concentrations, beyond that possible by the human eye. Another advantage of added tracers is the ability to co-inject a 'conservative' tracer that is not consumed by biogeochemical pathways with a nutrient or some biological element so that its uptake, reflected in changes in concentration, can be compared with that of the conservative tracer. For example, to determine whether nitrogen dissolved in stream water downwelling into the hyporheic zone of Little Lost Man Creek, California, USA, was undergoing transformation, the conservative tracer chloride (Cl^-) was co-injected with nitrate into the surface water for 20 days (Triska *et al.*, 1989a, b). Lateral outwelling of streamwater was indicated by chloride concentrations in water from mini-piezometers in the bank. Cl^- concentrations indicated >90% streamwater in wells <4 m from the wetted channel but fell to 0% at a well 10 m from the channel. Wells with elevated nitrate compared with that predicted from the Cl^- concentrations indicated subsurface nitrification whereas sites with less nitrate than predicted suggested denitrification during subsurface flow (Triska *et al.*, 1989b).

Tracers have also been used widely in lakes, sometimes as whole-lake experiments, to estimate groundwater inputs and outputs. Cole and Pace (1998) added lithium bromide to three small seepage lakes in Michigan, USA, and identified groundwater inputs averaging 57% of total water input. This enabled them to calculate the relative significance of SGW interactions versus input via precipitation for influx of solutes such as phosphorus and dissolved organic carbon. In the last decade, added tracer studies have become common in assessing SGW interactions in lakes and streams (Dahm *et al.* 1998; Harvey and Wagner, 2000; Constantz *et al.*, 2003), illustrating patterns of nutrient spiraling and transformation, interstitial retention and release zones, temporal changes in groundwater inputs, and providing fundamental data for most modeling activities. Nonetheless, some caution should be exercised in ecohydrological studies where chemical effects of the tracers might impact upon the biota as has been seen in salt dilution tracer studies in some surface streams (Wood and Dykes, 2002); this could be especially relevant for interstitial microbial processes.

9.5 Future Technical Challenges and Opportunities

Significant technical advances in testing and development of field methods for monitoring SGW interactions have occurred in the last decade (e.g., Baxter et al., 2003; Rosenberry, 2005), bringing them into the easy reach of ecohydrologists. With increasing recognition of the importance of SGW interactions in most aquatic ecosystems and their susceptibility to human impacts and global climate change (Hancock et al., 2005), the need for data sets at broader geographic scales is greater and there is scope for more use of GIS and remote-sensing methods (e.g., Zacharias et al., 2003). Studies integrating a suite of techniques have been particularly valuable, often because each complementary method illustrates further aspects of the complexity of SGW interactions. For example, the different scales at which each method might apply (Table 9.1) enable researchers to integrate measures at the sediment, reach or lake zone, and whole catchment scale, and seek ecohydrological interactions within and among these nested spatial scales.

There is also a pressing need for further technical advances in measuring SGW hydrological interactions at smaller spatial scales (<10 mm) because this is the scale at which many biogeochemical processes operate. With more powerful methods for assessing these, we will be able to identify how sediment-scale transformations in lakes and rivers act to govern reach or lake-zone water chemistry or ecosystem processes when we construct models of nutrient cycling and spiraling (Mulholland and DeAngelis, 2000). In turn, temporal changes in these processes may be modeled in association with changes in water level or flow regime so that hydroecologists can explore 'hot times' (as meant by Fisher et al., 2004) as well as hot spots of productivity or other ecosystem processes resulting from transient phases of SGW exchanges.

Like all technical developments, comparisons of instrument errors and sampling efficiency are essential, both between methods (Constantz et al., 2003) as well as in different geomorphic settings. Successful modeling of SGW interactions will only be as good as the input data, and hydroecologists are acutely aware of the effects on hydrological exchanges caused by layering of sediments in lake and river beds where variable hydraulic conductivities generate preferential flow paths that could be missed by point measurements (Woessner, 2000). Integration of several complementary methods and inclusion of geomorphologists within ecohydrological research teams would help resolve some of these dilemmas.

Ecologists working on benthic biota and processes in lakes and rivers now recognise the significant influence of groundwater inputs on algal productivity, invertebrate composition, and ecosystem processes. The suite of techniques described in this chapter equips ecohydrologists with tools to elucidate SGW interactions and their effects further. The relative simplicity of many of the methods lends itself to routine inclusion in sampling programs of benthic ecology in lakes and streams. Hydroecologists and ecohydrologists have major contributions to make in furthering our conceptual understanding of how SGW interactions underpin lake and stream ecosystem functioning. They will also find use for these tools in practical applications of how SGW interactions influence lake and river rehabilitation (Woessner, 2000; Boulton, 2007) or pollution and contaminant dynamics (Landmeyer *et al.*, 2003), guiding our successful management of the linked surface and groundwater resources (Winter *et al.*, 1998; Hancock *et al.*, 2005).

Acknowledgements

I appreciate the invitation to contribute to this timely book, and hope this perspective on methods encourages more ecologists and hydrologists to collaborate in ecohydrological and hydroecological studies of surface-groundwater interactions. I thank Emily Stanley and Maurie Valett for first introducing me to 'T-bar technology'; James LaBaugh, John Morrice, and Steve Thomas for patiently guiding me through groundwater hydrological theory and techniques; and Cecile Claret, Marie-Jo Olivier, Pierre Marmonier, and Florian Malard for demonstrating some of the European approaches in the PASCALIS program. The Australian Research Council provided funding for work presented here. I also thank Sarah Mika, Peter Hancock, Paul Wood and an anonymous referee for constructive comments on earlier drafts of the manuscript.

References

Alexander MD, Caissie D. 2003. Variability and comparison of hyporheic water temperatures and seepage fluxes in a small Atlantic salmon stream. *Ground Water*, **41**: 72–82.

Anderson MP, Cheng X. 1993. Long- and short-term transience in a groundwater/lake system in Wisconsin. *Journal of Hydrology*, **145**: 1–18.

Baxter C, Hauer FR, Woessner WW. 2003. Measuring groundwater-stream water exchange: New techniques for installing minipiezometers and estimating hydraulic conductivity. *Transactions of the American Fisheries Society*, **132**: 493–502.

Becker MW, Georgian T, Ambrose H, Siniscalchi J, Fredrick K. 2004. Estimating flow and flux of ground water discharge using water temperature and velocity. *Journal of Hydrology*, **296**: 221–233.

Benda LE, Poff NL, Tague C, Palmer MA, Pizzuto J, Cooper S, Stanley E, Moglen G. 2002. How to avoid train wrecks when using science in environmental problem solving. *BioScience*, **52**: 1127–1136.

Benson NG. 1953. The importance of groundwater to trout populations in the Pigeon River, Michigan. *Transactions of the North American Wildlife Conference*, **18**: 260–281.

Boulton AJ. 1993. Stream ecology and surface-hyporheic exchange: implications, techniques and limitations. *Australian Journal of Marine and Freshwater Research*, **44**: 553–564.

Boulton AJ. 2007. Hyporheic rehabilitation in rivers: restoring vertical connectivity. *Freshwater Biology* **54**: 133–144.

Boulton AJ, Hancock PJ. 2005. Rivers as groundwater dependent ecosystems: degrees of dependency, riverine processes, and management implications. *Australian Journal of Botany* (in press).

Boulton AJ, Stanley EH. 1995. Hyporheic processes during flooding and drying in a Sonoran Desert stream. II. Faunal dynamics. *Archiv für Hydrobiologie*, **134**: 27–52.

Boulton AJ, Findlay S, Marmonier P, Stanley EH, Valett HM. 1998. The functional significance of the hyporheic zone in streams and rivers. *Annual Review of Ecology and Systematics*, **29**: 59–81.

Brunke M, Gonser T. 1997. The ecological significance of exchange processes between rivers and groundwaters. *Freshwater Biology*, **37**: 1–33.

Buchanan TJ, Somers WP. 1969. *Discharge Measurements at Gaging Stations*. Techniques of Water Resource Investigations of the US Geological Survey, Book 3, Chapter A8. US Geological Survey, Washington, DC.

Burt TP, Hefting MM, Pinay G, Sabater S. 2007. The role of floodplains in mitigating diffuse nitrate pollution. In *Hydroecology and Ecohydrology: Past, Present and Future*, Wood PJ, Hannah DM, Sadler JP (eds). John Wiley & Sons: New York, pp. 253–268.

Claret C, Boulton AJ. 2003. Diel variation in surface and subsurface microbial activity along a gradient of drying in an Australian sand-bed stream. *Freshwater Biology*, **48**: 1739–1755.

Cleven E-J, Meyer EI. 2003. A sandy hyporheic zone limited vertically by a solid boundary. *Archiv für Hydrobiologie*, **157**: 267–288.

Cole JJ, Pace ML. 1998. Hydrologic variability of small, Northern Michigan lakes measured by the addition of tracers. *Ecosystems*, **1**: 310–320.

Coleman RL, Dahm CN. 1990. Stream geomorphology: effects on periphyton standing crop and primary production. *Journal of the North American Benthological Society*, **9**: 293–302.

Conant B. 2004. Delineating and quantifying ground water discharge zones using streambed temperatures. *Ground Water*, **42**: 243–257.

Constantz J, Cox MH, Su GW. 2003. Comparison of heat and bromide as ground water tracers near streams. *Ground Water*, **41**: 647–656.

Cooling MP, Boulton AJ. 1993. Aspects of the hyporheic zone below the terminus of a South Australian arid-zone stream. *Australian Journal of Marine and Freshwater Research*, **44**: 411–426.

Dahm CN, Valett HM. 1996. Hyporheic zones. In: *Methods in Stream Ecology*, Hauer FR, Lamberti GA (eds). Academic Press: San Diego, pp. 107–119.

Dahm CN, Grimm NB, Marmonier P, Valett HM, Vervier P. 1998. Nutrient dynamics at the interface between surface waters and groundwaters. *Freshwater Biology*, **40**: 427–453.

Danielopol DL. 1989. Groundwater fauna associated with riverine aquifers. *Journal of the North American Benthological Society*, **8**: 18–35.

Duff JH, Triska FJ. 2000. Nitrogen biogeochemistry and surface-subsurface exchange in streams. In *Streams and Ground Waters*, Jones J, Mulholland P (eds). Academic Press: New York, pp. 337–361.

Evans EC, Petts GE. 1997. Hyporheic patterns within riffles. *Hydrological Sciences Journal*, **42**: 199–213.

Fetter CW. 1988. *Applied Hydrogeology*. Merrill Publishing: Columbus, Ohio.

Fisher SG, Sponseller RA, Heffernan JB. 2004. Horizons in stream biogeochemistry: flowpaths to progress. *Ecology*, **85**: 2369–2379.

Freeze RA, Cherry JA. 1979. *Groundwater*. Prentice-Hall: Englewood Cliffs, New Jersey.

Gurrieri JT, Furniss G. 2004. Estimation of groundwater exchange in alpine lakes using non-steady mass-balance methods. *Journal of Hydrology*, **297**: 187–208.

Hancock PJ, Boulton AJ, Humphreys WF. 2005. The aquifer and its hyporheic zone: ecological aspects of hydrogeology. *Hydrogeological Journal*, **13**: 98–111.

Harvey JW, Wagner BJ. 2000. Quantifying hydrologic interactions between streams and their hyporheic zones. In *Streams and Ground Waters*, Jones J, Mulholland P (eds). Academic Press: New York, pp. 3–44.

Hubbert MK. 1940. The theory of groundwater motion. *Journal of Geology*, **48**: 785–944.

Hynes HBN. 1983. Groundwater and stream ecology. *Hydrobiologia*, **100**: 93–99.

Isiorho SA, Meyer JH. 1999. The effects of bag type and meter size on seepage meter measurements. *Ground Water*, **37**: 411–413.

Jones J, Mulholland P (eds). 2000. *Streams and Ground Waters*, Academic Press: New York.

Jones JB, Fisher SG, Grimm NB. 1995. Vertical hydrologic exchange and ecosystem metabolism in a Sonoran Desert stream. *Ecology*, **76**: 942–952.

Kaleris V. 1998. Quantifying the exchange rate between groundwater and small streams. *Journal of Hydraulic Research*, **36**: 913–932.

Kang WJ, Kolasa KV, Rials MW. 2005. Groundwater inflow and associated transport of phosphorus to a hypereutrophic lake. *Environmental Geology*, **47**: 565–575.

Kobayashi M, Kitaoka K, Yoshioka R, Horiuchi K. 2000. Groundwater flow discharging into a lake estimated by natural isotopes. *Internationale Vereinigung für Theoretische und Angewandte Limnologie, Verhandlungen*, **27**: 2283–2287.

Landmeyer JE, Bradley PM, Bullen TD. 2003. Stable lead isotopes reveal a natural source of high lead concentrations to gasoline-contaminated groundwater. *Environmental Geology*, **45**: 12–22.

Lee DR. 1977. A device for measuring seepage flux in lakes and estuaries. *Limnology and Oceanography*, **22**: 140–147.

Lee DR, Cherry J. 1978. A field exercise on groundwater flow using seepage meters and mini-piezometers. *Journal of Geological Education*, **27**: 6–10.

Lee DR, Hynes HBN. 1977. Identification of groundwater discharge zones in a reach of Hillman Creek in southern Ontario. *Water Pollution Research in Canada*, **13**: 121–133.

Lesack LFW. 1995. Seepage exchange in an Amazon floodplain lake. *Limnology and Oceanography*, **40**: 598–609.

Lock MA, John PH. 1978. The measurement of groundwater discharge into a lake by a direct method. *Internationale Revue der gesamten Hydrobiologie*, **63**: 271–275.

Malard F, Dole-Olivier M-J, Mathieu J, Stoch F. 2001. *Sampling Manual for the Assessment of Regional Groundwater Biodiversity.* http://www.pascalis-project.com/ (accessed April 2005).

Malard F, Tockner K, Dole-Olivier M-J, Ward JV. 2002. A landscape perspective of surface–subsurface hydrological exchanges in river corridors. *Freshwater Biology*, **47**: 621–640.

Malard F, Ferreira D, Dolédec S, Ward JV. 2003. Influence of groundwater upwelling on the distribution of the hyporheos in a headwater river flood plain. *Archiv für Hydrobiologie*, **157**: 89–116.

Malcolm IA, Soulsby C, Youngson AF, Hannah DM, McLaren IS, Thorne A. 2004. Hydrological influences on hyporheic water quality: implications for salmon egg survival. *Hydrological Processes*, **18**: 1543–1560.

Mulholland PJ, DeAngelis DL. 2000. Surface–subsurface exchange and nutrient spiraling. In *Streams and Ground Waters*, Jones J, Mulholland P (eds). Academic Press: New York, pp. 149–166.

Mutz M, Rohde A. 2003. Processes of surface–subsurface water exchange in a low energy sand-bed stream. *Internationale Revue der gesamten Hydrobiologie*, **88**: 290–303.

Nestler A, Goodwin J. 2007. A mathematical and conceptual framework for hydraulics. In *Hydroecology and Ecohydrology: Past, Present and Future*, Wood PJ, Hannah DM, Sadler JP (eds). John Wiley & Sons, Ltd: Chichester, 205–224.

Palmer MA. 1993. Experimentation in the hyporheic zone: Challenges and prospectus. *Journal of the North American Benthological Society*, **12**: 84–93.

Power G, Brown RS, Imhof JG. 1999. Groundwater and fish – insights from northern North America. *Hydrological Processes*, **13**: 401–422.

Riggs, H.C. 1972. *Low-flow Investigations. Techniques of Water Resource Investigations of the U.S. Geological Survey*. Book 4, Chapter B1. US Geological Survey: Washington, DC.

Rosenberry DO. 2005. Integrating seepage heterogeneity with the use of ganged seepage meters. *Limnology and Oceanography – Methods*, **3**: 131–142.

Rosenberry DO, Morin RH. 2004. Use of an electromagnetic seepage meter to investigate temporal variability in lake seepage. *Ground Water*, **42**: 68–77.

Rouch R. 1988. Sur la répartition spatiale des Crustacés dans le sous-écoulement d'un ruisseau des Pyrénées. *Annales de Limnologie*, **24**: 213–234.

Rutherford JE, Hynes HBN. 1987. Dissolved organic carbon in streams and groundwater. *Hydrobiologia*, **154**: 33–48.

Sanford W. 2002. Recharge and groundwater models: an overview. *Hydrogeology Journal*, **10**: 110–120.

Shaw RD, Prepas EE. 1990. Groundwater-lake interactions. I. Accuracy of seepage meter estimates of lake seepage. *Journal of Hydrology*, **119**: 105–120.

Silliman SE, Ramirez J, McCabe RL. 1995. Quantifying downflow through creek sediments using temperature time-series – one-dimensional solution incorporating measured surface temperature. *Journal of Hydrology*, **167**: 99–119.

Stanley EH, Boulton AJ. 1995. Hyporheic processes during flooding and drying in a Sonoran Desert stream. I. Hydrologic and chemical dynamics. *Archiv für Hydrobiologie*, **134**: 1–26.

Triska FJ, Kennedy VC, Avanzino RJ, Zellweger GW, Bencala KE. 1989a. Retention and transport of nutrients in a third-order stream in northwestern California: channel processes. *Ecology*, **70**: 1877–1892.

Triska FJ, Kennedy VC, Avanzino RJ, Zellweger GW, Bencala KE. 1989b. Retention and transport of nutrients in a third-order stream in northwestern California: hyporheic processes. *Ecology*, **70**: 1893–1905.

Triska FJ, Kennedy VC, Avanzino RJ, Zellweger GW, Bencala KE. 1990. In situ retention-transport response to nitrate loading and storm discharge in a third order stream. *Journal of the North American Benthological Society*, **9**: 229–239.

Valett HM, Hakenkamp C, Boulton AJ. 1993. Perspectives on the hyporheic zone: integrating hydrology and biology. Introduction. *Journal of the North American Benthological Society*, **12**: 40–43.

White DS. 1990. Biological relationships to convective flow patterns within stream beds. *Hydrobiologia*, **196**: 149–158.

Williams DD, Hynes HBN. 1974. The occurrence of benthos deep in the substratum of a stream. *Freshwater Biology*, **4**: 233–256.

Winter TC. 1999. Relation of streams, lakes, and wetlands to groundwater flow systems. *Hydrogeology Journal*, **7**: 28–45.

Winter TC, Harvey JW, Franke OL, Alley WM. 1998. *Ground Water and Surface Water – A Single Resource*. United States Geological Survey Circular 1139, Denver, Colorado.

Winter TC, LaBaugh JW, Rosenberry DO. 1988. The design and use of a hydraulic potentiomanometer for direct measurement of differences in hydraulic head between groundwater and surface water. *Limnology and Oceanography*, **33**: 1209–1214.

Woessner WW. 2000. Stream and fluvial plain ground-water interactions: rescaling hydrogeologic thought. *Ground Water*, **38**: 423–429.

Wondzell SM, Swanson FJ. 1996. Seasonal and storm dynamics of the hyporheic zone of a 4th-order mountain stream. I. Hydrological processes. *Journal of the North American Benthological Society*, **15**: 3–19.

Wood PJ, Dykes AP. 2002. The use of salt dilution gauging techniques: ecological considerations and insights. *Water Research*, **36**: 3054–3062.

Wood PJ, Hannah DM, Agnew MD, Petts GE. 2001. Scales of hydroecological variability within a groundwater-dominated stream. *Regulated Rivers: Research and Management*, **17**: 347–367.

Wright KK, Baxter CV, Li JL. 2005. Restricted hyporheic exchange in an alluvial river system: Implications for theory and management. *Journal of the North American Benthological Society*, **24**: 447–460.

Zacharias I, Dimitriou E, Koussouris T. 2003. Estimating groundwater discharge into a lake through underwater springs by using GIS technologies. *Environmental Geology*, **44**: 843–851.

10
Examining the Influence of Flow Regime Variability on Instream Ecology

Wendy A. Monk, Paul J. Wood and David M. Hannah

10.1 Introduction

River flow regimes demonstrate marked variability in space and time (Poff, 2002; Bower *et al.*, 2004). Flow variability is recognised as a valuable predictor of physical (abiotic) habitat conditions and exerts a significant influence upon biological processes within riverine ecosystems (see Richter *et al.*, 1996; Poff *et al.*, 1997; Wood and Armitage, 2004; Biggs *et al.*, 2005). However, while the impact of flow regime variability on instream communities is widely acknowledged by scientists and environmental managers (e.g., Smakhtin, 2001; Tharme, 2003), the integration of hydrological and ecological data sets and quantification of hydroecological associations is not straightforward. In part, this reflects the frequent mismatch in the nature and spatio-temporal resolution of available hydrological and ecological data, which stems from different primary purposes for data collection. River gauging (hydrometric) data are usually gathered to identify discharge fluctuations on a continuous basis, with records available for many rivers for several decades. In contrast, the limited medium- to long-term ecological data series available for analysis have largely been collected for single sites or as part of routine biomonitoring surveys at particular times of the year. Furthermore, demonstrating the influence of river flow on single organisms, populations, communities or ecological metrics is made more difficult by multiple, complex interactions that may occur between river flow and other instream characteristics, such as water quality and channel morphology (Effenberger *et al.*, 2006).

This chapter highlights the growing need for integrated hydrological and ecological data sets; and it explores analytical methods to examine the influence of flow regime variability on instream ecology. The results of a bibliographic survey of scientific journal publications that have examined (or alluded to) the influence of flow variability on riverine ecosystems are presented. The importance of understanding spatial and temporal scales of hydroecological variability is considered. Finally the hydrological and ecological time series are discussed, specifically the (mis)use and limitations of macroinvertebrate community data for pure and applied hydroecological research.

10.2 The Requirement for Hydroecological Data

Knowledge of the influence of physical processes upon water quality and the structure of biotic communities of streams and rivers underlies many national river system management strategies (Bunn and Arthington, 2002), and the majority of national and international water legislation (Black *et al.*, 2002). The quest to identify ecologically relevant hydrological indices has been driven by the need to identify and quantify variability in both ecological communities and individual populations that may be sensitive to natural changes or human modification of the river flow regime (Richter *et al.*, 1996). The potential impact of future climate change/variability and increasing anthropogenic influence(s) on river flow has focused attention on the need to understand the relationships between the flow regime, and instream biological processes and communities (Pringle, 2003).

The influence of natural flow regime variability on instream communities has been recognised increasingly by both ecologists and hydrologists (Richter *et al.*, 1996; Clausen and Biggs, 1997; Wood *et al.*, 2001; Snelder *et al.*, 2005). An ever-growing number of ecologically important flow regime characteristics have been identified, most of which are incorporated within the Indicators of Hydrologic Alteration (IHA) approach (Richter *et al.*, 1996). This methodology identified five facets of the flow regime that are potentially 'ecologically relevant': (i) magnitude of monthly water conditions; (ii) magnitude and duration of extreme water conditions; (iii) timing of annual extreme water conditions; (iv) frequency and timing of high and low pulses; and (v) rate and frequency of water condition change. It has been proposed that these facets of the flow regime (as characterised by hydrological indices) should be examined in association with ecological data in an attempt to quantify the instream community response to changes in flow (Richter *et al.*, 1996, 1997; Poff *et al.*, 1997). Significant effort has been made in recent years to identify ecologically relevant hydrological parameters and, to date, over 200 indices have been derived and utilised (Olden and Poff, 2003; Monk *et al.*, 2006). However, critical evaluation of the most appropriate hydrological descriptors of instream ecological response, and the nature and strength of hydroecological associations is lacking. This reflects a paucity of high-quality, medium- to long-term paired hydrological and ecological datasets for many geographical locations (exceptions, for example, Richter *et al.*, 1997; Wood *et al.*, 2001; Wright *et al.*, 2004).

10.3 Bibliographic Analysis

Given the interdisciplinary academic interest in the ecological response to river flow variability, a bibliographic survey was undertaken to examine where scientific papers examining this specific hydroecological link were published based upon standard library cataloguing systems: (i) geosciences (e.g., geography, geology and environmental sciences); (ii) biosciences (e.g., biology, biological sciences and ecology), and (iii) interdisciplinary journals, including water resources and engineering. In addition, the first author of each paper was used to determine the academic departments/institutions within which the research was conducted: (i) geosciences; (ii) biosciences; and (iii) engineering and water resources. The spatial scale and temporal resolution of the research was also examined based on data contained within each of the manuscripts reviewed. The survey was based upon papers listed in the ISI Web of Knowledge database (http://wok.mimas.ac.uk) using combinations of the following key words: hydrology and ecology, hydroecology, ecohydrology, flow variability, discharge and instream communities (including periphyton, macroinvertebrates and fish). The bibliographic search had an emphasis on papers published in the last ten years (unless widely cited in contemporary flow variability studies, for example, Hughes and James, 1989; Jowett and Duncan, 1990) to reflect current research trends.

As of June 2006, the survey results demonstrate that approximately 50% of the published research examining instream response to river flow variability appeared in ecological/biological sciences journals (Figure 10.1a) with 30% and 20% appearing in hydrological/geosciences and interdisciplinary journals, respectively. Examination of the spatial and temporal scales clearly indicated that micro-scale investigations were predominately published in ecological/biological sciences journals; while papers appearing in hydrological/geoscience and interdisciplinary journals were dominant at larger resolutions. However, when the affiliations of the first or corresponding author were examined, the vast majority were found to reside in biosciences departments and biological research institutes (73%) with much lower proportions from geosciences (16%) and engineering (water resources) (11%) (Figure 10.1b). Biologists and ecologists clearly dominated research at the micro-scale. Researchers from geosciences and engineering departments/institutes predominantly undertook research at the meso- and macro-scale. Interestingly, with respect to paper content, this survey indicated that a number of papers alluded to the importance of river flow variability for a range on instream physical habitat and biological processes but empirical data sets (including either hydrological and ecological observations) to validate the assumptions were limited in a number of instances.

10.4 Importance of Scale

The structure and patterns within river ecosystems are inextricably linked to the spatial and temporal resolution of study (Thorpe *et al.*, 2006). Theories and models linking processes operating at different spatial scales have a long tradition within the disciplines of hydrology and ecology (see Frissell *et al.*, 1986; Levin, 1992; Downes *et al.*, 1993).

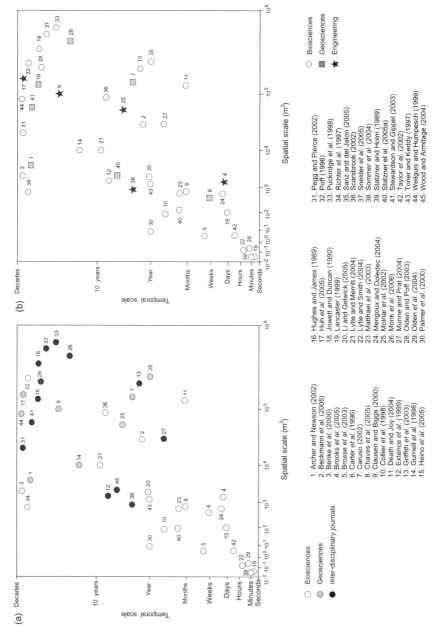

Figure 10.1 Spatial and temporal scale of research reported in scientific papers examining river flow variability based on (a) papers published in biosciences, geosciences and interdisciplinary journals, and (b) affiliation of the first/corresponding author

It has been argued that rivers should be viewed hierarchically as a set of nested filters whereby the large-scale processes (e.g., climatology and topography) constrains smaller scale hydrological (and hydraulic) and ecological processes (Poff, 1997; Snelder and Biggs, 2002). For example, instream community structure may be coarsely influenced by hydrological variability measured at the catchment scale as it directly determines local habitat structure and stability. Spatial hierarchical approaches attempt to conceptually and/or empirically explore the relationships between macro-scale features and processes, for example basin characteristics (Frissell *et al.*, 1986; Habersack, 2000), and those occurring at meso- (reach) and micro-scales (habitat) (Cannan and Armitage, 1999; Habersack, 2000; Poole, 2002; Benda *et al.*, 2004). However, research that integrates across and between different spatial and temporal scales is rare (Benda *et al.*, 2004; Lytle and Poff, 2004).

The natural flow regime of a river exhibits variability that reflects climate (first order driver) and basin characteristics (second order modifier; Bower *et al.*, 2004). It is widely acknowledged that many instream and riparian organisms display adaptations to flow regime dynamics (Lytle and Poff, 2004). This knowledge underpins research exploring the ecological responses to both short-term (sub-daily to monthly) and long-term (inter-annual and beyond) flow regime changes (Jowett and Duncan, 1990; Richter *et al.*, 1996; Poff *et al.*, 1997). Extreme hydrological events (floods and droughts) may result in significant modification to habitat and instream community structure, at multiple scales, that may persist for months/years or even reset successional trajectories (Lake, 2000). However, the way in which individual rivers, reaches and habitat patches, and their biota, respond to flow variability associated with climatological variability may be strongly influenced by basin characteristics. For example, the response of instream communities to low summer rainfall (seasonal drought) in basins dominated by surface runoff may be rapid (see Ledger and Hildrew, 2001) but the same conditions may result in no concurrent response in communities of groundwater-dominated rivers due to the buffering effect of aquifers (see Wood and Armitage, 2004). Thus, regional (inter-catchment) and local (intra-catchment) differences in flow regime characteristics clearly need to be recognised to understand fully the variable response of instream communities.

Appropriate measures of river 'flow' (i.e., descriptive statistics and indices of river discharge/runoff) used to quantify the influence upon instream communities or ecological processes may differ markedly between scales; hence, flow metrics are not always transferable. At small spatial scales, flow forces (e.g., flow velocity, shear forces and turbulence) may be better variables to describe 'flow' dynamics (Statzner *et al.*, 1988; Lancaster and Mole, 1999; Lytle and Smith, 2004; and see Figure 10.2). The ability of an individual benthic organism to hold station within the near-bed boundary layer whilst obtaining sufficient resources to maintain itself over several minutes is dependent on its ability to utilise and/or withstand small-scale flow forces (Crowder and Diplas, 2000). At spatial scales greater than habitat patches, it is frequently not possible or appropriate to collect point flow velocity readings and the most widely available 'flow' data is in the form of river discharge readings. This marks a major threshold in terms of both spatio-temporal sampling approaches (see break point in Figure 10.2) and disciplines – between (eco)hydraulics and (eco)hydrology. River discharge (the volume of water passing through a known cross section over a set time period) may have limited relevance to individ-

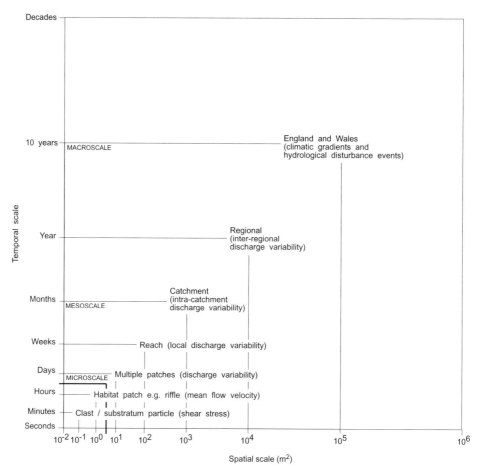

Figure 10.2 Conceptual model of the spatial and temporal scales used to examine the ecological influence of flow variability upon instream ecology indicating the spatial scale of investigation and exemplar ecologically relevant hydrological descriptors. The thick black line represents the 'break point' transition between scales of measurement using flow forces/velocity and river discharge series

ual taxa but provides a robust temporal measure of flow variability. River discharge may influence instream organisms and communities in multiple ways, for example through changes in channel morphology and habitat characteristics associated with flow disturbance, erosion and/or deposition of sediments and transport/delivery of food resources (see Lake, 2003; Matthaei *et al.*, 2003; Rabeni *et al.*, 2005). Events (floods and droughts) with returns periods in excess of 50 years may be the main structuring agents of regional and catchment-scale changes of instream communities (Lytle and Poff, 2004, and see Figure 10.2). More frequent events, such as the annual maximum flood, mean flow or baseflow conditions, may exert a strong control on reach- and habitat-scale instream communities and ecological processes (Armitage and Cannan, 2000).

10.5 River Flow Data: Collection and Analysis

River-flow time series provide the backbone to understanding the impacts of climate and land-use change on hydrology, and patterns of water utilisation (Marsh, 2002). Thus, flow data are vital for planning and management of sustainable water resources, particularly in the context of changing water legislation and increasing stakeholder engagement in water issues (see Petts, Chapter 13 in this book). Once automated hydrometric instrumentation is installed, river gauges constitute (arguably) non-resource intensive networks (Hart and Finelli, 1999). However, in reality, accurate and repeated ratings of river discharge for channel cross sections plus site maintenance remains a time-consuming and costly task. As a consequence, there is a global problem in supporting long-term hydrological data collection. There was a general increase in the number of river discharge stations worldwide during the 1980s related to concerns regarding human health and environmental impacts of pollution/water quality. However, drainage basins in many parts of the world remain ungauged (or poorly gauged), and river gauges have been decommissioned in many remote upland basins to allow focusing of funds on downstream operational gauges near population centres (Rodda, 1998) Hence, the ungauged basin problem remains one of the major challenges to hydrology (Sivapalan et al., 2003).

Thus, fluvial hydrology is strongly dependent upon observed 'flow' data. The most basic and fundamental measures of flow take two forms: (i) velocity – vector quantity equal to the rate that position changes with time (e.g., $m\,s^{-1}$), and (ii) discharge – as defined above (e.g., $m^3\,s^{-1}$). Flow velocity may be measured using various instruments and techniques, including current meters, floats, tracer dilution, ultrasonic and electromagnetic methods. Discharge is estimated using five main approaches: (i) the velocity–area method; (ii) tracer dilution; (iii) structures such as flumes and weirs; (iv) ultrasonic methods, and (v) electromagnetic methods. Further methodological details are provided by Herschy (1999) and Gordon et al. (2004).

In many industrialised nations, discharge data have been recorded since the mid-Twentieth century (e.g., Bergstroem and Carlsson, 1994; Maheshwari et al., 1995). In addition, river discharges have been reconstructed back to the Nineteenth century using historical climatological (rainfall) data (Jones and Lister, 1998; Jones et al., 2006). Some river-discharge time series have been extended even further using proxy palaeoclimatological data (e.g., Lamb, 1977). Palaeo-reconstruction has been used increasingly to improve estimates of flood and drought frequency and severity, for example analysis of sedimentary sequences (see Macklin et al., 2005). Recent research has explored the use of both epigraphic and documentary records to improve estimates of flood frequency (e.g., Macdonald et al., 2006).

Often, field data are digitised and stored in electronic archives to permit ready extraction. After sufficiently long river flow time-series have been collected (e.g., +30 years; WMO, 1966) and quality control checks applied, they may be analysed to detect trends and other patterns in various flow variables, including extremes such as floods and droughts (methods reviewed by Kundzewicz and Robson, 2004). Hydrometric time series should always be screened for quality assurance prior to analyses (Babovic, 2005). The integrity of individual time series needs to be considered by: (i) testing for inhomogeneity where there have been changes in the type, reliability and/or location of instrumentation; and (ii) identifying periods with missing or unreliable (erroneous) data. Missing values and erroneous data are the most common problems associated with hydrological time series. These problems are increased by the growing use of automated instrumentation that is prone to periodic failures, which may go undiscovered between site maintenance visits. However, if data are missing for short periods, it is often possible to interpolate values using mean values, incremental fills, regression (with other stations or variables) or autocorrelation methods. To explore the information contained within river flow time series a range of data mining techniques has been developed (Babovic, 2005).

There is a relatively long tradition of quantifying river flow characteristics within the hydrological sciences using numerous indices at event, monthly, seasonal and annual time scales (e.g., Nestler and Long, 1997; Bower et al., 2004). However, thought ('hydrologic') must be given to the form and spatio-temporal resolution of data for analyses to select the most appropriate and representative information to address the specific research question. To group rivers and identify spatio-temporal data structure, several methods for classification of river flow regime 'types' have been developed. Two techniques have been widely applied to provide the basis for subsequent hydroecological analysis: (i) principal components analysis (PCA); and (ii) cluster analysis (CA) techniques (e.g., Poff, 1996; Thomson et al., 2003). PCA is a data reduction technique that simplifies data structure by linear transformation of data onto orthogonal projections, termed principal components (PCs). The first PC explains the greatest variance within the data set, the second PC accounts for the second greatest variance, and so on. PCA can be employed to reduce dataset dimensionality while retaining those characteristics of the data set that contribute most to its statistical variance, by keeping lower-order principal components and ignoring higher-order ones. CA partitions a data set into groups (clusters), which possess some commonality as defined statistically by proximity according to some defined distance measure. There is a range of clustering algorithms (Bower et al., 2004), which fall into two broad types: (i) agglomerative – begins with each element as a separate cluster and merges them successively, and (ii) divisive – starts with the entire data set and divides it into smaller clusters (Legendre and Legendre, 1998).

Cluster analysis (sometimes using PCA as a precursor to reduce data set dimensionality) may be employed to group rivers with similar river flow characteristics. Figure 10.2 shows the results of classification of river flow regime shape (seasonality) for England and Wales (for further details see Monk et al., 2006). Three groups of rivers were identified representing differences in timing of peak flow (early, intermediate and late) and the rising and falling hydrograph limb (Figure 10.3). The river flow regionalisation can be explained by a climatological gradient (west to east) and differences in groundwater system buffering (major aquifers in the south-east) across England and Wales. In addition, it is possible to classify rivers based on the magnitude of flow (size) and combine the output from both of these independent classifications to identify hydrologically coherent groups (i.e., regions; see Hannah et al., 2000; Hannah et al., 2005). Therefore, river flow classifications may potentially provide a valuable precursor to structure between- and within-region analyses of hydroecological patterns (Monk et al., 2006).

10.6 Ecological Data: Collection and Analysis

The methods for collecting biological samples in river environments are very different to those employed for most hydrometric data, as samples are usually collected manually. Given this human sampling effort, in addition to the significant laboratory effort required for identifying

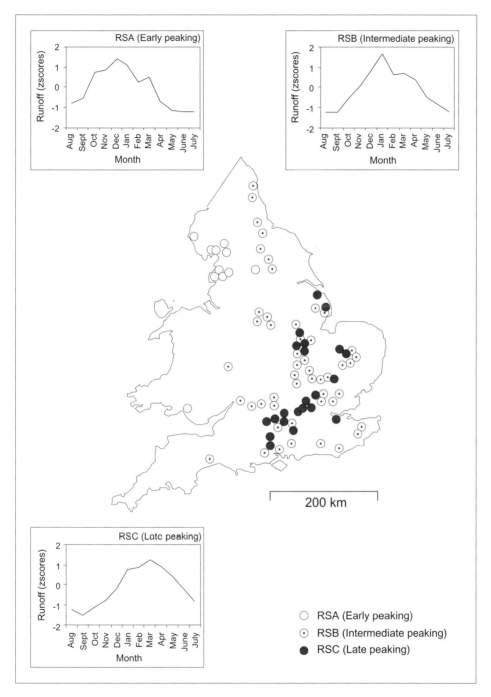

Figure 10.3 Location map of the 83 sites across England and Wales and flow regime shape classes displaying the spatial distribution of the regional shape classes and correspoinding average hydrographs for (a) RSA early peaking – December – January; (b) intermediate peaking – January, and (c) late peaking – March

samples and the lack of experienced taxonomists, very few long-term time series exist with a subseasonal resolution. For hydroecological analysis, careful consideration needs to be given to the nature of the hydrological phenomena to be investigated since this is likely to determine the most appropriate indicator organism(s). A range of biological indicators have been utilised within river systems, with a strong emphasis historically on algae, macroinvertebrates and fish (e.g., Fore and Grafe, 2002; Griffith *et al.*, 2005; Bonada *et al.*, 2006).

The quality of an ecological time-series depends on the accuracy and precision of the sampling process, laboratory procedures and data preparation methods used prior to data analysis. The primary objective of sampling is to collect an unbiased sample, which can be replicated consistently by following a stringent sampling procedure (Murray-Bligh, 1999; Metzeling *et al.*, 2003). Protocols for the assessment of freshwater ecosystems using biological indicators have been developed over several decades (Resh and McElravy, 1993). Samples can be collected quantitatively and semi-quantitatively. Quantitative sampling usually yields data in the form of: (i) counts (total individuals, number of taxa or species counts) per unit area (e.g. individuals per m^2), or (ii) counts per volume (e.g., individuals per litre). Semi-quantitative sampling is normally undertaken as part of routine monitoring programmes using so-called 'rapid' assessment techniques. These rely on a standardised sampling method that is employed to collect a sample over a set period of time, although the areal extent may vary. However, collecting representative samples that reflect the river community and its variability is extremely difficult because the distribution of many assemblages is patchy in space and time (Armitage *et al.*, 1995; Matthaei *et al.*, 2003; Effenberger *et al.*, 2006). In addition, it is particularly important to consider the life cycle of organisms (e.g., timing of emergence of aquatic insects) against the timing of sample collection to ensure community variability is accurately represented.

Sampling techniques and their resultant data quality should be considered before any analysis is undertaken, especially if utilising secondary data. A variety of biases and errors can be introduced into ecological data sets during collection and identification; for example: differences in sampling equipment (e.g., net mesh size), effort (time per unit area), habitat type, and potential operator error in both collection and taxonomic identification (Carter and Resh, 2001). Measuring and controlling data quality has become increasingly important for large-scale ecological studies where records have been compiled from a variety of sources (Clarke *et al.*, 2002). Data should always be assessed for input accuracy because without quality assurance it is difficult to have confidence in any subsequent results (James and McCulloch, 1990; Cao *et al.*, 2003).

Family-level biological data are widespread and although species-level counts may be desirable, it is usually only possible for short-term investigations of river flow effects (see Lancaster, 1999; Olden *et al.*, 2004). Examination of instream community response to long-term environmental variability (i.e., inter-annual) is severely limited by the absence of high-resolution species-level data collected for the specific purpose of examining variability (Jackson and Füreder, 2006). As a result, other secondary data sources, such as routine biomonitoring data, may have to be employed. Although these data may have been collected originally for a different primary purpose, biomonitoring samples may be the only available data source (Cao *et al.*, 2003; Statzner *et al.*, 2005b). Potential pitfalls of such data sets include changes in the taxonomic resolution of data (species/genus-level versus family-level records), the use of total counts and a variety of abundance classes (e.g., presence/absence or log-abundance categories). Where such problems occur, it is necessary to convert all data to the same format (usually the lowest common denominator/resolution) to allow comparison of the greatest number of sites and/or the longest time series possible (Cao et al., 2003). However, it is sometimes possible to analyse subsets of data for a more limited number of sites or shorter time periods to maximise the taxonomic resolution. In some cases, macroinvertebrate data may be reduced to biotic indices (e.g., water quality), family richness and diversity (Clausen and Biggs, 1997, 2000), or biological traits (Statzner et al., 2005b).

10.7 Integration of Hydrological and Ecological Data for Hydroecological Analysis

A range of organisms have been utilised in instream flow investigations with a strong emphasis on economically important fish (see Nestler et al., Chapter 12 in this book). However, for illustrative purposes, this section concentrates on the use of macroinvertebrate communities. Most benthic macroinvertebrate taxa have recognised species-specific flow tolerances and, therefore, they are widely utilised in the examination of flow variability of river ecosystems (Clausen and Biggs, 1997; Wood and Armitage, 2004). Because macroinvertebrates provide a strong indication of environmental conditions, they have been employed in studies at the scale of individual substrate clasts (Olden et al., 2004), habitat characteristics (Sommer et al., 2004), whole river systems (Wood et al., 2001), and national classifications (Clausen and Biggs, 1997; Monk et al., 2006). However, while it is possible to classify hydrological data at large temporal and spatial scales (Snelder et al., 2005; Poff et al., 2006), the availability of spatially comparable long-term benthic macroinvertebrate data is a major limitation to hydroecological analyses (Jackson and Füreder, 2006).

It is important that flow variability investigations should only consider macroinvertebrate data at sites with limited human impact, where water quality has not been a limiting factor, since biotic responses may reflect changes other than flow variability (e.g., organic pollution – Monk et al., 2006). Despite this need for caution, research using macroinvertebrate data originating from biomonitoring programmes has increased understanding of a range of abiotic factors on instream communities (e.g. Cao et al., 2003; Clarke et al., 2003; Statzner et al., 2005b; Monk et al., 2006). Biotic indices have been shown to vary in response to changes associated with groundwater abstraction (Bickerton et al., 1995) and diversity indices have been shown to display clear trends related to both flood and drought conditions (Clausen and Biggs, 2000; Wood and Armitage, 2004).

To date, only one metric has been specifically developed to examine macroinvertebrate community response to flow variability: the Lotic-invertebrate Index for Flow Evaluation (LIFE) (Extence et al., 1999). The LIFE methodology is based upon the sensitivity of river benthic macroinvertebrate species and families to particular flow velocity ranges. The LIFE score is sensitive to both artificial and natural flow changes (Extence et al., 1999; Monk et al., 2006). Using a paired hydroecological dataset of flow records (1980–1999) and macroinvertebrate family-level data (1990–2000) for 83 rivers across England and Wales (Figure 10.3), relationships between the LIFE score and 201 flow variables (as listed by Monk et al., 2006) were examined. When data for all 83 rivers were pooled, a single variable (SMED – median annual flow divided by the basin area) produced the best predictor of the LIFE score (Figure 10.4a). To assess spatial variability in hydroecological interaction, flow regimes were classified (using hierarchical agglomerative cluster analysis with Ward's method) to divide the 83 rivers between three classes/regions (Figure 10.3). Stepwise multiple linear regression models for the three regions indicated that between 14–41% of the variance in the LIFE score could be explained using one or two variables (Figure 10.4 and Table 10.1). It is perhaps not surprising that the hydrological variables (calculated from discharge) cannot explain a large amount of the ecological variation as they represent a large scale filter and an averaging of ecologically important small scale habitat characteristics. However, the predominance of the specific median discharge (SMED) in several models suggests that the size/area of the basin is a particularly important scaling factor that strongly influences ecological response to flow (cf. Biggs *et al.*, 2005). However, when rivers were examined on an individual basis, a wider range of hydrological variables was identified as the 'best' hydroecological predictors of the LIFE score (e.g., Figure 10.4c). This suggests that the classification (subdivision) of rivers into flow regions offers a means of increasing the predictive capacity, but at the scale of the individual river, site-specific characteristics (channel morphology and habitat structure) are important in determining faunal responses (Armitage *et al.*, 2001; Wood *et al.*, 2001; Monk *et al.*, 2006).

10.8 River Flow Variability and Ecological Response: Future Directions and Challenges

This chapter has demonstrated the pressing need for high quality hydrological and ecological series for both pure and applied hydroecological research. However, integrated hydroecological analysis (particularly over extended time periods and wide spatial domains) is not always straightforward or possible due to the markedly different methods of sampling and primary objectives of original data collection. The paucity of long-term ecological datasets is a significant impediment to the understanding of historical changes to instream communities (Jackson and Füreder, 2006), although it may be possible to make significant progress in some instances through the use of biomonitoring data (Monk *et al.*, 2006).

There is clearly a large volume of research currently being undertaken to examine the influence of flow regime variability upon riverine ecosystems. It is imperative that high quality baseline data sets are compiled to ensure that the impact of flow regime changes associated with anthropogenic impacts and climate change/variability can be identified (Ward and Stanford, 1979; Poff, 2002). Recent and ongoing research examining the influence of flow variability has been undertaken at a range of spatial and temporal resolutions. However, few researchers have attempted to bridge scales (e.g., Benda *et al.*, 2004) despite the need to understand the processes driving change at the small scale and manage river basins at large scales. In addition, while the ecological relevance and significance of flow regime variability is widely acknowledged, a number of the 'assumptions' stated in the scientific literature remain to be validated with empirical data.

The results of the bibliographic survey clearly demonstrated that most of the research undertaken to examine the influence of flow variability upon instream communities has

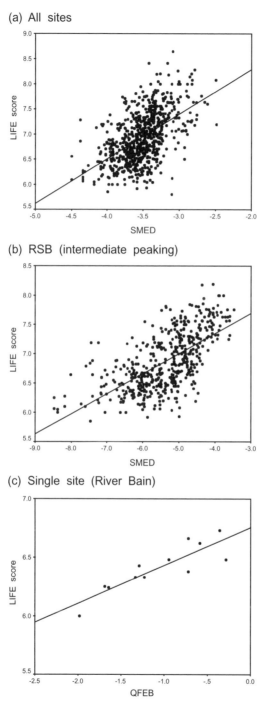

Figure 10.4 Scatter plots of the relationship between the LIFE score and flow regime indices with regression lines for: (a) all sites; (b) regime shape (RSB); and (c) an individual river (River Bain, Lincolnshire). SMED = specific median discharge, QFEB = February discharge

Table 10.1 *Stepwise multiple linear regression models for the LIFE score using hydrological variables for all sites and flow regime shape classes. SMED = specific median discharge, QFEB = February discharge, CVDF = coefficient of variation of daily discharge*

Model	Adjusted R^2	F	Number of rivers (samples)	Predictor variable and sign
(a) All Sites	0.381	442.62***	83 (719)	+SMED
(b) Shape Classes				
RSA	0.300	30.54***	11 (70)	−QFEB
RSB	0.411	334.01***	52 (479)	+SMED
RSC	0.137	14.45***	20 (170)	+SMED − CVDF

*** $p < 0.001$

been published in ecological/biological sciences journals and predominately undertaken by those in bioscience departments/research institutes. Research undertaken at smaller spatial scales and temporal resolutions is dominated by ecologists/biological scientists. These results are in marked contrast to a recent survey of research conducted under the banner of ecohydrology and hydroecology that demonstrated that half over of the research output was undertaken in physical sciences, dominated by geography, earth and environmental science departments (Hannah et al., 2004). Hydroecology/ecohydrology has been described as a new paradigm by some commentators, although others have questioned if it is perceived as anything particularly 'new' (Bond, 2003; Hannah et al., 2004; see also Wood et al., Chapter 1 in this book). This suggests that the label 'hydroecology' or 'ecohydrology' is not necessarily one that everyone recognises or necessarily wishes to subscribe to.

This chapter demonstrates the pressing need for high-quality, long-term, paired hydrological and ecological data sets. Clearly, generation of these important research and management resources (and appropriate methods for their analysis) requires true interdisciplinary collaboration within this rapidly changing research arena. This process has begun already, particularly at larger spatial resolutions, where patterns of ecological response to flow regime variability have been found. However, effective management of riverine ecosystems (at the basin scale) requires the characterisation of physical and biological processes (at the sub-reach scale). Thus, research is needed to examine the patterns and processes along the space–time transition between hydrological (discharge) and hydraulic (velocity) scales. The future challenge for river hydroecological research remains the integration between spatial and temporal scales, and providing understanding of both patterns observed within riverine environments and the underlying processes driving them.

References

Archer, D. and Newson, M. D. (2002). The use of indices of flow variability in assessing the hydrological and instream habitat impacts of upland afforestation and drainage. *Journal of Hydrology*, **268**: 244–258.

Armitage, P. D. and Cannan, C.E. (2000). Annual changes in the summer patterns of mesohabitat distribution and associated faunal assemblages. *Hydrological Processes*, **14**: 3161–3179.

Armitage, P. D., Pardo, I. and Brown, A. (1995). Temporal constancy of faunal assemblages in mesohabitats: application to management. *Archiv für Hydrobiologie*, **133**: 367–387.

Armitage, P. D., Lattmann, K., Kneebone, N. and Harris, I. (2001). Bank profile and structure as determinants of macroinvertebrate assemblages – seasonal changes and management. *Regulated Rivers: Research and Management*, **17**: 543–556.

Babovic, V. (2005). Data mining in hydrology. *Hydrological Processes*, **19**: 1511–1515.

Beckmann, M. C., Scholl, F. and Matthaei, C. D. (2005). Effects of increased flow in the main stem of the River Rhine on the invertebrate communities of its tributaries. *Freshwater Biology*, **50**: 10–26.

Benda, L., Poff, N. L., Miller, D., Dunne, T., Reeves, G., Pess, G. and Pollock, M. (2004). The network dynamics hypothesis: How channel networks structure riverine habitats. *BioScience*, **54**: 413–427.

Benke, A. C., Chaubey, I., Ward, G. M. and Dunn, E. L. (2000). Flood pulse dynamics of an unregulated river floodplain in the southeastern US coastal plain. *Ecology*, **81**: 2730–2741.

Bergstroem, S. and Carlsson, B. (1994). River runoff to the Baltic Sea: 1950–1990. *Ambio*, **23**: 280–287.

Bickerton, M., Petts, G., Armitage, P. D. and Castella, E. (1995). Assessing the ecological effects of groundwater abstraction on chalk streams: Three examples from eastern England. *Regulated Rivers: Research and Management*, **8**: 121–134.

Biggs, B. J. F., Nikora, V. I. and Snelder, T. H. (2005). Linking scales of flow variability to lotic ecosystem structure and function. *River Research and Applications*, **21**: 283–298.

Black, A. R., Bragg, O. M., Duck, R. W. and Rowan, J. S. (2002). Development of a method to assess ecological impact due to hydrological regime alteration of Scottish rivers. *IASH Publication*, **276**: 45–51.

Bonada, N., Prat, N., Resh, V. H. and Statzner, B. (2006). Developments in aquatic insect biomonitoring: a comparative analysis of recent approaches. *Annual Review of Entomology*, **51**: 495–523.

Bond, B. (2003). Hydrology and ecology meet – and the meeting is good. *Hydrological Processes*, **17**: 2087–2089.

Bower, D., Hannah, D. M. and McGregor, G. R. (2004). Techniques for assessing the climatic sensitivity of river flow regimes. *Hydrological Processes*, **18**: 2515–2543.

Brooks, A. J., Haeusler, T., Reinfelds, I. and Williams, S. (2005). Hydraulic microhabitats and the distribution of macroinvertebrate assemblages in riffles. *Freshwater Biology*, **50**: 331–344.

Brosse, S., Arbuckle, C. J. and Townsend, C. R. (2003). Habitat scale and biodiversity: influence of catchment, stream reach and bedform scales on local invertebrate diversity. *Biodiversity and Conservation*, **12**: 2057–2075.

Bunn, S. E. and Arthington, A. H. (2002). Basic principles and ecological consequences of altered flow regimes for aquatic biodiversity. *Environmental Management*, **30**: 492–507.

Cannan, C. E. and Armitage, P. D. (1999). The influence of catchment geology on the longitudinal distribution of macroinvertebrate assemblages in a groundwater dominated river. *Hydrological Processes*, **13**: 355–369.

Cao, Y., Hawkins, C. P. and Vinson, M. R. (2003). Measuring and controlling data quality in biological assemblage surveys with special reference to stream benthic macroinvertebrates. *Freshwater Biology*, **48**: 1898–1911.

Carter, J. L. and Resh, V. H. (2001). After site selection and before data analysis: sampling, sorting, and laboratory procedures used in stream benthic macroinvertebrate monitoring programs by USA state agencies. *Journal of the North American Benthological Society*, **20**: 658–682.

Carter, J. L., Fend, S. V. and Kennelly, S. S. (1996). The relationship among three habitat scales and stream benthic invertebrate community structure. *Freshwater Biology*, **35**: 109–124.

Caruso, B. S. (2002). Temporal and spatial patterns of extreme low flows and effects on stream ecosystems in Otago, New Zealand. *Journal of Hydrology*, **257**: 115–133.

Chaves, M. L., Chainho, P. M., Costa, J. L., Prat, N. and Costa, M. J. (2005). Regional and local environmental factors structuring undisturbed benthic macroinvertebrate communities in the Mondego River basin, Portugal. *Archiv für Hydrobiologie*, **163**: 497–523.

Clarke, R. T., Furse, M. T., Gunn, R. J. M., Winder, J. M. and Wright, J. F. (2002). Sampling variation in macroinvertebrate data and implications for river quality indices. *Freshwater Biology*, **47**: 1735–1751.

Clarke, R. T., Armitage, P. D., Hornby, D., Scarlett, P. and Davy-Bowker, J. (2003). *Investigation of the relationship between the LIFE Index and RIVPACS – Putting the LIFE into RIVPACS*. R & D Technical Report W6–044/TR1, Environment Agency: Bristol, UK.

Clausen, B. and Biggs, B. J. F. (1997). Relationships between benthic biota and hydrological indices in New Zealand streams. *Freshwater Biology*, **38**: 327–342.

Clausen, B. and Biggs, B. J. F. (2000). Flow variables for ecological studies in temperate streams: groupings based on covariance. *Journal of Hydrology*, **237**: 184–197.

Collier, K. J., Wilcock, R. J. and Meredith, A. S. (1998). Influence of substrate type and physico-chemical conditions on macroinvertebrate faunas and biotic indices of some lowland Waikato, New Zealand, streams. *New Zealand Journal of Marine and Freshwater Research*, **32**: 1–19.

Crowder, D. W. and Diplas, P. (2000). Using two-dimensional hydrodynamic models at scales of ecological importance. *Journal of Hydrology*, **230**: 172–191.

Death, R. G. and Joy, M. K. (2004). Invertebrate community structure in streams of the Manawatu–Wanganui region, New Zealand: the roles of catchment versus reach scale influences. *Freshwater Biology*, **49**: 982–997.

Downes, B. J., Lake, P. S. and Schreiber, E. S. G. (1993). Spatial variation in the distribution of stream invertebrates: implications of patchiness for models of community organization. *Freshwater Biology*, **30**: 119–132.

Effenberger, M., Sailer, G., Townsend, C. R. and Matthaei, C. D. (2006). Local disturbance history and habitat parameters influence the microdistribution of stream invertebrates. *Freshwater Biology*, **51**: 312–332.

Extence, C. A., Balbi, D. M. and Chadd, R. P. (1999). River flow indexing using British benthic macroinvertebrates: A framework for setting hydroecological objectives. *Regulated Rivers: Research and Management*, **15**: 543–574.

Fore, L. S. and Grafe, E. (2002). Using diatoms to assess the biological condition of large rivers in Idaho (USA). *Freshwater Biology*, **47**: 2015–2037.

Frissell, C. A., Liss, W. J., Warren, C. E. and Hurley, M. D. (1986). A hierarchical framework for stream habitat classification: viewing streams in a watershed context. *Environmental Management*, **10**: 199–214.

Gordon, N. D., McMahon, T. A., Finlayson, B. L., Gippel, C. J. and Nathan, R. J. (2004). *Stream Hydrology: An Introduction for Ecologists*. John Wiley & Sons, Ltd: Chichester.

Griffith, M. B., Husby, P., Hall, R. K., Kaufmann, P. R. and Hill, B. H. (2003). Analysis of macroinvertebrate assemblages in relation to environmental gradients among lotic habitats of California's Central Valley. *Environmental Monitoring and Assessment*, **82**: 281–309.

Griffith, M. B., Hill, B. H., McCormick, F. H., Kaufmann, P. R., Herlihy, A. T. and Delle, A. R. (2005). Comparative application of indices of biotic integrity based on periphyton, macroinvertebrates and fish to southern Rocky Mountain stream. *Ecological Indicators*, **5**: 117–136.

Gurnell, A. M., Bickerton, M. A., Angold, P., Bell, D., Morrissey, I., Petts, G. E. and Sadler, J. (1998). Morphological and ecological change on a meander bend: the role of hydrological processes and the application of GIS. *Hydrological Processes*, **12**: 981–993.

Habersack, H. (2000). The river-scaling concept (RSC): a basis for ecological assessments. *Hydrobiologia*, **422/423**: 49–60.

Hannah, D. M., Smith, B. P. G., Gurnell, A. M. and McGregor, G. R. (2000). An approach to hydrograph classification. *Hydrological Processes*, **14**: 317–338.

Hannah, D. M., Wood, P. J. and Sadler, J. P. (2004). Ecohydrology and hydroecology: A 'new paradigm'? *Hydrological Processes*, **18**: 3439–3445.

Hannah, D. M., Kansakar, S. R., Gerrard, A. J. and Rees, G. (2005). Flow regimes of Himalayan rivers of Nepal: nature and spatial patterns. *Journal of Hydrology*, **308**: 18–32.

Hart, D. D. and Finelli, C. M. (1999). Physical–biological coupling in streams: the pervasive effects of flow on benthic organisms. *Annual Review of Ecology and Systematics*, **30**: 363–395.

Heino, J., Parviainen, J., Paavola, R., Jehle, M., Louhi, P. and Muotka, T. (2005). Characterizing macroinvertebrate assemblage structure in relation to stream size and tributary position. *Hydrobiologia*, **539**: 121–130.

Herschy, R. W. (1999). *Hydrometry: Principles and Practices*. 2nd Edition. John Wiley & Sons, Ltd: Chichester.

Hughes, J. M. R. and James, B. (1989). A hydrological regionalization of streams in Victoria, Australia, with implications for stream ecology. *Australian Journal of Marine and Freshwater Research*, **40**: 303–326.

Huh, S. H., Dickey, D. A., Meador, M. R. and Ruhl, K. E. (2005). Temporal analysis of the frequency and duration of low and high streamflow: years of record needed to characterize streamflow variability. *Journal of Hydrology*, **310**: 78–94.

Jackson, J. K. and Füreder, L. (2006). Long-term studies of freshwater macroinvertebrates: a review of the frequency, duration and ecological significance. *Freshwater Biology*, **51**: 591–603.

James, F. C. and McCulloch, C. E. (1990). Multivariate analysis in ecology and systematics: Panacea or Pandora's box? *Annual Review of Ecology and Systematics*, **21**: 129–168.

Jones, P. D. and Lister, D. H. (1998). Riverflow reconstructions for 15 catchments over England and Wales and an assessment of hydrologic drought since 1865. *International Journal of Climatology*, **18**: 999–1013.

Jones, P. D., Lister, D. H., Wilby, R. L. and Kostopoulou, E. (2006). Extended riverflow reconstruction for England and Wales. *International Journal of Climatology*, **26**: 219–231.

Jowett, I. G. and Duncan, M. J. (1990). Flow variability in New Zealand rivers and its relationship to in-stream habitat and biota. *New Zealand Journal of Marine and Freshwater Research*, **24**: 305–317.

Kundzewicz, Z. W. and Robson, A. J. (2004). Change detection in hydrological records – a review of the methodology. *Hydrological Sciences Journal*, **49**: 7–19.

Lake, P. S. (2000). Disturbance, patchiness and diversity in streams. *Journal of the North American Benthological Society*, **19**: 573–592.

Lake, P. S. (2003). Ecological effects of perturbation by drought in flowing waters. *Freshwater Biology*, **48**: 1161–1172.

Lamb, H. H. (1977). *Climate: Present, Past and Future*. Volume 2. *Climatic History and the Future*. Methuen: London.

Lancaster, J. (1999). Small-scale movements of lotic macroinvertebrates with variations in flow. *Freshwater Biology*, **41**: 605–620.

Lancaster, J. and Mole, A. (1999). Interactive effects of near-bed flow and substratum texture on the microdistribution of lotic macroinvertebrates. *Archiv für Hydrobiologie*, **146**: 83–100.

Ledger, M. E. and Hildrew, A. G. (2001). Recolonization by the benthos of an acid stream following a drought. *Archiv für Hydrobiologie*, **152**: 1–17.

Legendre, P. and Legendre, L. (1998). *Numerical Ecology*. Elsevier: Amsterdam.

Levin, S. A. (1992). The problem of pattern and scale in ecology. *Ecology*, **73**: 1943–1967.

Li, R. Y. and Gelwick, F. P. (2005). The relationship of environmental factors to spatial and temporal variations of fish assemblages in a floodplain river in Texas, USA. *Ecology of Freshwater Fish*, **14**: 319–330.

Lytle, D. A. and Merritt, D. M. (2004). Hydrologic regimes and riparian forests: a structured population model for cottonwood. *Ecology*, **85**: 2493–2503.

Lytle, D. A. and Poff, N. L. (2004). Adaptation to natural flow regimes. *Trends in Ecology and Evolution*, **19**: 94–100.

Lytle, D. A. and Smith, R. L. (2004). Exaptation and flash flood escape in the Giant Water Bugs. *Journal of Insect Behavior*, **17**: 169–178.

Macdonald, N., Werritty, A., Black, A. R. and McEwen, L. J. (2006). Historical and pooled flood frequency analysis for the River Tay at Perth, Scotland. *Area*, **38**: 34–46.

Macklin, M. G., Johnstone, E. and Lewin, J. (2005). Pervasive and long-term forcing of Holocene river instability and flooding in Great Britain by centennial-scale climate change. *Holocene*, **15**: 15–38.

Maheshwari, B. L., Walker, K. F. and McMahon, T. A. (1995). Effects of regulation on the flow regimes of the River Murray, Australia. *Regulated Rivers: Research and Management*, **10**: 15–38.

Marsh, T. J. (2002). Capitalising on river flow data to meet changing national needs – a UK perspective. *Flow Measurement and Instrumentation*, **13**: 291–298.

Matthaei, C. D., Guggelberger, C. and Huber, H. (2003). Local disturbance history affects patchiness of benthic river algae. *Freshwater Biology*, **48**: 1514–1526.

Merigoux, S. and Doledec, S. (2004). Hydraulic requirements of stream communities: a case study on invertebrates. *Freshwater Biology*, **49**: 600–613.

Metzeling, L., Chessman, B. C., Hardwick, R. and Wong, V. (2003). Rapid assessment of rivers using macroinvertebrates: the role of experience, and comparisons with quantitative methods. *Hydrobiologia*, **510**: 39–52.

Molnar, P., Burlando, P. and Ruf, W. (2002). Integrated catchment assessment of riverine landscape dynamics. *Aquatic Sciences*, **64**: 129–140.

Monk, W. A., Wood, P. J., Hannah, D. M., Wilson, D. A., Extence, C. A. and Chadd, R. P. (2006). Flow variability and macroinvertebrate community response within riverine systems in England and Wales. *River Research and Applications*, **22**: 595–615.

Munne, A. and Prat, N. (2004). Defining river types in a Mediterranean area: a methodology for the implementation of the EU water framework directive. *Environmental Management*, **34**: 711–729.

Murray-Bligh, J. (1999). *Procedures for Collecting and Analysing Macroinvertebrate Samples – BT001*. The Environment Agency. Bristol, p. 176.

Nestler, J. M. and Long, K. S. (1997). Development of hydrological indices to aid cumulative impact analysis of riverine wetlands. *Regulated Rivers: Research and Management*, **13**: 317–334.

Olden, J. D. and Poff, N. L. (2003). Redundancy and the choice of hydrologic indices for characterizing streamflow regimes. *River Research and Applications*, **19**: 101–121.

Olden, J. D., Hoffman, A. L., Monroe, J. B. and Poff, N. L. (2004). Movement behaviour and dynamics of an aquatic insect in a stream benthic landscape. *Canadian Journal of Zoology*, **82**: 1135–1146.

Palmer, M. A., Swan, C. M., Nelson, K., Silver, P. and Alvestad, R. (2000). Streambed landscapes: evidence that stream invertebrates respond to the type and spatial arrangement of patches. *Landscape Ecology*, **15**: 563–576.

Pegg, M. A. and Pierce, C. L. (2002). Classification of reaches in the Missouri and lower Yellowstone rivers based on flow characteristics. *River Research and Applications*, **18**: 31–42.

Poff, N. L. (1996). A hydrogeography of unregulated streams in the United States and an examination of scale-dependence in some hydrological descriptors. *Freshwater Biology*, **36**: 71–91.

Poff, N. L. (1997). Landscape filters and species traits: towards mechanistic understanding and prediction in stream ecology. *Journal of the North American Benthological Society*, **16**: 391–409.

Poff, N. L. (2002). Ecological response to and management of increased flooding caused by climate change. *Philosophical Transactions of the Royal Society of London A*, **360**: 1497–1510.

Poff, N. L., Allan, J. D., Bain, M. B., Karr, J. R., Prestegaard, K. L., Richter, B. D., Sparks, R. E. and Stromberg, J. C. (1997). The natural flow regime: a paradigm for river conservation and restoration. *BioScience*, **47**: 769–784.

Poff, N. L., Olden, J. D., Pepin, D. M. and Bledsoe, B. P. (2006). Placing glocal stream flow variability in geographic and geomorphic contexts. *River Research and Applications*, **22**: 149–166.

Poole, G. C. (2002). Fluvial landscape ecology: addressing uniqueness within the river discontinuum. *Freshwater Biology*, **47**: 641–660.

Pringle, C. (2003). What is hydrologic connectivity and why is it ecologically important? *Hydrological Processes*, **17**: 2685–2689.

Puckridge, J. T., Sheldon, F., Walker, K. F. and Boulton, A. J. (1998). Flow variability and the ecology of large rivers. *Marine and Freshwater Research*, **49**: 55–72.

Rabeni, C. F., Doisy, K. E. and Zweig, L. D. (2005). Stream invertebrate community functional responses to deposited sediment. *Aquatic Sciences*, **67**: 395–402.
Resh, V. H. and McElravy, E. P. (1993). Contemporary quantitative approaches to biomonitoring using benthic macroinvertebrates. In D. M. Rosenberg and V. H. Resh (Eds). *Freshwater Biomonitoring and Benthic Macroinvertebrates*. Routledge, Chapman and Hall Inc.: New York, pp. 159–194.
Richter, B. D., Baumgartner, J. V., Powell, J. and Braun, D. P. (1996). A method for assessing hydrologic alteration within ecosystems. *Conservation Biology*, **10**: 1163–1174.
Richter, B. D., Baumgartner, J. V., Wigington, R. and Braun, D. P. (1997). How much water does a river need? *Freshwater Biology*, **37**: 231–249.
Rodda, J. C. (1998). Hydrological networks need improving! *Proceedings of the International Conference on World Water Resources at the Beginning of the 21st Century*. IHP-V Technical Documents in Hydrology. UNESCO, Paris. **18**: 91–102.
Sanz, D. B. and del Jalon, D. G. (2005). Characterisation of streamflow regimes in central Spain, based on relevant hydrobiological parameters. *Journal of Hydrology*, **310**: 266–279.
Scarsbrook, M. R. (2005). Persistence and stability of lotic invertebrate communities in New Zealand. *Freshwater Biology*, **47**: 417–431.
Sivapalan, M., Takeuchi, K., Franks, S. W., Gupta, V. K., Karambiri, H., Lakshmi, V., Liang, X., McDonnell, J. J., Mendiondo, E. M., O'Connell, P. E., Oki, T., Pomeroy, J. W., Schertzer, D., Uhlenbrook, S. and Zehe, E. (2003). IAHS decade on Predictions in Ungauged Basins (PUB), 2003–2012: Shaping an exciting future for the hydrological sciences. *Hydrological Sciences Journal*, **48**: 857–880.
Smakhtin, V. U. (2001). Low flow hydrology: a review. *Journal of Hydrology*, **240**: 147–186.
Snelder, T. H. and Biggs, B. J. F. (2002). Multiscale river environment classification for water resources management. *Journal of the American Water Resources Association*, **38**: 1225–1239.
Snelder, T. H., Biggs, B. J. F. and Woods, R. H. (2005). Improved eco-hydrological classification of rivers. *River Research and Applications*, **21**: 609–628.
Sommer, T. R., Harrell, W. C., Solger, A. M., Tom, B. and Kimmerer, W. (2004). Effects of flow variation on channel and floodplain biota and habitats of the Sacramento River, California, USA. *Aquatic Conservation: Marine and Freshwater Ecosystems*, **14**: 247–261.
Statzner, B. and Holm, T. F. (1989). Morphological adaptations of benthic invertebrates to stream flow – an old question studied by means of a new technique (Laser Doppler Anemometry). *Oecologia*, **53**: 290–292.
Statzner, B., Gore, J. A. and Resh, V. H. (1988). Hydraulic stream ecology – observed patterns and potential applications. *Journal of the North American Benthological Society*, **7**: 307–360.
Statzner, B., Merigoux, S. and Leichtfried, M. (2005a). Mineral grains in caddisfly pupal cases and streambed sediments: resource use and its limitation through conflicting resource requirements. *Limnology and Oceanography*, **50**: 713–721.
Statzner, B., Bady, P., Dolédec, S. and Schöll, F. (2005b). Invertebrate traits for biomonitoring of large European rivers: an intial assessment of trait patterns in least impacted river reaches. *Freshwater Biology*, **50**: 2136–2161.
Stewardson, M. J. and Gippel, C. J. (2003). Incorporating flow variability into environmental flow regimes using the flow events method. *River Research and Applications*, **19**: 459–472.
Taylor, B. W., McIntosh, A. R. and Peckarsky, B. L. (2002). Reach-scale manipulations show invertebrate grazers depress algal resources in streams. *Limnology and Oceanography*, **47**: 893–899.
Tharme, R. E. (2003). A global perspective on environmental flow assessment: emerging trends in the development and application of environmental flow methodologies for rivers. *River Research and Applications*, **19**: 397–441.
Thomson, J. R., Taylor, M. P. and Brierley, G. J. (2003). Are river styles ecologically meaningful? A test of the ecological significance of a geomorphic river characterization scheme. *Aquatic Conservation: Marine and Freshwater Ecosystems*, **14**: 25–48.
Thorpe, J. H., Thoms, M. C. and Delong, M. D. (2006). The riverine ecosystem synthesis: biocomplexity in river networks across space and time. *River Research and Applications*, **22**: 123–147.

Toner, M. and Keddy, P. (1997). River hydrology and riparian wetlands: a predictive model for ecological assembly. *Ecological Applications*, **7**: 236–246.

Ward, J. V. and Stanford, J. A. (Eds) (1979). *The Ecology of Regulated Streams*. Plenum: New York.

Weilguni, H. and Humpesch, U. H. (1999). Long-term trends of physical, chemical and biological variables in the River Danube 1957–1995: A statistical approach. *Aquatic Sciences*, **61**: 234–259.

WMO (1966). *Climate Change, Technical Note No. 79, WMO No. 195-TP-100*. World Meteorological Organisation. Geneva, p. 53.

Wood, P. J., Hannah, D. M., Agnew, M. D. and Petts, G. E. (2001). Scales of hydroecological variability within a groundwater-dominated stream. *Regulated Rivers: Research and Management*, **17**: 347–367.

Wood, P. J. and Armitage, P. D. (2004). The response of the macroinvertebrate community to low-flow variability and supra-seasonal drought within a groundwater dominated stream. *Archiv für Hydrobiologie*, **161**: 1–20.

Wright, J. F., Clarke, R. T., Gunn, R. J. M., Kneebone, N. T. and Davy-Bowker, J. (2004). Impact of major changes in flow regime on macroinvertebrate assemblages of four chalk stream sites, 1997–2001. *River Research and Applications*, **20**: 775–794.

11

High Resolution Remote Sensing for Understanding Instream Habitat

Stuart N. Lane and Patrice E. Carbonneau

11.1 Introduction

This chapter is concerned with the use of high resolution remotely sensed data for understanding instream habitat. The focus is primarily on parameters commonly labelled as first order (e.g., Petts, 1984): depth and substrate. Remote sensing of higher order parameters (e.g., water temperature) is becoming better established (see Torgersen *et al.*, 1999, 2001; Mertes, 2002; Whited *et al.*, 2002; Kay *et al.*, 2005), but this is beyond the chapter's scope. First, we reflect upon why it is important to develop high resolution remote sensing in relation to understanding instream habitat and focus upon the issue of scale in relation to the estimation of habitat suitability. We show that the range of scales that is relevant to habitat estimation extends from small scales (sub-metre) to much larger (tens of km) scales and this requires a commensurate development of technologies that can measure this scale range. Two main sections follow: remote sensing of (i) depth (and by implication channel morphology); and (ii) substrate. For each methodology identified, we introduce the technique, identify key literature, and illustrate each technique with a specific example.

11.2 Scale, the Grain of Instream Habitat and the Need for Remotely Sensed Data

The ecologically available habitat in a river depends on several key parameters, notably flow velocity and depth, wetted perimeter, substrate, water temperature and pH (see Elso and Giller, 2001; Maddock *et al.*, 2001; Leclerc, 2005). Traditionally, emphasis has been placed upon hydraulic variables, notably velocity, depth and wetted perimeter. This is for two reasons: (i) they are important controls on organism metabolism, both directly and indirectly, controlling the balance between access to food and expenditure on swimming; and (ii) the spatial and temporal changes in higher order parameters (such as pH and temperature) should track changes in velocity and depth to some extent. It is not surprising that width, depth and velocity are commonly incorporated into some form of habitat score such as PHABSIM (see Milhous *et al.*, 1984), which can be used to determine habitat suitability in situations where flow is uniform, or approximately uniform (Milhous *et al.*, 1989). PHABSIM has a quasi two-dimensional hydraulic treatment, using a basic flow division method to apportion the wetted perimeter into different habitat suitabilities. The PHABSIM approach has been questioned, notably for smaller streams or where spatial variation in habitat is high (e.g., Leclerc *et al.*, 1995; Ghanem *et al.*, 1996; Crowder and Diplas, 2002).

The main challenge for hydraulic habitat analyses like PHABSIM is that they do not represent the within-reach scale variability in habitat availability. In conceptual terms, by including a basic division of flow, PHABSIM is recognising that habitat studies based upon setting instream *flow* requirements alone will be insufficient. How a given flow is manifest in terms of available instream habitat will be conditioned by slope (strictly of the energy line), channel geometry and bed roughness. The importance of channel geometry is shown for a gravel patch of 0.70 m by 0.20 m (Figure 11.1a), based upon a flume dataset in which digital photogrammetry was used to generate 0.002 m digital elevation models precise to ±0.001 m using two media methods (Butler *et al.*, 1998, 2002). Figure 11.1b shows the predictions of depth and depth-averaged velocity for a flow of approximately $0.07\,m^3\,s^{-1}$ obtained using a three-dimensional hydraulic model (Lane *et al.*, 2004a; Hardy *et al.*, 2005). In terms of field survey, a 0.70 m by 0.20 m patch is considerably smaller than would be considered in the application of a model like PHABSIM. In using a cumulative excedance, the distributions in Figure 11.1b do not address the critical question of what the minimum spatial scale of depth and velocity must be for a part of the flow domain to be usable. However, they do demonstrate the very substantial variability in habitat relevant parameters over very small spatial scales. This spatial variability will also change as a function of discharge, such that optimum habitat locations will change as a function of time.

In relation to instream ecology, a spatial and temporal patchiness in habitat suitability will result and this has been shown to matter. For instance, in relation to salmonids, small scale patchiness has been identified as important at almost all life stages, including spawning (e.g., Bardonnet and Baglinière, 2000), alevin emergence (e.g., Rubin, 1998), juveniles (e.g., Lonzarich *et al.*, 2000) and Kelt survival (e.g., Cunjak *et al.*, 1998). For these reasons, salmonids, along with other instream organisms, have been linked to local scale habitat patchiness and aspects of population dynamics emergent in fisheries populations at larger scales are likely to be strongly dependent upon smaller scale processes (Lammert and Allen, 1999; Folt *et al.*, 1998).

There is now an increasing incorporation of system patchiness into habitat simulation models (see Leclerc, 2005). Much of this is based upon a move to two-dimensional habitat modelling. This is logical as organisms, fish in particular, may be relatively mobile and hence move over a range of spatial scales. Examples include Leclerc *et al.*'s (1995, 1996) finite element modelling of habitat suitability; Tiffan *et al.*'s (2002) simulation of flow

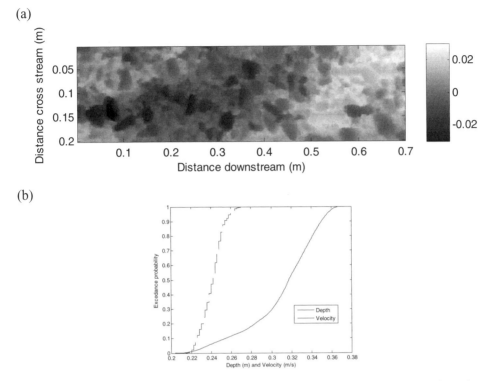

Figure 11.1 A digital elevation model of a patch of gravel (a, shown as deviations from the surface mean) and the associated distributions of depth and velocity (b), the latter determined by depth-averaging model predictions from a three-dimensional numerical model (Lane et al., 2004a)

depths and depth-averaged velocities at different steady state discharges; and habitat probabilistic modelling by Guay *et al.* (2000). Crowder and Diplas (2000) used hydraulic model output to determine energy gradients and 'velocity shelters'/ refugia in gravel-bed streams and showed, by linking these to observed fish behaviour (Crowder and Diplas, 2002) that boulders enhanced the potential availability of the right habitat.

Central to the habitat modelling described above is a minimum scale of variability on flow and depth that must be determined in order to determine habitat suitability. This corresponds with the minimum size of patch that an organism can make use of at a given point in its life cycle. At the same time, analysis must capture the range over which an organism might migrate. This is not only a substantive issue, in relation to what instream organisms need but also a methodological challenge. For instance, both the two-dimensional hydraulic models and the Crowder and Diplas analyses described above are based upon evaluations of gradients in depth and/or velocity. Their evaluation will depend upon the spatial scale over which calculations are made, and this will need to be evaluated in relation to the spatial resolution of the data available to populate biological models. The analysis in Figure 11.1 is underpinned by high resolution measurement of channel geometry to the point that variability normally described as roughness (i.e., the effects of grains and grain organisations) is being measured explicitly as topography. There is no sense in which such data could be routinely used in habitat suitability assessment as a result of the effort that would be required to acquire and to manage the topographic information over the spatial scales of relevance to instream organisms. In practice, the collection of field data on river bed geometry inevitably involves a trade off between (Lane, 1998, 2000): (i) improving the resolution of data collection to allow measurement of smaller scales of topographic variability; and (ii) increasing the spatial extent of the study. This mirrors the challenge for instream habitat assessment. Remote sensing is one approach to addressing this challenge as it offers the possibility of measurement over a very wide range of spatial scales. Whilst the last 15 years have demonstrated the potential for remote sensing in the measurement of river topography in general (see Lane, 2000, for review), it is only more recently that this has been extended to include instream habitat assessment. In this chapter, we focus upon how developments in remote sensing are facilitating the acquisition of distributed data on parameters such as depth and morphology and substrate information, over a range of spatial scales, ranging from the grain (the spatial patch occupied by an organism) through to river reach and tributary scales.

11.3 Depth and Morphology

In this chapter, we combine consideration of depth and morphology, primarily because determination of depth is often a component that allows eventual determination of morphology. Using depth as a surrogate for morphology is not necessarily appropriate as water surface slopes are rarely zero, and this means that a depth map may look different to an elevation map and also vary as stage changes. However, for application to hydraulic models, in order to determine the two dimensional patterns of velocity and ultimately other parameters like water temperature and substrate, it is necessary to have bed elevations. We illustrate below how depth information may be combined with known water surface information in order to acquire morphological information. There are three primary methods for acquiring high resolution depth and morphological data for the purpose of understanding instream habitat: (i) image processing; (ii) photogrammetry; and (iii) laser scanning.

11.3.1 Image Processing

Principles The use of remotely sensed pixel information to map water depth in fluvial environments by the application of calibrated depth–pixel relationships is well documented. This has used multispectral (e.g., Lyon *et al.*, 1992; Marcus, 2002; Whited *et al.*, 2002; Winterbottom and Gilvear, 1997), colour (e.g., Westaway *et al.*, 2003) and greyscale (e.g., Winterbottom and Gilvear, 1997) data sources. All of these approaches rely upon the same theoretical basis: if a beam of light with an incoming intensity I_{in} passes through a transparent medium of thickness x, the remaining outgoing intensity I_{out} can be written as:

$$I_{out} = I_{in} e^{-cx} \tag{11.1}$$

where c is the rate of absorption of the medium, which varies according to properties of the medium such as turbidity and the frequency of the incident light. In the case of imagery, the outgoing intensity is captured as brightness levels and, as a result of the exponential dependence of depth, results in shallow submerged areas having a brighter colour than deep submerged areas, all other things being equal.

Application of Equation (11.1) is crucially dependent upon knowledge of c and I_{in} and there are two broad approaches to their parameterisation: (i) empirical methods, which combine geolocated field measurements with rectified measurements of brightness at the same locations on the image, in order to determine c and I_{in} (e.g., Marcus, 2002; Winterbottom and Gilvear, 1997; Westaway et al., 2003); and (ii) physically based approaches in which sampled properties of the water column are used to model c (e.g., Lyon et al., 1992; Legleiter et al., 2004; Voss et al., 2003).

Strictly, Equation (11.1) does not hold in the presence of a reflective bottom. The presence of a reflective bottom acts as a second light source and the upwelling radiation is not attenuated at the same rate as the downwelling radiation (Legleiter et al., 2004). Rigorous treatment of light attenuation as a function of depth requires a complete radiative transfer model that takes into account all direct and scattered sources of radiation. In turn, this rests upon determination of parameters that require auxiliary assumptions. Notably, field measurements of the spectral reflectance of the bottom are required. Technical advances in the area of field spectroscopy are making this more feasible, notably in coastal waters (Carter et al., 2003; Zhang et al., 2003). With these additional parameters and models, the quality of the results that can be obtained is excellent (Lyon et al., 1992; Legleiter et al., 2004).

Given the parameter dependence of these fully rigorous treatments, empirical approaches based upon inference of the unknowns in Equation (11.1) using regression analysis have been more widely adopted. The primary control upon the successful application of image processing to depth measurement is the type of imagery available. Generally, studies using multispectral data produce the best results. For example, Winterbottom and Gilvear (1997) compared multispectral and grey-scale imagery, obtaining a calibration fit (level of explained variance in depth measurements) of 67 % in the multispectral case and 55 % in the grey scale case. However, standard colour imagery can retain certain advantages over multispectral imagery. The cost of a multispectral image survey is commonly greater than that of a standard colour equivalent (Roberts and Anderson, 1999). Furthermore, standard colour imagery can attain centimetre scale resolution whilst multispectral imaging sensors are still generally limited to meter scale resolution (Carbonneau et al., 2004, 2005).

The main remaining challenge is in handling the effects of spatial and/or temporal illumination condition and bottom reflectance. Commonly, base illumination will vary between images within a survey as well as between surveys. This means that either calibration information will be required for each image and/or each survey, or a correction must be applied that rescales the spectrum of brightness values for each image to those of the image(s) with calibration data. Carbonneau et al. (2006) developed and evaluated an automated feature-based algorithm for this purpose and this is described below. Unfortunately spatial variability in bottom reflectance remains much more of a challenge. Bottom reflectance will vary with grain size, periphyton cover and potentially grain colour in situations where there are differences in geology. The spatial structure of these variables is generally not known *a priori*. Spatial variation in bottom reflectance will commonly be confused with spatial variability in water column absorption of light. Thus, even if calibration relationships can be developed for different bottom reflectance scenarios, implementation of such relationships is problematic, which emphasises one of the fundamental assumptions and limitations of an empirically based depth mapping approach.

Example application: depth mapping in an 80 km salmon spawning river As stated earlier, the application of depth mapping to high resolution imagery of fluvial environments brings additional difficulties. In particular, if centimetre resolution airborne imagery is collected at the catchment scale, this necessarily implies that a very large number of photographs, hundreds to thousands, will be required to cover the whole study area. In such cases, the use of digital images to measure light intensity values can be problematic (Fonstad and Marcus, 2005), and aperture and exposure times need particular consideration as these factors control the conversion of actual light intensity in the field to digital image numbers. Almost always, there will be spatial variability in illumination due to lighting effects rather than depth changes.

Carbonneau *et al.* (2006) used features in the image to correct for between-image illumination differences. The use of image classification data that distinguishes dry, wet and vegetated areas, can allow for an automated identification of the wet/dry interface. If we assume that wet pixels immediately adjacent to the dry area are at near-zero depth, we can use the brightness levels of these pixels to equalise the illumination levels in the image data set and correct the depth mapping results. Since this process can be automated, it can allow for the large scale production of depth information. Carbonneau *et al.* (2006) applied this approach to the 80-km main branch of the Sainte-Marguerite, a gravel bed salmon river in Quebec, Canada. This resulted in 2092 images, each having a ground resolution of 3 cm. Ground truth was secured with 1500 GPS measurements taken in the wadable area of the flow. The ground truth data and the feature based illumination correction method were then used to derive bathymetric maps (see Figure 11.2) with a final resolution of $4 m^2$. The final precision of the process was evaluated at ± 15 cm which shows that empirical methods yield results with an inherent amount of noise. Figure 11.3 shows a depth distribution profile for the 40 km upstream half of the St-Marguerite river: each vertical shows the distribution of depths at each 2-m location down the river. Structures in the depth profile remain clearly visible such as persistent shallow reaches (blue) alternating with deep reaches (red) associated to the riffle–pool sequence. Such high resolution, continuous and large scale measurements of flow depth can only be achieved with remote sensing methods.

11.3.2 Photogrammetry

Principles Photogrammetry has now been used in a number of river measurement studies (e.g., Barker *et al.*, 1997; Butler *et al.*, 1998, 2002; Carbonneau *et al.*, 2003; Dixon *et al.*, 1998; Heritage *et al.*, 1998; Lane *et al.*, 1994, 2003, 2004b; Lane, 1998, 2000; Westaway *et al.*, 2000, 2001, 2003). Much of this follows from developments in digital photogrammetry that have made it possible to automate data collection fully through digital stereomatching. In turn, this has reduced the dependence of photogrammetry upon

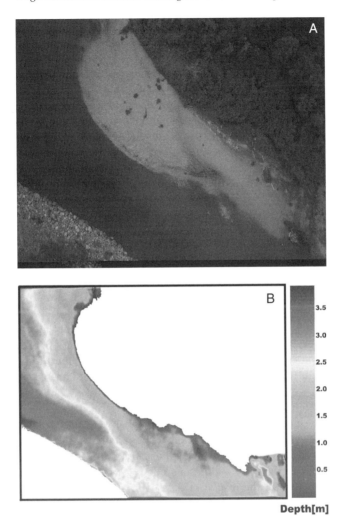

Figure 11.2 (See also colour plates) A sample image used for both automated depth mapping and grain size mapping (A), image resolution is 3 cm and scale is 1:350, and the depth image (B)

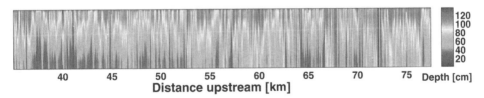

Figure 11.3 (See also colour plates) Distribution profile of depth values for the upper 40 km of the main branch of the St-Marguerite river, Quebec, Canada. The profile, displayed as an image, is composed of 4000 concatenated cumulative distributions of the depth values for each successive 10 m reach of the river. Distance upstream is given by the x axis. A preponderance of blue indicates shallow areas with actual depth values given by the colour scale

expensive hardware, making it a more viable option for many applications. Lane et al. (1993) summarise the basic principles of photogrammetry. In summary, two images of known position and geometry and with overlapping coverage of the study area of interest are digitised. Central to the digitising is the matching of a corresponding (or homologous) point pair, one point on each image, on the basis of local image properties. The latter may be areas with similar patterns of brightness (area-based matching) or features identifiable on both images (feature-based matching). Imagery can either be acquired using digital sensors, or by scanning of conventional film-based imagery using a photogrammetric standard scanner. Digital sensors are cheap to operate and yield the required information quickly, but may limit spatial resolution due to limits on the maximum size of the sensing array used to store data and may also require greater levels of camera calibration in order to obtain reliable data. Regardless of data source, these methods can allow rapid generation of high resolution digital elevation models.

Photogrammetry has one fundamental advantage over many survey techniques. There is a one-to-one correspondence between image pixel dimensions (which is a function of scale, defined as the ratio of focal length to the distance of the sensor from the object of interest) and the precision of derived surface elevations. The best possible spatial resolution is about five-times the pixel dimension. Thus, a survey can be designed to deliver a chosen resolution and precision through acquiring imagery at the right scale. Thus, it is not surprising that photogrammetric applications to the measurement of river bed surfaces have included high resolution laboratory-based measurement (e.g., Butler et al., 1998, 2002; Lane et al., 2002), close range field-based measurement (e.g., Lane et al., 1994; Barker et al., 1997) and airborne measurement (e.g., Westaway et al., 2000, 2001, 2003; Lane et al. 2003, 2004b). High resolution spaceborne platforms are unlikely to yield the imagery that is required for dealing with the relatively low relative relief of river surfaces.

The application of digital photogrammetry has been extensively described in the references given above. In summary, the stages in a photogrammetric analysis are: (i) project design in which sensor options are evaluated and used to identify the appropriate scale of data collection, both imagery and ground control; (ii), if the sensor is not of known position and orientation at the instant of data collection, acquisition of ground control using either specially laid out targets or recovery of ground control from information visible on the imagery (e.g., buildings or road markings); (iii) determination of sensor position and orientation using information in (ii), and extending the analysis to sensor calibration if sensors with unknown geometric properties have been used; (iv) identification of homologous point pairs, almost invariably using some form of stereomatching algorithm, to determine elevations; and (v) managing the individual data points that arise to control possible point error. The latter is particularly important as a result of the transition to using automated identification of homologous point pairs using stereomatching. Manual identification of point pairs inevitably involves a degree of user evaluation of the success or otherwise of a particular match. Manual identification may involve relatively subjective user decisions over exactly where to place a data point and this has been shown to impact upon surface representation (Lane, 1998). Automated identification can lead to many points that are not correctly matched, with associated error resulting. Applications of photogrammetry to rivers have shown that digital methods require very careful post-processing to manage these errors; a review of the methods available for handling error in digital photogrammetric morphological data is provided by Lane et al. (2004a).

One major issue dominates all photogrammetric measurements of rivers: handling inundated zones. In some situations, identification of homologous point pairs may be possible as the water is sufficiently clear. This allows application of two media photogrammetric methods, which have been developed for both close range (e.g., Butler *et al.*, 2002) and airborne (e.g., Westaway *et al.*, 2000, 2001) applications. It uses a refraction correction based on methods developed during the 1970s and 1980s for analytical photogrammetric applications (e.g., Harris and Umbach, 1972; Fryer, 1983). This correction requires knowledge of the elevation of the water surface at which refraction occurs. In the close range applications, this has commonly been obtained through using a perspex viewing tank (e.g., Butler *et al.*, 2002), which also reduces water surface waviness that can cause a serious increase in the level of refraction. In airborne applications, elevation data have been acquired photogrammetrically for water adjacent points and used to interpolate a water surface (e.g., Westaway *et al.*, 2000, 2001) from these points using a series of image processing steps.

This approach works largely when the water is clear. In theory, it should to work in any situation where the optical depth of the imagery exceeds the maximum water depth, allowing application to situations where the water has some suspended load or washload. However, even small amounts of turbidity may reduce the textural variability of the bed, making the identification of homologous point pairs difficult, either automatically or manually. Thus, a second approach has been adopted for situations where homologous point pair identification is impeded (Westaway *et al.*, 2003). This has coupled the depth measurement techniques described above to water surfaces interpolated from water adjacent points, allowing depth maps obtained from image processing to be transformed into digital elevation models.

Example application: close range determination of gravel habitat   Carbonneau *et al.* (2003) demonstrate the use of a close-range digital photogrammetric methodology to acquire topographic information on coarse gravel river habitat at minimal cost. Digital elevation models (DEMs) were derived from 1:165 scale imagery obtained with a standard 35-mm film single lens reflex (SLR) camera, a commercial desktop scanner and suitable photogrammetry software.

Traditionally, the use of commercial SLR cameras as sensors in photogrammetric applications was considered impossible as a result of the severe geometric distortion associated with the camera lens and the unstable geometry associated with the relationship between the image, the position of the lens, and the object field of interest. However, the development of software-based photogrammetry packages in the late 1990s allowed implementation of mathematical models (bundle adjustments) to model the effects of distortion and geometry upon the relationship between image and ground point positions. Carbonneau *et al.* (2003) found that standard commercial photography equipment could be used to produce digital elevation models, provided certain steps were taken to offset the lesser optical quality of commercial cameras produced for public use. This research showed that 15 to 20 ground control points (i.e., points on the terrain that are visible in the overlapping imagery and have measured spatial positions) are necessary to model the terrain surface successfully. The end results were DEMs having a spatial resolution of 1-mm of approximately 500 × 500 pixels in size. Figure 11.4 shows an example. In this case, several cobbles are embedded in a matrix of sand. Despite the presence of some

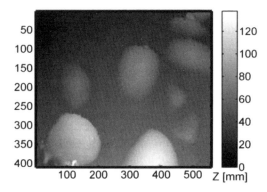

Figure 11.4 Example of a close range DEM obtained from a commercially available SLR film camera. DEM resolution is 1 mm. Precision is ±2.1 mm

errors, notably in the bottom left corner, the DEM clearly outlines the clasts and thus gives relevant information for investigations into small scale bed structures. Quality assessment showed that the precision of the resulting DEMs was consistently less than 10% of the D_{50} of the bed particles. This translates into sub-centimetric precision. Such close range applications have potential for instream habitat research. For instance, topography exerts a dominant influence on the flow velocity field, which is a key parameter in swimming energy expenditure. Additionally, juvenile salmon can use the gravel void spaces as shelter during winter periods (Rimmer *et al.*, 1983). This implies that the structure and compactness of the surface bed material can influence the amount of available shelter and this can be quantified from these digital elevation models.

11.3.3 Laser Scanning

Principles Unlike the two techniques for depth and morphology measurement described thus far, which rely upon capturing the signal that comes from radiation emitted from ground surface points of interest, laser scanning is an active form of remote sensing. In the simplest of terms, an airborne or ground-based laser emits a pulse of electromagnetic radiation. This is then reflected from the surface of interest and is detected upon its return to the laser. The time of travel is converted to a distance of travel. If the position and orientation of the laser is known with sufficient precision, this distance can be used to determine a three-dimensional coordinate. By changing the orientation of the laser by a small amount, additional coordinates can be determined. A full explanation of the principles of this technology can be found in Wehr and Lohr (1999) and Baltsavias (1999). The design of laser scanning surveys is not dissimilar to that associated with photogrammetry, in that it is necessary to trade between the spatial extent of the area of interest and the quality and resolution of data required, in relation to the characteristics of the sensor. Hicks *et al.* (2002) and Lane *et al.* (2003) illustrate this for an application to braided river morphology. The reach was flown three times to give an average cross-stream point spacing of 4.6 m and downstream point spacing of 1.7 m. As with the photogrammetry, post-processing is required in order to improve the quality of the data collected. This is now a very well-established component of laser scanner data collection, not least because of a working group of the ISPRS Commission 3, concerned with how to filter and to interpolate laser-scanned data (Sithole and Vosselman, 2003).

The use of laser scanning for the determination of river corridor morphology is now routine, not only in terms of research applications (e.g., Marks and Bates, 2000; Gomes Pereira and Wicherson, 1999; Bates and de Roo, 2000; Horritt and Bates, 2001; Yu and Lane, 2006a,b) but also in terms of practical floodplain mapping. For instance, most of the UK's floodplains have been measured using airborne laser scanners (Light Detection and Range or LiDAR). More recently, braided rivers, where a significant percentage of the active width may be visible at low flow, have also been measured using laser scanning (e.g., Hicks *et al.*, 2002; Charlton *et al.*, 2003; Lane *et al.*, 2003). These approaches have two limitations. First, they have typically used laser scanners at wavelengths that lead to negligible water surface penetration and so have relied upon filling in the residual inundated areas using methods such the depth estimation using image processing described above (e.g., Westaway *et al.*, 2003). However, by using a different wavelength, it is possible to get some water penetration and possibly subaerial bottom reflectance (e.g., Irish and Lillycrop, 1999). Second, airborne applications can be limited as a result of both the costs and logistics of survey. An equally exciting development in the simultaneous measurement of subaerial and submerged topography is terrestrial oblique laser scanning. Heritage and Hetherington (2006) report on one of the first applications of this to the fluvial environment, obtaining 0.01 m resolution data for a reach of river approximately 120 m in length and 12 m in width, through a total of 21 scans from different locations and taking about two days to acquire. This certainly represents a crucial method for the measurement of river bed topography at the reach scale. Heritage and Hetherington (2006) note difficulties in measuring the morphology of individual clasts as a result of direction-dependent shadowing effects and hence its potential for measuring grain scale morphology will be more restricted.

Example application: braided river morphology   Figure 11.5 shows a digital elevation model (after Hicks *et al.*, 2002; Westaway *et al.*, 2003) obtained through a combination of laser scanning and the image processing techniques described above. These data have

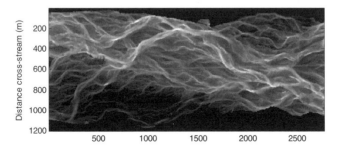

Figure 11.5   *Detrended digital elevation model of the Waimakariri River, South Island, New Zealand, plotted to 2.0 m resolution. Colour scale extends from +2.0 m (black) to −2.0 m (white) with respect to local valley slope.*

been interpolated to 2-m resolution. The laser scanned data for the dry areas were precise to ±0.100 m (Lane *et al.*, 2003) on the basis of comparison with independent check data. Generally, this study showed that the laser scanned data gave better results in terms of data precision than did the equivalent photogrammetric approach in terms of both precision (precise to ±0.131 m after appropriate survey design and filtering), and elimination of the problem of propagation of small random error in the photogrammetric bundle adjustment into systematic surface error in the final DEMs (Lane *et al.*, 2004b). Photogrammetry remains preferable if multiple repeat surveys are required, as the mobilisation cost is lower, especially if the sensor is positioned and oriented on board an aircraft, eliminating the need for any ground survey.

11.4 Substrate

Substrate has been established as an important habitat-determining factor for both fish (e.g., Cunjak, 1988; Heggenes, 1990; Rimmer *et al.*, 1983) and macroinvertebrates (e.g., Rice *et al.*, 2001). Methods for substrate characterisation using remote sensing can be divided into two different approaches. The first is based upon the identification of discrete grains and then the synthesis of a particle size distribution and hence substrate parameters from measurements of each discretised grain. This is commonly labelled 'photo-sieving'. As it requires the clear identification of the boundaries between adjacent grains, the image pixel resolution must be small as compared with the grain being discretised, and hence it is commonly applied at close range. The second method is based upon the textural signature of many grains and the expectation that the greater the typical size of grains in a sediment patch, the greater the textural variability in images of those grains. This does not require the discretisation of grain boundaries, but the image pixel resolution must still be sufficient to include some of the textural variability caused by grain boundaries. This means that the method is suitable with high resolution airborne imagery.

11.5 Discrete Grain Identification

11.5.1 Principles

Fully automated discrete grain identification (e.g., Butler *et al.*, 2001; Sime, 2003; Graham *et al.*, 2005a,b) builds upon earlier work (e.g., Iriondo, 1972; Adams, 1979) that shows that photography could be used to acquire grain size. This method has much appeal as it is rapid, involves minimal if any surface destruction and, in theory, can be applied to both exposed and inundated areas. The basic principle of automation is that a grain's boundaries are well defined with respect to the rest of the image (McEwan *et al.*, 2000). McEwan *et al.* (2000) developed a successful methodology for extracting data from a high resolution digital elevation model (DEM). Application to test cases, of known grain-size distributions, produced extremely encouraging results. However, it is preferable to use unprocessed two-dimensional imagery. The basic stages in automated grain size identification involve some form of image segmentation in order to identify possible grain size boundaries. The boundaries then have to be grown in order to connect such that individual grains are discretely identified. The grains then need to be measured, commonly in terms of the visible short and long axes (e.g., Butler *et al.*, 2001). Graham *et al.* (2005a,b) provide an overview of the best methods for doing this and software is now available that fully automates the process (http://www.sedimetrics.com/). It is important to note (e.g., Adams, 1979) that there is not necessarily an equivalence between actual grain-size and the projection of grain-size onto a two-dimensional image, not least because of particle imbrication, armouring and burial affects, but it is possible to determine calibration relationships to correct for such effects, should correction be required.

11.5.2 Example Application: Exposed Gravel Grain-size Distributions

Butler *et al.* (2001) report one of the first attempts to extract grain-size data automatically from conventional photographic data. This was based upon thresholding a scanned raw photograph followed by watershed segmentation and the fitting of ellipses to each particle. From the ellipses, long and short axes were identified. They explored the effects of various aspects of the data collection process on the quality of the results obtained. Although the process resulted in both over- and under-segmentation of individual grains, these effects were insignificant when judged at the population level, by comparison with manually traced results. When expressed in phi units, median grain size estimates for the automated method were −4.80 and for the manually traced estimates were −4.82; D_{84} estimates were −4.23 and −4.40 respectively, and D_{95} estimates were −3.30 and −3.34. Subsequent tests showed that the primary control upon data quality was the image scale, which controlled the minimum acceptable grain size that could be identified. In relation to the surface grain size, there will be an image scale at which discrete identification of individual grains is no longer possible and ensemble methods (see below) will be required. Other researchers have fully assessed this approach and Graham *et al.* (2005a,b) provide full guidance on this method.

11.6 Ensemble Grain Size Parameter Determination

11.6.1 Principles

One of the areas where recent progress has been considerable has been in the application of automated image processing methods to high resolution airborne imagery in determining ensemble grain size parameters (e.g., Carbonneau *et al.*, 2004, 2005; Carbonneau, 2005). The principle of this method rests upon their being an expression of surface grain-size variability in the image, such as in the textural composition of the image.

In the case of fluvial gravels, the textural composition of the image is controlled by the shading effect of the particles, which creates light–dark contact zones in proportion to the actual size of the particles. An empirical approach can then be used to relate the local image texture generated by the shading to the actual particle size. This approach is conceptually simple and, if properly applied, could allow for automated grain size measurements from image data. However, appropriate image texture properties must be identified. Furthermore, calibration data are required to establish the functional relationship between image information and grain size. Examination of the available literature reveals two image properties that could potentially be used for grain size mapping: (i) co-occurrence based image texture; and (ii) semivariance. Image texture, as defined by the co-occurrence matrix, was among the first local image properties to be developed capable of segmenting image areas that appear visually distinct to a human observer (Haralick *et al.*, 1973; Haralick, 1979; Conners *et al.*, 1984). Thus, image texture may allow for grain size determination since patches of different grain sizes appear distinct to a human observer, provided image resolution is sufficient. Both image semivariance and texture were tested in Carbonneau *et al.* (2004) as possible image properties capable of deriving grain size from airborne digital imagery.

11.7 Example Application: Substrate Mapping in a Salmon River

The work of Carbonneau *et al.* (2004) used the same image data set as described previously for the bathymetry mapping. As stated earlier, the theoretical underpinning for grain size mapping needs to be supported by appropriate calibration data. Carbonneau *et al.* (2004) used geolocated field measurements of grain size to establish a grain size correlation with two-dimensional semivariance calculations for 33×33 pixel regions of 3-cm resolution imagery. This analysis over a 33×33 pixel region yields grain size information with a resolution of 1 m. Figure 11.6 shows an example of a grain size map showing the local D_{50} of the dry exposed gravel for the reach shown in Figure 11.2(a). Examination of the bed material present in Figure 11.2(a) shows that in addition to the gravel bar, there is an area of bank stabilisation where boulders have been used to protect a nearby road. Figure 11.5(a) shows that the grain size mapping algorithm has correctly measured the different bed material types in this image. In this case, the authors found an error of 13 mm on their automated grain size estimates. Crucially, once calibrated, this method is fully automated, thus allowing for the processing of the full image data set for the extraction of median surface grain sizes of dry exposed bed material. A significant part of the channel area is occupied by water. However, medium sized rivers often have significant shallow areas where some visible bed material texture remains. Carbonneau *et al.* (2005) demonstrated that the grain size mapping method developed for the exposed dry area could be extended to the wetted, shallow, area of the bed. Analysing the airborne imagery of the St-Marguerite river in Canada, it was found that 67% of the channel area had a shallow visible bottom and 19% was dry exposed bed material. Therefore, the extension of grain size mapping to the shallow areas allowed for 86% of the channel to be measured with meter scale resolution. This equates to roughly 3×10^6 median grain size measurements equally distributed along both the streamwise and lateral dimension of the channel. The only limitation of the method is in the deep part of the wetted perimeter where the bed is not visible. In Figure 11.5b, the deep area shows a uniform grain size of between 5 and 10 cm. In fact, this is due to texture produced by waves at the water surface. Such waves will obviously be a source of error when grain size mapping is applied to the whole wetted channel and in case of white water rapids to errors can be quite large. Another source of error to be considered is also shown in Figure 11.5(b). Examination of the edges of the measured area shows lines of coarse material that are obviously not present. The light/dark contact zone created by the wet/dry interface produces additional image texture, which is falsely interpreted as coarse material.

These types of error show that the use of such automated grain size mapping methods necessarily implies that the spatial coverage that can be achieved will be slightly offset by a lesser overall data quality. The same observation also applies to the automated depth

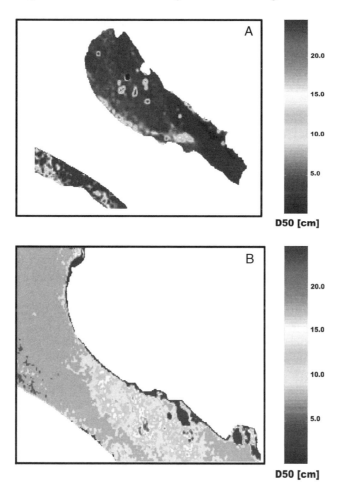

Figure 11.6 (See also colour plates) *Example of a dry area median grain size map (A), resolution is 1 m and precision is ±13 mm, and (B) an example of a shallow area median grain size map, resolution is 1 m and precision is ±28 mm, both produced from the image shown in Figure 11.2(A)*

mapping method. However, two points emerge. First, we must maintain research efforts to improve the intrinsic data quality of these methods. Second, whilst there has been a traditional emphasis upon establishing the quality of grain-size estimates at a point, capturing the spatial variability of grain size over the reach and entire tributary length scales has required careful sampling in order to avoid contamination of derived grain-size estimates by local scale sample variability (e.g., Rice and Church, 1998). This is not necessary with the methods reported above.

Thus, whilst the kinds of remotely-sensed grain size estimates we describe here may cause some error, this may well be compensated for by the reduction in error associated with sampling the full spatial extent of available habitat. As these datasets span the a wider range of scales of grain size variability, and when taken with the depth data above, it means that we can begin to examine, in a quantitative fashion, modern concepts of landscape ecology such as patch dynamics and habitat connectivity (Cadenasso et al., 2006) in a fluvial context. This avenue of research could potentially yield very significant advances to our knowledge and understanding of fluvial ecosystems.

11.8 Future Developments

There is no doubt that the last 20 years has seen a revolution in our ability to measure depth and substrate using remote sensing technology. Much of this has been in the context of research applications, although it is notable that at least some of these research applications are being adopted as part of practical assessments of river habitat. This will continue and as it does, so remotely sensed data may become routinely incorporated in practical habitat assessment. However, and in turn, this will raise new biological challenges for two reasons. First, it is quite possible that the resolution and extent over which depth and substrate can now be measured may actually exceed our ability to measure the biota that make use of it. This leads into a second problem: the development of ecologically effective parameters (Lane et al., 2006). In conventional modelling of river hydraulics those scales that cannot be incorporated explicitly into analyses (e.g., microtopography) are commonly represented through parameters (e.g., roughness) that represent the effects of unrepresented or partially represented processes upon model predictions (Lane, 2005). From a biological perspective, we need a similar research effort that seeks to couple biological observations to these new scales of measurement of both depth and substrate. This kind of research could lead to a new suite of biological parameters that allow the increasingly routine remote sensing of depth and substrate to be interpreted in habitat terms.

References

Adams, J. 1979. Gravel size analysis from photographs. American Society of Civil Engineers, *ASCE Journal of the Hydraulics Division*, **104**, 1247–55.

Baltsavias, E.P. 1999. Airborne laser scanning: basic relations and formulas. *ISPRS Journal of Photogrammetry and Remote Sensing*, **54**, 199–214.

Bardonnet, A. and Baglinière, J.-L. 2000. Freshwater habitat of Atlantic salmon (*Salmo salar*). *Canadian Journal of Fisheries and Aquatic Science*, **57**, 497–506.

Barker, R., Dixon, L. and Hooke, J. 1997. Use of terrestrial photogrammetry for monitoring and measuring bank erosion. *Earth Surface Processes and Landforms*, **22**, 1217–1227.

Bates, P.D. and De Roo, A.P.J. 2000. A simple raster-based model for flood inundation simulation. *Journal of Hydrology*, **236**, 54–57.

Butler, J.B., Lane, S.N. and Chandler, J.H. 1998. Assessment of DEM quality characterising surface roughness using close range digital photogrammetry. *Photogrammetric Record*, **16**, 271–291.

Butler, J.B., Lane, S.N. and Chandler, J.H. 2001. Automated extraction of grain-size data from gravel surfaces using digital image processing for hydraulic research. *Journal of Hydraulic Research*, **39**, 1–11.

Butler, J.B., Lane, S.N., Chandler, J.H. and Porfiri, E. 2002. Through-water close range digital photogrammetry in flume and field environments. *Photogrammetric Record*, **17**, 419–439.

COLOR PLATE

Figure 3.1 A complex ERS habitat on the River Otter in Devon (UK)

Figure 3.6 Downstream organisation of ERS patches on the River Severn at Caersws (mid-Wales, UK)

COLOR PLATE

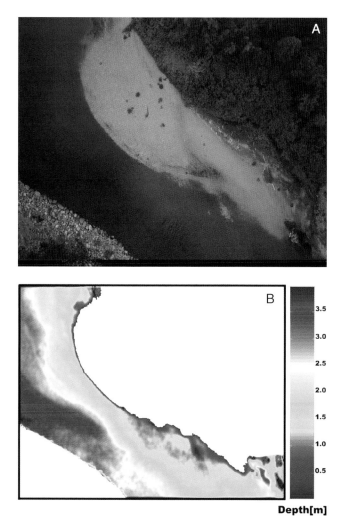

Figure 11.2 A sample image used for both automated depth mapping and grain size mapping (A), image resolution is 3 cm and scale is 1:350, and the depth image (B)

COLOR PLATE

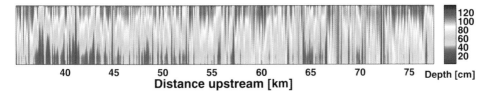

Figure 11.3 Distribution profile of depth values for the upper 40 km of the main branch of the St-Marguerite river, Quebec, Canada. The profile, displayed as an image, is composed of 4000 concatenated cumulative distributions of the depth values for each successive 10 m reach of the river. Distance upstream is given by the x axis. A preponderance of blue indicates shallow areas with actual depth values given by the colour scale

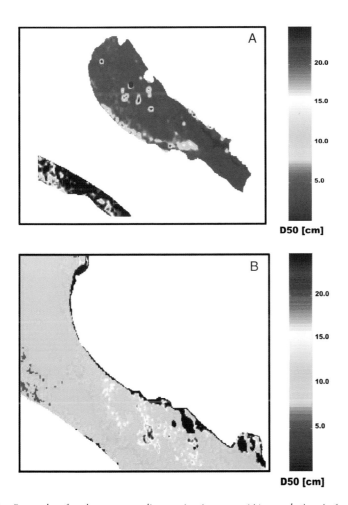

Figure 11.6 Example of a dry area median grain size map (A), resolution is 1 m and precision is ±13 mm, and (B) an example of a shallow area median grain size map, resolution is 1 m and precision is ±28 mm, both produced from the image shown in Figure 11.2(A)

COLOR PLATE

Figure 18.5 The confluence of a small karstic groundwater tributary (bottom; width approximately 1m) and the Taillon glacier stream, French Pyrénées (flow right to left), illustrating the marked contrast in suspended sediment concentrations

Cadenasso, M.L., Prickett, S.T.A. and Grove J.M. 2006. Dimensions of ecosystem complexity: Heterogeneity, connectivity and history. *Ecological Complexity*, **3**, 1–12.

Carbonneau, P.E. 2005. The threshold effect of image resolution on image-based automated grain size mapping in fluvial environments. *Earth Surface Processes and Landforms*, **30**, 1687–1693.

Carbonneau P.E., Lane, S.N. and Bergeron, N.E. 2003. Low cost close range digital photogrammetry for the quantification of fish habitat. *International Journal of Remote Sensing*, **24**, 2837–2854.

Carbonneau P.E., Lane, S.N. and Bergeron, N.E. 2004. Catchment-scale mapping of surface grain size in gravel-bed rivers using airborne digital imagery. *Water Resources Research.*, **40**, W07202.

Carbonneau, P.E., Bergeron, N.E. and Lane, S.N. 2005. Automated grain size measurements from airborne remote sensing with an application to granulometric long profile measurements in fluvial environments. *Water Resources Research*, **41**, W11426.

Carbonneau P.E., Lane, S.N. and Bergeron, N.E. 2006. Feature based image processing methods applied to bathymetric measurements from airborne remote sensing in fluvial environments. *Earth Surface Processes and Landforms*, **31**, 1413–1423.

Carter, K.L., Liu, C.C., Lee, Z., English, D.C., Patten, J., Chen, F.R., Ivey E. and Davis, C.O. 2003. Illumination and turbidity effects on observing faceted bottom elements with uniform Lambertian albedos, *Limnology and Oceanography*, **48** (Part 2), 355–363.

Charlton, M.E., Large, A.R.G. and Fuller, I.C. 2003. Application of airborne lidar in river environments: The River Coquet, Northumberland, UK. *Earth Surface Processes and Landforms*, **28**, 299–306.

Conners, R.W., Trivedi, M.M. and Harlow, C.A. 1984. Segmentation of a high resolution urban scene using texture operators. *Computer Vision, Graphics and Image Processing*, **25**, 273–310.

Crowder, D.W. and Diplas, P. 2000. Evaluating spatially explicit metrics of stream energy gradients using hydrodynamic model simulations. *Canadian Journal of Fisheries and Aquatic Science*, **57**, 1497–1507.

Crowder, D.W. and Diplas, P. 2002. Assessing changes in watershed flow models with spatially explicit hydraulic models. *Journal of the American Water Resources Association*, **38**, 397–408.

Cunjak, R.A. 1988. Behaviour and microhabitat of young atlantic salmon (*Salmo salar*) during winter. *Canadian Journal of Fisheries and Aquatic Science*, **45**, 2156–2160.

Cunjak, R.A., Prowse, T.D. and Parrish, D.L. 1998. Atlantic salmon (*Salmo salar*) in winter: 'the season of parr discontent'? *Canadian Journal of Fisheries and Aquatic Science*, **55** (Suppl. 1), 161–180.

Dixon, L.F.J., Barker, R., Bray, M., Farres, P., Hooke, J., Inkpen, R., Merel, A., Payne, D. and Shelford, A. 1998. Analytical photogrammetry for geomorphological research. Chapter 4 in Lane, S.N., Richards, K.S. and Chandler, J.H. (eds) *Landform Monitoring, Modelling and Analysis*. John Wiley & Sons, Ltd, Chichester, pp. 63–94.

Elso, J.I. and Giller, P.S. 2001. Physical characteristics influencing the utilization of pools by brown trout in an afforested catchment in Southern Ireland. *Journal of Fish Biology*, **58**, 201–221.

Folt, C.L., Nislow, K.H. and Power, M. 1998. Implications of temporal and spatial scale for Atlantic salmon (*Salmo salar*) research. *Canadian Journal of Fish and Aquatic Science*, **55**(Supp. 1), 9–21.

Fonstad, M.A. and Marcus, W.A. 2005, Remote sensing of stream depths with hydraulically assisted bathymetry (HAB) models. *Geomorphology*, **72**, 320–329.

Fryer, J.G. 1983. A simple system for photogrammetric mapping in shallow-water. *Photogrammetric Record*, **11**, 203–208.

Ghanem, A., Steffler, P. and Hicks, F. 1996. Two-dimensional hydraulic simulation of physical habitat conditions in flowing streams. *Regulated Rivers: Research and Management*, **12**, 185–200.

Gomes Pcreira, L.M. and Wicherson, R.J. 1999. Suitability of laser data for deriving geographical information: A case study in the context of management of fluvial zones. *ISPRS Journal of Photogrammetry and Remote Sensing*, **54**, 105–114.

Graham, D.J., Rice, S.P., Reid, I. 2005a. A transferable method for the automated grain sizing of river gravels. *Water Resources Research*, **41**, Art. No. W07020 JUL 21 2005.

Graham, D.J., Reid, I., Rice, S.P. 2005b. Automated sizing of coarse-grained sediments: Image-processing procedures. *Mathematical Geology*, **37**, 1–28.

Guay, J.C., Boisclair, D., Rioux, D., Leclerc, M., Lapointe, M. and Legendre, P. 2000. Development and validation of numerical habitat models for juveniles of Atlantic salmon (*Salmo salar*). *Canadian Journal of Fisheries and Aquatic Science*, **57**, 2065–2075.

Haralick, R.M. 1979. Statistical and structural approaches to texture. *Proceedings of the IEEE*, **67**, 786–804.

Haralick, R.M., Shanmugan, K. and Dinstein, I. 1973. Textural features for image classification. *IEEE Transactions on Systems, Man and Cybernetics*, **3**, 610–621.

Hardy, R.J., Lane, S.N., Lawless, M.R., Best, J.L., Elliott, L. and Ingham, D.B. 2005. Development and testing of a numerical code for treatment of complex river channel topography in three-dimensional CFD models with structured grids. *Journal of Hydraulic Research*, **43**, 468–480.

Harris, W.D. and Umbach, M.J. 1972. Underwater mapping. *Photogrammetric Engineering*, **38**, 765–772.

Heggenes, J. 1990. Habitat utilization and preferences in juvenile Atlantic salmon (*Salmo salar*) in streams. *Regulated Rivers: Research and Management*, **5**, 341–354.

Heritage, G. and Hetherington, D.E. 2006. Towards a protocol for laser scanning in fluvial geomorphology. *Earth Surface Processes and Landforms*, in press.

Heritage, G.L., Fuller, I.C., Charlton, M.E., Brewer, P.A. and Passmore, D.P. 1998. CDW photogrammetry of low relief fluvial features: Accuracy and implications for reach-scale sediment budgeting. *Earth Surface Processes and Landforms*, **23**, 1219–1233.

Hicks, D.M., Duncan, M.J., Walsh, J.M., Westaway, R.M. and Lane, S.N. 2002. New views of the morphodynamics of large braided rivers from high-resolution topographic surveys and time-lapse video. In Dyer, F.J., Thoms, M.C., Olley, J.M. (eds), *The Structure, Function and Management Implications of Fluvial Sedimentary Systems*. IAHS Publication No. 276, pp. 373–380.

Horritt, M.S. and Bates, P.D. 2001. Effects of spatial resolution on a raster based model of flood flow. *Journal of Hydrology*, **253**, 239–249.

Iriondo, M.H. 1972. A rapid method for size analysis of coarse sediments. *Journal of Sedimentary Petrology*, **42**, 985–986.

Irish, J.L. and Lillycrop, W.J., 1999. Scanning laser mapping of the coastal zone: the SHOALS system. *ISPRS Journal of Photogrammetry and Remote Sensing*, **54**, 123–129.

Kay, J.E., Kampf, S.K., Handcock, R.N., Cherkauer, K.A., Gillespie, A.R. and Burges, S.J. 2005. Accuracy of lake and stream temperatures estimated from thermal infrared images. *Journal of the American Water Resources Association*, **41**, 1161–1175.

Lammert, M. and Allan, J.D. 1999. Assessing biotic integrity of streams: effects of scale in measuring the influence of land use/cover and habitat structure on fish and macroinvertebrates. *Environmental Management*, **23**, 257–270.

Lane, S.N. 1998. The use of digital terrain modelling in the understanding of dynamic river systems. Chapter 14 in Lane, S.N., Richards, K.S. and Chandler, J.H. (eds) *Landform Monitoring, Modelling and Analysis*, John Wiley & Sons, Ltd, Chichester, pp. 311–342.

Lane, S.N. 2000. The measurement of river channel morphology. *Photogrammetric Record*, **16**, 937–961.

Lane, S.N. 2005. Roughness – time for a re-evaluation? *Earth Surface Processes and Landforms*, **30**, 251–253.

Lane, S.N., Richards, K.S. and Chandler, J.H. 1993. Developments in photogrammetry: the geomorphological potential. *Progress in Physical Geography*, **17**, 306–328.

Lane, S.N., Chandler, J.H. and Richards, K.S. 1994. Developments in monitoring and terrain modelling small-scale river-bed topography. *Earth Surface Processes and Landforms*, **19**, 349–368.

Lane, S.N., Hardy, R.J., Elliot, L. and Ingham, D.B. 2002. High-resolution numerical modelling of three-dimensional flows over complex river bed topography. *Hydrological Processes*, **16**, 2261–2272.

Lane, S.N., Westaway, R.M. and Hicks, D.M. 2003. Estimation of erosion and deposition volumes in a large gravel-bed, braided river using synoptic remote sensing. *Earth Surface Processes and Landforms*, **28**, 249–271.

Lane, S.N., Hardy, R.J., Elliott, L. and Ingham, D.B. 2004a. Numerical modelling of flow processes over gravelly-surfaces using structured grids and a numerical porosity treatment. *Water Resources Research*, W01302 JAN 8 2004.

Lane, S.N., Reid, S.C., Westaway, R.M. and Hicks, D.M. 2004b. Remotely sensed topographic data for river channel research: the identification, explanation and management of error. In Kelly, R.E.J., Drake, N.A. and Barr, S.L. *Spatial Modelling of the Terrestrial Environment*, John Wiley & Sons, Ltd, Chichester, pp. 157–174.

Lane, S.N., Mould, D.C., Carbonneau, P.E., Hardy, R.J. and Bergeron, N. 2006. Fuzzy modelling of habitat suitability using 2D and 3D hydrodynamic models: biological challenges. In Ferreira, R.M.L., Alves, E.C.T.L., Leal, J.G.A.B. and Cardoso, A.H. (eds) *River Flow 2006*, Volume 2, Taylor and Francis, London, pp. 2043–2053.

Leclerc, M. 2005. Ecohydraulics: a new interdisciplinary frontier for CFD. In Bates, P.D., Lane, S.N. and Ferguson, R.I. (eds) *Computational Fluid Dynamics: Applications in Environmental Hydraulics*, John Wiley & Sons, Ltd, Chichester, pp. 429–460.

Leclerc, M., Boudreault, A., Bechara, J.A. and Corfa, G. 1995. Two-dimensional hydrodynamic modelling: a neglected tool in the Instream Flow Incremental Methodology. *Transactions of the American Fisheries Society*, **124**, 645–662.

Leclerc, M., Boudreault, A., Bechara, J.A. and Belzile, L. 1996. Numerical method for modelling spawning habitat dynamics of landlocked salmon, *Salmo salar*. *Regulated Rivers: Research and Management*, **12**, 273–285.

Legleiter, C.J., Roberts, D.A., Marcus, W.A. and Fonstad, M.A. 2004. Passive optical remote sensing of river channel morphology and in-stream habitat: Physical basis and feasibility, *Remote Sensing of Environment*, **93**, 493–510.

Lonzarich, D.G., Lonzarich, M.R. and Warren, M.L., Jr. 2000. Effects of riffle length on the short-term movement of fishes among stream pools. *Canadian Journal of Fisheries and Aquatic Science*, **57**, 1508–1514.

Lyon, J.G., Lunetta, R.S. and Williams, D.C. 1992. Airborne multispectral scanner data for evaluating bottom sediment types and water depths of the St. Marys river, Michigan. *Photogrammetric Engineering and Remote Sensing*, **58**, 951–956.

Maddock, I.P., Bickerton, M.A. and Spence, R. 2001. Reallocation of compensation releases to restore river flows and improve instream habitat availability in the Upper Derwent catchment, Derbyshire, UK. *Regulated Rivers: Research and Management*, **17**, 417–441.

Marcus, W.A. 2002. Mapping of stream microhabitats with high spatial resolution hyperspectral imagery. *Journal of Geographical Systems*, **4**, 113–126.

Marks, K. and Bates, P.D. 2000. Integration of high-resolution topographic models with floodplain flow models. *Hydrological Processes*, **14**, 2109–2122.

McEwan, I.K., Sheen, T.M., Cunningham, G.J., and Allen, A.R. 2000. Estimating the size composition of sediment surfaces through image analysis. *Proceedings of the Institution of Civil Engineers – Water, Maritime and Energy*, **142**, 189–195.

Mertes, L.A.K. 2002. Remote sensing of riverine landscapes. *Freshwater Biology*, **47**, 799–816.

Milhous, R.T., Wegner, D.L. and Waddle, T. 1984. Users guide to the physical habitat simulation system. US Fish and Wildlife Service Biological Services Program. FWS/OBS-81/43.

Milhous, R., Updike, M. and Snyder, D. 1989. PHABSIM system reference manual: Version 2. US Fish and Wildlife Service FWS/OBS 89/16. (Instream Flow Information Paper 26).

Petts, G.E. 1984. *Impounded Rivers: Perspectives for Ecological Management*, John Wiley & Sons, Ltd, Chichester.

Rice, S.P. and Church M. 1998. Grain size along two gravel-bed rivers: Statistical variations, spatial patterns and sedimentary links, *Earth Surface Processes and Landforms*, **23**, 345–363.

Rice, S.P., Greenwood, M.T. and Joyce, C.B. 2001. Tributaries, sediment sources, and the longitudinal organisation of macroinvertebrate fauna along river systems. *Canadian Journal of Fish and Aquatic Science*, **58**, 824–840.

Rimmer, D.M., Paim, U. and Saunders, R.L. 1983. Changes in the selection of microhabitat by juvenile Atlantic salmon (*Salmo salar*) at the summer–autumn transition in a small river. *Canadian Journal of Fisheries and Aquatic Sciences*, **41**, 469–475.

Roberts, A.C.B. and Anderson, J.M. 1999. Shallow water bathymetry using integrated airnorne multi-spectral remote sensing. *International Journal of Remote Sensing*, **20**, 497–510.

Rubin, J.F. 1998. Survival and emergence pattern of the sea trout fry in substrata of different compositions. *Journal of Fish Biology*, **53**, 84–92.

Serway, R.A. 1983. *Physics for Scientists and Engineers*. CBS College Publishing, ••, 132–133.

Sime, L.C. 2003. Information on grain sizes in gravel-bed rivers by automated image analysis. *Journal of Sedimentary Research*, **73**, 630.

Sithole, G. and Vosselman, G. 2003. *Report: ISPRS Comparison of Filters,* Working Group 3 of ISPRS Commission III (http://www.commission3.isprs.org/wg3/, accessed 06-10-06).

Tiffan, K.F., Garland, R.D. and Rondorf, D.W. 2002. Quantifying flow-dependent changes in subyearling fall Chinook salmon rearing habitat using two-dimensional spatially explicit modelling. *North American Journal of Fisheries Management*, **22**, 713–726.

Torgersen, C.E., Price, D.M., Li, H.W. and McIntosh, B.A. 1999. Multiscale thermal refugia and stream habitat associations of chinook salmon in northeastern Oregon. *Ecological Applications*, **9**, 301–319.

Torgersen, C.E., Faux, R.N., McIntosh, B.A., Poage, N.J. and Norton, D.J. 2001. Airborne thermal remote sensing for water temperature assessment in rivers and streams. *Remote Sensing of Environment*, **76**, 386–398.

Voss, K.J., Mobley, C.D., Sundman L.K., Ivey, J.E. and Mazel C.H. 2003. The spectral upwelling radiance distribution in optically shallow waters. *Limnology and Oceanography*, **48** (Part 2), 364–373.

Wehr, A. and Lohr, U. 1999. Airborne laser scanning – an introduction and overview. *ISPRS Journal of Photogrammetry and Remote Sensing*, **54**, 68–82.

Westaway, R.M., Lane, S.N., and Hicks, D.M. 2000. Development of an automated correction procedure for digital photogrammetry for the study of wide, shallow gravel-bed rivers. *Earth Surface Processes and Landforms,* **25**, 200–226.

Westaway, R.M., Lane, S.N. and Hicks, D.M. 2001. Airborne remote sensing of clear water, shallow, gravel-bed rivers using digital photogrammetry and image analysis. *Photogrammetric Engineering and Remote Sensing*, **67**, 1271–1281.

Westaway, R.M., Lane, S.N. and Hicks, D.M. 2003. Remote survey of large-scale braided rivers using digital photogrammetry and image analysis. *International Journal of Remote Sensing*, **24**, 795–816.

Whited, D., Stanford, J.A. and Kimball, J.S. 2002. Application of airborne multispectral digital imagery to quantify riverine habitats at different base flows. *Rivers Research and Applications*, **18**, 583–594.

Winterbottom, S.J. and Gilvear, D.J. 1997. Quantification of channel bed morphology in gravel-bed rivers using airborne multispectral imagery and aerial photography. *Regulated Rivers: Research and Management*, **13**, 489–499.

Yu, D. and Lane, S.N. 2006a. Urban fluvial flood modelling using a two-dimensional diffusion-wave treatment. Part 1: mesh resolution effects. *Hydrological Processes*, **20**, 1541–1565.

Yu, D. and Lane, S.N. 2006b. Urban fluvial flood modelling using a two-dimensional diffusion-wave treatment. Part 2: development of a sub-grid-scale treatment. *Hydrological Processes*, **20**, 1567–1583.

Zhang, H., Voss, K.J., Reid, R.P. and Louchard E.M. 2003. Bidirectional reflectance measurements of sediments in the vicinity of Lee Stockings Island, Bahamas, *Limnology and Oceanography*, **48** (Part 2), 380–389.

12
A Mathematical and Conceptual Framework for Ecohydraulics

John M. Nestler, R. Andrew Goodwin, David L. Smith, and James J. Anderson

12.1 Introduction

The basic premise underlying ecohydraulics is deceptively simple – to meld basic principles of biology and ecology (for brevity, both biology and ecology will be collectively termed ecology from here forward) and hydraulic engineering. However, different disciplines can have very different traditions and conventions. This is particularly true for hydraulic engineering and ecology. For example, many conventional engineering tools trace their origin back to the conservation principles of Newton (i.e., conservation of mass, momentum, and energy), whereas many ecological tools trace their origin to Malthusian population growth and the ideas and concepts of Darwin. While there are many classroom and research examples of how the concepts of Newton can be integrated into ecology (e.g., Vogel, 1983; Pennycuick, 1992), the schism between engineering and ecology largely remains in water resources management tools and continues to impede the integration of hydraulic engineering and ecology into the new discipline of ecohydraulics. Reconciling this schism is the theme of this chapter.

The need to accommodate disparate traditions and to study processes that vary over wide ranges of spatial and temporal scales separates ecohydraulics from other disciplines and could be an impasse that prevents its further development. The different approaches of hydraulic engineers and ecologists: (i) are applied to processes that differ substantially in spatial and/or temporal scales; (ii) can be traced to different modeling traditions; and (iii) utilize different sets of mathematical formulations, concepts, and assumptions. The full promise of ecohydraulics modeling and analysis may remain elusive because of the difficulty in reconciling the separate scale ranges that hydraulic engineers and ecologists each use to study the world. This difficulty has two consequences, best expressed as two interrelated questions:

- First at a science level, how can researchers develop fundamental principles for ecohydraulics and thereby advance this field as a discipline when the underlying principles of its component disciplines appear so disparate?
- Subsequently at a practical level, how can resource managers implement sustainable development and biodiversity preservation that depend, ultimately, on an accurate understanding, integration, and prediction of the causal relationship between changes in the physico-chemical environment (the focus of engineers) with response of individuals, populations, and communities (the focus of ecologists) without developing such fundamental principles?

The way forward may lie in the recognition that processes associated with each different hierarchical level of an ecosystem also have associated scale ranges (Levin, 1992). Neither the traditions nor conventions of engineers and biologists separately can adequately describe the different processes operating over the entire range of scales that typically characterize ecosystem dynamics and natural complexity. Neither discipline by itself can address discontinuity and mismatch of scales at which hydraulics are typically modeled with scales at which fish respond to their environment (Kondolf *et al*., 2000; Bult *et al*., 1999, Railsback, 1999). Therefore, a conceptual or mathematical bias is introduced when either engineering or ecology is used by itself to represent the dynamics and complexity of an aquatic ecosystem. For example, ecosystem processes may vary in temporal and spatial scale from fine-scale hydrodynamics associated with habitat selection by stream macroinvertebrates, to intermediate scales associated with chemical transformations typical of water quality dynamics, to large-scale ecological population dynamics in which one cycle may last decades and extend over thousands of kilometers. Although importance of scale is known, there are limited approaches to including scale quantitatively as a metric to describe ecosystem processes (Nestler and Sutton, 2000) even though scale-associated issues are known substantially to affect prediction accuracy and reduce usefulness of single-discipline models for decision making (Nuttle, 2000).

It is in this science gray zone between engineering and ecology that the new interdisciplinary field of ecohydraulics is emerging. Ecohydraulics offers the promise of spanning the tools, concepts, and traditions of its two component disciplines. By so doing, this new discipline can address many of the most important resource management issues facing the world. However, before the new field of ecohydraulics can be elevated to the same stature as its more established components of ecology and hydraulics, it is first necessary to develop a scientific foundation for ecohydraulics comprising guiding, fundamental principles. Simply applying standard hydraulic engineering tools to address ecological issues is an insufficient theoretical basis for ecohydraulics and will not support the development of ecohydraulics as a separate discipline. Ecohydraulics must be built on a scientific 'common denominator' that allows hydraulic engineering and ecology to be melded into a new discipline.

Ecohydraulics should be an integrated discipline that honors the conventions and traditions of both ecologists and hydraulic engineers. It should recognize that the two component disciplines of engineering and ecology have different concepts and approaches, and that each focuses on certain processes over limited ranges of scales. We believe these differences can be distilled to the dominant reference frameworks used by each of the two component disciplines. To be elevated to the same stature as either aquatic ecology or hydraulic engineering, ecohydraulics must integrate the approaches of the component disciplines and, through the resulting synergy, develop new tools and approaches that are presently beyond the reach of either ecology or hydraulic engineering separately. We propose that the specialized goal of ecohydraulics should be to 'integrate hydraulic and biological tools to improve the analysis and prediction of ecological response to physicochemical change in aquatic settings in support of water resources management'.

The objectives of this chapter are: (i) to provide a brief historical background on ecohydraulics; (ii) to provide a suite of unifying concepts that can be used by both ecologists and hydraulic engineers so that each can better understand the field of the other; (iii) to relate natural processes to an appropriate modeling approach used either by engineers or ecologists, and (iv) to illustrate how hydraulics and ecology can be integrated using two examples. These organizing concepts can be employed to couple existing tools of ecologists and hydraulic engineers to provide a parsimonious and useful representation of aquatic ecosystems that may describe natural complexity at extremes of spatial and temporal scales. By so doing, a firm foundation can be created upon which the discipline of ecohydraulics can be built.

12.2 Ecohydraulics: Where Do the Ideas Come From?

The ideas underpinning ecohydraulics can be traced to two separate beginnings; one centered in hydraulic engineering and the other centered in ecology. Ecologists in academic settings have a long history of studying the relationship between fluid flow and ecological response (e.g., Ambühl, 1959). This history is elegantly presented and reviewed by Vogel (1983) in his book *Life in Moving Fluids*. Every student of either aquatic ecology or hydraulic engineering should read this book in order to understand the importance of fluid dynamics to aquatic biota at a first principles level. Since this book was first published there have been many examples in which scientists describe the relationship between flow fields and organismic response (Statzner and Higler, 1986; Statzner *et al.*, 1988; Carling, 1992; Pavlov *et al.*, 2000; Smith *et al.*, 2005). However, while scientifically interesting, these investigations remained primarily in the realm of academia and were seldom used to support water resources management and decision making.

Hydraulic engineers first systematically attempted to include organismic response into their work during efforts to develop design criteria for fish passage facilities at dams. The body of work can be traced back about 300 years (Odeh, 1999) with many of the most important works described by Clay (1961, 1995). Examples of an engineering approach to fish passage include Bell (1973), Powers and Osborne (1985), and Bell (1991). However, these studies tended to treat fish as engines that exhibited different categories of swimming performance important to passage design. As fish passage technology studies became more multidisciplinary, ecologists working primarily in Europe began developing natural fish ways. Their efforts began to integrate principles of fluvial geomorphology, river ecology, population dynamics, and behavior to supplement development of hydraulic design criteria (e.g., see works in Jungwirth *et al.*, 1998, and Newbury and Gaboury, 1993). However, the detailed understanding of the movement strategies used by fish to 'hydro-navigate' through the river in search of different habitats was beyond reach.

The second major pathway in which hydraulic engineering contributed to ecohydraulics was through the development of aquatic habitat assessment methods as exemplified by the Instream Flow Incremental Methodology (Bovee, 1982) and similar methods. This methodology was an outgrowth of stream gauging techniques (Pierce, 1941; Viessman and Lewis, 1996) that divides the river into cells and allows the calculation of discharge. The analysis was performed by relating average depth, average velocity, and cover in each cell to previously determined habitat suitability curves. Cell specific values of habitat were then integrated over time and space to describe habitat dynamics of target species as part of assessing the effects of different flow alternatives. This methodology was widely applied (Tharme, 2003) but often criticized for its lack of ecological realism (Orth and Maughan, 1982; Gore and Nestler, 1988) because it relied so extensively on engineering and hydrologic methods.

The general lack of success in developing fish passage technology based solely on simple hydraulic parameters, like average velocity and average depth, is mirrored by the criticisms faced by users of aquatic habitat assessment methods. Increasingly, there is interest in closing this gap between hydraulic engineers and ecologists as both disciplines better understand their own limitations. In both cases, critical reviewers bemoan the lack of 'first principles' to underpin the development of new tools and new concepts in fish passage and improved habitat assessment methods.

12.3 Reference Frameworks of Engineering and Ecology

As disciplines, hydraulic engineering and ecology can be decomposed into 'first principles' by progressively reducing their concepts until they are irreducible (as in Aristotle, 350 BC). Identifying first principles of the two disciplines should lead to the scientific common denominator that can be used to integrate them to generate the new discipline of ecohydraulics. Mathematical models are abstractions of the guiding traditions and conventions of any discipline and, therefore, an examination of the attributes of their respective models should point to a discipline's first principles. The most fundamental attribute of a mathematical model is the manner in which it represents space–time and scale (Nestler *et al.*, 2005). Three modeling reference frameworks are typically encountered in hydraulics and ecology, each of which deals with space–time and scale differently: Eulerian, Lagrangian, and Agent (Goodwin *et al.*, 2006). We know of no other frameworks for handling spatial and temporal dynamics of entities in ecosystems (Parrish and Edelstein-Keshet, 1999). Typically, one of these frameworks is used by itself to formalize and simulate natural processes, although features of the environment are inherently neither Eulerian, Lagrangian, nor Agent. As described below, each framework has specific strengths and weaknesses and appears to be optimally suited to address certain ranges of scale relative to the size of the physical domain to which the model is applied (Goodwin *et al.*, 2006). These frameworks embody the fundamental principles underpinning hydraulic engineering and ecology, and we believe that an understanding of them is necessary to in order continue the evolution of a concept set that can serve as the theoretical and mathematical foundation for ecohydraulics.

12.3.1 Eulerian Reference Framework

In this framework the physical domain is discretized into a mesh of interconnected compartments or cells (Figure 12.1A). Using established conservation (in a physics sense) equations, mass, energy, and momentum are transported through the mesh and balanced at compartment interfaces (Thomann and Mueller, 1987; Cassell *et al.*, 1998). The Eulerian framework is useful for simulating processes occurring over times and distances that are short compared with time and distance intervals used to update transfers across cell boundaries. The Eulerian framework is typically used when entities of modeling interest are very small in size relative to the discretization of the physical domain of the system. Such processes can be averaged across an individual grid cell and their products transported passively by fluid flow without propagating substantial error. For example, bulk flow of water depends on molecular properties such as density and viscosity, which vary with temperature. Over the discrete time and distance intervals commonly employed in environmental models, variation of fluid properties is sufficiently gradual that they can be averaged into control volumes. Although larger scales must be considered, they are typically addressed as boundary conditions or initial conditions and not addressed in the governing equations. The Eulerian framework has also proven useful for simulating lower trophic levels (e.g., Gin *et al.*, 1998) that are defined by short temporal and limited spatial scales.

12.3.2 Lagrangian Reference Framework

Lagrangian frames are preferable to Eulerian frames when aggregation of constituents into control volumes results in an unacceptable accumulation of error. This typically occurs when entities of modeling interest exhibit dynamics that are intermediate or large in scale relative to discretization of the physical domain of the system. Lagrangian schemes retain individual identities of constituent particles or discrete volumes (both referred to as 'particles' for brevity) and track them separately as they move throughout the computational grid (Figure 12.1B). The Lagrangian reference frame imposes no conservation principles other than the preservation of particle identity. Therefore, conservation principles, if required, must be imposed as additional constraints on the model system (Parrish and Edelstein-Keshet, 1999; Gravel and Staniforth, 1994). The movement of fish eggs or drogues (i.e. a passively drifting flow field instrument) in a flow field is best represented using the Lagrangian framework (e.g., North *et al.*, 2005).

12.3.3 Agent Reference Framework

The agent framework is required when an entity of interest exhibits complex behaviors due to its internal state or external signals (Figure 12.1C). This integration of internal state, external signals, and resultant complex behaviors cannot be represented adequately with either the Eulerian or Lagrangian frameworks or a combination of the two. Ideally, such entities are best represented using all three reference frameworks together. A simple example is a fish holding position in a river to feed. Its perception of the flow field is best defined using a Lagrangian frame (based on its specific location in the flow field) but its habitat (a block of water with uniform hydraulic conditions within which it occurs) is typically defined by a Eulerian frame. The fish's impetus to move depends on its internal

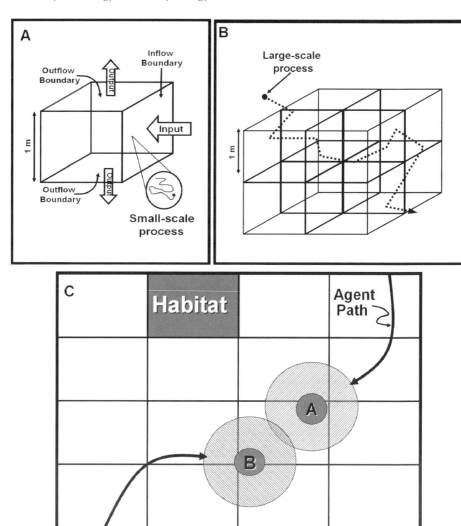

Figure 12.1 Scale relationships between the three reference frameworks. (A) Symbolic representation of the Eulerian framework in which entities associated with small scales are aggregated into a control volume (cell) and simplified as masses and fluxes. Note that particle identity is lost. (B) Symbolic representation of the Lagrangian framework where an entity (particle) exhibiting larger scale dynamics passively moves through a number of Eulerian cells. Such particles cannot be aggregated into cells without accumulation of substantial error and must, therefore, maintain their individual identities. The 3-D Eulerian cells are shown only for comparison. (C) Symbolic representation of the agent framework in which an entity A moves through space and interacts with other agents B via rules that govern agent–agent interactions and agent–system interactions based on information acquired within the agent response envelop (cross-hatched circles)

state, typically a time-varying trade-off between the need to feed and avoid predators, and the signals that it acquires providing information about the condition of its immediate surroundings. Fundamental to an agent framework is the concept of memory in which past events determine how the agent will respond to its present surroundings. In such a setting, the governing equations used in the Eulerian and Lagrangian frames are inadequate to describe the dynamics of this class of entity. Instead, the entity must be represented with its own separate set of governing equations. In ecological settings, detailed population dynamics or movement behaviors of organisms such as fish in aquatic environments, or moose in terrestrial environments, are typically simulated using individual-based models (IBM), a type of agent-based model.

12.4 Concepts for Ecohydraulics

If the goal of ecohydraulics is to relate accurately the biotic response to the physicochemical environment, then the predominantly Eulerian world of engineering must be reconciled and integrated with the predominantly Lagrangian/Agent world of ecology. To this end, we propose that ecohydraulics be founded on the concept of integrated reference frameworks. A quantitative method of implementing this concept is the Eulerian–Lagrangian–Agent method (ELAM) (Goodwin et al., 2006). In their fullest embodiment, integrated reference frameworks take advantage of the strengths of each of the reference frameworks described above to create a single, unified knowledge base in which information can be rotated, translated, or transformed to meet the information needs of any of the three reference frameworks. In such a framework, spatially and temporally incremental physicochemical information (such as hydraulics or water quality information) is stored in an Eulerian framework at discrete points within the grid (mesh). However, individual organisms exist in continuous time and space and make directed movements or execute other behaviors based on cues and gradients in important stimulus variables and their internal state. The mismatch between incremental and continuous space–time must be addressed before ecology and hydraulic engineering can be integrated. To bridge the gap between the information needs of individuals or groups of biota and the way information is stored in an Eulerian framework, interpolation methods can be used to shift information to points of interest that do not fall on grid points where Eulerian information is stored. Once information has been shifted, agent-based methods can be used to describe how individual organisms such as fish or shellfish interface with the physico-chemico environment (Figure 12.1C) or with other individual organisms of their own or other species. In this unified scheme, the Eulerian framework is the domain of the hydraulic engineer and the agent framework is the domain of the ecologist. The Eulerian and agent frameworks are linked together via the Lagrangian framework (Figure 12.2).

Integrated reference frameworks can be used to address a wide variety of simulation challenges because each frame can be applied at the scale for which it is best suited – the ideal foundation for ecohydraulics where two apparently divergent sets of concepts and tools must be integrated. We offer two complementary example applications of integrated reference frameworks to illustrate how engineering and ecological models can be coupled to create a greater synthesis. The first example implicitly couples the Eulerian and Lagrangian frameworks to create a hydraulically realistic description of fish habitat

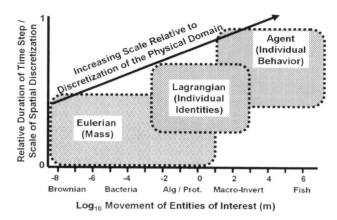

Figure 12.2 Symbolic relationships between optimum reference frameworks and inherent entity scale attributes. Note hydraulic engineering methods typically (with exceptions) used in water resource management employ tools and concepts founded on Eulerian representations, and that biologists and ecologists employ tools best classified as agent-based tools and concepts. Lumped parameter population growth models, competition models, and other types of ecological model designed to study biological populations can be envisioned as simplifications of agent-based frameworks

selection. The second example is the Numerical Fish Surrogate (NFS) (Goodwin et al., 2006), which illustrates a complete coupling in which all three reference frameworks are explicitly integrated into one system. The NFS is an example of an integrated reference frameworks model based on ELAMs and exemplifies a single, integrated knowledge engine in which information can be rotated, translated, converted, or rescaled, as needed, to be used by any one of the frameworks.

12.5 Two Examples of Ecohydraulics

12.5.1 Example 1: Semi-quantitatively Describing Habitat of Drift Feeding Salmonids

Background While many factors can impact abundance of salmonids in streams, hydraulic pattern is often the first feature of a stream to be simulated and analyzed. Drift feeding salmonids occupy a focal position located in relatively low velocity water adjacent to a faster reach of stream. From the focal position, a fish can dart out into the faster current to feed. This behavior allows the fish to have the bio-energetic benefits of swimming in slow water while having access to the increased food delivery rate of fast water. Although this conceptual model is widely applied, rarely has it been noted that it is an inherently shear based description of habitat occupancy. Most habitat analyses are based on an Eulerian representation of the flow field in which average velocity is measured at the focal position of the fish (a Lagrangian representation), or at some arbitrary point in the water column near the fish. The result of this Eulerian–Lagrangian conceptual mismatch (i.e., an average cell velocity used to characterize a point location) is that hydraulically based habitat descriptions are unable to replicate the distribution of fish in streams and, therefore, cannot predict changes in abundance as a function of changes in hydraulics.

Approach The following procedures reconcile the reference framework mismatch and illustrate how integrated reference framework concepts can be used to guide a relatively simple hydraulic habitat analysis. We consider this procedure to be an excellent example of a semi-quantitative ecohydraulics approach in which hydraulic habitat analysis is integrated with ecologically based fish behavior. Microhabitat data from the Yakima River, Washington State USA (Allen, 2000) was used to calculate the exposure strain rate (Neitzel *et al.*, 2004) of juvenile Chinook salmon. An example of using a spatial velocity gradient based approach to describing habitat (Figure 12.3) is found in Smith *et al.*, (in review). The exposure spatial velocity gradient, or 'exposure strain rate', (*e*) is:

$$e = \frac{\partial \bar{u}}{\partial y}$$

where $\bar{u}$ is the average water velocity (cm/s), and y is the characteristic length (cm) resulting in e having units of cm/s/cm. The characteristic length was taken to be 0.4 cm, or the

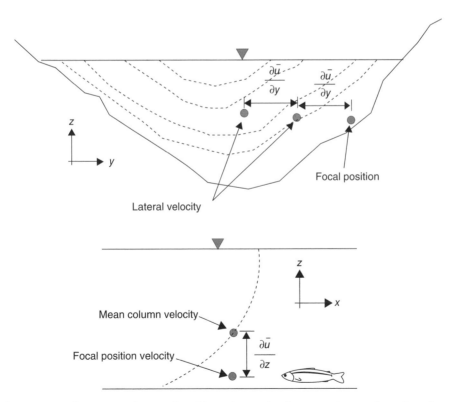

Figure 12.3 Illustration of vertical and lateral spatial velocity gradients, also referred to as vertical and lateral 'exposure strain rates', respectively

minimum head width of fish observed during the study. A common characteristic length is needed to allow comparisons between all calculated values of e.

Results Qualitatively, the exposure strain rate describes the Lagrangian conceptual drift feeding model using a metric (spatial velocity gradients) that has physical meaning and maps back to an Eulerian habitat representation. The exposure strain rate was calculated for spring, summer, and fall (data unavailable for winter) and five reach scale geomorphic habitat types: deep pools, deep runs, low gradient riffles, run glides, and shallow pools. Microhabitat characteristics were measured through direct observation and sample sizes ranged between 241 and 393 fish. In each habitat type, three types of exposure strain rate were calculated. The vertical strain rate was calculated as the difference between the focal velocity and the mean column velocity measured 0.6 m depth from the surface. Two lateral strain rates were calculated as the difference between the focal position average velocity and the mean column velocity 0.6 and 1.2 m toward the center of the channel.

It was thought that a high or low spatial velocity gradient, or 'exposure strain rate', would represent poor quality habitat since if the spatial velocity gradient was high, a fish darting out to feed would be swept downstream and thus have to struggle to regain its former focal position or acquire a new one. Conversely, a focal position with a low spatial velocity gradient might represent a location where food delivery rates were low. Therefore, the vertical strain rates should fall in a fairly narrow range for a given size class of fish. Statistical analysis supported the hypothesis that vertical strain rates were similar across the range of reach scale habitat types. In other words, juvenile Chinook salmon were occupying focal positions that have similar vertical spatial velocity gradients across all habitat types. A deep pool focal position was similar to a low gradient focal position in terms of vertical spatial velocity gradients. The lateral strain rates calculated at 0.6 and 1.2 m from the focal positions showed differences in the lateral spatial velocity gradients. Deep pool lateral strain rates were lower than in low gradient riffles.

Significance of the example This approach to describing habitat use is consistent with the ecological model of drift feeding; however it is different than the normal application of spatial velocity gradients to describe rivers. For example, spatial velocity gradients referenced from the fish's focal position for the fish increase as distance from the fish increases if average velocity increases over the distance between the two measurement points. If, however, spatial velocity gradients are calculated on smaller scales, and the fish is near a boundary, gradients decrease on moving away from the boundaries, and increase as one moves closer to the boundaries. Since most fish observations were near a boundary, it is possible that variation of spatial velocity gradients integrated over the fish length serve as guidance for drift feeding fish seeking suitable focal positions.

There are three conclusions that can be drawn from this.

(i) Focal positions are statistically similar in terms of a velocity gradient across different habitat types. This implies that fish were selecting focal positions independent of reach scale habitat.
(ii) Velocity gradients were different between habitat types at scales of 0.6 and 1.2 m. Taken together, it appeared that juvenile Chinook were selecting focal positions with

similar levels of gradient intensity independent of overall reach scale hydraulics associated with different reach scale habitats.

(iii) By conceptually integrating the separate approach of the engineer and biologist in this example, a new insight was gained that could not have been obtained from the exclusive use of one of the disciplines.

12.5.2 Example 2: Quantitatively Describing Fish Swim Path Selection in Complex Flow Fields

Background Hydropower dams on the Snake and Columbia Rivers in the Pacific Northwest of the USA block the out-migration of juvenile salmon (migrants). Bypass systems are constructed to intercept as many of these fish as possible and thereby prevent them from entering the turbines where they can be potentially injured or killed (Figure 12.4). To work effectively, migrants must be attracted to the vicinity of the bypasses and the hydrodynamic characteristics of bypass entrances must encourage the entry of migrants. However, the hydraulic design criteria for neither the approach nor entrances were known, leading to the construction of large, expensive systems of variable performance with concomitant negative impacts on migrants.

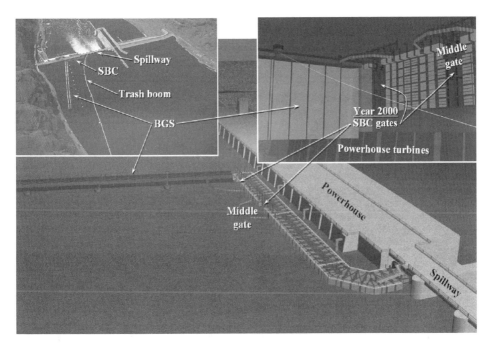

Figure 12.4 Illustration of project features at Lower Granite Dam on the Snake River, WA, USA. A Surface Bypass Collector (SBC) is designed to pass fish around the dam, and the Behavioral Guidance Structure (BGS) is intended to guide migrants to the SBC and occlude them from the three turbine intakes nearest the shore

Approach The Numerical Fish Surrogate (NFS) is an example of how the three reference frameworks can be integrated into a single consistent mathematical tool to address a problem that is presently beyond the reach of either hydraulic engineers or biologists (Goodwin *et al.*, 2006). The Eulerian reference framework is represented by a computational fluid dynamics (CFD) model that outputs hydraulic information at millions of nodes to describe the flow field in the dam forebay up to approximately 1200 m from the dam that is encountered by migrants approaching the dam. The Lagrangian framework is represented by the passive particle traces made by interpolating information from the nodes of the CFD model mesh to points that represent the path made by a neutrally buoyant passive particle. The agent framework is represented by the behavior rules that can be applied to the passive ('dumb') particles that allow them to acquire information from the CFD model mesh, to have a memory to help define their inner state, and to use process information from the environment relative to its inner state as the basis of swim behaviors (become 'smart' particles) (Figure 12.5). The parameters of the behavioral model can be recursively adjusted to minimize the difference between traces made by virtual and real fish. By interpreting the best behavioral rules, it is possible to gain insight into the hydraulic navigation strategy of real fish, how fish use hydrodynamic cues to make movement decisions, and to improve dam design and operation to minimize impacts on fish.

Results As described in detail in Goodwin *et al.* (2006), NFS predictions generally match observed passage proportions with goodness of fit (slope/R-square) decreasing from spillway (0.95/0.85), bypass (1.07/0.80), to powerhouse (1.15/0.61). The reduced goodness of fit for the powerhouse is likely related to the difficulty of maintaining constant operation of the powerhouse during the collection of fish tracks. Total powerhouse discharge usually remains constant, but the units in operation and the distribution of load across those units typically changes during a test. However, the CFD model is run at

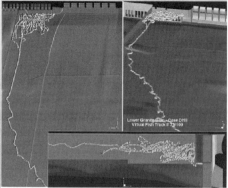

Figure 12.5 Comparison of observed and predicted swim paths made by out-migrating juvenile salmon. Acoustically tagged migrant telemetry data from Cash et al. (2002)

steady state and consequently does not capture changes in operation. Spillway operation is held constant during a test and therefore meets the steady-state assumption. Bypass system operation is also held constant, but its location nearer the powerhouse than the spillway for most cases reduces its goodness of fit. Turning the behavioral rules off (passive transport) reduces the goodness of fit (slope/R-square) of the spillway to 0.49/0.50, bypass 0.10/0.10, and powerhouse 0.24/0.08 (Goodwin *et al.*, 2006). Improvement in goodness of fit provided by the rules represents the contribution of the behavioral rules towards the quality of the predictions.

The 'traffic rule' used by migrants is derived from the context of fluvial geomorphology of free flowing rivers (Figure 12.6 described more fully in Nestler *et al.*, 2007). We describe total flow field distortion using a metric called 'total hydraulic strain' because it intrinsically conveys an awareness of the underlying distortion mechanisms of: (i) linear deformation (whose tensor metric components are normal strain rates); (ii) rotation (whose tensor metric components are angular velocities), and (iii) angular deformation (whose tensor metric components are one-half the true shearing strain rates). Although rotation is not due to normal or shearing strain rates, the same spatial velocity gradients induce both angular deformation (shearing strain) and rotation. In free flowing rivers, flow field pattern results from flow resistance (Leopold *et al.*, 1964). Without flow resistance there is no force to distort a unit volume of water

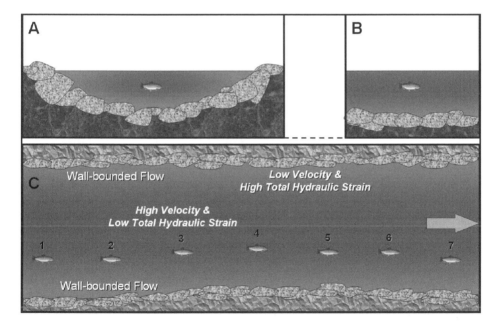

Figure 12.6 Channel contains both friction resistance (wall-bounded flow, illustrated) and form resistance (free-shear flow, not illustrated). Areas of elevated total hydraulic strain are illustrated with darker red and identify areas of impending turbulence in both wall-bounded and free-shear flows. By avoiding these areas a migrant can navigate a channel in a way that minimizes risks of turbulence exposure, delay, predation, and inefficient swimming conditions

once it is set into motion by the force of gravity (Ojha and Singh, 2002). Flow resistance can be separated into two categories for sub-critical, steady flow: friction resistance and form resistance. Friction resistance in a simple, straight, uniform channel produces a flow pattern in which average velocities are lowest nearest a source of friction (such as the channel bottom and edges) with a zero water velocity occurring at the water–channel interface. Pattern in the total hydraulic strain field is the inverse of pattern in the velocity field, with lowest total hydraulic strain occurring furthest from sources of friction resistance and highest near the sources.

Form friction is created by large woody debris or rock outcrops projecting into the channel. As in the case of friction resistance, total hydraulic strain associated with form resistance increases towards the signal source. In contrast to bed friction (where water velocity decreases towards the friction source), water velocity increases towards the signal source for form resistance because of local reduction in conveyance area and increased travel distance of water flowing around an obstruction. A fish approaching a stump from the upstream direction will sense an increase in total hydraulic strain and an increase in water velocity until solid boundary effects very close to the obstruction are encountered. By integrating information between the total hydraulic strain and velocity fields, fish have sufficient information to separate channel structures associated with friction or form resistance, thereby creating a hydrodynamic 'image' of their immediate surroundings.

Significance of the example   There are three significant findings from this work:

(i) The NFS can accurately predict response of fish to the hydrodynamic fields created by different dam designs and operations, allowing water resource managers to design bypass systems optimally and take other steps to minimize the impact of water resource development on out-migrating juvenile salmon. This tool could not have been developed using concepts restricted to either hydraulic engineering or ecology.
(ii) In natural, free flowing rivers there is a dynamic equilibrium between flow field and bed form. Fish have acquired behaviors and sensing systems over geologic time to navigate such systems hydraulically. Unfortunately, flow pattern at dams is determined by size, orientation, and operation of gates, valves, and orifices and not by a dynamic equilibrium between discharge and channel bed form. The creation of flow features, such as high energy intake plumes that are uncommon in free flowing rivers causes fish to become 'confused' at dams when they apply behaviors to hydrodynamic cues evolved in free flowing rivers over geologic time to features at dams that do not exhibit this relationship. Dams can be designed to incorporate 'natural hydrodynamic signals' into their design to reduce their impact on fish. By so doing, the hydraulic 'foot print' of the dam can be minimized, perhaps even made invisible, to migrating fish with attendant benefits to society such as the elimination of costly but ineffectual bypass construction projects and the reduction of impacts to migrating fish. This solution to the vexing challenge of managing impacts of dams could not have been developed from either hydraulic engineering or ecology as individual disciplines.
(iii) At an application level perspective, ELAM models like the Numerical Fish Surrogate address several needs in ecohydraulics approaches: (a) conversion of information

from sources that differ in metric, range, scale, and dimensionality to a form of computer script (agents) consistent with animal sensory capabilities (Bian, 2003); (b) ability systematically to organize and evaluate behavior hierarchies from the integration of information from various sensory modalities that may take varying precedence during the changing phases of a behavioral sequence (Sogard and Olla, 1993; New *et al.*, 2001); (c) decentralized computer script for adding, eliminating, or modifying components without affecting the rest of the model (Ginot *et al.*, 2002); (d) the theoretical and computational basis to elicit vector-based virtual movement of individuals responding to abiotic and biotic stimulus data provided in either Eulerian (Tischendorf, 1997) or Lagrangian form (Nestler *et al.*, 2005), and (e) ability easily to compare model output to field-collected data (e.g., Figure 12.5) (Hastings and Palmer, 2003).

12.6 Discussion

12.6.1 An Opportunity for Engineers and Ecologists

Developing tools that can be used to understand and predict natural complexity and preserve biodiversity is a major challenge in water/aquatic resource management. The concept of integrated reference frameworks is a paradigm that can be used by both hydraulic engineers and ecologists for building tools to address the challenge of sustainable development. For engineers, this concept leads to new areas for technology growth and new applications of established tools. For ecologists, the concept allows them to provide clear and unequivocal mathematical formulations analogous to many physics and engineering applications that relate physiochemical change to ecological response. Each discipline can expand into new areas of study and research, and ultimately develop the tools that water resource managers require.

The concept of integrated reference frameworks allows engineering and ecology based approaches to be integrated either qualitatively (similar to the first example) or quantitatively (similar to the second example) to address difficult environmental water-related issues. The significance of both examples lies in the fact that they are both built on integrated reference framework concepts and that without this foundation, neither example could be possible. In addition, the concept allows both hydraulic engineers and aquatic ecologists to communicate better when budget and schedule realities limit investigative scope.

Integrated reference frameworks offer a number of advantages over single framework modeling approaches because of their ability to realistically to simulate ecological processes that occur across a wide range of scales. ELAM models allow the three frameworks to be mathematically coupled, and conceptually integrated to simulate accurately processes that occur across a wide range of scales typically encountered in ecosystem analysis and simulation. ELAMs have the potential partially to address the problems identified by Alewell and Manderscheid (1998) that some ecological processes are inherently too difficult to simulate, and by Turchin (1997) that full spatiotemporal analysis is conceptually difficult. In particular, the Lagrangian module of ELAMs can be used to simulate animal movement behavior, a difficult but critical element in simulating and managing larger aquatic organisms because they often exhibit integrated responses to complex situations (Schilt and Norris, 1997), well outside the simulation capabilities of Eulerian-based models.

12.6.2 Challenges and Limits for Ecohydraulics

Below, we identify three of the challenges and limitations of ecohydraulics that can be the source of future research and development:

The challenge of the habitat mosaic: the habitat mosaic concept (Pringle et al., 1988; Townsend, 1989; Townsend and Hildrew, 1994), used to characterize aquatic ecosystem dynamics, requires that rivers be considered as a dynamic patchwork of interconnected habitats that dynamically appear and disappear along the river corridor in response to flow and channel change. However, most of the tools used by both engineers and ecologists often require the steady state assumption for flow and almost always assumes a rigid channel. In the future, ecohydraulics must embrace both the perspectives and tools used by river geomorphologists to address more fully river management issues such as short-term bedform variations, long-term reach-scale cycles of aggradation and degradation, and channel evolution, all of which are simultaneously at play.

The challenge of multi-scale features: almost all engineering tools used to describe channel morphology use a relatively limited range of cell or mesh sizes to discretize the channel, and they assume that the Cartesian coordinate location of physical features within the channel is important. However, river channels are probably self-similar structures so that methods borrowed from fractal geometry (e.g., Nestler and Sutton, 2000) may provide more accurate, useful, and realistic depiction of the structure of river channels that can be applied better to understand habitat dynamics for many different sized biota. We envisage future tools whose output can be simultaneously queried at multiple scales. For example, information at a relatively large scale may be needed to understand salmon habitat dynamics, but information at a much smaller scale may be required to understand habitat dynamics of aquatic macroinvertebrates.

The challenge of the real world – budgets and schedules: most resource managers are not interested in supporting lengthy, expensive, high-resolution, academic investigations of streams and rivers. Instead they favor the application of relatively simple, fast, and inexpensive screening methods. We recommend that ecological model similitude analysis (Petts et al., 2006) be conducted to reconcile this tension between basic science and resource management. In this approach, high-resolution, first-principles based, studies are conducted to describe processes of interest. Once completed, the high-resolution methods are progressively simplified by coarsening time and space scales, by simplifying equations through combing or eliminating coefficients, or by reducing dimensionality. Using a combination of sensitivity and divergence analysis, the high-resolution methods are simplified until the answer they give differs substantially from the high-resolution methods. The point immediately prior to divergence represents the simplest model to provide scientifically defensible output. Ecological model similitude analysis provides a comprehensive framework where basic scientists, applied scientists, and decision makers can integrate their contributions toward sustainable development. Basic scientists using ELAMs can explore and define causal relationships between environmental change and ecological response, applied scientists can develop simpler, utilitarian methods based on solid science

12.7 Conclusions

ELAMs realistically simulate ecological processes that occur across a wide range of scales and therefore integrate the tools and principles of multiple disciplines. For the example in this manuscript, the principles underlying the engineering computational fluid dynamics (CFD) model were not compromised in any way to simulate the movement dynamics of salmon. Concomitantly, no guiding ecological principles were compromised to accommodate the fluid dynamics simulation. Unlike single framework modeling approaches, ELAMs allow multiple, diverse, disciplines to collaborate in a way that does not require any individual discipline to compromise its guiding principles. This capability can be expanded to include other modeling approaches in the future, such as individual-based models or geomorphology models, to develop ever more realistic tools to guide water resource management.

In a practical sense, ecohydraulics as the discipline, and ELAM as the tool, integrates the point of view of the resource manager, who sees the habitat from the fixed perspective of the Eulerian framework, and the point of view of the fish, which sees a varying world to which it must respond.

Acknowledgements

The authors thank the many colleagues whose ideas and concepts were freely shared over many years that culminated in the suite of ideas presented herein. We particularly acknowledge the contributions of Larry Weber and his students at IIHR – Hydroscience and Engineering, University of Iowa. We thank Jim Pennington and Toni Toney both of the US Army Engineer Research and Development Center, Vicksburg, MS. We would also like to thank Tracy Hopkins for her assistance with document and figure preparation. The tests described and the resulting data presented herein, unless otherwise noted, were obtained from research conducted under the sponsorship of the US Army Engineer District Walla Walla, the Grant County Public Utility District, and the System-Wide Water Resources Program (SWWRP), a US Army Corps of Engineers research and development initiative. 3-D computer virtual fish animations and further information on Eulerian–Lagrangian–Agent methods (ELAM), including the Numerical Fish Surrogate, can be found at http://el.erdc.usace.army.mil/emrrp/nfs/

Permission was granted by the Chief of Engineers to publish this information.

References

Alewell C, Manderscheid B. 1998. Use of objective criteria for the assessment of biogeochemical ecosystem models. *Ecological Modelling*, **107**: 213–224.
Allen MA. 2000. Seasonal microhabitat use by juvenile spring chinook salmon in the Yakima River Basin, Washington. *Rivers*, **7**: 314–332.

Ambühl H. 1959. Die Bedeutung der Strömung als ökologischer Faktor. *Schweizerische Zeitschrift fur Hydrologie*, **21**: 133–264.

Aristotle. 350 BCE. *Physics*, Book 1. Translated by RP Hardie and RK Gaye, MIT web site http://classics.mit.edu/Aristotle/physics.1.i.html.

Bell MC. 1973. *Fisheries Handbook of Engineering Requirements and Biological Criteria*. US Army Corps of Engineers. Fish Passage Development and Evaluation Program, North Pacific Division, Portland, Oregon.

Bell M. 1991. *Fisheries Handbook of Engineering Requirements and Biological Criteria*. US Army Corps of Engineers, Fish Passage Development and Evaluation Program, North Pacific Division, Portland, Oregon.

Bian L. 2003. The representation of the environment in the context of individual-based modeling. *Ecological Modelling*, **159**: 279–296.

Bovee KD. 1982. A guide to stream habitat analysis using the instream flow incremental methodology. *Instream Flow Information Paper No. 12*, FWS/OBS-82/26, US Fish and Wildlife Service, Washington, DC.

Bult TP, Riley SC, Haedrich RL, Gibson RJ, Jeggenes J. 1999. Density-dependent habitat selection by juvenile Atlantic salmon (*Salmo salar*) in experimental riverine habitats. *Canadian Journal of Fisheries and Aquatic Sciences*, **56**: 1298–1306.

Carling PA. 1992. The nature of the fluid boundary layer and the selection of parameters for benthic ecology. *Freshwater Biology*, **28**: 273–284.

Cash KM, Adams NS, Hatton TW, Jones EC, Rondorf DW. 2002. Three-dimensional fish tracking to evaluate the operation of the Lower Granite surface bypass collector and behavioral guidance structure during 2000. Final report prepared by US Geological Survey Columbia River Research Laboratory for the US Army Corps of Engineers, Walla Walla District, Walla Walla, WA.

Cassell EA, Dorioz JM, Kort RL, Hoffmann JP, Meals DW, Kirschtel D, Braun DC. 1998. Modeling phosphorus dynamics in ecosystems: mass balance and dynamic simulation approaches. *Journal of Environmental Quality*, **27**: 293–298.

Clay CH. 1961. *Design of Fish Ways and Other Fish Facilities*. Department of Fisheries and Oceans, Canada, Ottawa, Ontario.

Clay CH. 1995. *Design of Fishways and Other Fish Facilities*, second edition Lewis Publishers, Boca Raton, Ann Arbor, London, Tokyo.

Gin KY, Guo HJ, Cheong H-F. 1998. A size-based ecosystem model for pelagic waters. *Ecological Modeling*, **112**: 53–72.

Ginot V, Le Page C, Souissi S. 2002. A multi-agents architecture to enhance end-user individual-based modelling. *Ecological Modelling*, **157**: 23–41.

Goodwin RA, Nestler JM, Anderson JJ, Weber LJ, Loucks DP. 2006. Forecasting 3-D fish movement behavior using an Eulerian–Lagrangian-agent method (ELAM). *Ecological Modelling*, **192**: 197–223.

Gore JA, Nestler JM. 1988. Instream flow studies in perspective. *Regulated Rivers: Research and Management*, **2**: 93–101.

Gravel S, Staniforth A. 1994. A mass-conserving semi-Langrangian scheme for the shallow-water equations. *Monthly Weather Review, American Meteorological Society*, **122**: 243–248.

Hastings A, Palmer MA. 2003. A bright future for biologists and mathematicians? *Science*, **299**: 2003–2004.

Jungwirth M, Schmutz S, Weiss S. 1998. *Fish Migration and Fish Bypasses*. Fishing News Books. London.

Kondolf GM, Larsen EW, Williams JG. 2000. Measuring and modeling the hydraulic environment for assessing in-stream flows. *North American Journal of Fisheries Management*, **20**: 1016–1028.

Leopold LB, Wolman MG, Miller JP. 1964. *Fluvial Processes in Geomorphology*. W. H. Freeman and Company, San Francisco, CA.

Levin SA. 1992. The problem of pattern and scale in ecology. *Ecology*, **73**: 1943–1967.

Neitzel DA, Dauble DD, Čada GF, Richmond MC, Guensch GR, Mueller RP, Abernethy CS, Amidan B. 2004. Survival estimates for juvenile fish subjected to a laboratory-generated shear environment. *Transactions of the American Fisheries Society*, **133**: 447–454.

Nestler JM and Sutton VK. 2000. Describing scales of features in river channels using fractal geometry concepts. *Regulated Rivers: Research and Management*, **16**: 1–22.

Nestler JM, Goodwin RA, Loucks DP. 2005. Coupling of engineering and biological models for ecosystem analysis. *Journal of Water Resources Planning and Management*, **131**(2): 101–109.

Nestler JM, Goodwin RA, Smith DL, Anderson JJ, Li S. Optimum fish passage and guidance designs are based in the hydrogeomorphology of natural rivers. *River Research and Applications*. In press.

New JG, Fewkes LA, Khan AN. 2001. Strike feeding behavior in the muskellunge, *Esox masquinongy*: contributions of the lateral line and visual sensory systems. *Journal of Experimental Biology*, **204**: 1207–1221.

Newbury RW, Gaboury MN. 1993. Exploration and rehabilitation of hydraulic habitats in streams using principals of fluvial behavior. *Freshwater Biology*, **29**: 195–210.

North EW, Hood RR, Chao S-Y, Sanford LP. 2005. The influence of episodic events on transport of striped bass eggs to the estuarine turbidity maximum nursery area. *Estuaries*, **28**(1): 108–123.

Nuttle WK. 2000. Ecosystem managers can learn from past successes. *EOS: Transactions of the American Geophysical Union*, pp. 278, 284.

Odeh M (editor). 1999. Innovations in fish passage technology. *American Fisheries Society*, Bethesda, Maryland.

Ojha CSP, Singh RP. 2002. Flow distribution parameters in relation to flow resistance in an up-flow anaerobic sludge blanket reactor system. *Journal of Environmental Engineering*, **128**(2): 196–200.

Orth DJ, Maughan OE. 1982. Evaluation of the incremental methodology for recommending instream flow for fishes. *Transactions of the American Fisheries Society*, **3**: 413–445.

Parrish J, Edelstein-Keshet L. 1999. Complexity, pattern, and evolutionary trade-offs in animal aggregation. *Science*, **284**: 99–101.

Pavlov DS, Lupandin AI, Skorobogatov MA. 2000. The effects of flow turbulence on the behavior and distribution of fish. *Journal of Ichthyology*, **40**: S232-S261.

Pennycuick CJ. 1992. *Newton Rules Biology*. New York, Oxford University Press.

Petts GE, Nestler J, Kennedy R. 2006. Advancing science for water resources management. *Hydrobiologia/Developments in Hydrobiology, Special Issue*. In press.

Pierce CH. 1941. Investigations of methods and equipment used in stream gauging. *Water Supply Paper 868-A*, US Geological Survey, Washington, DC.

Powers P, Osborn JF. 1985. An investigation of the physical and biological conditions affecting fish passage success at culverts and waterfalls. *Bonneville Power Administration Final Report*, BPA Project number 198201400.

Pringle CM, Naiman RJ, Bretschko G, Karr JR, Oswood MW, Webster JR, Welcomme RL, Winterbourn MJ. 1988. Patch dynamics in lotic systems – the stream as a mosaic. *Journal of the North American Benthological Society*, **7**(4): 503–524.

Railsback S. 1999. Reducing uncertainties in in-stream flow studies. *Fisheries*, **24**(4): 24–26.

Schilt CR, Norris KS. 1997. Perspectives on sensory integration systems: problems, opportunities, and predictions. In *Animal Groups in Three Dimensions*, JK Parrish and WM Hamner (eds) Cambridge University Press, New York, pp. 225–244.

Smith DL, Brannon EL, Odeh M. 2005. Response of juvenile rainbow trout to turbulence produced by prismatoidal shapes. *Transactions of the American Fisheries Society*, **134**: 741–753.

Smith DL, Allen MA, Brannon EL. In press. Characterization of velocity gradients inhabited by juvenile Chinook salmon by habitat type and season. *American Fisheries Society Bioengineering Symposium*.

Sogard SM, Olla BL. 1993. Effects of light, thermoclines and predator presence on vertical distribution and behavioral interactions of juvenile walleye pollock, *Theragra chalcogramma* Pallas. *Journal of Experimental Marine Biology and Ecology*, **167**: 179–195.

Statzner B, Higler B. 1986. Stream hydraulics as a major determinant of benthic invertebrates. *Freshwater Biology*, **16**: 127–139.

Statzner B, Gore JA, Resh VH. 1988. Hydraulic stream ecology: observed patterns and potential applications. *Journal of North American Benthological Society*, **7**(4): 307–360.

Tharme RE. 2003. Global perspective on environmental flow assessment: emerging trends in the development and applications of environmental flow methodologies for rivers. *Rivers Research and Applications*, **19**: 397–441.

Thomann RV, Mueller JA. 1987. *Principles of Surface Water Quality Modeling and Control*. Harper & Row Publishers, Inc., New York.

Tischendorf L. 1997. Modelling individual movements in heterogeneous landscapes: potentials of a new approach. *Ecological Modelling*, **103**: 33–42.

Townsend CR. 1989. The patch dynamics concept of stream community ecology. *Journal of the North American Benthological Society*, **8**: 36–50.

Townsend CR, Hildrew AG. 1994. Species traits in relation to a habitat templet for river systems. *Freshwater Biology*, **31**: 265–276.

Turchin P. 1997. Quantitative analysis of animal movements in congregations. *Animal Groups in Three Dimensions*, JK Parrish and WM Hammer (eds) Cambridge University Press, New York, pp. 107–112.

Viessman Jr W, Lewis GL. 1996. *Introduction to Hydrology*, fourth edition Harper Collins College Publishers, New York.

Vogel S. 1983. *Life in Moving Fluids: The Physical Biology of Flow*. Princeton University Press, p. 352.

13

Hydroecology: the Scientific Basis for Water Resources Management and River Regulation

Geoffrey E. Petts

13.1 Introduction

The challenge of 21st century river management is to better balance human water demands with the needs of rivers themselves.

(Postel and Richter, 2003, p. 4)

Water shortage in the face of increasing human demands and uncertainties in resource assessment caused by predictions of climate change are key issues for this century. Flood protection and waterway regulation for navigation are also high priorities. The conservation of biodiversity, improvement in ecosystem health, and restoration of ecosystem integrity are now embedded in strategy documents, such as the European Union's Water Framework Directive (Barth and Fawell, 2001), but in practice they are rarely given the investment required by the sustainability agenda. Too often, nature conservation issues along river corridors are limited (i) to protecting relatively small areas of wetland or floodplain forest as wildlife sanctuaries or sites of special scientific interest, or (ii) to enhancing habitat by installing artificial structures and to maintaining fisheries by stocking with hatchery-reared fish. The Twenty-first century sustainability agenda requires that we implement actions to conserve, protect and restore river corridors and their water-dependent ecosystems.

Table 13.1 Examples of Mega-projects to secure water resources constructed between 1935 and 1970. References to work on hydroecological impacts [1] Petts (1984); [2] Stanford and Ward (1986); [3] Davies (1986); [4] Davies and Wishart (2000); [5] Petr (1986); [6] Prowse et al. (2002); [7] Prowse and Conly (2002); [8] Dumont (1986); [9] Wishart et al. (2000)

Date	Project	River	Comment
1935	Hoover Dam	Colorado, USA	First of 19 high dams to control flows along the river [1, 2]
1942	Grand Coulee	Columbia, USA	5494 MW capacity hydro-power dam and largest of the 23 dams regulating the river and its major tributaries for hydro-power and irrigation supplies [1]
1959	Kariba Dam	Zambezi, Zimbabwe	A $160 \times 10^9 \, m^3$ reservoir with 1200 MW hydropower capacity [1, 3, 4]
1964	Bratsk	Angara (Yenisey)	$169 \times 10^9 \, m^3$ reservoir and 4600 MW capacity hydro-power dam
1964	Akosombo	Volta, Ghana	$148 \times 10^9 \, m^3$ reservoir with a 912 MW capacity dam; drowned area $8480 \, km^2$ [1, 5]
1967	W.A.C. Bennett	Peace, Canada	$74 \times 10^9 \, m^3$ reservoir was first large project in US part of MacKenzie basin [1, 6, 7]
1970	High Aswan	Nile, Egypt	$164 \times 10^9 \, m^3$ reservoir supporting hydro-power generation and 2.5 million ha of irrigated land [1, 8, 9]

Success in achieving sustainable water resources management will require lessons to be learned from the legacies of the control-by-construction agenda of the Twentieth century. The era of the megaproject in water resources management opened with the completion of the Hoover Dam on the Colorado River, USA, in 1935. The project was planned as the hub of a large region supplying hydroelectric power and irrigation water and led to unprecedented growth in economy and population (Thomas, 1956, p. 29). A fear of drought led to aggressive policies using an armoury of 'weapons' to combat water problems, improve national security, and realise the 'American dream' – a limitless economy (p. 408). Despite the growing awareness of the environmental damage that could be done to river systems by impoundments and inter-basin transfers (e.g., Petts, 1984) the development of water resources focused on 'control by construction' (Table 13.1). By 1990, there were over 40 000 large dams over 15 m high; 400 mega-projects[†] had been constructed or were planned (Gleick, 1998). At the start of this century, large dams contributed nearly 20% of the world's electricity supply (WCD, 2000) and irrigation agriculture consumed some $2500 \, km^3$ of water in producing 40% of the world's food (Gleick, 1998).

[†]Megaprojects include those with dams over 150 m high, a dam volume of over 25 million m^3, a reservoir volume exceeding 25 billion m^3, or an installed capacity of more than 1000 megawatts.

Globally, the natural services provided by river ecosystems are threatened and in some specific cases, are already over exploited (Postel and Carpenter, 1997; Naiman *et al.*, 2002). These services include not only water for drinking, irrigation and other human needs, but also other goods such as fish, waterfowl and riparian plants, and nonextractive benefits such as recreation, transportation, energy, flood regulation and self-purification. Along the world's great rivers, 60% of the flows have already been diverted and large dryland rivers such as the Nile, Colorado and Yellow rivers no longer reach the sea (Gleick, 1998). In Australia, water abstraction from the Murray–Darling system is approaching 90% of mean annual runoff with flow to the ocean being a rare event (Schofield *et al.*, 2000), and most of the floodplain wetlands have been lost (Reid and Brooks, 2000). Even where abstractions do not severely deplete river flows, series of dams now regulate entire river systems. Along the Volga between Rybinsk and Volvograd, 50 years of dam construction has turned the river into a series of eight shallow lakes with a combined volume of 150 km^3 and a surface area of 20000 km^2 (Khaiter *et al.*, 2000). In general terms, the lessons for sustainable river regulation from 'control by construction' are that ecological degradation results from: (i) loss of flood flows to reset instream, riparian and floodplain habitats; (ii) reduction of low flows causing habitat limitation and change, e.g. by siltation, plant encroachment or water-quality changes; (iii) loss of hydrological cues of life-cycle behaviour such as migration and spawning; (iv) unnatural seasonal flow variations, e.g. below irrigation-supply dams; and (v) unnatural rates of flow rise and recession, e.g. below power-peaking dams.

By 2025, 40% of the world's population could face problems related to water shortage. Sustaining or restoring the natural functioning of water-dependent ecosystems in the face of continuing growth of human demands is a major challenge (Postel and Richter, 2003) that requires acceptance of a multiuse ethic to provide flexible water budgets to support ecological functions (Stalnaker, 1994). Its solution requires the allocation of water to protect water-dependent ecosystems along river corridors from the impacts of abstraction and flow regulation. Governing objectives are included in policy statements. The EU Water Framework Directive, for example, requires the achievement of a 'good ecological status' in all water bodies by 2015 through the application of an integrated, catchment-based approach to water management (Achleitner *et al.*, 2005; Table 13.2), but sustainable water management still appears to be no more than an aspiration. The interdisciplinary science of hydroecology offers water resource managers and river regulators new knowledge of ecological responses to hydrological change, new tools to assess water development scenarios, and a framework to improve communication of ecological issues in adaptive management. Its application to planning, policy development and operational management offers optimism that the next generation may have the approaches and tools necessary to realise this aspiration. This chapter provides a perspective that places this optimism in context.

13.2 A Scientific Basis for Water Resources Management

Hydroecology provides an interdisciplinary approach to managing river flows. It involves the integration of hydrology, geomorphology and ecology to advance knowledge of lotic systems in order to improve the management of abstractions and river regulation schemes,

Table 13.2 Elements of the European Union's Water Framework Directive that relate to protecting and restoring flows for rivers and their water-dependent ecosystems (based on Barth and Flawell, 2001)

Defined 'reference' conditions are required to represent the natural state of rivers, streams and any floodplain lakes (denoted as 'high status'). These will be based upon river types within ecoregions. Given the high degree of human influence, definitions of the degree of modification are critical as are indications of restoration potential.

The classification for rivers shall include:
- Composition and abundance of aquatic flora, benthic invertebrate fauna and abundance and age structure of fish
- Hydrological regime; quantity and dynamics of water flow and connection to groundwater aquifers, river continuity and morphological conditions including variations in river depth and width, structure and substrate of the river bed, and structure of the riparian zone
- General thermal conditions, oxygenation, salinity, acidification, and nutrient conditions; specific pollutants and other nonpriority substances being discharged to the river.

Water management is to be based on river basins, over-riding administrative structures and boundaries.

---

and to protect water-dependent ecosystems along river corridors. It includes basic science for application to the challenges of balancing human water needs and the needs of water-dependent ecosystems across a range of spatial scales, and of seeking solutions to regional water resource concerns and reach-scale river management problems. The recent growth in popularity of the broader discipline of 'ecohydrology' (Zalewski, 2000) as 'the study of the functional interactions between hydrology and biota at the catchment scale' (p. 1), advanced by the UNESCO International Hydrological Programme in response to the challenge of converging hydrological, hydraulic and ecological research to develop truly integrated physical-biological models (Janauer, 2000), has created an environment of opportunity to embed hydroecological perspectives within water resources management.

The evolution of 'hydroecology' has been intimately associated with water resources problems. Its roots may be found in case studies of the environmental effects of large dams during the late 1960s and early 1970s. Gill's (1971) theoretical assessment of the long-term influence of river impoundment on the ecology of the Mackenzie River Delta and the detailed analysis of the influence of the Vir Valley reservoir on the ecology of the Svratka river, Czech Republic (Penaz et al., 1968) clearly established the need for an interdisciplinary approach to problems of river regulation. During the second half of the 1970s advances in three key themes expanded activity in linking hydrology and ecology: (i) increasingly quantitative studies of the ecological effects of dams upon the downstream river (e.g., Armitage, 1976; Ward, 1976); (ii) new methods for predicting the instream flow needs of biota (Bovee, 1978; Gore, 1978), and (iii) demonstration of the importance of the flood regime in sustaining the fisheries of large rivers (Welcomme, 1979).

Ward and Stanford (1979) provided a powerful catalyst for interdisciplinary advances involving specialists in physical sciences and biology. Their edited volume *The Ecology of Regulated Streams* not only demonstrated the magnitude of world-wide stream regulation but also proposed directions for future scientific investigations on stream ecosystems altered by upstream impoundments. They highlighted many of the key principles that were to be elaborated over the following 25 years. These included: that pristine rivers are heterogeneous but relatively predictable and sustain high species diversity because of the temporal and spatial heterogeneity of their environment; that major impacts of dams are caused by disrupting the stream continuum and the loss of the 'reset mechanism' of flooding; and that predictive models and management tools for river regulation require knowledge of life-histories of species and their environmental requirements.

From a physical-science perspective, subsequent developments may be considered as either (i) hydrological or (ii) hydraulic investigations. Hydrological studies have focused on the 'flow regime' and its inter-annual variability; groundwater and surface-water interactions especially in relation to water abstraction from major water-supply aquifers and river interactions with its alluvial aquifer; and channel-floodplain connectivity through inundation and drainage, with particular emphasis on nutrient dynamics. The timing, in relation to species life-cycles, and predictability of the seasonal high-flow stages have been shown to be significant, especially in relation to terrestrial and aquatic productivity (Junk *et al.*, 1989). Where the flow regime is predictable, biota may have adapted to this regime and developed profitable strategies for occupying the seasonally inundated riparian and floodplain areas. Thus, Bayley (1991) describes the 'flood pulse advantage' as the degree by which the annual multi-species fish yield exceeds that from a system with constant water level. An 'advantage' is also seen in the subsidy to production due to seasonal flooding of the riparian zone of higher vegetation (Stromberg, 1993; Ahn *et al.*, 2004), amphibians (Galat and Lipkin, 2000) and birds (Galat and Lipkin, 2000; Kingsford and Auld, 2005).

Hydraulic investigations have focused on channel-forming discharges, channel dynamics and habitat turnover within the channel-floodplain mosaic, and on instream flows and hydraulic stream ecology. The major driver of channel form, the structure of the floodplain habitat mosaic, and the frequency of habitat turnover is the range of high flows experienced for less than 2% of the time. These include the bankfull discharge, which is often taken to have a recurrence interval of 1.5 years on the annual series. More frequent, lower magnitude, 'spates' are important for maintaining instream habitats, not least by acting as flushing flows preventing the build-up of silt and organic debris (Reiser *et al.*, 1989; Milhous, 1998) and vegetation encroachment into a channel (Johnson, 1997; Merritt and Cooper, 2000). However, it is the action of high flows in connecting riparian and floodplain processes with in-channel processes, not least through the delivery of living and dead wood (Gurnell *et al.*, 2005), that sustains the high biocomplexity of river corridors.

Medium and low flows, together experienced for about 90% of the time along most rivers, sustain a diversity of aquatic ('hydraulic') habitats. The concept of 'hydraulic stream ecology' is centred on the principle that the relative difference in speed between an organism and the medium in which it lives affects the energy budget of that organism (Statzner *et al.*, 1988). Current velocity is significant for lotic organisms, influencing respiration and other measures of metabolism, feeding biology, and behavioural characteristics including rheotaxis, locomotory activity, schooling and territoriality. In natural channels, the complex channel morphology determined by floods, creates a heterogeneous hydraulic environment. Different species of animals have preferred 'hydraulic' habitats that may be simply defined, in terms of velocities, depths and substratum. Thus, habitat suitability criteria show the probability-of-use of microhabitats by a specific life stage or a particular species (e.g. Vilizzi *et al.*, 2000). In recent years, 'ecohydraulics' has emerged as a major discipline, following the First specialist International Symposium on Habitat Hydraulics in 1994, under the auspices of the International Association of Hydraulic research (Saltveit, 1996).

13.2.1 Principles for Sustainable River Regulation

Fundamentally, the ecological integrity of riverine ecosystems depends on their natural dynamic character (Poff *et al.*, 1997). Flow regimes reflect their hydro-climatic setting (Figure 13.1a) but both the seasonal pattern and inter-annual variability can be markedly

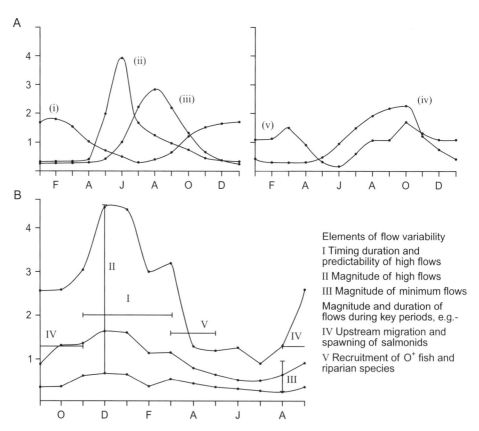

Figure 13.1 Analysis of a flow regime. (a) Typical flow regimes for the main ecoregions (i – temperate maritime, winter rainfall dominated; ii – nival, snowmelt dominated; iii – tropical, summer rainfall dominated) and the impact of flow regulation by Hume Dam on the flow regime of the River Murray, Australia (iv – before and v – after regulation; after Baker and Wright, 1978). (b) The maximum, mean and minimum monthly flows for a 35-year record from temperate river in Scotland with examples of the ecologically significant flows

altered by impoundments and abstractions from surface waters and groundwaters. Naiman *et al.* (2002) summarize the fundamental ecological principles for the sustainable management of water resources, focusing on the need to sustain *flow variability* that mimics the natural, climatically driven variability of flows at least from season to season and from year to year, if not from day to day (Figure 13.1b). There are three fundamental principles: the natural flow regime shapes the evolution of aquatic biota and ecological processes; every river has a characteristic flow regime and an associated biotic community; and every river sector has a channel form determined by the interaction of the flow regime with the available sediment load, within a valley of particular slope and width, modified by the type of riparian woody vegetation.

From these were elaborated three specific principles for advancing the provision of environmental flows (after Bunn and Arthington, 2002):

(i) Flow is a major determinant of physical habitat in rivers, which in turn is a major determinant of biotic composition.
(ii) Maintenance of the natural patterns of connectivity between habitats (a) along a river and (b) between a river and its riparian zone and floodplain, is essential to the viability of populations of many riverine species.
(iii) Aquatic species have evolved life history strategies primarily in response to the natural flow regime and the habitats that are available at different times of the year and in wet and dry years.

They added a fourth principle that has specific relevance to regulated rivers:

(iv) The invasion and success of exotic and introduced species along river corridors is facilitated by regulation of the flow regime, especially with the loss of natural wet–dry cycles.

Thus, the basic environmental knowledge needed to formulate policy decisions and management approaches on water allocations to meet environmental needs along rivers – the theoretical and conceptual basis of hydroecology – has been established. In practice, integrating (a) hydrology and biology, and (b) hydraulics and biology, and then the two groups for water resources management and river regulation, remains a major challenge (Petts *et al.*, 2006a).

13.3 Hydroecology in Water Management

The *principle* of including ecological issues in regulating river flows, particularly for populations of migratory fish, has been embedded in legislation for many years. In the UK, private acts towards the end of the Nineteenth century made provision for flows below dams, taking account not only of navigation, public health and the rights of downstream users, but also the protection of fisheries (Sheail, 1984). The Water Resources Act 1963 required the River Authorities to set 'minimum acceptable flows' (MAFS) and since then all new abstraction licenses have contained conditions to protect the water environment where necessary (Petts, 1996). The conditions include 'hands-off' flows (or 'flow reservations') that require abstractions to cease when flows fall below a specified discharge or water level and 'maintained flows' that under extreme low flow conditions require river support by groundwater pumping, reservoir releases, or inter-basin transfers.

In the UK, a definition of sustainable development has become embedded in water policy (DEFRA, 2002) but lacking scientifically sound techniques, no formal MAFS have been set. In England and Wales the Environment Agency (2001, 2004) is the statutory body with a duty for strategic water resource management – charged with managing the various demands on water so as to protect the long-term future of the water environment while encouraging sustainable development. However, at present 'the Agency is unable to fully meet all of its policy requirements . . . mainly because hydrological assessments are not sufficiently consistent or robust' (Environment Agency, 2004, p. 5). Specifically, the lack of appropriate, scientifically sound tools is impacting upon the needs of the European Union Habitats Directive and the UK Biodiversity Action Plan (BAP) as well as the agency's Catchment Abstraction Management Strategies (CAMS). The Habitats Directive and UKBAP seek to safeguard valuable nature conservation sites and threatened species, including wetlands, and to enhance biodiversity. CAMS are the agency's mechanism for managing abstractions through licensing and provide the framework for integrating ecological considerations within water resource regulation across England and Wales. Furthermore, water allocation to protect environmental needs still has relatively low priority among agencies and the public, not least in the aftermath of natural disasters, especially floods, and threats to national security. In the USA, for example, Hurricane Katrine (see the *Economist*, September 3rd 2005) is likely to have a lasting impact on political and social judgements. However, progress must still be made in advancing tools for water management (Petts *et al.*, 2006a).

13.3.1 Water Allocation to Protect Riverine Systems

The need to determine environmental flows to protect riverine ecosystems is a major driver of hydroecology (Petts *et al.*, 2006b). Early attempts to set instream flows for rivers were targeted at sport or game fisheries and focused on measures of dry weather flow (DWF). Until the late 1970s, river management remained more of an 'art' than a science. The developments during the late 1970s and 1980s were reviewed by Petts and Maddock (1994) and the promise of hydroecology in developing more sustainable approaches was promoted by Petts *et al.* (1995). By the early 1990s, the science and management of regulated rivers had expanded from the determination of instream flows that focused on a target species to *environmental flows*. These new approaches addressed the sustainability of communities and ecosystems. They recognised that a set of minimum flow constraints does not provide sufficient protection for river ecosystems; they incorporated flows not only for different times of the year but also for different years to meet the flow needs of the different species (and their life stages) within a riverine-dependent community (Petts, 1996; Richter *et al.*, 2006; Stalnaker, 1996).

13.3.2 Defining Ecologically Acceptable Flow Regimes

The management of water resources requires determination of the water volume to be reserved to meet the ecological needs of water-dependent communities along a river corridor. One approach to incorporating environmental flows into water resources management (Petts, 1996; Petts et al., 1999) requires the determination of (i) environmental flows to inform short-term and local operational rules, (ii) ecologically acceptable hydrographs to manage seasonal and short-series of inter-annual flows, and (iii) ecologically acceptable flow duration curves to aid long-term water resources planning. Fundamentally, the ecologically acceptable flow regime (EAFR) gives explicit recognition to the role of flow and water quality (especially temperature) *variability* and channel *dynamics* (i.e., changing patterns of hydraulic habitats) in evaluating water resource scenarios to manage riverine ecosystems.

The approach (Figure 13.2) requires consideration of the fundamental hydrological variables: (a) flow magnitude and timing, and (b) flow frequency and duration. First, the determination of environmental flows to meet specific ecological targets recognises that species will experience average years, poor years and good years, with varying flow conditions, or different lifestages will experience conditions that are more or less favourable at critical times. These environmental flows for target species can be determined using a range of hydroecological tools reviewed below. Each tool involves certain assumptions that introduce uncertainties into the determinations. However, the environmental flows may be used to define the basic operational rules for regulating rivers, for example, to set conditions attached to abstraction licences governing when and how much water may be taken.

The second step is as yet more uncertain and draws attention to the need for long-term coupled datasets (Petts et al., 2006a). Each environmental flow must be assigned an 'acceptable' frequency. This allows the specification of 'typical', wet-year and dry-year hydrographs. The process can be informed by combining flow time series with habitat-suitability curves for the target species to create habitat suitability time series. A set of habitat suitability time series representing the range of species within a river reach or

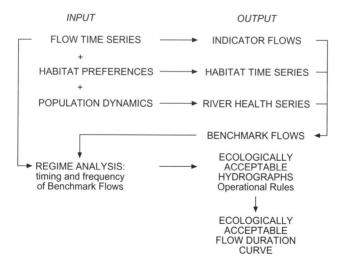

Figure 13.2 A process for establishing ecologically acceptable flow regimes

sector under consideration can be useful in illustrating the complex sequence of hydraulic habitat utilisation by a biological community. In the most sophisticated approaches, habitat time series and population dynamics are integrated to develop chronologies of 'river health' (e.g., Capra *et al.*, 1995).

The third step is to combine the 'ecologically acceptable hydrographs' into a flow duration curve for determining the long-term average allocation of the water volume required to meet the ecological targets. This could be compared to the historical series of gauged and naturalized flows as evidence in support of discussions about trade-offs between water users.

The above provides a hydroecological approach to allocating water but decisions about trade-offs require expert judgement. Riverine ecosystems are adapted to the variability of flows from year to year but at this time assigning frequencies and durations to environmental flows is largely arbitrary. Case studies of the ecological impacts of rare, high, floods or prolonged droughts may be informative but knowledge is required of the habitat and population recovery mechanisms if frequencies and durations are to be defined. A second problem is the definition of ecological targets for a catchment or river reach. These must be defined as precisely as possible to ensure that the available tools provide the 'best estimates' of the environmental flows. For example, 'end-of-summer flows maintaining at least 20 % of the optimum usable habitat for juvenile trout' or 'river levels ensuring that the riparian wetlands are inundated for at least 40 days between March and May'. However, it is inevitable that final decisions are based upon expert judgement. Is it acceptable to lose or gain habitats or species, or change the age structure of a population, or the density of an organism? And then there are questions about the location and spatial extent of the impact. Is the change acceptable in some reaches? Is it acceptable if it is restricted to one reach? Relatively little is known about the impact of spatial fragmentation along the river continuum upon the integrity of catchment ecosystems. Nevertheless, the evaluation of the EAFR can be an iterative process allowing the substitution of different ecological targets, and evaluation of flow-allocation scenarios in order to balance the needs of all users on the basis of the best available scientific information.

13.3.3 Determining Environmental Flows

River regulation rules informed by environmental-flow determinations have achieved significant water savings (e.g., Harman and Stewardson, 2005) but there is no single best method or approach to defining environmental flows (IUCN, 2003). Given the variety of water resource contexts, the range of environmental settings and variety of species concerned, there has been an increasing tendency to use evidence-based expert judgement. In reality, all environmental flow assessments provide the evidence for 'expert panels' – the decision-making process that evaluates tradeoffs between water users. From a flow-management perspective, these trade-offs include those between magnitude, frequency, duration, timing and rate of change (Poff and Ward, 1989) and the evidence is often hierarchically structured to include three orders (Petts, 1980) or levels (Young *et al.*, 2004), broadly linking primary fluvial processes, to habitat structure, and then to biota.

Tharme (2003) identifies over 200 approaches that have been described for advising on environmental flows in 44 countries. Broadly, environmental flow methods fall into four types (Table 13.3). These range from reconnaissance-level assessments relying on

Table 13.3 Summary of environmental-flow methods as four general types

Method	Examples	Source/selected references
Hydrological indices	Baxter method recommended 20% ADF to protect Atlantic salmon juveniles.	Baxter (1961)
	Montana Method recommended absolute minimum of 10% ADF with good habitat protected at 30% ADF.	Tennant (1976)
	Indicator of hydrological alteration method and regime indicators analyse the significance of descriptors of the annual flow regime.	Richter et al. (1996, 1997, 1998), Monk et al. (2006)
Habitat assessment	Channel form assessment using inflection in relationship of depth, wetted perimeter or velocity with discharge; also known as hydraulic geometry approach.	Petts et al. (1995) Stewardson and Gippel (2003), Jowett (1998)
	'Meso-habitat' or 'physical biotope assessment' provides a rapid assessment method and seeks to maximize 'habitat diversity'.	Newson and Newson (2000), Parasiewicz (2001), Dyer and Thoms (2006)
Biological response models	Relate channel hydraulics, substratum type and cover to habitat suitability criteria for target species (or life stage) to establish weighted usable area (WUA) available at different flows.	1-D hydraulic model (PHABSIM), Bovee (1978), Gore (1978) used world-wide, Tharme (2003), 2-D hydraulic models: Parasiewicz (2003), Stewart et al. (2005)
	Variants have sought to integrate population data for target species.	Capra et al. (1995)
Habitat-inclusive biological models.	Model community development as a function of biological processes (reproduction, growth, competition) and physical habitat dynamics.	Capra et al. (2003), Morales et al. (2006a, b)

ecologically informed hydrological, geomorphological, and hydraulic approaches to complex hydrodynamic habitat modelling and biological modelling. The first seek surrogate indices for conditions assumed to be suitable for target biota or limiting for 'ecosystem health'. The more advanced 'biological-response' approaches use 'habitat suitability' indices to represent preferences of different species and life-stages for physical habitats defined in terms of one or more micro-habitat variables (velocity, depth, substratum, cover). Recent advances (e.g., Morales et al., 2006a,b), however, suggest that major developments in the future may focus on models of biological processes of mortality, dispersal, reproduction, growth and competition that incorporate physical components.

Hydrological methods These rely on the statistical analysis of historical daily flow records. In their most simple form, environmental flows can be expressed as a hydrological statistic to define the Dry Weather Flow (DWF) most commonly for setting minimum flows to protect fish. A fixed percentage of the average daily flow (ADF) has been used, with some studies (Baxter, 1961; Tennant 1976) proposing 20% ADF to protect aquatic habitat in streams; severe habitat loss occurring below 10% ADF. In the UK, the preferred approach has been to use a flow duration statistic (such as the 95th percentile flow) or the mean annual minimum 7-day flow frequency statistic. Recognising that the DWF should be applied to natural, most sensitive rivers, Drake and Sheriff (1987) suggested that lower flows, such as 0.5 DWF, might be applied to rivers with degraded habitats that are less sensitive to low flows.

Recognition that flow variability is a primary determinant of species distribution between and within riverine systems has led researchers to analyse flow regimes in order to define 'ecologically relevant' hydrological variables. Thus, Richter *et al.* (1997) advanced the 'Indicators of Hydrologic Alteration' method (IHA; Richter *et al.*, 1996, 1998), using 32 hydrologic parameters for each year of flow record to characterize inter-annual variation before (reference period) and after flow regulation/abstraction. Using this approach, Galat and Lipkin (2000), for example, demonstrated that the impact of mainstream impoundment on the Missouri River was (i) a reduction in the magnitude and duration of the annual flood pulse; (ii) an increase in the magnitude and duration of annual discharge minima; (iii) a reduction in frequency of low-flow pulses; (iv) earlier timing of March–October low-flow pulses, and (v) a general increase in the frequency of flow reversals with a reduction in the range of flows.

Olden and Poff (2003) reviewed 171 hydrologic indices from 420 sites across continental USA to show that the IHA method successfully characterises all of the major components of the flow regime. The IHA provides a framework to identify the relative risks of ecological degradation, enabling the focusing of reach-based assessment and mitigation measures, but effective hydroecological calibration and validation requires long-term datasets and analyses of reference conditions. Nevertheless, in exploring the biological significance of the indices, specific alternative or additional indices can be added to the IHA. Thus, Galat and Lipkin (2000) used an additional variable specifically to include the Julian date of the vegetation growing season 1-day minimum flow (defining the growing season on the Missouri River, USA, as March 1st to October 31st). The timing of minimum flows during this period is relevant to the reproductive success of particular bird and turtle species that nest on exposed sand islands, to the germination of mud-flat plants, and for nursery areas for many fishes. For aquatic fauna, Wood *et al.* (2001a,b) isolated the significance of late-winter/spring high flows – especially the absence of high flows in drought years – for the summer, low-flow macroinvertebrate community within a groundwater-dominated, chalk stream in England. Using stepwise multiple regression Wood *et al.* (2001a) found that high flows (or their absence) 4–7 months prior to sampling (i.e., winter–spring) were most important in describing the summer, low-flow community. By developing a classification system of annual hydrograph shape and 'magnitude', they demonstrated that for this particular river, the timing of the hydrograph peak appears to be critical with an early (before February) and low magnitude hydrograph peak being correlated with relatively low invertebrate community abundance (Wood *et al.*, 2001b).

In Europe, the IHA is being evaluated for classifying river flow regime alterations to meet the requirements of the European Community Water Framework Directive (Black *et al.*, 2005). Monk *et al.* (2006) used a 20-year paired flow and macroinvertebrate survey record from 83 rivers in England and Wales to demonstrate the ecological importance of the 'flow regime' indices for macroinvertebrate community metrics at the regional scale, and highlighted the ecological importance of (i) monthly flows, and (ii) the magnitude and duration of annual extreme flow conditions within the temperate maritime climatic region of England and Wales.

One application of the statistical characterization of ecologically relevant hydrograph parameters would be to define the variability of the dimensions of the flow regime within which artificial influences should be contained (Richter *et al.*, 1997). Thus, Galat and Lipkin (2000) recommended changes in reservoir management to return the regulated flows to within the pattern of natural variability, thereby simulating a natural riverine ecosystem. They argued that naturalization of the flow regime would not only benefit the ecological system but also the economic value of the Missouri River, once the products of agriculture, electric-power generation and transportation are integrated with the socio-ecological benefits of a naturalized flow regime.

Habitat surveys   Advances continue to be made in developing simple systems to assess the condition of river reaches in terms of meso-scale habitat typologies using qualitative substratum types (Armitage *et al.*, 1995), physical biotopes (Newson and Newson, 2000) and meso-habitats (Parasiewicz, 2001). Biotopes (habitat units) are inferred from visual observation of surface flow character. Each habitat unit is assumed to be an area where an animal might be observed for a significant proportion of the time, which for a fish may correspond in size to the fundamental hydromorphologic units such as a pool, riffle or run and that these can be defined by hydraulic attributes and cover. The use of physical habitat mapping as a basis for analysing riverine ecosystems requires the establishment of the biological significance of the 'meso-habitats'. In this context, meso-habitats are equivalent to the 'functional units' of the fluvial hydrosystem model (see Amoros and Petts, 1993) within which groups of functional units (functional sets) characterise channel types (meandering, braided, etc.). Such approaches assume that biological communities have evolved to exploit the full range of meso-habitats; the variability of flows determining when and for how long meso-habitats are available to different species at different locations throughout the stream network. Thus, Dyer and Thoms (2006) concluded that a range of flows is required to optimise habitat diversity. Habitat duration curves provide summary statistics on average habitat availability and these could be developed to consider periods of habitat persistence related to key biological time-windows. Combining flow regime analyses (e.g., Harris *et al.*, 2000; Monk *et al.*, 2006) with reach scale channel classification systems, such as the template provided by Rosgen (1996) that includes not only a range of channel types but also degraded forms of those channel types, may provide a profitable direction for future research to establish a new tool for channel management and restoration. Furthermore, as noted by Parasiewicz (2001), if community structure reflects habitat structure then securing habitat for the most common species might preserve the most profound characteristics of the ecosystem and provide survival conditions for the majority of the aquatic community.

There is much debate about the identification and parameterization of 'physical habitat' at the 'meso-scale'. The attractiveness of this 'meso-scale' for managers is its practicality (Newson *et al.*, 1998). The cost of micro-scale surveys is prohibitive in many cases or where micro-scale approaches are used, transect spacing is often inadequate for characterizing habitat variability, especially in morphologically complex channels. Maddock (1999) concluded that questions remained about the appropriate spatial scales for such surveys to yield cost-effective returns for managers. Similarly, Clifford *et al.* (2006) consider a key question to be that of which distinct combinations of physical attributes may be recognized by field survey and mapped onto discrete biotic assemblages. A detailed analysis of flow biotopes and functional habitats in relation to physical habitat delimiters (flow depth, velocity and Froude number) failed to demonstrate linkages between physical units and biological performance (Clifford *et al.*, 2006). A significant element of the problem is the dynamic relationship between physical attributes and discharge, which reduces the coherence of particular biotope units. Thus, Emery *et al.* (2003) isolated four to six statistically distinct clusters that could be mapped back into the river reach under lower flows but this changed to a threefold division of zones around the channel centreline, channel margins and intermediate areas as stage increased; not only did the number of clusters reduce but their arrangement changed with increasing flow. Consequently, attempts to argue the biological significance of meso-scale hydraulic habitat surveys appear premature. Nevertheless, meso-scale studies can reflect 'real world' habitat complexity and provide a multidisciplinary tool for informing management strategies. Thus, such surveys have been included in the UK River Habitat Survey (Raven *et al.*, 1997) and could provide an important tool in respect of habitat audit issues within the European Water Framework Directive (Raven *et al.*, 2002).

Ecohydraulic models These include physical habitat models and biological response models. In the case of the first, attention has focused on implications of sediment transport (Milhous, 1982, 1998; Carling, 1995) and stable channel design (Bettess, 1996; Hey, 2006). Particular attention has been given to providing good quality habitat for fish spawning, to maintain the intra-gravel environment for egg development over winter, and for fry at emergence in late winter. One common problem associated with flow regulation and abstractions is siltation of the channel bed and this is a particular problem at salmonid spawning sites (Milhous, 1982; Reiser *et al.*, 1989). Over the past 20 years there has been a sustained effort to improve scientific understanding of channel siltation and to derive tools for determining the flows needed to flush fine sediments. These 'sediment-maintenance flows' are designed to flush fines without eroding the important underlying gravels. Wu and Chou (2004) present a simulation approach that allows systematic evaluation of flushing options incorporating (i) the initial bed sediment condition, (ii) the ultimate goal of flushing, i.e. the final bed sediment condition, including the loss of gravels, and (iii) the flushing discharge defined in terms of volume, magnitude, duration and timing. In another example, Baptist *et al.* (2004) proposed the Cyclic Floodplain Rejuvenation strategy to provide flood protection and biodiversity enhancement using management intervention to remove deposited sediment and reset vegetation succession. For the highly regulated River Waal, they used a one-dimensional hydraulic model together with rule-based models for sedimentation and vegetation succession, to show that safe flood levels and a vegetation mosaic including softwood forest, similar to the historic reference for the river, could be sustained when about 15% of the total floodplain area is rejuvenated with a return period of 25–35 years.

The most widely used biological-response model is Physical HABitat SIMulation (PHABSIM). This provides a framework for coupling the changing hydraulic conditions with discharge and the habitat preferences of one or more selected species (Bovee, 1978; Gore, 1978). It requires quality input data and this is often time consuming and expensive to obtain. Furthermore, the output is location and species/life-stage specific. Nevertheless, PHABSIM has been supported in a legal context, has had widespread application as a management tool, and has provided ecologists with a voice in water-resource decision making (Tharme, 2003). There is also no doubt that the accumulated experience means that its strengths and weaknesses are well understood.

PHABSIM relies on two principles: each species chosen exhibits preferences within a range of habitat conditions that it can tolerate, and the area of stream providing these conditions can be quantified as a function of discharge and channel structure. Considerable efforts have been spent on attempts to assess the ecological integrity of PHABSIM by demonstrating the biological significance of 'carrying capacity' as a limiting factor of population size. However validation of the approach in biological terms has proved difficult (Lamouroux et al., 1999; Kondolf et al., 2000) and problems have arisen in establishing discrete relationships between biological populations and the Weighted Usable Area (WUA) from empirically derived habitat suitability curves. Microhabitat availability may describe the potential for a particular species or life stage but the biomass of that species or life stage within a community can be reduced by a range of biological processes such as reproduction, energetics and mortality that may be influenced by one or more unspecified environmental factors.

One-dimensional hydraulic models such as the step-backwater hydraulic models that are most commonly used in PHABSIM, generalise hydraulic variability and fail to describe the variety of specific habitats by breaking a river reach into independent cells, each being described by a single velocity, depth or substratum value. Careful selection of cross sections within and between mesohabitats minimise difficulties. However, 1-D flow models cannot accurately predict spatial patterns of velocity in natural rivers that have significant lateral flow components, although they are useful for determining average velocity variations with changing discharge. Thus, some researchers have argued for the use of 2-D hydraulic models because of their potential accurately and explicitly to quantify spatial variations of flow patterns (Crowder and Diplas, 2000). Although most studies have sought to model large-scale flow patterns, recent studies have focused on local, biologically important flow complexity (Crowder and Diplas, 2006) and these may provide a link to meso-scale habitats and 'biotopes'. Two-dimensional models that calculate downstream and lateral flow components (and 3-D models that include vertical velocities) describe the spatial and temporal heterogeneity of hydraulic conditions, and therefore, habitat heterogeneity and diversity. Bovee (1996) and Hardy (1998) recommend 2-D approaches for computing landscape-ecology metrics across a variety of scales especially for simulating meso-habitat patterns. Thus, for rivers with high species richness and generalised fish habitat use patterns, Stewart et al. (2005) used a 2-D flow model, meso-habitat abundance, and biomass estimates from multiple sites to predict the effect of discharge on adult standing stocks of two native fish species, showing a very strong association with indices of habitat heterogeneity.

Parasiewicz (2003) advanced the approach to modelling mesohabitats using 'Meso-HABSIM'. This maps mesohabitats at different flows along extensive sections of a river, establishes the suitability of each mesohabitat for the dominant members of the fish community, and derives rating curves to describe changes in relative areas of suitable habitat in response to flow. The rating curves can be developed for a mesohabitat, reach or sector, and it is this scaling that enables easy translation of model outputs into management. Mesohabitat approaches also address the problem of needing to consider multiple species and community structure.

Habitat suitability criteria   Notwithstanding the strengths and weaknesses of the hydraulic models, in practice, the quality of the suitability criteria may have the strongest influence on the output quality of any of the biological-response models. Simple indices are based on frequency of occurrence of actual habitat conditions used by a target organism in a particular reach. More complex indices combine these with additional information on the availability of habitat. In the latter, the ratio of the proportion of habitat utilised to available habitat area within the reach defines the habitat preference. Habitat use or preference indices for a number of physical variables may be combined to obtain composite indices (e.g., Vadas and Orth, 2001) but these involve a number of assumptions (Bovee, 1986): multiplication of suitability indices for individual habitat variables assumes that the target species selects each physical habitat variable independently of other variables; using the arithmetic mean of the suitability indices assumes that good habitat conditions on one variable can compensate for bad habitat conditions on another, and using the lowest suitability index assumes that the most limiting factor determines habitat suitability. Composite suitability indices also assume that all physical variables are equally important and independent. The statistical methodologies for analysing species–environment relationships have been reviewed by Ahmadi-Nedushan *et al.* (2006) who focus on multivariate approaches for defining composite indices and the analysis of multi-species data.

New research is needed to refine our basic knowledge of biological responses to flow conditions, including the moisture requirements of specific plant species and inundation tolerances for floodplain fauna. For example, an important food source for waterfowl during their spring and autumn migrations along the Mississippi River is provided by seeds of millets (*Echinochloa* sp.) that grow on floodplain mudflats exposed when seasonal floods recede. Ahn *et al.* (2004) demonstrated that increased water levels that drown the plants during the summer growing season had caused a decline in productivity. Using a moist-soil plant model incorporating daily water depth, flood timing within the growing season, and flood duration, they demonstrated that a dry period of more than 85 days is required during the growing season to achieve at least 50% of maximum production. In another recent example, Kingsford and Auld (2005) used 25 years of breeding data for ten species of colonial waterbirds in the Macquarie Marshes, Australia, to show that the number of waterbird nests was positively related to the flow prior to breeding and area inundated, and that breeding was triggered by a threshold flow. Such information has immediate value by informing river regulation policies.

It is clear that major advances require long-term ecological datasets coupled with hydrological and water-quality datasets, together with 'reference' data from relatively undisturbed catchments. Attempts have been made to couple the characterisation of flow and temperature regimes (e.g., Harris *et al.*, 2000) but these must be linked to flow- and temperature-sensitive biological processes, such as the timing of smolting in fish. Thus, Schramm and Eggleton (2006), with reference to catfish growth in the lower Mississippi river, USA, demonstrated that the flood-pulse concept applies more strongly to temperate floodplain–river ecosystems when thermal aspects of flood pulses are considered. Particular complications can arise where streams are influenced by groundwater upwelling, which not only influences flows but also water quality, for example producing deoxygenated conditions in spawning gravels and poor egg survival (Malcolm et al., 2005).

Habitat-inclusive biological models From a biological perspective, understanding responses of biota to habitat temporal variability is needed to identify the magnitude, duration and frequency of habitat limiting periods or carrying capacity (Capra et al., 1995) and this requires long-term and coupled hydrological, hydraulic (reflecting the dynamics of channel morphology) and biological time series. Much research effort has focused on summer minimum flows but recent research on responses of trout populations to flow variability has suggested the importance of winter flows (Cattaneo et al., 2002; Lobon-Cervia, 2003; Mitro et al., 2003) in determining recruitment. Sabaton et al. (1997) and Gouraud et al. (2001) used 'Modypop' – a matrix model based on age classes – to simulate changes in age class of trout using habitat-dependent parameters: summer low flows that limit adult trout biomass and spring flows that limit the young-of-the-year numbers between emergence and their first summer. Capra et al. (2003) showed that post-emergence high flows have a major impact on the density of 0+ fishes. However, whilst such models are useful for developing and testing concepts, their role as management tools remains limited until they can be validated by empirical studies involving multisite sampling. Halls and Welcomme (2004) used an age-structured population dynamics model, incorporating density-dependent growth, mortality and recruitment to show that exploitable biomass of a common floodplain fish species is maximised by minimising the rate of flood recession and maximising the flood duration and area inundated. However, they also highlighted the need for additional research on spawning behaviour, system primary production and critical habitat availability.

In a most promising development, Morales et al. (2006a) advance a hydroinformatics tool to integrate principles from biology and river hydraulics. They illustrate their approach by considering freshwater unionid mussels that have a complex life cycle, including a period as obligate parasites of fish. The life-cycle stages – egg/larvae, parasitic, juvenile, adult – depend on regional-scale (flow regime and water quality) and reach-scale (hydraulic habitats) environmental conditions over decadal timescales (their life spans). The 'mussel dynamics model' predicts community development as a function of individual growth and reproduction, and simulates interactions between individuals to estimate population dynamics. Biotic interactions include relationships with host fish and intra- and inter-species food competition. Mussel-environment interactions for the different life stages are captured at a scale of 1 m^2 using habitat suitability rules for the various parameters considered. In this case, hydraulic habitat was defined by a dimensionless shear–stress ratio – a measure of substrate stability – the shear stress at given flow rate as a function of the shear stress at onset of sediment motion (Morales et al., 2006b). Habitat suitability was evaluated for various flows covering the expected yearly hydrograph, recognising that each life stage is temperature dependent. Classically, the model demonstrated that for low-density species, even a small level of habitat modification could have a substantial impact on population survival.

13.4 Applications to Water Resource Problems

Once the environmental flows have been determined to meet the ecological targets, they can be combined (Figure 13.2) to define the EAFR. Thus, Petts (1996) and Petts *et al.* (1999) used five fundamental environmental flows to define the ecologically acceptable hydrographs and flow duration curve for the River Babingley, UK (Figure 13.3). Habitat surveys and studies of Brown trout (*Salmo trutta*) and six species of invertebrate led to the definition of: (i) a channel maintenance flow (the bankfull discharge); (ii) a flow that

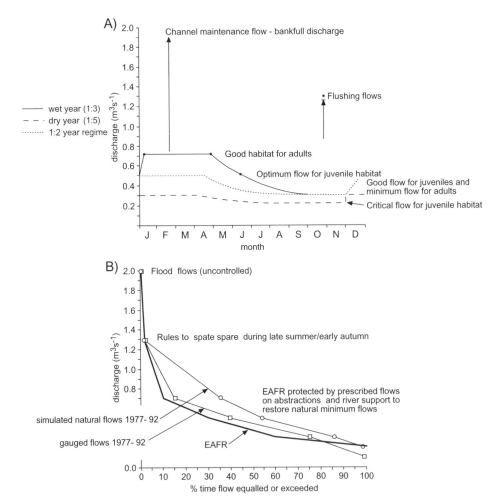

Figure 13.3 An example of an ecologically acceptable flow regime derived from the process outlined in Figure 13.2 applied to the River Babingley, UK (after Petts, 1996; Petts et al., 1999). (A) ecologically acceptable hydrographs, and (B) the ecologically acceptable flow duration curve compared with the duration curves for gauged flows and simulated natural flows taking into account all abstractions and discharges (e.g., from waste-water treatment works)

maximizes usable habitat for adult trout; (iii) a flow that sustains connectivity between, and usable habitats for trout in, all reaches; (iv) the flow that sustains 'normal' low flow habitats, and (v) the flow that sustains a few habitat refuges and below which all habitat for the target species will be lost. The EAFR defines (a) the water volume required to protect the river as a trout stream, and (b) the minimum maintained flow. The area between the simulated natural flow duration curve and the EAFR defines the volume of water available to be licensed for abstraction. Prescribed flow conditions in abstraction licences protect the ecologically acceptable hydrographs using 'hands-off' flows to stop abstraction. The extent to which the gauged regime fell below the EAFR is a measure of the need for flow alleviation during dry years.

Approaches to integrating water resource and river management considerations within the process of determining environmental flows include a range of 'holistic methodologies'. All are inter-disciplinary and science based, and use an adaptive management framework to implement strategic, incremental actions to reduce environmental risk in the face other human demands for water. A common feature of all 'holistic' approaches is that they allow environmental and ecological benefits from water allocations to be weighed against those to be derived from other environmental management actions, such as morphological restoration, riparian and floodplain re-establishment, and catchment land-use management.

In data-rich areas, suites of models may be integrated to achieve the holistic outcomes. Thus, Bratrich *et al.* (2004) advanced an environmental management matrix to address five management domains: instream flows; hydropeaking; reservoir management; bedload management, and power-plant structure. Their matrix includes hydrological character, drainage-network connectivity, channel morphology, river corridor landscape and biological communities. They used dynamic habitat models to quantify the effects of different instream flow regulation schemes at morphologically distinct sites.

In other areas, such as Australia and southern Africa, a lack of ecological data and process models, and political pressure to deliver environmental flow recommendations in short time frames, often less than one year, has led to reliance on multidisciplinary expert panels (Cottingham *et al.*, 2002; Young *et al.*, 2004). King *et al.* (2003) developed a value-based system (DRIFT). Their approach links the productivity of large floodplain rivers to their flow characteristics and was produced for a developing region with severe water shortage, and where riparian subsistence human populations are a significant factor. DRIFT (downstream response to imposed flow transformation) provides a data-management tool for many types and sources of information, predictive models, theoretical principles and 'expert knowledge' of a panel of scientists. DRIFT is based around four modules: (i) the biophysical module describes the present nature and functioning of the ecosystem; (ii) the sociological module identifies subsistence users at risk from flow abstraction or regulation; (iii) a module that combines the outputs from the first two to develop biophysical and subsistence scenarios, and (iv) a module to address mitigation and compensation costs. Arthington *et al.* (2003) applied DRIFT to establish environmental flow requirements of fish in Lesotho rivers and proposed that the methodology can provide a best practice framework for conducting scientific panel studies, although they acknowledged a number of risks with the approach.

Richter *et al.* (2006) present a five-step process for developing environmental flow recommendations, at the heart of which is the expert judgement of the range of stakeholders – best available scientific knowledge – but with the emphasis on trials and monitoring; an iterative process that seeks refinement as a result of experiment. Each flow allocation is seen as an opportunity for scientific experiment, the outcomes of which will further inform the flow allocation process. Thus, in their case study of the Savannah River, the movements of target fish species were monitored during the implementation of a high-pulse release from a dam to verify that the fish were able to move upstream past a lock-and-dam structure that had previously impeded spawning migrations. In a second 'experiment', tracking seed establishment during growing seasons in which flows were adjusted for this purpose was used to monitor the recruitment of floodplain tree species.

13.4.1 Communication and Policy Development

In parallel with advances in hydroecology as a multidisciplinary science, advances must be made in embedding hydroecology in water resources management and policy making. Of course, policy making and decision making require the best available evidence-based information founded on sound science, but there remains a need to promote the view of riverine ecosystems as legitimate users of water, requiring the same level of respect, advocacy, and protection as that allocated to other societal needs.

At the Earth Summit of 1992 *The Rio Declaration* and *Agenda 21* (UNCED, 1992) placed biodiversity, ecosystems and public and stakeholder participation at the head of the 'development' agenda. Scientifically based tools for water management have been available for some years (e.g., Calow and Petts, 1992, 1994) but we still need to close the debate about unsustainable exploitation or scientifically based stewardship. The reality is that environmental management is mostly about people and their behaviours, not about ecosystems. The role of science is to provide society with the best assessment of ecological reality (Lackey, 2001). Society's values and priorities drive water policy. Furthermore, the short decision time frames employed by the hierarchy of politicians and officials in natural resource management constrains the role that science plays. The holistic approaches to establishing environmental flows, such as DRIFT, that involve multidisciplinary panels of experts, provide an important framework for promoting hydroecological knowledge and tools in water allocation. Indeed, DRIFT was designed to run in parallel to a public participation process (King *et al.*, 2003) that provides (i) information on river concerns, which need to be described in each scenario, such as recreational fishery, a conservation area, a riverside business or a flood hazard, and (ii) a continuing process of information exchange and education allowing all parties to understand and evaluate options. Hydroecology as an integrative and applied science has an important role to play in communicating knowledge about the dynamics and sustainability of river ecosystems to ensure that the decision-making process is fully informed.

By the start of this millennium there had developed a vanguard movement to restore regulated rivers and to reconnect rivers with their floodplains, to advance decision-making processes to balance consumptive water use with that needed to maintain a healthy river, and to allocating our limited water resources to meet the needs of both humans and riverine ecosystems (Postel and Richter, 2003). However, a realism had evolved in river management, not least among the developed nations, that neither supports the stewardship model of environmental responsibility nor addresses the call for society to balance human demands with the needs of the world's rivers (Postel and Richter, 2003). A vision of contemporary ecological reality, presented for the western United States by Lackey (2003), is adaptable and applicable to western Europe. Here, wild Atlantic salmon are close to attaining a status enjoyed by wolves, brown bears and beaver, wild animals that are unlikely to disappear entirely, but struggle to survive as remnants of once flourishing species in limited areas of their original range, and where fisheries are sustained by hatchery-bred fish. In these managed rivers, the concept of ecological health relates to societal values, not science.

Hydroecology has an important role in supporting the development of science to calibrate indicators of health but there are limited opportunities to develop new understanding of natural or pristine systems having a high degree of ecological integrity to define 'benchmark' or 'reference' conditions (Tockner *et al*., 2003). Such opportunities must be taken by supporting hydroecological research on reference rivers. However, the underlying issue is that scientifically informed management interventions are futile without the political will to tackle the core drivers. Using the example of wild salmon in the Pacific Northwest, Lackey (2003) argued that despite significant advances in science that have demonstrated the direct causes of the population decline, and growth in public support to reverse the 150-year downward trend, the recovery prognosis for wild salmon is poor. A number of factors influence how societies prioritize water use and water use remains, at best, only one of a number of priorities that society ranks high. Lackey (2003) argued that substantial and pervasive changes in individual and collective preferences must take place if sustainable water management is to be realised. These preferences are driven by many factors including trends in international trade to improve economic efficiency and continued growth in population levels and their aggregate demands to support chosen life styles leading to increasing competition for increasingly scarce water resources.

13.5 Conclusions

'How much water must be allocated to protect river ecosystems?' remains a key question that we are far from answering. What we now know is that it requires understanding of the direct and indirect interactions between flows and biota over a range of time and space scales. Ten years ago, Stanford *et al*. (1996) presented a general protocol for restoring regulated rivers that requires a management system founded on natural habitat restoration and maintenance. Hydrological variability in discharge and water temperature, and changes of channel and floodplain form that determine the mosaic of hydraulic (velocity dependent) and hydrological (inundation dependent) patches, are now recognised as key drivers of river ecosystems. This chapter presents the interdisciplinary science of hydroecology not only as a conceptual and theoretical framework but also as a structured toolbox of approaches to manage water resources and achieve sustainable river regulation. The toolbox comprises: the four types of approach to defining environmental flows; the ecologically acceptable flow regime that provides a template for integrating environmental flows into water resources assessment and abstraction management; and the holistic approaches that provide the vehicle for incorporating flow assessments into planning and the decision-making process. However, to improve the hydroecological models requires coupled datasets linking habitat and biota over timescales of decades.

Our immediate challenge is to place water resource development within the context of fundamental ecological principles in order to maintain the ecological vitality (i.e., goods and services) of the fluvial system. This requires:

- more information about how habitat fragmentation affects organisms, and how the effects vary with location in a catchment;
- more information about how the fluvial system responds to *accelerated rates* of flow variability;
- more information about how biota respond to *sequences* of physical habitat changes and the timescales of these responses;
- interdisciplinary research on the few remaining 'nearly natural' large rivers to establish reference conditions;
- coupled physical–biological monitoring of management actions on rivers over significant timescales (at least 10 years) to provide data for more sophisticated habitat-inclusive biological models;
- active manipulations and experiments to measure ecosystem responses in order to distinguish the effects of direct human impacts on flow regimes (e.g., by dams, land-use change) from the effects of climate change.

There have been major advances in science over the past decade but limited developments in sustainable water and river regulation. Holistic strategies are being developed in the context of adaptive management but a continuing failure by policy makers to give due recognition to the array of goods, services and other benefits provided by aquatic ecosystems remains a major constraint to progress. There is still an infatuation with maximizing economic yield, a belief that technology provides the solution to environmental risks, and a danger that adaptive management could come to focus more on participant satisfaction with the process than on the environmental sustainability of the outcome. Hydroecology provides the collective interdisciplinary expertise to provide improved analyses of ecological problems, stronger tools to aid the decision-making process and better evidence to inform stakeholder participation. Hydroecologists must engage more closely with the policy agendas to ensure that this potential is realized in practice.

References

Achleitner S., de Toffol S., Engelhard C. and W. Rauch. 2005. The European Water Framework Directive: water quality classification and implications to engineering planning. *Environmental Management*, **35**, 517–525.

Ahmadi-Nedushan, B., St-Hilaire A., Berube M., Robichaud E., Thiemonge N. and B. Bobee. 2006. A review of statistical methods for the evaluation of aquatic habitat suitability for instream flow assessment. *River Research and Applications*, **22**, 503–524.

Ahn C., Sparks R.E. and D.C White. 2004. A dynamic model to predict responses of millets (*Echinochloa* sp.) to differen hydrological conditions for the Illinois floodplain-river. *River Research and Applications*, **20**, 485–498.

Amoros C. and G.E. Petts (eds). 1993. *Hydrosystemes Fluviaux*. Masson, Paris.

Armitage P.D. 1976. A quantitative study of the invertebrate fauna of the River Tees below Cow Green reservoir. *Freshwater Biology*, **6**, 229–240.

Armitage P.D., Pardo I. and A. Brown. 1995. Temporal constancy of faunal assemblages in meso-habitats: application to management. *Archive für Hydrobiologie*, **133**, 367–387.

Arthington A.H., Rall J.L., Kennard M.J. and B.J. Pusey. 2003. Environmental flow requirements of fish in Lesotho Rivers using the DRIFT methodology, *River Research and Applications*, **19**, 641–666.

Baker B.W. and G.L. Wright. 1978. The Murray Valley: its hydrologic regime and the effects of water development on the river. *Proceedings of the Royal Society of Victoria*, **90**, 103–110.

Baptist M.J., Penning W.E., Duel H., Smits A.J.M., Geerling G.W., van der Lee G.E.M. and J.S.L. van Alphen. 2004. Assessment of the effects of cyclic floodplain rejuvenation on flood levels and biodiversity along the Rhine River. *River Research and Applications*, **20**, 285–298.

Barth F. and J. Flawell. 2001. The Water Framework Directive and European Water Policy. *Ecotoxicology and Environmental Safety*, **50**, 103–105.

Baxter G. 1961. River utilization and the preservation of migratory fish life. *Proceedings of the Institution of Civil Engineers*, **18**, 225–244.

Bayley P.B. 1991. The flood-pulse advantage and the restoration of river-floodplain systems. *Regulated Rivers*, **6**, 75–86.

Bettess R. 1996. Sediment transport and channel stability. In Calow P. and G.E. Petts (eds) *The Rivers Handbook*, Blackwell Scientific, Oxford, Vol. 2, pp. 227–253.

Black A.R., Rowan J.S., Duck R.W., Bragg O.M. and B.E. Clelland. 2005. DHRAM: a method for classifying river flow regime alterations for the EC Water Framework Directive. *Aquatic Conservation*, **15**, 427–446.

Bovee K.D. 1978. The incremental method of assessing habitat potential for coolwater species, with management implications. *American Fisheries Society Special Publication*, **11**, 340–346.

Bovee K.D. 1986. Development and evaluation of habitat suitability criteria for use in the instream flow incremental methodology. *Instream Flow Information Paper*, **21**, US Fish and Wildlife Service Biological Report No. 86.

Bovee K.D. 1996. Perspectives on two-dimensional river habitat models: the PHABSIM experience. *Proceedings of the Second International Symposium on Habitat Hydraulics*, INRS-Eau, Quebec, Canada, B149–162.

Bratrich C., Truffer B., Jorde K., Markard J., Meier W., Peter A., Schneider M. and B. Wehrli. 2004. Green hydropower: a new assessment procedure for river management. *River Research and Applications*, **20**, 865–882.

Bunn S.E. and A.H. Arthington. 2002. Basic principles and ecological consequences of altered flow regimes for aquatic biodiversity. *Environmental Management*, **30**, 492–507.

Calow P. and G.E. Petts (eds). 1992. *The Rivers Handbook*, Volume 1. Blackwell Scientific Publications, Oxford.

Calow P. and G.E. Petts (eds). 1994. *The Rivers Handbook*, Volume 2. Blackwell Scientific Publications, Oxford.

Capra H., Breil P. and Y. Souchon. 1995. A new tool to interpret magnitude and duration of fish habitat variations. *Regulated Rivers*, **10**, 291–289.

Capra H., Sabaton C., Gouraud V., Souchon Y. and P. Lim. 2003. A population dynamics model and habitat simulation as a tool to predict Brown Trout demography in natural and bypassed stream reaches. *River Research and Applications*, **19**, 551–568.

Carling P. 1995. Implications of sediment transport for instream modelling of aquatic habitat. In Harper D.M. and A.J.D. Ferguson (eds) *The Ecological Basis for River Management*, John Wiley & Sons, Ltd, Chichester, pp. 17–32.

Cattanéo F., N. Lamouroux, P. Breil and H. Capra. 2002. The influence of hydrological and biotic processes on brown trout (*Salmo trutta*) population dynamics. *Canadian Journal of Fisheries and Aquatic Sciences*, **59**, 12–22.

Clifford N.J., Harmar O.P., Harvey G. and G.E. Petts 2006. Physical habitat, eco-hydraulics and river design: a review and re0-evaluation of some popular concepts and methods. Aquatic Conservation, in press.

Cottingham P., Thoms M.C. and G.P. Quinn. 2002. Scientific panels and their use in environmental flow assessment in Australia. *Australian Journal of Water Resources*, **5**, 103–111.

Crowder D.W. and P. Diplas. 2000. Using two-dimensional hydrodynamic models at scales of ecological importance. *Journal of Hydrology*, **230**, 172–191.

Crowder D.W. and P. Diplas. 2006. Applying spatial hydraulic principles to quantify stream habitat. *River Research and Applications*, **22**, 79–90.

Davies B.R. 1986. The Zambezi River System. In Davies B.R. and K.F. Walker (eds) *The Ecology of River Systems*, Dr W. Junk, Dordrecht, pp. 225–266.

Davies B.R. and M.J. Wishart. 2000. River conservation in the countries of the Southern African Development Community (SADC). In Boon P.J., Davies B.R. and G.E. Petts (eds) *Global Perspectives on River Conservation*. John Wiley & Sons, Ltd, Chichester, 179–204.

DEFRA. 2002. *Directing the Flow: Priorities for Future Water Policy*. Department of the Environment, Food and Rural Affairs, London.

Drake P.J. and J.D.F. Sherriff. 1987. A method for managing river abstractions and protecting the environment. *Journal of Water and Environmental Management*, **1**, 371–386.

Dumont H.J. 1986. The Nile River System. In Davies B.R. and K.F. Walker (eds) *The Ecology of River Systems*. Dr W. Junk, Dordrecht, pp. 61–74.

Dyer F.J. and M.C. Thoms. 2006. Managing river flows for hydraulic diversity: an example of an upland regulated gravel-bed river. *River Research and Applications*, **22**, 257–268.

Economist. 2005. A city cilenced. *The Economist*, September 3rd, 11.

Emery J.C., Gurnell A.M., Clifford N.J., Petts G.E. and P.J. Soar. 2003. Classifying the performance of riffle-pool bedforms for habitat assessment and river rehabilitation design. *River Research and Applications*, **19**, 533–549.

Environment Agency. 2001. *Our vision: Making it Happen*. Environment Agency, Bristol, UK.

Environment Agency. 2004. *Hydroecology: Integration for Modern Regulation*. Environment Agency, Bristol, UK.

Galat, D.L. and R. Lipkin. 2000. Restoring the ecological integrity of great rivers: historical hydrographs aid in defining reference conditions for the Missouri River. In Jungwirth M., Muhar S. and S. Schmutz (eds) *Assessing the Ecological Integrity of Running Waters, Development in Hydrobiology* **149**, Kluwer Academic Dordrecht, The Netherlands, pp. 29–48.

Gill M.A. 1971. Damming the McKenzie: a theoretical assessment of the long-term influence of impoundment on the ecology of the MacKenzie River Delta. In *Proceedings Peace–Athabaska Delta Symposium*, Water Resources Centre, University of Alberta, Edmonton, pp. 204–222.

Gleick P.H. 1998. *The World's Water*. Island Press, Washington DC.

Gore J.A. 1978. Technique for predicting instream flow requirements of benthic macroinvertebrates. *Freshwater Biology*, **8**, 141–151.

Gouraud V., Baglinière J.L., Baran P., Sabaton C., Lim P. and D. Ombredane. 2001. Factors regulating brown trout populations in two French rivers: application of a dynamic population model. *Regulated Rivers*, **17**, 557–569.

Gurnell A.M. and G.E. Petts. 2002. Island-dominated landscapes of large floodplain rivers, a European perspective. *Freshwater Biology*, **47**, 581–600.

Gurnell A.M., Tockner K., Edwards P. and G.E. Petts. 2005. Effects of deposited wood on biocomplexity of river corridors. *Frontiers in Ecology and Environment*, **3**, 377–382.

Halls A.S. and R.L. Welcomme. 2004. Dynamics of river fish populations in response to hydrological conditions: a simulation study. *River Research and Applications*, **20**, 985–1000.

Hardy T.B. 1998. The future of habitat modelling and instream flow assessment techniques. *Regulated Rivers*, **14**, 405–420.

Harman C. and M. Stewardson. 2005. Optimizing dam release rules to meet environmental flow targets. *River Research and Applications*, **21**, 113–130.

Harris N., Gurnell A.M., Hannah D.M, and G.E. Petts. 2000. Classification of river regimes: a context for hydroecology. *Hydrological Processes*, **14**, 2831–2848.

Hey R.D. 2006. Fluvial geomorphological methodology for natural stable channel design. *Journal of the American Water Resources Association*, 357–374.

IUCN. 2003. *Flow. The Essentials of Environmental Flows*. International Union for Conservation of Nature and Natural Resources, Gland Switzerland.

Janauer G.A. 2000. Ecohydrology: fusing concepts and scales. *Ecological Engineering*, **16**, 9–16.
Johnson W.C. 1997. Equilibrium response of riparian vegetation to flow regulation in the Platte River Nebraska. *Regulated Rivers*, **13**, 403–416.
Jowett I.G. 1998. Hydraulic geometry of New Zealand rivers and its use as a preliminary method of habitat assessment. *Regulated Rivers*, **14**, 451–466.
Junk W.J., Bayley P.B. and R. E. Sparks. 1989. The flood pulse concept in river-floodplain systems. In D.P. Dodge (ed.) *Proceedings of the International Large River Symposium*. Canadian Special Publication Fisheries Aquatic Sciences, **106**, 110–127.
Khaiter P.A., Nikanorov A.M., Yereschukova M.G., Prach K., Vadineanu A., Oldfield J. and G.E. Petts. 2000. River conservation in central and eastern Europe (incorporating the European parts of the Russian Federation). In Boon P.J., Davies B.R. and G.E. Petts (eds) *Global Perspectives on River Conservation*, John Wiley & Sons, Ltd, Chichester, pp. 105–126.
King J.M., Brown C. and H. Sabet. 2003. A scenario-based holistic approach to environmental flow assessments for rivers. *River Research and Applications*, **19**, 619–639.
Kingsford R.T. and K.M. Auld. 2005. Waterbird breeding and environmental flow management in the Macquarie Marshes, arid Australia. *River Research and Applications*, **21**, 187–200.
Kondolf G.M., Larsen E.W. and J.G. Williams. 2000. Measuring and modeling the hydraulic environment for assessing instream flows. *North American Journal of Fisheries Management*, **20**, 1016–1028.
Lackey R.T. 2001. Defending reality. *Fisheries*, **26**, 26–27.
Lackey R.T. 2003. A salmon-centric view of the 21st Century in the western United States. *Renewable Resources Journal*, **21**, 11–15.
Lamouroux N., Capra H., Pouilly M. and Y. Souchon. 1999. Fish habitat preferences in large streams of southern France. *Freshwater Biology*, **42**, 673–687.
Lobon-Cervia J. 2003. Spatio-temporal dynamics of brown trout production in a Cantabrian stream: effects of density and habitat quality. *Transactions of the American Fisheries Society*, **132**, 621–637.
Maddock I. 1999. The importance of physical habitat assessment for evaluating river health. *Freshwater Biology*, **41**, 373–391.
Malcolm I.A., Soulsby C., Youngson A.F. and D.M. Hannah. 2005. Catchment-scale controls on groundwater–surface water interactions in the hydorheic zone: implications for salmon embryo survival. *River Research and Applications*, **21**, 977–990.
Merritt D.M. and D.J. Cooper. 2000. Riparian vegetation and channel change in response to river regulation: a comparative study of regulated and unregulated streams in the Green River basin, USA. *Regulated Rivers*, **16**, 543–564.
Milhous R.T. 1982. Effect of sediment transport and flow regulation on the ecology of gravel-bed rivers. In Hey R.D., Bathurst J.C. and C.R. Thorne (eds) *Gravel-bed Rivers: Fluvial Processes, Engineering and Management*. John Wiley & Sons, Ltd, Chichester, pp. 819–842.
Milhous R.T. 1998. Modelling of instream flow needs: the link between sediment and aquatic habitat. *Regulated Rivers*, **14**, 79–94.
Mitro M.G., Zale A.V. and B.A. Rich. 2003. The relation between age-0 rainbow trout *(Oncorhynchus mykiss)* abundance and winter discharge in a regulated river. *Canadian Journal of Fisheries and Aquatic Sciences*, **60**, 135–139.
Monk, W.A., Wood, P.J., Hannah, D.M., Wilson A., Extence C.A. and R.P. Chadd. 2006. Flow variability and macroinvertebrate community response within riverine systems. *River Research and Applications*, **22**, 595–615.
Morales Y., Weber L.J., Mynett A.E. and T.J. Newton. 2006a. Mussel dynamics model: a hydroinformatics tool for analysing the effects of different stressors on the dynamics of freshwater mussel communities. *Ecological Modelling*. **197**, 448–460.
Morales Y., Weber L.J., Mynett A.E. and T.J. Newton. 2006b. Effects of substrate and hydrodynamic conditions on the formation of mussel beds in a large river. *Journal of the North American Benthological Society*. **25**, 664–676.
Naiman R.J., Bunn S.E., Nilsson C., Petts G.E., Pinay G. and L. Thompson. 2002. Legitimizing fluvial ecosystems as users of water: an overview. *Environment Management*, **30**, 455–467.

Newson M.D. and C.L.Newson. 2000. Geomorphology, ecology and river channel habitat: mesoscale to basin-scale challenges. *Progress in Physical Geography*, **24**, 195–217.

Newson, M.D., Harper D.M., Padmore C.L., Kemp J.L. and B. Vogel. 1998. A cost-effective approach for linking habitats, flow types and species requirements. *Aquatic Conservation*, **8**, 431–446.

Olden J. and N.L. Poff. 2003. Redundancy and the choice of hydrologic indices for characterizing streamflow regimes. *River Research and Applications*, **19**, 101–121.

Parasiewicz P. 2001. MesoHABSIM: A concept for application of instream flow models in river restoration planning. *Fisheries*, **26**, 6–13.

Parasiewicz P. 2003. Upscaling: integrating habitat model into river management. *Canadian Water Resources Journal*, Special Issue on Habitat Modelling and Conservation of Flows, **28**, 283–300.

Penaz M., Kubicek F., Marvan P. and M. Zelinka. 1968. Influence of the Vir River Valley Reservoir on the hydrobiogiocal and ichthyological conditions in the River Svratka. *Acta Scientiarum naturalium Academiae Scientiarum bohemoslovacae-Brno*, **2**, 1–60.

Petr T. 1986. The Volta River System. In Davies B.R. and K.F. Walker (eds) *The Ecology of River Systems*. Dr W. Junk, Dordrecht, pp. 163–184.

Petts G.E. 1980. Long-term consequences of upstream impoundment. *Environmental Conservation*, **7**, 325–332.

Petts G.E. 1984. *Impounded Rivers*. John Wiley & Sons, Ltd, Chichester.

Petts G.E. 1996. Water allocation to protect river ecosystems. *Regulated Rivers*, **12**, 353–367.

Petts G.E. and Maddock I. 1994. Flow allocation for in-river needs. In Calow P. and G.E. Petts (eds) *Rivers Handbook*, Vol. 2, Blackwell Scientific Publications, Oxford, pp. 289–307.

Petts G.E. (ed.) 2003. Environmental flows for river systems. *River Research and Applications*, Special Issue, **19**, 5–6.

Petts G.E., Maddock I., Bickerton M. and A.J. Ferguson. 1995. Linking hydrology and ecology: the scientific basis for river management. In Harper D.M. and A.J.D. Ferguson (eds) *The Ecological Basis for River Management*. John Wiley & Sons, Ltd, Chichester.

Petts G.E., Bickerton M.A., Crawford C., Lerner D.N. and D. Evans. 1999. Flow management to sustain groundwater-dominated stream ecosystems. *Hydrological Processes*, **13**: 497–513.

Petts G.E., Nestler J. and R. Kennedy. 2006a. Advancing science for water resources management. *Hydrobiologia*, **565**, 277–288.

Petts G.E., Morales Y. and J. Sadler. 2006b. Linking hydrology and biology to assess the water needs of river ecosystems. *Hydrological Processes*, **20**, 2247–2251.

Poff N.L. and J.V. Ward. 1989. Implication of streamflow variability and predictability for lotic community structure: a regional analysis of streamflow patterns. *Canadian Journal of Fisheries Aquatic Science*, **46**, 1805–1818.

Poff N.L., Allan J.D., Bain M.B., Karr J.R., Prestegaard K.L., Richter B., Sparks R. and J. Stromberg. 1997. The natural flow regime: a new paradigm for riverine conservation and restoration. *BioScience*, **47**,769–784.

Postel S. and S. Carpenter. 1997. Freshwater ecosystem services. In Dailly G.C. (ed.) *Nature's Services: Societal Dependence on Natural Ecosystems*. Washington, DC, Island Press.

Postel S.L. and Richter B. 2003. *Rivers for Life*. Island Press, Washington, DC.

Prowse T.D. and F.M. Conly. 2002. A review of hydroecological results of the Northern River Basins Study, Canada. Part 2. Peace-Athabaska delta. *River Research and Applications*, **18**, 447–460.

Prowse T.D., Conly F.M., Church M. and M.C. English. 2002. A review of hydroecological results of the Northern River Basins Study, Canada. Part 1. Peace and Slave rivers. *River Research and Applications*, **18**, 429–446.

Raven P.J., Fox P.A., Everard M., Holmes N.T.H. and F.H. Dawson. 1997. River Habitat Survey: a new system for classifying rivers according to their habitat quality. In Boon P.J. and D.L. Howell (eds) *Freshwater Quality: Defining the Indefinable?* The Stationery Office, Edinburgh, pp. 215–234.

Raven P.J., Holmes N.T.H., Charrier P., Dawson F.H., Naura M. and P.J. Boon. 2002. Towards a harmonized approach for hydromorphological assessment of rivers in Europe: a qualitative comparison of three survey methods. *Aquatic Conservation*, **12**, 405–424.

Reid M.A. and J.J. Brooks. 2000. Detecting effects of environmental water allocations in wetlands of the Murray-Darling Basin, Australia. *Regulated Rivers*, **16**, 479–496.

Reiser D.W., Ramey M.P. and T.A. Wesche, 1989. Flushing flows. In Gore J.A. and G.E. Petts (eds) *Alternatives in Regulated River Management*. CRC Press, Boca Raton, Florida, 91–138.

Richter, B.D., Baumgartner J.V., Powell J. and P.D. Braun. 1996. A method for assessing hydrologic alteration within ecosystems. *Conservation Biology*, **10**, 1163–1174.

Richter, B.D., Baumgartner J.V., Wingington R. and P.D. Braun, 1997. How much water does a river need? *Freshwater Biology*, **37**, 231–249

Richter B.D., Baumgartner J.V., Braun D.P. and Powell J. 1998. A spatial assessment of hydrological alteration within a river network. *Regulated Rivers*, **14**, 329–340.

Richter B.D., Warner A.T., Meyer J.L. and K. Lutz. 2006. A collaborative and adaptive process for developing environmental flow recommendations. *River Research and Applications*. In press.

Rosgen D. 1996. *Applied River Morphology*. Wildland Hydrology, Pagosa Springs, Colorado.

Sabaton C., Siegler L., Gouraud V., Bagliniere J.L. and S. Manne. 1997. Presentation and first applications of a dynamic population model for brown trout (*Salmo trutta* L.): aid to river management. *Fisheries Management and Ecology*, **4**, 425–438.

Saltveit S.J. (ed.). 1996. First International Symposium on Habitat Hydraulics. *Regulated Rivers*, special issue, **12**, 2 & 3.

Schofield N.J., Collier K.J., Quinn J., Sheldon F. and M.C. Thoms. 2000. River conservation in Australia and New Zealand. In Boon P.J., Davies B.R. and G.E. Petts (eds) *Global Perspectives on River Conservation*. John Wiley & Sons, Ltd, Chichester, 311–333.

Schramm H.L. Jr. and M.A. Eggleton. 2006. Applicability of the flood-pulse concept in a temperate floodplain river ecosystem: thermal and temporal components. *River Research and Applications*, **22**, 543–554.

Sheail J. 1984. Constraints on water-resource development in England and Wales, concept and management of compensation flows. *Journal of Environmental Management*, **19**, 351–361.

Stalnaker C.B. 1994. Evolution of instream flow habitat modelling. In Calow P. and Petts G.E. (eds) *Rivers Handbook*, **2**, 276–288.

Stalnaker C.B. 1996. Evolution of instream flow habitat modelling. In Calow P. and G.E. Petts (eds) *The Rivers Handbook*. Blackwell Scientific, Oxford, Vol. 2, pp. 276–286.

Stanford J.A. and J.V. Ward. 1986. The Colorado River System. In Davies B.R. and K.F. Walker (eds) *The Ecology of River Systems*. Dr W. Junk, Dordrecht, pp. 353–384.

Stanford, J.A., Ward J.V., Liss W.J., Frissell C.A., William R.N., Lichatowich J.A. and C.C. Coutant. 1996. A general protocol for restoration of regulated rivers. *Regulated Rivers: Research and Management*, **12**, 391–413.

Statzner B., Gore J.A., and V.H. Resh. 1988. Hydraulic stream ecology: observed patterns and potential applications. *Journal of the North American Benthological Society*, **7**, 307–360.

Stewardson M.J. and C.J.Gippel. 2003. Incorporating flow variability into environmental flow regimes using the Flow Events Method. *River Research and Applications*, **19**, 459–472.

Stewart G., Anderson R. and E. Wohl. 2005. Two-dimensional modelling of habitat suitability as a function of discharge on two Colorado rivers. *River Research and Applications*, **21**, 1061–1074.

Stromberg J.C. 1993. Instream flow models for mixed deciduous riparian vegetation within a semi-arid region. *Regulated Rivers*, **8**, 225–236.

Tennant D.L. 1976. Instream flow requirements for fish, wildlife, recreation and related environmental resources. *Fisheries*, **1**, 6–10.

Tharme R.E. 2003. A global perspective on environmental flow assessment: emerging trends in the development and application of environmental flow methodologies for rivers. *River Research and Applications*, **19**, 397–442.

Thomas, W.L. Jr. (ed.). 1956. *Man's Role in Changing the Face of the Earth*. University of Chicago Press, Chicago.

Tockner K., Ward J.V., Arscott B.A., Edwards P.J., Kollmann J., Gurnell A.M., Petts G.E. and B. Maiolini. 2003. The Tagliamento River: a model ecosystem of European importance. *Aquatic Sciences*, **65**, 239–253.

UNCED. 1992. *Agenda 21: The United Nations Programme of Action from Rio*, United Nations Department of Public Information, New York.

Vadas R.L. and D.J. Orth. 2001. Formulation of habitat suitability models for stream fish guilds: do the standard methods work. *Transactions of the American Fisheries Society*, **130**, 217–235.

Vilizzi L., Copp G.H. and J.-M. Roussel. 2000. Assessing variation in suitability curves and electivity profiles in temporal studies of fish habitat use. *River Research and Applications*, **20**, 605–618.

Ward J.V. 1976. Comparative limnology of differentially regulated sections of a Colorado mountain river. *Archiv für Hydrobiologie*, **78**, 319–342.

Ward J.V. and J.A. Stanford (eds). 1979. *The Ecology of Regulated Streams*. Plenum Press, New York.

WCD. 2000. *Dams and Development*. World Commission on Dams, Earthscan, London.

Welcomme R.L. 1979. *Fisheries Ecology of Floodplain Rivers*, Longman, London.

Wishart M.J., Ganneur and H.T. El-Zanfaly. 2000. River conservation in North Africa and the Middle East. In Boon P.J., Davies B.R. and G.E. Petts (eds) *Global Perspectives on River Conservation*, John Wiley & Sons, Ltd, Chichester, pp. 127–154.

Wood P.J., Agnew M.D. and G.E. Petts. 2001a. Hydro-ecological variability within a groundwater dominated stream. In *Hydro-ecology: Linking Hydrology and Aquatic Ecology*, IAHS Publ. Vol. 266, pp. 151–160.

Wood, P.J., Hannah D.M., Agnew M.D. and G.E. Petts. 2001b. Scales of hydroecological variability within a groundwater-dominated stream. *Regulated Rivers*, **17**, 347–367.

Wu Fu-Chun and Yi-Ju Chou. 2004. Tradeoffs associated with sediment maintenance flushing flows: a simulation approach to exploring non-inferior options. *River Research and Applications*, **20**, 591–604.

Young W.J., Chessman B.C., Erskine W.D., Raadik T.A., Wimbush D.J., Tilleard J., Jakeman A. J., Varley I. and J. Verhoeven. 2004. Improving expert panel assessments through the use of a composite river consition index – the case of the rivers affected by the Snowy Mountains hydro-electric scheme, Australia. *River Research and Applications*, **20**, 733–750.

Zalewski M. 2000. Ecohydrology – the scientific background to use ecosystem properties as management tools toward sustainability of water resources. *Ecological Engineering*, **16**, 1–8.

14

The Role of Floodplains in Mitigating Diffuse Nitrate Pollution

Tim Burt, Mariet M. Hefting, Gilles Pinay and Sergi Sabater

14.1 Context

The chemistry of stream water reflects the biogeochemical processes operating within a catchment. However, there is more to understanding the changing composition of stream water than just chemistry and biology: the functional behaviour of a catchment system also depends on the hydrological processes operating. In this regard, the riparian zone, as the landscape element linking terrestrial and aquatic ecosystems, may well be the most important component of the catchment system. Even in intensively managed landscapes, the interaction between ecological communities and the hydrological cycle within the riparian zone can result in the river system being buffered from the worst effects of diffuse pollution from the catchment area. In the case of nitrate, whilst uptake by the riparian vegetation may be important (Peterjohn and Correll, 1984), most studies have concluded that the significance of vegetation relates more to its interaction with microbial processes and denitrification (e.g., Groffman *et al.*, 1992; Haycock and Pinay, 1993; Hill, 1996; Addy *et al.*, 1999). The ecohydrological significance of the near-stream zone is, therefore, far greater than its surface area would suggest, arising from distinctive biogeochemical conditions present within a waterlogged soil–vegetation system bordering the river channel.

Diffuse nutrient inputs to surface and ground waters are a major problem in developed countries (Burt *et al.*, 1993). In Europe, increasing contamination by nitrate is a common trend in intensive agricultural areas, the problem having become worse in recent years (European Commission, 2002). The reduction of diffuse nitrate contamination through appropriate agricultural practices was the objective behind the EU Nitrate Directive (91/676), but successful amelioration requires a landscape-based approach. It is within this context that the importance of riparian zones in mitigating diffuse nitrate fluxes from adjacent upland areas has been examined over the last two decades, beginning with the pioneering studies of Peterjohn and Correll (1984) and Lowrance et al. (1984). It is now widely accepted that preservation of a fully functioning riparian zone is an effective way to remove pollutants before runoff enters the watercourse (Sweeney et al., 2004), as well as helping to protect channel integrity and river biodiversity.

It is nevertheless true that the efficiency of these areas in retaining nutrients has remained much under discussion, their recognised role ranging from being a sink or a net source of nitrate (Bedard-Haughn et al., 2004; Fisher and Acreman, 2004). Not unrelated to this uncertainty, riparian zones are very varied in their size (presence or absence of a floodplain; if present, width of floodplain), vegetation and land use (forested or farmland) and the nature of the upslope catchment (area, land use, topography; Burt et al., 2002; Vidon and Hill, 2004a). What riparian zones have in common is that they form a biogeochemical boundary, or ecotone, between terrestrial and aquatic ecosystems (Naiman and Décamps, 1997). Conditions within the riparian zone create the potential for nitrate removal and so change the water quality before it enters the river. Bishop et al. (2004) argue that, since the chemistry of water moving downslope is determined at any given point by soil chemistry, the particular role of the riparian soil is to set the stream water chemistry, since this is the last soil in contact with the water before it becomes runoff. Thus, the riparian zone is perhaps the most important element of the hydrological landscape, given that it can decouple the linkage between the major landscape elements, hillslope and channel. In a sense therefore, the riparian zone must, paradoxically, act both as a conduit and a barrier, linking the terrestrial and aquatic environments but also acting as a barrier between them. (Burt, 2005)

By strict definition, the riparian zone comprises just the vegetation along the bed and banks of the river channel (Tansley, 1911). However, this definition has been broadened more recently to encompass a wider area alongside the channel; very often, the floodplain is taken to be consistent with this more broadly defined zone. The riparian landscape is unique among environments because it is a terrestrial habitat strongly affecting and affected by aquatic environments (Malanson, 1993). Here we review the role of floodplain buffer zones in mitigating nitrate pollution; this requires a focus on subsurface flow and on relevant nitrogen cycling processes. Riparian zones may buffer from other kinds of pollutant, in particular eroded soil and sediment-bound pollutants such as phosphate and pesticides, but these are not considered here, since a different hydrological focus – on surface runoff – is needed.

14.2 Nitrogen Removal by Riparian Buffers: Results of a Pan-European Experiment

As already noted, riparian zones are spatially and temporally heterogeneous with respect to hydrology (Lowrance et al., 1997), soil characteristics (Jacinthe et al., 1998) and biogeochemical processes (Hedin et al., 1998; Hill et al., 2000). This variability affects the rate of nitrate removal in the riparian zone since the major pathway for nitrate movement is through subsurface flow (Hill, 1996). This means that the removal capacity of riparian zones is controlled both by their hydrological characteristics (water residence time and degree of contact between soil and groundwater; Gold *et al.*, 1998), and the biological processes (plant uptake and denitrification; Groffman *et al.*, 1992; 1996b). The relative influence of these factors depends both on soil characteristics (Groffman *et al.*, 1992; Pinay *et al.*, 1995; Flite *et al.*, 2001) and on the nitrogen input to the riparian zone (Hanson *et al.*, 1994). Moreover, vegetation type (e.g., forest, grass, arable crop) seems also to affect the nitrate removal efficiency of riparian zones (Groffman *et al.*, 1996a). However, such conclusions remain controversial since some studies have shown that meadow sites cannot remove nitrate (Osborne and Kovacic, 1993; Hubbard and Lowrance, 1997), whilst others have shown the exact opposite (Haycock and Pinay, 1993; Schnabel *et al.*, 1996; Addy *et al.*, 1999).

14.2.1 The NICOLAS Project

A comprehensive analysis of the variations in the nitrogen removal capacity of riparian zones located across a wide range of climatic, hydrological, land cover and diffuse nitrogen input fluxes was the objective of a joint European research project (NICOLAS: **Ni**trogen **Co**ntrol by **La**ndscape **S**tructures in Agricultural Environments). The project aimed to evaluate the natural performance of riparian zones in buffering waterborne fluxes of diffuse nitrogen pollution from agricultural to aquatic environments. The study sites (Figure 14.1) covered a range of climatic and soil conditions from the Mediterranean to Central and Northern Europe. Some of the sites were characterised by a continental climate (the Polish and Romanian sites). The Swiss site had typical alpine climate, while the Dutch, British and French sites were characterised by humid maritime climate. Finally, the Spanish site was typically Mediterranean. As a whole, the sites extended from 41° to 53°N latitude. Mean annual temperature ranged from 7.8 to 12°C, and potential evapotranspiration ranged from low (Swiss site) to very high (Spanish site). Six of the study sites were covered by herbaceous or shrubby vegetation, while the remaining eight were forested. Upslope agricultural fields (usually cereals or pasture) bordered these areas. The riparian zones had a wide range of widths (from 10 to 60 m), slopes (from 0% to 22%) and soil characteristics. Soils ranged from organic rich to organic poor, and from sandy to clayey. There was significant variability in soil water content and saturation, both in terms of water table height within the topsoil and the extent of time that it remained saturated (Sabater *et al.*, 2003).

14.2.2 Climatic and Hydrological Controls on the Efficiency of Riparian Buffers

Climatic characteristics can affect the water table level and its dynamics (Burt *et al.*, 2002), as well as temperature and soil moisture, which determine the rates of processes such as soil respiration, mineralization and plant productivity (Rustad *et al.*, 2001). An obvious question in a multisite study is to determine whether climatic factors could affect the efficiency of the riparian area in nitrogen removal. The NICOLAS study showed that there was significant nitrate removal within the riparian zone at most of the study sites (Figure 14.2). Annual nitrate removal efficiency was close to 30% per metre of riparian strip at the French forested site, while at four other sites (the herbaceous Romanian, Dutch

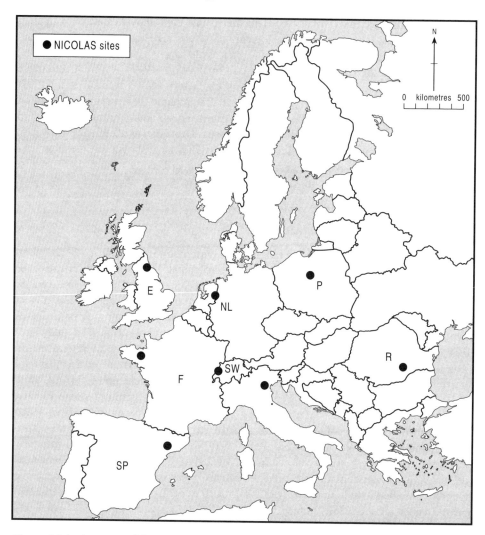

Figure 14.1 Location of the study sites included in the NICOLAS project. Most sites included both a forested and herbaceous buffer zone

and one of the French sites, and the forested Spanish site) the efficiency was about $10\% \, m^{-1}$. The nitrate removal efficiency decreased to around $5\% \, m^{-1}$ for four other sites (the two English herbaceous sites, one herbaceous French site and the Swiss site). For the remaining sites, nitrate removal efficiency was close to zero (the Dutch forested site, the two forested Polish sites, the forested British, and the Romanian sites).

Nitrate is removed in riparian zones by means of the microbial process of denitrification in soils, and by plant uptake (Peterjohn and Correll, 1984; Hill, 1996). The first process results in a net loss of nitrogen to the atmosphere, while the second may only represent a transient fate for nitrogen, unless vegetation harvest is effective. The two processes are closely linked to climatic conditions. Denitrification can account for 50–90% of the total

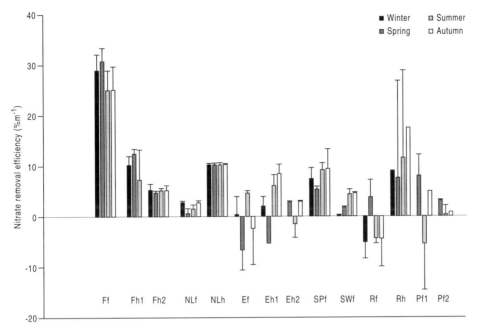

Figure 14.2 *Mean seasonal nitrate removal rates and standard deviation of the means, for the different sites studied. Acronyms are those detailed for Figure 14.1. 'f' and 'h' define the respective forested and herbaceous study site for each country. In cases where more than one study site occurs for each category they appear defined with numbers*

nitrate elimination when soils are usually water saturated (Nelson et al., 1995; Kruk, 2003). Denitrification may be constrained by low soil temperature (continental climates of central Europe) or low soil moisture (Mediterranean climate). However, in most riparian buffers analysed in NICOLAS there was considerable nitrogen removal (Figure 14.2) with rates ranging from $5\% m^{-1}$ to $30\% m^{-1}$. There was no significant climatic pattern associated with this variation. Instead, the large variation in nitrate removal efficiency between the riparian study areas related to the particular characteristics of individual riparian buffers (Hill et al., 2000).

Geomorphological and hydrological conditions tend to be the most important controls of the nitrogen removal capacity of riparian zones (Lowrance et al., 1997). It has been found that riparian zones with their water table close to the soil surface are more effective at removing nitrogen compounds than other sites where water tables are lower (Cooper, 1990; Haycock and Pinay, 1993; Hefting et al., 2004). Across the NICOLAS sites, those with a flat riparian zone (floodplain), which allows a high water table to be maintained, were more effective than those sites where the riparian zone was sloping (Burt et al., 2002; Sabater et al., 2003). Studying hydrological flow paths is crucial in riparian buffer zone research, since the actual pathway of water flow through substrates is often complex (Clément et al., 2003b; Hefting et al., 2005). Hence the riparian zone may contain a mosaic of suitable and unsuitable flow paths for nitrate removal and, as a result, process rates are not a simple function of the riparian surface area but rather of the length and period of hydrological contact.

The two main processes involved in nitrogen removal (i.e., denitrification and plant uptake) can operate under different hydrological and temperature conditions, and other sources of variation may mask the varying influence of season in each of these processes (Pinay and Décamps, 1988; Nelson et al., 1995). For instance, at the French and the Dutch sites, which have mild climatic conditions, there was no significant difference between seasons within each site despite significant differences between sites (Figure 14.2). Nitrate removal at these sites was a combined result of denitrification and plant uptake (Clément et al., 2003a; Hefting et al., 2003), the relevance of each factor shifting according to variations in temperature, water table and nitrate input, which masked any pattern potentially identifiable with season. Likewise, high nitrogen removal rates were measured at the Spanish site, even in summer when the riparian soils were dry (Bernal et al., 2003). This indicates that plant uptake (either by shrubs or trees, depending on the time of year) was involved in nitrate depletion at that site, even though other processes such as microbial assimilation could also be involved in nitrate removal from groundwater (Butturini et al., 2003). Plant uptake was most probably the main removal process at that site during dry soil conditions. This apparent lack of seasonality in nitrate depletion reinforces the fact that within the range of climatic conditions tested, a wide range of riparian zones can significantly remove nitrogen inputs. In another of the NICOLAS studies, plant uptake was also found to be more important than denitrification in the drier sites (Hefting et al., 2005). The conclusion arising from the NICOLAS research is that plant uptake operates either simultaneously with denitrification or in isolation when hydrological conditions preclude denitrification.

14.2.3 The Effect of the Riparian Vegetation Type on Nitrate Removal

One of the most conflicting questions in the literature has been the respective efficiency of herbaceous and forested riparian zones (e.g., Haycock and Pinay, 1993; Osborne and Kovacic, 1993; Correll, 1997). In our study, significant differences in nitrogen removal capacity were measured between forested and meadow riparian zones at several sites (Sabater et al., 2003; Hefting et al., 2005). Yet, there was no consistent difference between regions. For instance, within the French site, the nitrogen removal rate of the forested zone was significantly higher than the meadow zone, while at the Dutch (Figure 14.3) and the Romanian sites the meadow riparian zones were significantly more efficient. The average removal rates of herbaceous and forested zones in the different regions were $4.43\%\,m^{-1}$ and $4.21\%\,m^{-1}$ for herbaceous and forested zones respectively. These results show that there are not *a priori* significant differences between herbaceous and forested strips in their function as nitrogen removal zones.

It has been suggested that variation in denitrification for different herbaceous and forested riparian wetlands is fundamentally related to the organic matter quality of the site (Groffman et al., 1991). However, Parry et al. (1999) demonstrated that a difference in denitrification rate between a pasture and a cultivated field was, instead, related to differences in the pore space structure resulting from agricultural practices. On the other hand, trees have deep roots that make it possible to capture deep groundwater and therefore may have a significant role even under considerable water stress (Schade et al., 2001). Our results suggest that both types of vegetation cover can provide similar nitrogen removal rates, and that both could be used to mitigate diffuse nitrogen pollution over a

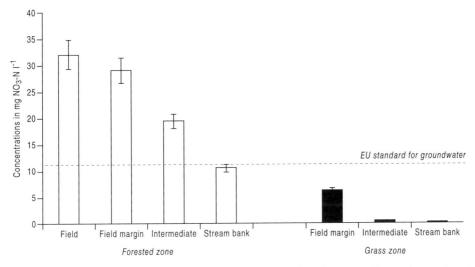

Figure 14.3 *Annual mean nitrate concentrations in subsurface runoff in a forested and grass-covered buffer zone along first-order streams in Twente, The Netherlands (open bars indicate the forested zone, black bars indicate the grassland zone). Data after Hefting et al (2006a)*

wide range of climatic conditions. Of course, one of the problems in comparing the effect of vegetation type on nitrate depletion is that other variables may be more important than plant cover in the riparian zone, in particular the characteristics of the upslope area (area, soil permeability, crops grown) and of the riparian zone itself (water table dynamics, flow paths, sediment types, topography). Some of these factors are further explored in Section 14.3. Another complicating effect is that nitrate removal may be influenced by the legacy of past vegetation communities rather than by present plant cover (Addy *et al.*, 1999; Hill *et al.*, 2004).

14.2.4 Nitrogen Saturation Effect

The increasing rate of fertilizer application on intensively farmed land has meant that about 20% of EU groundwater monitoring stations register nitrate concentrations of over 50 mg NO_3/l, and 40% over 25 mg NO_3/l (European Commission, 2002). These are levels that might approach or exceed those causing nitrogen saturation in the system (Aber *et al.*, 1989), therefore affecting the process of plant uptake and mineralisation of organic matter, potentially increasing the nitrogen available for leaching. The results of the pan-European study analysed here showed that nitrate load was one of the main factors controlling variation in nitrate removal rates between the riparian zones. The significance of nitrogen load for nitrate removal was only seen for nitrate concentration inputs higher than 5 mg N/l, when nitrate removal efficiency was negatively correlated with nitrate input ($r = -0.59$, $p < 0.05$). This relationship followed a pattern of negative exponential decay (Figure 14.3). Hydraulic gradient was also negatively correlated with nitrate removal ($r = -0.27$, $p < 0.05$) at sites receiving water with concentrations above 5 mg/l nitrate-N. This implies the importance of prolonged saturation within the riparian zone for

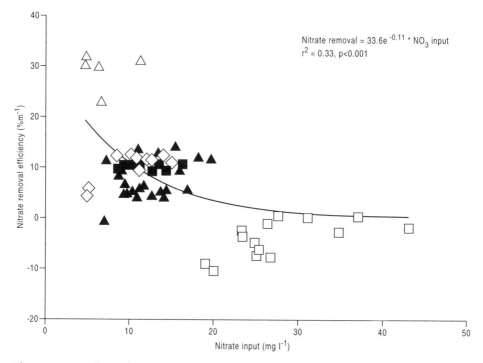

Figure 14.4 *Relationship between nitrate removal and nitrate input when nitrate concentrations were less than 5 ppm in the groundwater input, in the forested and herbaceous sites. Solid symbols = herbaceous sites, open symbols = forested sites. Triangles = French, squares = Dutch, rhombus = Spanish sites*

denitrification to be effective, but only up to a certain limit (Burt *et al.*, 1999). There is an apparent null or negative nitrogen removal by the riparian buffers when receiving nitrogen inputs up to 20 mg/l nitrate-N (Figure 14.4, white squares). The negative relationship between nitrate load and riparian zone removal efficiency found in some sites suggests also a saturation effect of long-term nitrate loading, which exceeds the buffering capacity (Aber, 1992) of the riparian zones. Hanson *et al.* (1994) observed clear symptoms of nitrogen saturation in a forested riparian zone subjected to long-term enrichment. These symptoms consisted of enrichment of total plant and microbial nitrogen pools as well as an increase in soil nitrogen processes rates, namely mineralisation and nitrification. The most representative case in our study for nitrogen saturation is the Dutch forested site, which is located in an area of long-lasting nitrogen enrichment, and soil process rates (decomposition and mineralisation) were significantly higher than in all the other sites included in this study (Pinay and Burt, 2000).

14.2.5 N$_2$O Emissions

Riparian buffer zones are hotspots for denitrification but may also exhibit considerable N$_2$O emission, specifically if nitrate loading rates are high or pH values are low (Hefting *et al.*, 2003). N$_2$O is a trace gas that contributes to the enhanced greenhouse effect (Wang *et al.*, 1976) and plays a role in the stratospheric ozone depletion (Crutzen, 1970). Estimates of the global N_2O budget indicate that soils are globally a major source of N_2O with 1.8–5.3 Tg y^{-1} (Prather *et al.*, 1995). N_2O is a by-product of nitrification and denitrification; production of N_2O by nitrifying bacteria results from reduction of NO_2 when oxygen is limiting. Production of N_2O during denitrification is affected by the relative availability of electron donors and acceptors; yet the relative contribution of these processes to the N_2O emission is still unclear. It is likely that denitrification is the major process of N_2O formation in nitrate-loaded and permanently wet riparian zones. To optimize nitrate mitigation by riparian zones, it is useful to determine which environmental conditions are favourable for denitrification without increasing N_2O emissions (i.e., the reduction process should produce N_2 not N_2O).

Large spatial variability in denitrification and N_2O fluxes under field conditions often complicates the quantification of the total N_2O emissions and $N_2O:N_2$ ratio during denitrification. Moreover, groundwater flow patterns in riparian zones are often complex and create additional spatial differences. The NICOLAS study site on a forested buffer zone in the Netherlands revealed significant differences in nitrate removal capacity and N_2O emissions between different groundwater flow paths (Hefting *et al.*, 2006a,b). In some places there was clear evidence that nitrate moved completely along the shallow groundwater flow path while other flow paths showed high removal rates in the same period. Distinct spatial patterns of denitrification and N_2O emission were observed along the high nitrate removal transect compared with no clear pattern along the low nitrate removal transect, where denitrification activity was very low. N_2O emissions ranged from 2 to 13 mg N m^{-2} d^{-1}, while denitrification rates ranged from 36 to 133 mg N m^{-2} d^{-1}. As expected, the ratio of N_2O to N_2 was higher along the flow path with low nitrate-removal efficiency and high nitrate loading rates (winter: 0.93; summer: 0.22) compared with the ratios found along the transect with high nitrate removal efficiency (winter: 0.29, summer: 0.0032). However, N_2O flux measurements from a groundwater flow path with high nitrate-removal capacity indicated a significant N_2O emission from effective riparian zones under suboptimal temperatures. Analysing the spatial patterns found in this detailed study (Hefting *et al.*, 2006a,b), it was concluded that N_2O emission was higher whenever any significant factor was reducing the denitrification rate, i.e. temperature, water-filled pore space or pH.

14.3 Landscape Perspectives

Agricultural intensification in Europe in the Twentieth century has fundamentally changed landscape functioning in terms of hydrology and nutrient fluxes. Drastic changes in hydrological pathways have occurred as a result of draining of agricultural land, straightening channels and regulating stream discharge. These hydrological changes have generally led to a decrease in the overall residence time of water within the landscape, consequently limiting the spatial extent and shortening the duration of saturated soil conditions in riparian areas. Since groundwater levels exert a strong control over the nitrate removal from riparian zones (Hefting *et al.*, 2004), these hydrological changes may have significantly reduced the effectiveness of riparian zones for water quality improvement.

14.3.1 Upslope–Riparian Zone–Channel Linkage

The general characteristics of the surrounding landscape, notably geology and geomorphology, control the magnitude and fluctuations of groundwater flow into and through riparian zones. Most studies of riparian zone processes have either examined single sites (e.g., Burt *et al*., 1999) or have compared adjacent sites (e.g., Haycock and Burt, 1993, compared neighbouring grassland and woodland sites). Only two studies have conducted comparative studies across a range of sites:

- Burt *et al*. (2002) examined water table fluctuations at the NICOLAS study sites described above. An identical experimental design was employed at each site: a 10-metre grid of dipwells was employed, with monthly measurements for at least a year. The existence of a floodplain proved crucial in providing a low hydraulic gradient: the floodplain needed to be sufficiently wide to sustain a high water table and allow buffer zone processes to operate effectively, but not so wide as to completely decouple upslope areas from the river channel. Soil hydraulic conductivity was also important in optimising soil water residence times within the riparian zone. If the alluvial sediments are too permeable, water tables are not sustained for long periods; on the other hand, if the alluvium is impermeable, too little water flows through to optimise denitrification. Although all sites displayed a clear annual cycle of water table fluctuation, climate seemed no more important than geomorphology in controlling the range of variation. Where there is direct slope-channel linkage, the water table is usually too deep to favour denitrification. The combination of a (flat) floodplain and soils of moderate permeability sustains a high water table and allows buffer zone processes to operate effectively.
- Vidon and Hill (2004a) considered several characteristics of landscapes that determine the nitrate removal capacity of riparian areas. They examined the size of the upland aquifer, and the topography and sediment lithology of the riparian zone. A stable water table was maintained in riparian zones well connected to large upland aquifers by slopes of >5% gradient. Where slopes were lower and/or the aquifer smaller, riparian sites were seasonally disconnected from upslope contributing areas. In the worst case, level areas underlain by clay soils yield only very small subsurface fluxes; runoff either occurs as overland flow or, if there has been under-drainage, bypasses the soil through mole and tile drains. As in the NICOLAS study, Vidon and Hill (2004a) conclude that even the most favourable situation requires a flat riparian zone to optimise water table conditions; a slope extending to the stream decreases the buffer role of the riparian zone. Vidon and Hill (2004b) extended their analysis to estimate the overall nitrate removal efficiency of their study sites. They concluded that as the upland permeable sediment depth and the slope gradient at the perimeter increase, riparian zones shift from small to large nitrate sinks within the southern Ontario landscape.

As noted in the introduction, Bishop *et al*. (2004) argue that the particular role of the riparian soil is to set the stream water chemistry, since this is the last soil in contact with the water before it becomes runoff. The degree of biogeochemical decoupling between hillslope and channel depends on what opportunity there is for different source waters to mix within the near-stream zone, and on the residence times of these different waters as they flow through the riparian zone, which in turn is controlled by near-stream topography and sedimentary architecture. As far as nitrate is concerned, the optimal condition would seem to be a floodplain with alluvium of intermediate permeability where residence times are sufficient to allow significant amounts of both subsurface flow *and* denitrification. To operate effectively, the riparian 'buffer zone' therefore requires a balance between the functions of conduit and barrier (Burt, 2005).

14.3.2 Catchment-scale Considerations

Typically, studies of riparian zones tend to focus at the site or catena scale (Park and van de Giesen, 2004). However, the scale of planning and water management is invariably at the landscape or catchment level (Viaud *et al.*, 2004). At that scale, spatial analysis of landscape characteristics is much less tractable and a 'lumped' approach is usually adopted (Burt and Pinay, 2005). For example, N removal has been found to correlate with the percentage area occupied by riparian wetlands or with the percentage area of hydromorphic soils (Jones *et al.*, 1976; Curmi *et al.*, 1998). A recent review of wetlands and water quality revealed that catchment-wide studies from very different regions around the world all indicate that a minimum wetland area covering at least 2–5% of the total catchment is needed for stream water quality improvement (Verhoeven *et al.*, 2006).

Burt and Pinay (2005) have extended the essentially two-dimensional analysis of slope–floodplain–channel linkage to the three-dimensional catchment perspective. In steep headwater basins, areas of soil saturation will be limited in extent and while they may dominate discharge production, their ability to buffer hillslope runoff will be limited; in effect, slope-channel linkage is too efficient. At the other extreme, in very large river valleys, slopes may be entirely decoupled from the channel with very wide floodplains (often artificially drained) in between. Even if there is some linkage between slope and channel, this may well be limited in time and space. Burt and Pinay (2005) conclude that riparian zones may be most effective in the middle-order sections of river basins where there is a sufficiently wide floodplain to sustain a high water table and allow buffer zone processes to operate effectively, but not so wide as to completely decouple upland areas from the channel.

14.3.3 N Loading

Hydrogeological setting not only determines nitrate removal effectiveness in terms of water residence time, water table level and the extent of anaerobic soil conditions, it also determines the intensity of nitrogen loading of riparian areas. Groundwater N inputs were found by Vidon and Hill (2004a) to be correlated with the depth of the upland permeable sediments (contributing area) and the slope gradient at the upland–riparian boundary. Slope length between the riparian zone and interfluve also had a significant positive correlation with N loading. This slope length parameter is an indicator of upland areas contributing to the subsurface flow, given that the upslope contributing area can often be difficult to assess (although the use of digital elevation models can help greatly in this respect – see Lane, 2004).

N loading in rural catchments is obviously also related strongly to agricultural practice. The ever-increasing amount of fertilizer use causes considerable amounts of nutrients to leach from the agricultural fields into the groundwater and then to flow towards lower levels and aquatic systems in the landscape. In the Netherlands, riparian areas in agricultural catchments are subject to increasingly high nitrate inputs. Average nitrate loadings were high in a forested buffer zone in Twente, NE Netherlands at $87\,g\,NO_3\text{-}N\,m^{-2}\,y^{-1}$. As noted above, nitrate loading rates at these sites have reached such high levels that nitrate retention has levelled off (Sabater et al., 2003; Hefting et al., 2006a). Tracing flow paths and nitrate removal in chronically N loaded riparian zones showed that these buffer zones were still mitigating diffuse nitrate fluxes (38% of the incoming nitrate load), although groundwater nitrate concentrations after riparian zone passage were barely lower than the EU groundwater standard. In addition, as already discussed, N transformations in riparian zones receiving water very rich in nitrate can have high emissions of the greenhouse gas N_2O (Hefting et al., 2003). This is in accordance with the overall high nitrous oxide emissions from rivers and estuaries which was found to rise from 0.3 to 6% of the denitrification rate with increasing N loading (Kroeze and Seitzinger, 1998). In such cases, N transformation within riparian buffer zones may result in an unfavourable shift from mitigating water pollution to an increase in greenhouse gas emission.

14.4 Future Perspectives

As science and policy develop in the early Twenty-first century, several challenging opportunities present themselves:

- From an experimental point of view, one of the most important matters is to quantify the effect of water regime change, due to climatic change or human activity, on the functioning of the riparian zones and, more specifically, on their N removal ability. For instance, a reduction in the duration of the high water period along low stream orders would lead to downslope movement of the dry–wet interface within the riparian zones, thereby favouring further intrusion of upland allochthonous nitrate within the riparian zones. Moreover, it is expected that this intrusion would lead to further incursion of pesticides in riparian soils as well, since these xenobiotic molecules are often found together with nitrate. To what extent does this intrusion reduce bacterial denitrification of newly contaminated soils where pesticide decomposition would be less effective since the bacterial population is not adapted to repeated pesticide application? Another aspect of this challenge is related to an increase in the occurrence extreme flood events. This could reduce the amount of fine-textured sediment deposited, especially in constrained floodplains, thereby reducing floodplain soil productivity. Lack of fine-textured sediment could also reduce the heavy metal adsorption capacity of alluvial soils. A key question is that of how a lack of heavy metal sequestration would affect nitrogen cycling processes downstream.
- Integrated catchment management, if it is to be successful, must be conducted at a scale of hundreds or even thousands of square kilometres. If the hypothesis of the riparian buffer zone is accepted, catchment managers will require an appropriate framework for assessing the N removal capacities of riparian zones in different landscapes. A geographical information system (GIS) provides a means of upscaling results of field-scale experiments to the catchment scale. Used in combination with a model such as REMM – the **R**iparian **E**cosystem **M**anagement **M**odel (Inamdar et al., 2000), a GIS can

provide a practical basis for identifying riparian buffer zones, actual or potential (i.e., those requiring restoration). As an example of this approach in relation to diffuse pollution, Quinn et al. (2004) consider the minimum information requirements for integrated river basin management at a scale of thousands of square kilometres, but in so doing, also recognise the need to downscale results back to the end-user at the field/farm scale.

- Given that many floodplains have been greatly altered by various forms of human activity, it will be necessary to determine the potential restoration possibilities of riparian zones under various geographical and socio-economical contexts. The willingness of farmers actively to embrace the notion of a riparian buffer zone, by making significant changes to their management of floodplain land, depends on a large number of social, economic and political factors. Farmer responses may vary widely regionally and nationally, depending for example, on levels of financial and technical assistance available.

- This in turn raises wider questions about land management and the protection of the water environment. There is a range of both supranational (e.g., EU Water Framework and Habitats Directives) and national legislation that is impacting upon the relationship between land management, hydrology and the water environment. The legislation is and will change how we use the landscape. What are the science and social science challenges arising from new legislation? How does land management as driven by agricultural policy interface with the needs of land management to alleviate diffuse pollution? Conversely, to what extent does the implementation of policy to lessen diffuse pollution reflect the science of diffuse pollution? With floodplains specifically in mind, to what extent may hydrological restoration of riparian zones be in conflict with other Water Framework Directive drivers, in particular the 'good ecological status' of in-stream ecosystems? And looking to the future, will climate change undermine current attempts to manage riparian zones and mitigate diffuse pollution? In many ways therefore, the biggest challenge in researching diffuse pollution lies at the interface between science, policy and land management. In this regard, riparian buffer zones take on special significance, forming as they do the biogeochemical boundary between terrestrial and aquatic ecosystems.

References

Aber, J.D. (1992). Nitrogen cycling and nitrogen saturation in temperate forest ecosystems. *Trends in Ecology and Evolution*, **7**: 220–224.

Aber, J.D., Nadelhoffer, K.J., Steudler, P., and Melillo, J.M. (1989). Nitrogen saturation in northern forest ecosystems. *Bioscience*, **39**: 378–386.

Addy, K.L., Gold, A.J., Groffman, P.M., and Jacinthe, P.A. (1999). Ground water nitrate removal in subsoil of forested and mowed riparian buffer zones. *Journal of Environmental Quality*, **28**: 962–970.

Bedard-Haughn, A., Tate, K.W., and van Kessel, C. (2004) Using nitrogen-15 to quantify vegetative buffer effectiveness for sequestering nitrogen in runoff. *Journal of Environmental Quality*, **33**: 2252–2262.

Bernal, S., Butturini, A., Nin, E., Sabater, F., and Sabater, S. (2003) Leaf litter dynamics and nitrous oxide emission in a Mediterranean riparian forest: Implications for soil nitrogen dynamics. *Journal of Environmental Quality*, **32**: 191–197.

Bishop, K., Seibert, J., Kohler, S., and Laudon, H. (2004). Resolving the Double Paradox of rapidly mobilised old water with highly variable responses in runoff chemistry. *Hydrological Processes*, **18**: 185–189.

Burt, T.P. (2005). A third paradox in catchment hydrology and biogeochemistry – decoupling in the riparian zone. *Hydrological Processes*, **19**: 2087–2089.

Burt, T.P., and Pinay, G. (2005). Linking hydrology and biogeochemistry in complex landscapes. *Progress in Physical Geography*. In press.

Burt, T.P., Heathwaite, A.L., and Trudgill, S.T. (1993). *Nitrate: Process, Pattern and Management*. John Wiley & Sons, Ltd: Chichester.

Burt, T.P., Matchett, L.S., Goulding, K.W.T., Webster, C.P., and Haycock, N.E. (1999). Denitrification in riparian buffer zones: the role of floodplain sediments. *Hydrological Processes*, **13**: 1451–1463.

Burt, T.P., Pinay, G., Matheson, F.E., Haycock, N.E., Butturini, A., Clément, J.C., Danielescu, S., Dowrick, D.J., Hefting, M.M., Hillbricht-Ilkowska, A., and Maitre, V. (2002). Water table fluctuations in the riparian zone: comparative results from a pan-European experiment. *Journal of Hydrology*, **265**: 129–148.

Butturini, A., Bernal, S., Nin, E., Hellin, C., Rivero, L., Sabater, S., and Sabater, F. (2003). Influences of the stream groundwater hydrology on nitrate concentration in unsaturated riparian area bounded by an intermittent Mediterranean stream. *Water Resources Research*, **39**: 1–13.

Clément, J.C., Holmes, R.M, Peterson, B.J. and Pinay, G. (2003a). Isotopic investigation of denitrification in a riparian ecosystem in western France. *Journal of Applied Ecology*, **40**: 1035–1048.

Clément, J.C., Aquilina, L., Bour, O., Plaine, K., Burt, T.P., and Pinay, G. (2003b). Hydrological flowpaths and nitrate removal rates within a riparian floodplain along a fourth-order stream in Brittany (France). *Hydrological Processes*, **17**: 1177–1195.

Cooper, A.B. (1990). Nitrate depletion in the riparian zone and stream channel of a small headwater catchment. *Hydrobiologia*, **202**: 13–26.

Correll, D.L. (1997). Buffer zones and water quality protection: general principles. In: Haycock, N.E., Burt, T.P., Goulding, K.W.T, and Pinay, G. (eds) *Buffer Zones: Their Processes and Potential in Water Protection*, Quest Environmental: Harpenden, pp. 7–20.

Crutzen, P.J. (1970). The influence of nitrogen oxides on the atmospheric ozone content. *Quarterly Journal of the Royal Meteorological Society*, **96**: 320–325.

Curmi, P., Durand, P., Gascuel-Odoux, C., Merot, P., Walter, C., and Taha, A. (1998) Hydromorphic soils, hydrology and water quality: spatial distribution and functional modelling at different scales. *Nutrient Cycling in Agroecosystems*, **50**: 127–142.

European Commission. (2002). Implementation of Council Directive 91/676/EEC concerning the protection of waters against pollution caused by nitrates from agricultural sources. Report.

Fisher, J., and Acreman, M.C. (2004). Wetland nutrient removal: a review of the evidence. *Hydrology and Earth System Sciences*, **8**: 673–685.

Flite, O.P., Shannon, R.D., Schnabel, R.R., and Parizek, R.R. (2001). Nitrate removal in a riparian wetland of the Appalachian valley and ridge physiographic province. *Journal of Environmental Quality*, **30**: 254–261.

Gold, A.J., Jacinthe, P.A., Groffman, P.M., Wright, W.R., and Puffer, R.H. (1998). Patchiness in groundwater nitrate removal in a riparian forest. *Journal of Environmental Quality*, **27**: 146–155.

Groffman, P.M., Axelrod, E.A., Lemunyon, J.L., and Sullivan, W.M. (1991). Denitrification in grass and forest vegetated filter strips. *Journal of Environmental Quality*, **20**: 671–674.

Groffman, P.M., Gold, A.J., and Simmons, R.C. (1992). Nitrate dynamics in riparian forests: microbial studies. *Journal of Environmental Quality*, **21**: 666–671.

Groffman, P.M., Hanson, G.C., Kiviat, E., and Stevens, G. (1996a). Variation in microbial biomass and activity in four different wetland types. *Soil Science Society of America Journal*, **60**: 622–629.

Groffman, P.M., Howard, G., Gold, A.J., and Nelson, W.M. (1996b). Microbial nitrate processing in shallow groundwater in a riparian forest. *Journal of Environmental Quality*, **25**: 1309–1316.

Hanson, G.C., Groffman, P.M., and Gold, A.J. (1994). Symptoms of nitrogen saturation in a riparian wetland. *Ecological Applications*, **4**: 750–756.

Haycock, N.E., and Burt, T.P. (1993). The role of floodplain sediments in reducing the nitrate concentration of subsurface runoff: a case study in the Cotswolds, England. *Hydrological Processes*, **7**: 287–295.

Haycock, N.E., and Pinay, G. (1993). Groundwater nitrate dynamics in grass and poplar vegetated riparian buffer strips during the winter. *Journal of Environmental Quality*, **22**: 273–278.

Hedin, L.O., Vonfischer, J.C., Ostrom, N.E., Kennedy, B.P., Brown, M.G., and Robertson, G.P. (1998). Thermodynamic constraints on nitrogen transformations and other biochemical processes at soil-stream interfaces. *Ecology*, **79**: 684–703.

Hefting, M.M., Bobbink, R., and de Caluwe, H. (2003). Nitrous oxide emission and denitrification in chronically nitrate-loaded riparian buffer zones. *Journal of Environmental Quality*, **32**: 1194–1203.

Hefting, M., Clément, J.C., Dowrick, D., Cosandey, A.C., Bernal, S., Cimpian, C., Tatur, A., Burt, T.P., and Pinay, G. (2004). Water table elevation controls on soil nitrogen cycling in riparian wetlands along a European climatic gradient. *Biogeochemistry*, **67**: 113–134.

Hefting, M.M., Clément, J.C., Bienkowski, P., Dowrick, D., Guenat, C., Butturini, A., Topa, S., Pinay, G., and Verhoeven, J.T.A. (2005). The role of vegetation and litter in the nitrogen dynamics of riparian buffer zones in Europe. *Ecological Engineering*, **24**: 465–482.

Hefting M.M., Beltman B., Karssenberg, D., Rebel, K., Riessen M., and Spijker M. (2006a). Water quality dynamics and hydrology in nitrate loaded riparian zones in the Netherlands. *Environmental Pollution*, **139**: 143–156.

Hefting M.M., Bobbink, R., and Janssens M.P. (2006b). Spatial variation in denitrification and N_2O emission in relation to nitrate removal efficiency in an N-stressed riparian buffer zone. *Ecosystems*, **9**: 550–563.

Hill, A.R. (1996). Nitrate removal in stream riparian zones. *Journal of Environmental Quality*, **25**: 743–755.

Hill, A.R., Devito, K.J., Campagnolo, S., and Sanmugadas, K. (2000). Subsurface denitrification in a forest riparian zone: interactions between hydrology and supplies of nitrate and organic carbon. *Biogeochemistry*, **51**: 193–223.

Hill, A.R., Vidon, P., and Langat, J. (2004). Denitrification potential in relation to lithology in five headwater riparian zones. *Journal of Environmental Quality*, **33**: 911–919.

Hubbard, R.K., and R. Lowrance. (1997). Assessment of forest management effects on nitrate removal by riparian buffer systems. *Trans ASAE*, **40**: 383–391.

Inamdar, S.P., Sheridan, J.M., Williams, R.G., Bosch, D.D., Lowrance, R.R., Altier, L.S., and Thomas, D.L. (2000). Evaluation of the Riparian Ecosystem Management Model (REMM). I. Hydrology. *Transactions of the American Society of Agricultural Engineers*, **42**: 1679–1689.

Jacinthe, P.A., Groffman, P.M., Gold, A.J., and Mosier, A. (1998). Patchiness in microbial nitrogen transformations in groundwater in a riparian forest. *Journal of Environmental Quality*, **27**: 156–164.

Jones, J.R., Bokofka, B.P., and Bachmann, R.W. (1976). Factors affecting nutrient loads in some Iowa streams. *Water Research*, **10**: 117–122.

Kruk, M. (2003). Biogeochemical multifunctionality of wetland ecotones in lakeland agricultural landscape. *Polish Journal of Ecology*, **51**: 247–254.

Kroeze, C., and Seitzinger, S.P. (1998). The impact of land use on N_2O emissions from watersheds draining into the northeastern Atlantic Ocean and European seas. *Environmental Pollution*, **102** (Si): 149–158.

Lane, S.N., Brookes, C.J., Kirkby, M.J., and Holden, J. (2004). A network index based version of TOPMODEL for use with high resolution digital topographic data. *Hydrological Processes*, **18**: 191–201.

Lowrance, R.R., Todd, R.L., and Asmussen, L.E. (1984). Nutrient cycling in an agricultural watershed. II. Streamflow and artificial drainage. *Journal of Environmental Quality*, **13**: 27–32.

Lowrance, R., Altier, L.S., Newbold, J.D., Schnabel, R.R., Groffman, P.M., Denver, J.M., Correll, D.L., Gilliam, J.W., Robinson, J.L., Brinsfield, R.B., Staver, K.W., Lucas, W., and Todd, A.H.

(1997). Water quality functions of riparian forest buffers in Chesapeake Bay watersheds. *Environmental Management*, **21**: 687–712.

Malanson, G.P. (1993). *Riparian Landscapes*. Cambridge University Press: Cambridge.

Naiman, R.J., and Decamps, H. (1997). The ecology of interfaces: riparian zones. *Annual Review of Ecology and Systematics*, **28**: 621–658.

Nelson, W.M., Gold, A.J., and Groffman, P.M. (1995). Spatial and temporal variation in groundwater nitrate removal in a riparian forest. *Journal of Environmental Quality*, **24**: 691–699.

Osborne, L.L., and Kovacic, D.A. (1993). Riparian vegetated buffer strips in water-quality restoration and stream management. *Freshwater Biology*, **29**: 243–258.

Park, S.J., and van de Giesen, N. (2004). Soil-landscape delineation to define spatial sampling domains for hillslope hydrology. *Journal of Hydrology*, **295**: 28–46.

Parry, S., Renault, P., Chenu, C., and Lensi, R. (1999). Denitrification in pasture and cropped soil clods as affected by pore space structure. *Soil Biology and Biochemistry*, **31**: 493–501.

Peterjohn, W.T., and Correll, D.L. (1984). Nutrient dynamics in an agricultural watershed: Observations on the role of a riparian forest. *Ecology*, **65**: 1466–1475.

Pinay, G., and Décamps, H. (1988). The role of riparian woods in regulating nitrogen fluxes between the alluvial aquifer and surface water: A conceptual model. *Regulated Rivers*, **2**: 507–516.

Pinay, G., Ruffinoni, C., and Fabre, A. (1995). Nitrogen cycling in two riparian forest soils under different geomorphic conditions. *Biogeochemistry*, **30**: 9–29.

Pinay, G., and Burt, T.P. (2000). *Nitrogen Control in Agricultural Landscapes*. Final report. European Commission DG XII. NICOLAS – ENV4 – CT 97–0395.

Prather, M., Derwent, R., Ehhalt, D., Fraser, P., Sanhueza, E., and Zhou, X. (1995). Other trace gases and atmospheric chemistry. In J. Houghton, L.G. Meira, E. Haites, N. Harris and K. Maskell (eds), *Climate Change*. New York: Cambridge University Press.

Quinn, P.F., Hewett, C.J.M., and Doyle, A. (2004). Scale appropriate modelling: from mechanisms to management. In Tchiguirinskaia, I., Bonell, M., and Hubert, P. (eds) *Scales in Hydrology and Water Management*.Wallingford: International Association of Hydrological Sciences Press. IAHS Publication 287, pp. 17–37.

Rustad, L., Campbell, J., Marion, G., Norby, R., Mitchell, M., Hartley, A., Cornelissen, J., and Gurevitch, J. (2001). A meta-analysis of the response of soil respiration, net nitrogen mineralization, and aboveground plant growth to experimental ecosystem warming. *Oecologia*, **126**: 543–562.

Sabater, S., Butturini, A., Clement, J.C., Burt, T., Dowrick, D., Hefting, M., Maitre, V., Pinay, G., Postolache, C., Rzepecki, M., and Sabater, F. (2003). Nitrogen removal by riparian buffers along a European climatic gradient: Patterns and factors of variation. *Ecosystems*, **6**: 20–30.

Schade, J.D., Fisher, S.G., Grimm, N.B., and Seddon, J.A. (2001). The influence of a riparian shrub on nitrogen cycling in a Sonoran Desert stream. *Ecology*, **82**: 3363–3376.

Schnabel, R.R., Cornish, L.F., Stout, W.L., and Shaffer, J.A. (1996). Denitrification in a grassed and a wooded, valley and ridge, riparian ecotone. *Journal of Environmental Quality*, **25**: 1230–1235.

Sweeney, B.W., Bott, T.L., Jackson, J.K., Kaplan, L.A., Newbold, J.D., Standley, L.J., Hession, W.C., and Horwitz, R.J. (2004). Riparian deforestation, stream narrowing, and loss of stream ecosystem services. *Proceedings of the National Academy of Sciences of the United States of America*, **101**: 14132–14137.

Tansley, A.G. (1911). *Types of British Vegetation*. Cambridge: Cambridge University Press.

Verhoeven, J.T.A., Arheimer, B., Yin, C., and Hefting M.M. (2006). Global and regional concerns over the purification function of wetlands. *Trends in Ecology and Evolution*, **21**: 96–103.

Viaud, V., Merot, P., and Baudry, J. (2004). Hydrochemical buffer assessment in agricultural landscapes: From local to catchment scale. *Environmental Management*, **34**: 559–573.

Vidon, P.G.F., and Hill, A.R. (2004a). Landscape controls on the hydrology of stream riparian zones. *Journal of Hydrology*, **292**: 210–228.

Vidon, P.G.F., and Hill, A.R. (2004b). Landscape controls on nitrate removal in stream riparian zones. *Water Resources Research*, **40**: W03201, doi:10.1029/2003WR002473.

Wang, W., Lacis, Y., Mo, T., and Hansen, J. (1976). Greenhouse effects due to man made perturbations of trace gases. *Science*, **194**: 685–690.

15
Flow–Vegetation Interactions in Restored Floodplain Environments

Rachel Horn and Keith Richards

15.1 The Need for Ecohydraulics

Channels, riparian zones and floodplains are dynamic locations subjected to disturbance and fluctuating water and sediment inputs during floods. Species inhabiting these environments have evolved tolerance to natural disturbance, and may even depend on disturbance to complete their life cycles. Naturally, the ecosystems of these environments at the interface between terrestrial and lotic systems have high levels of biodiversity (Petts, 1990), and the disturbance by floods creates conditions for a heterogeneity of habitats that support a varied mix of species (Richards *et al.*, 2002; Hughes *et al.*, 2005). These ecosystems have intrinsic values associated with their high levels of biodiversity, but can also be valued for their contributions to fisheries, groundwater recharge and discharge, navigation and transport, recreation, timber production, carbon sequestration, nutrient and sediment retention, floodwater storage, and pollution control (Costanza *et al.*, 1997).

Unfortunately, hydraulic engineers have long sought to manage rivers and their flow regimes, the aim often being to stabilise them, and to reduce flooding. Engineered interventions include straightening and dredging rivers (channelisation); steepening, raising and protecting banks; and removing aquatic, bank and floodplain vegetation ('snagging'). The purpose has commonly been to convey floodwater efficiently and rapidly, but the consequence has been to reduce severely the very disturbance regimes that sustain the ecological health of these environments, and to restrict lateral connectivity between channel and floodplain. These practices have destroyed many aquatic and riparian ecosystems, and in western Europe in particular, natural floodplain ecosystems (for example) have largely disappeared, with remaining patches in a critical condition or, in fact, entirely managed as plantations. Floodplain forest is therefore a priority habitat type in the European Habitats Directive (92/43/EEC, 1992).

In recent years there has been a growing global awareness of the ecological importance of river and wetland environments, leading to an emphasis on restoration, renaturalisation, and the maintenance of the functions of channel and floodplain ecosystems (Ward et al., 2001). This has also sought to restore the potential for natural functions to deliver the range of natural services whose valuation Costanza et al. (1997) have reviewed. Inevitably, this has required research to underpin this valuation. One area that has seen a rapid growth of research has been the interaction of vegetation with the hydraulics of flow, which must be understood so that the consequences of restoring natural aquatic, riparian and floodplain vegetation can be modelled, and shown to be acceptable and even beneficial in some locations, without increasing flood risk in sensitive or populated locations. The need for this research is illustrated by the two cases of aquatic weed cutting, and restoring floodplain woodland.

In the former case, management of aquatic macrophytes by manual or mechanical (weed cutting) or chemical (herbicide) means has been practised for many years, both to maintain navigability and to reduce flood risk. Such practices have a severe ecological impact, since the macrophytes are crucial elements in the aquatic ecology, providing food, energy, nutrients and shelter for aquatic invertebrate fauna. Removing macrophytes may reduce flow resistance at low flow, and lower the water depth, but this can be deleterious for fish populations, whose habitat preferences include a minimum depth. However, it is by no means clear that such management practices are necessary for flood control. Macrophyte control that fails to remove the rhizomes is followed by rapid regrowth that may even be stimulated by cutting. Reduction of low flow water levels after cutting only provides more light and speeds photosynthesis (Dawson, 1978), so regrowth is usually rapid, and involves multiple, vigorous, plants whose roughness effect may be even greater (and that immediately require expensive recutting). Summer cutting encourages regrowth in the autumn, so that channels in the highest risk period have a higher biomass of young plants, compared with the thinner, larger, older plants that were cut. However, it is also possible that the effects of macrophytes on flow resistance and water levels (and flood risk) has been exaggerated.

Although water levels may decline noticeably after cutting at times of low flow, this is a minor effect in relation to flood levels, when serious risks only begin to arise when flow goes out-of-bank and the effect of the in-channel macrophytes may be limited. Complete blockage of channels by plants is rare, since the effect of such blockage on velocities, gas and nutrient exchanges, and plant health, could in essence self-limit increased vegetative density. Instead, macrophytes tend to aggregate in ways that maintain intervening channels of faster flow, which limit the overall resistance; the maintenance of such channels through macrophytes is therefore a practice to be recommended in preference to complete clearance (Dawson, 1989). Most importantly, the streamlining of flexible aquatic plants in high velocity flows (see Section 15.4.1) implies that the flow resistance associated with the aquatic vegetation decreases at high flows, and the risk of increased water levels and flood risk is therefore less severe than historically assumed (Pitlo and Dawson, 1990; Larsen et al., 1990).

In the second example, the restoration of naturally wooded floodplains runs counter to conventional flood management practices, where removal of woody vegetation is encouraged to improve conveyance. Woody vegetation reduces overbank velocities, raises upstream water levels, and is traditionally assumed to increase flood risk. However, if transmission of flood water can be delayed in an undeveloped or naturalised floodplain where increased water levels are acceptable, the floodplain storage could attenuate flood peaks downstream in populated reaches (Hughes, 1980; Naef *et al.*, 2000). In integrated catchment management, the previously assumed conflict between vegetation and flooding could therefore be turned to advantage using selected, appropriately located, areas as flood storage and attenuation zones.

The effectiveness of such storage depends on whether the retention is of standing water (temporary pondage) or flowing water (Naef *et al.*, 2000), and the effectiveness of the latter depends on the flow resistance of the floodplain vegetation. In order to plan for such retention, it is necessary to understand and predict this resistance effect. Naef *et al.* (2000) suggest that for long-duration hydrographs, floodplain storage fills before the peak arrives, and the effectiveness is accordingly limited. However, this conclusion was based on 1-D modelling, and more sophisticated 2- or 3-D modelling with spatially varied roughness could produce different conclusions. Certainly such modelling would be necessary in order to evaluate an additional issue, namely the management of floodplain storage through systems of drainage channels with gates and flaps that encourage drainage when the main channel flow levels drop, so that storage capacity is recovered quickly enough to accommodate a second hydrograph.

These two examples show the need to be able to model river and floodplain flows, and flow interactions between channels and floodplains, in quite subtle manners in order to optimise the joint benefit of biological restoration and flood control. This highlights the need to reduce the uncertainty with which flow conductance through vegetation and upstream water levels can be modelled.

15.2 The Basic Hydraulics of Flow–Vegetation Interaction

Resistance to flow arises because of energy dissipation generated by frictional forces exerted both by the channel boundary and by obstacles to the flow (including vegetation), as well those forces arising within the flow itself. The total resistance reflects many factors that contribute to the overall energy loss, including the bed roughness (determined by the size, shape, and distribution of the particles forming the solid boundary), irregularity of the bed surface, obstructions, the type and density of vegetation, and the degree of meandering. These are not always present, and vary in relative importance at different locations and times. Where resistance due to vegetation or other large obstacles occurs, the bed friction is often of minor importance, as the internal flow distortion due to the vegetation becomes dominant. Energy loss is also affected by seasonal changes in vegetation character; by floodplain flow that crosses the main channel in a meander bend; by transport and jamming of woody debris; and by shear stresses at the interface between the floodplain and the main channel (Trieste and Jarrett, 1987).

Einstein and Banks (1950) crucially showed, experimentally, that the total resistance is the sum of component resistances. As a result of this finding, much effort has been devoted to examining the contribution of individual components to a total resistance coefficient, such as Manning's n. Cowan (1956) summarised this additivity principle by suggesting that n is given by:

$$n = (n_b + n_1 + n_2 + n_3 + n_4)m \qquad (15.1)$$

where n_b is a base value of n for a straight, uniform, smooth channel in natural materials, n_1 is a factor for the effect of surface irregularities; n_2 is a value for variations in shape and size of the channel cross-section; n_3 is a value for obstructions; n_4 is a value for vegetation; and m is a correction factor for meandering of the flow. To be valid, Equation (15.1) requires that each component coefficient is independent of the others. Tables 15.1 and 15.2 illustrate some examples of vegetation roughness values.

This has led to research to quantify component roughness coefficients. For example, in an unvegetated open channel the bed material is a major determinant of flow resistance, and even in vegetated channels it can be significant, depending on its size and form relative to the vegetation, and the density, or rather sparseness, of the vegetation. Early work on flow resistance in pipes and channels focused on a characteristic grain size of the channel bed, and in Strickler's widely referenced study (Henderson, 1966; Smart, 1999), median grain diameter D is related to Manning's n:

$$n = 0.034\, D^{1/6} \qquad (15.2)$$

Other factors, such as flow depth, bed topography, bed slope and boundary mobility may all affect the roughness coefficient. The bed roughness coefficient decreases as the depth increases, so n_b is not just dependent on D, but also depends on grain size relative to depth. Nevertheless, this example shows that a component roughness can be correlated to a measurable property of a controlling factor.

15.2.1 Roughness Properties of Vegetation

The question therefore arises as to what properties of vegetation must be measured in order to characterise its contribution to the resistance to flow through it. Clearly the density and arrangement of stems, and the volume of canopy, will be important properties, but a large number of parameters may be needed to specify the shapes and distribution fully. These should include the shape, diameter, d (m), height, h_v (m) and momentum absorbing area (*MAA*) as defined by Fathi-Moghadam (1996) as well as the arrangement of the array of vegetation (whether it is random, in parallel or staggered patterns), and the density of the 'stems'. Vegetation height is sometimes described in terms of a submergence ratio (Stone and Shen, 2002) or relative height. This parameter, denoted by h^*, has a value between 0 and 1 and is the ratio of the height of the plants (h_v) to the flow depth (y_n). For emergent vegetation, the submerged height is equal to the flow depth and $h^* = 1$.

Natural trunks or stems do not form a regular pattern so a common approach (Kouwen, 1992) is to characterise their arrangement by the density or number of stems per unit area of ground, M (number/m^2). Other researchers (Li and Shen, 1973; Nepf, 1999; Ree and Crow, 1977) show that the spatial arrangement of stems or of cylinders (representing

Table 15.1 Extract from Chow (1959): values of the roughness coefficient n.

D-1. Minor Natural Streams	Minimum	Normal	Maximum
(a) Streams on plain (top width at flood stage < 100 ft)			
(1) Clean, straight, full stage, no rifts or deep pools	0.025	0.030	0.033
(4) Clean, winding, some pools and shoals, some weeds and stones	0.035	0.045	0.050
(7) Sluggish reaches, weedy deep pools	0.050	0.070	0.080
(8) Very weedy reaches, deep pools or floodways with heavy stand of timber and underbrush	0.075	0.100	0.150
(b) Mountain streams, no vegetation in channel, banks usually steep, trees and brush along banks submerged at high stages			
(1) Bottom: gravels, cobbles and few boulders	0.030	0.040	0.050
(2) Bottom: cobbles with large boulders	0.040	0.050	0.070
D-2. Floodplains			
(a) Pasture, no brush			
(1) Short grass	0.025	0.030	0.035
(2) High grass	0.030	0.035	0.050
(b) Cultivated areas			
(1) No crop	0.020	0.030	0.040
(2) Mature row crops	0.025	0.035	0.045
(3) Mature field crops	0.030	0.040	0.050
(c) Brush			
(1) Scattered brush, heavy weeds	0.035	0.050	0.070
(2) Light brush and trees in winter	0.035	0.050	0.060
(3) Light brush and trees in summer	0.040	0.060	0.080
(4) Medium to dense brush in winter	0.045	0.070	0.110
(5) Medium to dense brush in summer	0.070	0.100	0.160
(d) Trees			
(1) Dense willows, summer, straight	0.110	0.150	0.200
(2) Cleared land with tree stumps, no sprouts	0.030	0.040	0.050
(3) Same as above, but with heavy growth of sprouts	0.050	0.060	0.080
(4) Heavy stand of timber, a few down trees, little undergrowth, flood stage below branches	0.080	0.100	0.120
(5) Same as above, but with flood stage reaching branches	0.100	0.120	0.160
D-3. Major natural streams (top width at flood stage > 100 ft). The *n* value is less than that for minor streams of similar description because banks offer less effective resistance.			
(a) Regular section with no boulders or brush	0.025	....	0.060
(b) Irregular and rough section	0.035	....	0.100

Reprinted with permission of the estate of V.T. Chow.

Table 15.2 Adjustment values for factors that affect roughness of floodplains (from Arcement and Schneider, 1989)

Amount of vegetation (n_4)	n value adjustment	Example
Small	0.001–0.010	Dense growth of flexible turf grass ... or weeds, where the average depth of flow is at least twice the height of the vegetation, or supple tree seedlings such as willow ... where the average depth of flow is at least three times the height of the vegetation
Medium	0.011–0.025	Turf grass where the average depth of flow is from one to two times the height of the vegetation, or moderately dense stemmy grass, weeds, or tree seedlings growing where the average depth of flow is from two to three times the height of the vegetation; brushy moderately dense vegetation, similar to 1- to 2-year-old willow trees in the dormant season
Large	0.025–0.050	Turf grass where the average depth of flow is about equal to the height of the vegetation, or 8- to 10-year-old willow trees intergrown with some weeds and brush (no foliage) where the hydraulic radius exceeds 2 feet, or mature row crops such as small vegetables, or mature field crops where depth of flow is at least twice the height of the vegetation
Very large	0.050–0.100	Turf grass where the average depth of flow is less than half the height of the vegetation, or moderate to dense brush, or heavy stand of timber with few down trees and little undergrowth where depth of flow is below branches, or mature field crops where depth of flow is less than the vegetation height
Extreme	0.100–0.200	Dense bushy willow ... (all vegetation in full foliage), or heavy stand of timber, few down trees, depth of flow reaching branches.

Courtesy of the United States Geological Survey.

vegetation stems) can affect the overall drag and therefore vegetative resistance. Parallel and staggered arrays often show lower bulk drag coefficients than do random arrays, as a result of sheltering from upstream stems (see Section 15.3). Stem spacing is denoted by s (m), and spacings in the flow-parallel and flow-transverse directions may be indicated by s_x and s_y. The mean spacing, s, and the number of plants per unit area, M, are linked by $M = 1/s^2$, providing the mean spacing directions are measured orthogonally. Measurement of the density of vegetation is addressed in several papers (Arcement and Schneider, 1989; Fathi-Moghadam and Kouwen, 1997), which use various different metrics: the area of ground occupied by the vegetation per unit surface area, expressed as a percentage (m^2/m^2); the projected plant area (frontal area) opposing the flow, either per unit area of ground (m^2/m^2) (Herbich and Shulits, 1964; Wooding et al., 1973), or per unit volume of flow (in m^2/m^3) (Nepf, 1999; Petryk and Bosmajian, 1975). The surface area occupied by the vegetation is usually estimated assuming circular stems:

$$\lambda_a = \frac{M\pi d^2}{4}(\%) \qquad (15.3)$$

where λ_a is the area occupied by vegetation per unit area of the bed (%), M is the number of cylinders/stems per unit area (m^{-2}) and d is the cylinder/stem diameter. Using the projected plant area per unit volume is similar to the method of representing vegetation density suggested by Petryk and Bosmajian (1975) where vegetation with projected area in a streamwise direction A_i, in a channel of length L, with cross sectional area of flow A, has a vegetation density of:

$$\frac{\sum A_i}{AL} \qquad (15.4)$$

Given the range of measures of density, care is required when using relationships with roughness that the measure employed is both consistent and appropriate for the purpose.

Fathi-Moghadam (1996) has introduced an additional concept, that of the momentum absorbing area (*MAA*) of a plant community, which is the projection of the area of the leaves and stems in the cross-sectional area of the flow.

$$MAA = \phi^2 A_0 \qquad (15.5)$$

where $\phi = (2/\pi)$ accounts for random, three-dimensional orientation of the leaves and stems of a tree, and A_0 is the total one-sided area of the submerged biomass. Järvelä (2004) also presented a method for estimating the flow resistance of woody vegetation in both its leafless and leafy condition, based on easily measurable plant characteristics: for leafless vegetation the bulk drag coefficient is based on a characteristic diameter that is derived from theory on the mechanical design of trees; for leafy vegetation, a species-specific drag coefficient is introduced, which along with the leaf area index accounts for the plant flexibility and shape in flow.

Vegetation not only affects the flow through its geometric properties: its stiffness also has a significant effect on flow resistance. Flexible vegetation can become streamlined in high velocity flow, and its submerged height is then reduced by bending, further reducing the resistance to flow. It may seem simple to categorise vegetation into two distinct types: flexible and rigid. Flexible vegetation, such as grasses and aquatic species, is able to bend, streamline and even become flattened towards the bed depending on the flow conditions. Rigid vegetation (e.g., tree trunks) remains upright whatever the flow conditions. But branches and leaves, and young saplings, are flexible in high velocity flow; and grass stems can remain rigid in low flow. Thus, individual species or groups of vegetation may not always fall clearly into defined categories (Darby, 1999); different portions of the same plant can behave in distinctly different ways; and the same plant may belong to different groups at different stages of its life cycle.

Kouwen and Unny (1973) and Kouwen and Li (1980) undertook studies into the biomechanics of flexible vegetation lining channels, and defined relative roughness as the ratio of the mean height of the deflected vegetation to its undeflected height, and introduced a stiffness parameter *MEI*, which takes into account the density M, and the rigidity EI (elasticity E, and moment of inertia, or slenderness I) of the vegetation.

$$h_\mathrm{d} = 0.14 h_v \left[\frac{\left(\dfrac{MEI}{\tau_0}\right)^{0.25}}{h_v} \right]^{1.59} \quad (15.6)$$

Here h_d is the deflected (or effective) roughness height of the vegetation, MEI is the stiffness parameter, τ_0 is the mean boundary shear stress ($= \rho g y_n S_f$), y_n is depth of channel flow ($y_n > h_v$) and h_v is height of vegetation. Some values of MEI for vegetation are tabulated in Kouwen and Li (1980), and Darby (1999) suggested that stems with stiffness parameters greater than about 200 N/m^2 should be considered nonflexible.

Kouwen and Li (1980), Kouwen and Unny (1973) and Kouwen et al. (1969) studied flow velocity as a function of the relative roughness height, density and flexibility of vegetation, and its erectness. They identified three flow regimes, in the first of which grasses remain upright in the flow. In the second, the vegetation oscillates in the flow, and in the third, the vegetation lies prone and there is a significant (about fivefold) reduction in the roughness. The third phase requires a high velocity to flatten the vegetation, but then the reduction of roughness allows an even higher velocity. Kouwen and Li (1980) suggested that the limiting friction velocity (U^*) between the erect and prone states of vegetation is a function of the flexibility of the vegetation, with the vegetation becoming prone when:

$$U^* > U^*\text{critical} = \text{minimum}\ (0.028 + 6.33\ MEI^2, 0.023\ MEI^{0.106}) \quad \text{(SI units)} \quad (15.7)$$

Kouwen and Fathi-Moghadam (2000) have also investigated the effect of flexibility on the roughness of vegetation, specifically that of conifers. A velocity increase caused bending and streamlining of the plants, but the friction factors (and relation to flow velocity) depended on the species of tree and characteristics of shape, MAA and material properties (Fathi-Moghadam, 1996; McMahon, 1975). Another characteristic, the vegetation index, ζE, which relates to the resonant frequency, mass and length of a tree, was found to reduce the difference between the friction factors of tree species tested.

One final property of vegetation that can be an important control of flow is the simple volumetric displacement arising from the presence of the vegetation. If a cross section of flow is occupied by a significant volume of vegetation, then the flow has to be carried between the plants. This effect is reflected in a 'blockage factor' (Green, 2005). Such blockage can result in local flow velocities that are much faster than would be expected given the overall roughness, and implies that it can be extremely difficult to define the mean velocity in the cross section.

15.2.2 Nonlinearities

Although roughness components may be additive (compare Equation 15.1), the actual value of a component roughness coefficient may be nonlinearly dependent on some attribute of the controlling factor. This was originally shown in experiments in which wooden strips were placed perpendicular to the flow in a flume (Johnson, 1944). When the strips are very widely spaced, their roughness is low, and the same is true when they are close together because the flow skims across the gaps between them. At a certain spacing, the eddies shed from one strip impinge on the next strip downstream, and the roughness is maximised. This phenomenon may be expected to occur in rigid-stemmed vegetation, where in some densities and spatial arrangements, the turbulence induced by each stem may dissipate before the next stem, while in others, there is mutual interference and high roughness.

Another nonlinearity that complicates the analysis of roughness is revealed by the role of flow depth. Notwithstanding Equation (15.2), the friction coefficient for bed material is dependent on the ratio of bed material grain size to flow depth. This means that an iterative approach is necessary to determine the depth and roughness of a given discharge over a given bed, since the roughness both determines, and is partly controlled by, the depth. If the resistance to flow is provided by both bed material and rigid vegetation, then the bed roughness depends on the grain size and the water depth, but the latter is itself dependent on the total roughness. Furthermore, the roughness of emergent vegetation depends on its frontal density, which also depends on the water depth. Thus, determining the component roughness coefficients is again an iterative problem.

15.3 Drag Coefficients and Vegetation

One way of characterising the frictional effect of vegetation is through consideration of the drag coefficient required to balance the force driving the flow against the resisting force of the vegetation. In the case of a single cylinder of diameter d, and height h_v, isolated in a flow of depth y_n and mean velocity U, the force F_D (N) on the cylinder is expressed by:

$$F_D = C_d' \left(\frac{\rho g U^2}{2g} \right) dh_v \quad (15.8)$$

where C_d' is the drag coefficient of a single cylinder. The best estimate of this drag coefficient in fully turbulent flow around a cylinder with a rough surface is that it takes a value of 1.2 across a wide range of Reynolds numbers.

Things become more complicated in *arrays* of cylinders since, as is apparent from Equation (15.8), with all else being equal, the drag coefficient is inversely related to the velocity. In an array of cylinders, there is a complex pattern of velocity variation between the cylinders, and the array drag coefficient depends on the way in which the velocity to which it relates is defined (another illustration of the nonlinearity discussed in Section 15.2.2). Considering a series of cylinders (i.e., rigid stems) aligned along the flow path, the approach velocity to a downstream cylinder will depend on how far it is from the upstream cylinder, in the same manner as shown in Johnson's (1944) experiments with wooden strips. The approach velocity on the downstream cylinder is reduced due to a wake velocity defect, and also because the turbulence caused by the wake may delay the point of flow separation on the circumference of the downstream cylinder, reducing the size of its wake and therefore further suppressing the drag. Thus, there is a sheltering effect that reduces the drag on downstream objects, implying that the longer the flow path through vegetation, the lower the array drag coefficient. Experiments by Li and Shen

(1973) suggest that this is asymptotic as the velocity and drag equilibriate at array coefficients of between 0.75 for parallel cylindrical arrays and 1.10 for staggered arrays.

Thus, the reduction in drag for an array can be represented using an 'average' drag coefficient that differs from the value for a single stem. In fact, there are two ways of achieving this. One is to require the velocity in Equation (15.8) to reflect the average approach velocity to each stem (which is lower than the maximum or mean velocity), with C_d' retaining the individual stem value; the other is for the drag coefficient to be modified to account for the reduced drag force of an array compared with the sum of individual cylinder drag forces. Neither is straightforward, however. For example, in an array there may be locally higher velocity because of the blockage effect in dense arrays, and consequent acceleration of flow between stems, even if the average velocity is reduced. However, it may be very difficult to determine what the average velocity is (without a dense pattern of point measurements), or at what location it is appropriate to measure a point velocity in order to capture the mean velocity accurately.

15.4 Velocity, Velocity Profiles and Vegetation Character

Early experimental investigations into the design of vegetated channels by the US Department of Agriculture Soil Conservation Service (1937–1946) culminated in the production of the *Handbook of Channel Design for Soil and Water Conservation* (Stillwater Outdoor Hydraulic Laboratory, 1954). This design guide provided a series of curves developed from experiments on channels covered with submerged, flexible, vegetation such as grass and crops, which related Manning's n to the flow properties of the channel, VR (product of velocity and hydraulic radius); these are commonly referred to as n–VR curves. These did not distinguish between different flexibilities, simply classifying levels of vegetal retardance (see Figure 15.1). Here, the degree of retardance depended mainly on the height, density and uniformity of the vegetation, but above a minimum level of vegetation cover, the height of the vegetation outweighed the effect of density on the retardance. These observations are of limited applicability (they do not apply to emergent vegetation), and embody the iterative problem that n and VR are mutually dependent; it is difficult to predefine a value of n in order to estimate velocity.

Furthermore, this early analysis only provided insight into the mean velocity, when it is often necessary to understand how vegetation and flow interact to produce complex velocity profile shapes, which can then be integrated across the flow depth to provide estimates of mean velocity. In order to characterise these more detailed properties of flow over and through vegetated surfaces, convenient distinctions are between submerged and emergent vegetation, and rigid and flexible vegetation (even though the latter are rather fuzzier differences in practice than in theory).

Within a stand of emergent rigid vegetation, the velocity profile is widely held to be more or less vertical (apart from a shallow region close to the bed, where the bed roughness defines a logarithmic profile). Experimental results reported by Stone and Shen (2002) and Tsujimoto *et al*. (1992) confirm that the velocity in the vegetation layer is almost uniform and that the effect of bed friction on the velocity profile is important only very near the bed where the velocity decreases to zero (Figure 15.2a). With submerged

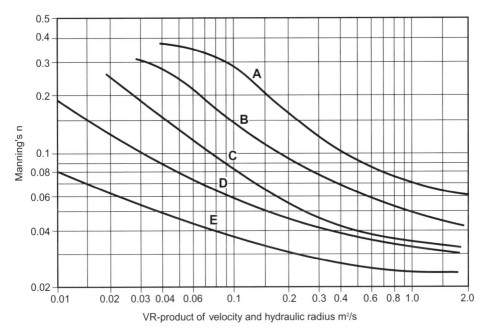

Figure 15.1 *n-VR curves for different degrees of vegetal retardance (from the Handbook of Design for Soil and Water Conservation, Stillwater Outdoor Hydraulic Laboratory, 1954). A = very high degree of vegetal retardance, B = high, C = moderate, D = low, E = very low degree. Courtesy of the United States Department of Agriculture.*

rigid vegetation of uniform height, the flow is considered to have three layers: the vegetation layer, a shear layer and a layer of surface flow (Figure 15.2b). In general the flow velocity in the vegetation layer is significantly smaller than that in the surface layer, because of the drag imparted by the vegetation. The difference in velocity between these two layers causes a shear force at the interface between the vegetation layer flow and that in the free stream above, with associated turbulent mixing close to the boundary between the layers. This creates a third, intervening shear layer.

Further complications are introduced with flexible vegetation (such as in-channel submerged, rooted, streamlined plants like *Elodea* and *Potamogeton*), whose inclination and change in height result in drag characteristics that change with the flow conditions. Similar behaviour occurs in flexible parts of woody vegetation, such as tree branches, leaves, and saplings, and in flexible ground-cover floodplain vegetation. Kutija and Hong (1996) addressed this behaviour in a numerical model where bending depends on cantilever beam theory, and the drag force is proportional to the height of the vegetation. Using a similar model, Tsujimoto *et al.* (1996) concluded that the velocity distribution in submerged flexible vegetation was roughly uniform, and proposed a resistance law by iterative coupling of sub-models for the velocity distribution in deformed vegetation (in which flexible plants are negative momentum sources and additional turbulent energy sources), and for plant deformation by shear flow (the plants being cantilevers with finite deformation).

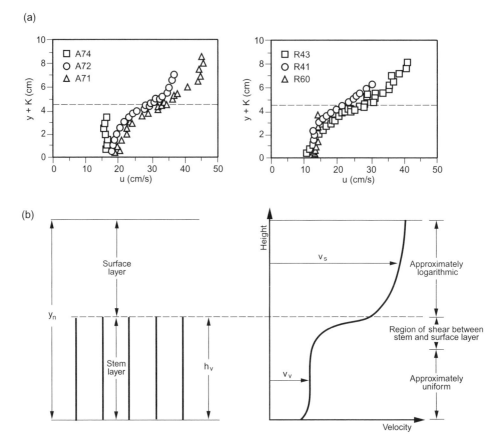

Figure 15.2 (a) Velocity profiles in emergent and submerged vegetation (after Tsujimoto et al., 1992). The dotted horizontal line indicates the vegetation height – vegetation was emergent for experiments A74 and R60. (b) Velocity profile within and above submerged vegetation. V_s = mean surface-layer velocity; V_v = mean stem-layer velocity.

In large rivers with significant stage changes during floods, the flow can inundate firstly the rigid trunks of trees, then the more flexible canopy above. Very few data exist to describe velocity profiles in such cases, but Figure 15.3a illustrates the general expectation. This implies that high velocity jets occur beneath the canopy, once the flow encounters the high drag of the canopy layer. Measurements with upward-pointing acoustic Doppler velocimeters on the floodplain of the River Rhine (Figure 15.3b) have shown that velocity profiles in areas of grass cover are logarithmic above the grass, while between willows, the velocity increases from the bed to a peak at a height of about 1 m (below the canopy), then declines above this as the flow is retarded in the canopy layer (Bölscher et al., 2005). At a flow depth of 2.75 m, the velocity at about 1 m above the bed is about $1\,\mathrm{m\,s^{-1}}$, while at 2.75 m is has decreased to about $0.66\,\mathrm{m\,s^{-1}}$.

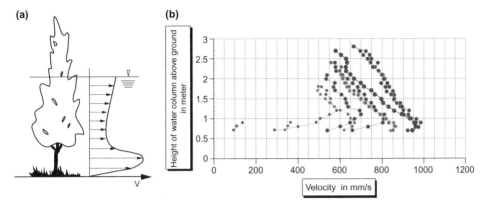

Figure 15.3 (a) Hypothetical velocity profile when flow occurs through a tree canopy (after Freeman et al., 2000). (b) Velocity profiles measured using a fixed, upward-looking acoustic Doppler profiler on a low floodplain area of the River Rhine near Breisach, located amongst a stand of willows (after Bölscher et al., 2005). At a flow depth of about 1 m, the velocity profile is controlled by the roughness of grass at the surface (which appears to become flattened and smoother by the second profile). Once the deeper flow penetrates into the lower canopy of the trees, the velocity is retarded above 1 m, and accelerates beneath the canopy. With permission of Berliner Geographische Abhandlungen.

15.5 Dimensionality: Flow Velocity in Compound Channels with Vegetation

Thus far, discussion of flow–vegetation interactions has been based on one- or two-dimensional considerations, but in a real case, flow and vegetation both occur in a channel and/or on a floodplain in a three-dimensional context. There may be flexible, submerged aquatic vegetation in the channel, and a mixture of flexible and rigid, submerged and emergent vegetation on the floodplain. In such a context the patterns of flow in and around the vegetation are more complex, and are macro-scale in their significance (that is, not restricted to flow structures around individual plants, but extending their impact to the overall flow cross section).

When overbank flow occurs in a river, the flow cross section becomes a two-stage channel, and complex patterns of turbulent flow interaction occur between the flow within the channel and the flow across the floodplain. If the channel meanders, the direction of overbank flow will be continually changing relative to the underlying current in the channel, resulting in energy losses due to the interaction. Compound channel flows are characterised by turbulent shear at the interface between the channel and floodplain, with banks of vortices transferring momentum between the main channel and floodplain flows (Knight and Shiono, 1996; van Prooijen *et al.*, 2005). If the floodplain is vegetated, these interactions are even more pronounced.

Figure 15.4 shows the typical pattern of interface vortices and secondary flow structures associated with channel–floodplain flow interaction. The strength of these momentum

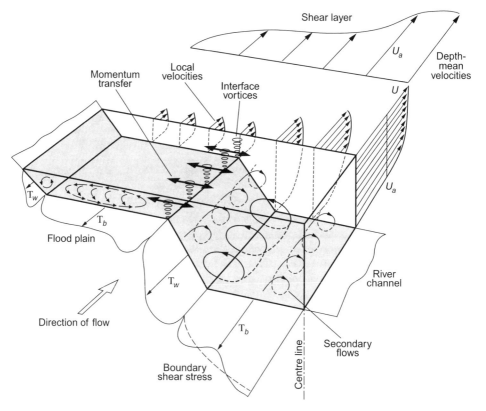

Figure 15.4 *Vortex structures and hydraulic parameters associated with overbank flow (after Knight and Shiono, 1996)*

exchange processes varies with the geometry of the system (sloping banks, for example, ease the flow of water onto the floodplain relative to vertical banks), and with the relative depths of floodplain and channel flow. Roughness differences between the channel and floodplain are particularly important, and are strongly dependent on the presence of vegetation. Vigorous boils arise periodically from the foot of a vegetated bank and strong transverse flows into the vegetation area alternate with weaker ones in the opposite direction. Vegetation stems vibrate vigorously where the surface flow extends into the vegetation (Hasegawa *et al.*, 1999). The higher the floodplain vegetation density and roughness, the more the discharge is displaced from the vegetated floodplain into the main channel, and strong shear layers develop at the edge of the vegetated zone, with secondary currents extending into the main channel (Naot *et al.*, 1996). Flume experiments by Schnauder (2004) illustrate these effects (Figure 15.5); this example is based on low-density rigid cylindrical roughness elements on the floodplain, representing trees.

In compound flow involving a floodplain and a channel, calculating and modelling discharge is difficult because roughness estimation needs to allow for the energy losses associated with the turbulent momentum exchanges described above. If these are not represented correctly, and roughness is underestimated, estimated velocities will be too

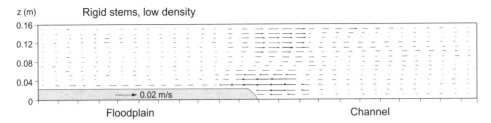

Figure 15.5 Secondary circulation at floodplain–channel interface (after Schnauder, 2004) with arrows showing vectors of longitudinally averaged, transverse flow components. Reprinted with permission from Dr-Ing Ingo Schnauder

high, flow depths too low, and the water surface elevation will be too low. Alternatively, if a known water surface elevation is fitted, the modelled partition of total discharge between the channel and the overbank flow will be incorrect, as will be the relative flow velocities between the channel and the floodplain.

15.6 Some Empirical Illustrations of Flow–Vegetation Interactions

This section illustrates the foregoing general and theoretical discussion with results from experimental studies. The majority of these were conducted in a 15 m, 0.3 m-wide flume, both with rigid cylindrical roughness elements designed to mimic trees, and with flexible, buoyant neoprene tubing. The subsections show that progressively more complex vegetation arrays result in velocity characteristics that are systematically explicable, and consistent with theoretical expectations. In addition, however, some results are reported of experiments in a field-scale flume facility on the River Wien in Vienna, where controlled flow releases of the order of 28–30 m^{3}s^{-1} can take place in a channel with an artificial floodplain on which willows are growing, inundating this floodplain to depths of 1–1.5 m.

15.6.1 Velocity Profiles in Submerged Rigid Vegetation

Figure 15.6 shows velocity profiles measured in a field of submerged rigid stems 0.15 m high in the laboratory flume, in flows of two depths and in staggered arrays of two different densities, created at different spacings by dowels 0.03 m and 0.08 m in diameter. The profiles are all essentially uniform within the stem layer, as suggested in Section 15.4, and the mean velocities for the profiles in the higher density array of 'stems' are lower than the equivalents in the low-density array. However, velocities are higher amongst stems of a given density, represented by smaller diameter, more closely spaced dowels (presumably because of the acceleration due to blockage). There is a systematic variation in velocity in the stem layer depending on the measurement location. The slowest velocity is always measured in the wake of a dowel, half-way to the next one downstream. Measurements ahead of a dowel and between dowels are quite similar. One property of all of these profiles not often noticed is that the shear layer between the stem

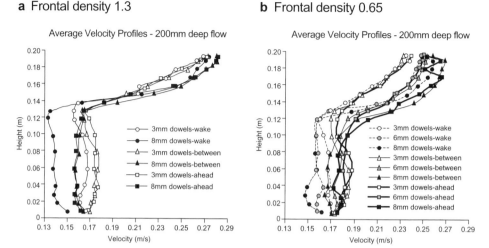

Figure 15.6 Velocity profiles for a submerged, staggered array of stems 0.15 m tall

and surface flow layers penetrates significantly into the upper part of the stem layer. This implies that models that simply assume two layers and a constant velocity in the stem layer will not capture the mean stem layer velocity correctly. It also implies that estimating the mean velocity in an array of rigid submerged stems will be problematic, because velocity profiles vary with location in the array and are more complex than a simple uniform profile.

15.6.2 Velocity Profiles in Emergent Rigid Vegetation

A similar effect is noticed in velocity profiles in emergent rigid stems, and is associated with proximity to the free surface. Although profiles are approximately uniform, close to the surface there is an acceleration of flow, rather like the surface layer above a layer of submerged stems. This is shown in Figure 15.7, which also shows that in emergent cases, the same relationships with density, stem diameter and measurement location occur. Again, however, the implication is that estimating mean velocity on the basis of a spatial pattern of point measurements is likely to be problematic.

15.6.3 Velocity Profiles in a Mixture of Submerged and Emergent Vegetation

In a staggered array of tall (emergent) and short (submerged) stems, velocity profiles have exactly the characteristics one would expect from consideration of Sections 15.6.1 and 15.6.2. In the example shown in Figure 15.8, the velocity profiles are measured amongst a staggered array of just-emergent (0.15 m) stems, with shorter (0.05 m stems) also in a staggered array between them. Within the basal layer, the stem density is therefore higher, and the vertically uniform velocity is low as a result. Near the top of the submerged stems, there is a shear layer where velocity increases to the higher (again uniform) velocity in the upper stem layer where the stem density is lower. This shear layer penetrates into the

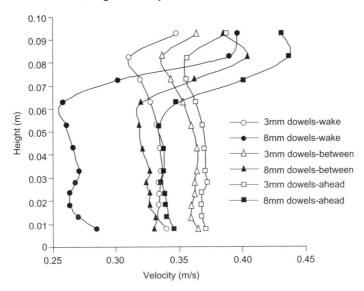

Figure 15.7 *Velocity profiles within a staggered array of emergent stems 0.15 m high in a flow depth of 0.10 m, and with a frontal density of 1.3*

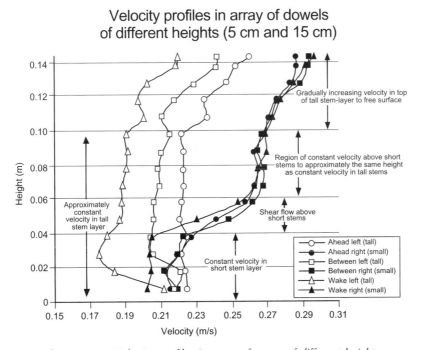

Figure 15.8 *Velocity profiles in array of stems of different heights*

upper part of the submerged stem layer, however. Near the water surface, there is an acceleration of flow amongst the emergent stems. The slowest overall velocity profile is in the wake of a tall stem; the fastest is measured ahead of a short stem, where it is only retarded in the basal layer. What this example shows is that where there is a complex vegetation array, it will result in complex velocity profiles that may be difficult to predict quantitatively, but which are qualitatively explicable.

15.6.4 Velocity Profiles in Submerged, Flexible Vegetation

Flexible vegetation becomes streamlined by the flow, and as a result 'interferes' with the velocity profile at the heights to which it is bent; it also oscillates in the flow, the more so as flow velocity increases, and the oscillation extracts momentum from the flow. Given these observations, the profiles in Figure 15.9, measured vertically through a stand of flexible, buoyant neoprene tubes between rows 5 and 6 of the 'rooted' tubes, are quite explicable. The strands in rows 3 and 4 ('rooted' upstream of rows 5 and 6) were partially streamlined, so that they were bent to cross the vertical through which measurements were made, and distorted the velocity at the heights at which they did so.

15.6.5 Complex Velocity Patterns in Staggered Arrays

The empirical examples discussed above illustrate that point velocities in flow through vegetation will vary spatially (transversely and with height above the bed) in complex

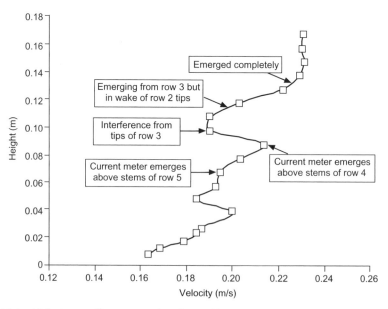

Figure 15.9 Velocity profile measured with a midget current meter between rows 5 and 6 of a stand of flexible strands of buoyant neoprene tube

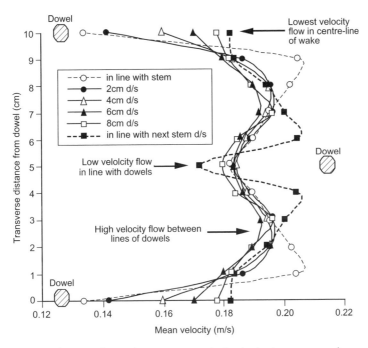

Figure 15.10 Distribution of stem-layer mean vertical velocity in a staggered array of 6.3 mm dowels at 10 cm centres

ways, making estimation of mean velocity very time-consuming and uncertain. Figure 15.10 reinforces this, showing a detailed set of vertically averaged velocities in plan view. This reveals that the maximum velocity across the channel in measurements made between the upstream stems occurs close to those stems, and that accordingly when measurements are made in the cross-section containing the next row of stems, the maximum is close to the stem in that section (this arises as a result of acceleration around the obstacle represented by the stem). The implication is that sinuosity in the maximum velocity flow thread is induced by the existence of stems. In addition, there is a reduction in velocity in the wake of upstream stems, and this persists for some distance downstream.

15.6.6 Velocity Variation Across a Partially Vegetated Channel

If measurements of velocity are made across the flume with one half of the width of the flume occupied by submerged rigid stems, the velocities in the 'vegetated' and 'unvegetated' halves of the flume may be compared. In addition, the total depth can be separated into that up to the height of the submerged stems, and that above (the stem layer and the surface layer, but continued across the section at the level of the tops of the rigid stems). Figure 15.11 shows two dimensions of the effect of rigid submerged vegetation on the flow. Firstly, the surface layer velocities are approximately uniform across the entire

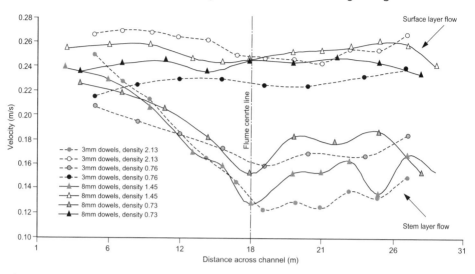

Figure 15.11 Variation in depth-averaged mean velocity across a channel half occupied by submerged rigid stems to mimic a vegetated floodplain

section. Secondly, the stem layer velocities in the 'vegetated' half of the flume are also approximately uniform across this half of the section, but at a much lower velocity. Thirdly, there is, however, a slight suggestion of a minimum velocity at the interface between the 'vegetated' and 'unvegetated' halves of the channel. Finally, in the 'unvegetated' half of the channel, the stem-layer velocity increases away from the 'vegetated' half, until it is almost the same as that in the surface layer above; the rate of increase decreases with distance, and the behaviour is almost comparable to a vertical velocity profile rotated through 90°. What is particularly noticeable is that the submerged stems have an effect on stem-layer velocity in the 'unvegetated' half of the channel for a distance almost equal to the lateral extent of the 'vegetated' half itself.

It is possible to compare the conveyance of the total flow as it is apportioned between both the 'vegetated' and the 'unvegetated' halves of the channel, and between the stem-layer and the surface layer (across the whole channel). In Figure 15.11, each pair of curves (stem-layer and surface) relates to an experiment with a different density of stems in the 'vegetated' stem layer. As the density decreases between these four experiments, the proportion of the flow carried in the 'vegetated' half of the channel increases, but is always less than half. At the same time, the proportion carried by the stem layer gradually increases, and in these experiments (with a depth of 0.20 m and stems 0.15 m high), averages approximately two-thirds. This shows that vegetation density has a significant and systematic influence on the conveyance of flow in a cross section, and a region of dense vegetation will displace flow and raise the water level and increase the velocity in adjacent regions.

15.6.7 An Alternative Means of Assessing Vegetative Roughness: The Water Surface Slope

Because of the evident complexity of measuring the mean velocity of flow through vegetation when the velocity profile is very variable spatially, alternative strategies for assessing the roughness associated with vegetation are needed in preference to the conventional one of back-calculating roughness from measured velocity and slope. In the flume experiments conducted on different densities and spacings of arrays, an alternative approach has been to measure the friction slope over the stretch of flume occupied by 'vegetation', and to solve the gradually varied flow equation using:

$$S_f = S_{fv} + S_{fb} = \frac{Mdh^* \hat{C}_d U_v^2}{2g(1-\lambda_a h^*)} + \frac{f}{8} \frac{U_v^2}{gR(1-\lambda_a h^*)} \qquad (15.9)$$

where the friction slope (S_f) is separated into components due to the vegetative drag (S_{fv}) and the bed (S_{fb}) roughness. The former depends on the stem density (M) and diameter (d), the proportion of the bed area occupied by stems (λ_a), and the relative submergence (h^*). Partial Manning coefficients due to the vegetative component of energy loss are then derived, and are correlated to the properties of the stem array included in Equation (15.9), and those that measure longitudinal (X) and transverse (Y) spacing, but which are modified spacing measurements applicable to random as well as regular arrays (for example, defining the average spacing to the nearest in-line stem upstream, X_{mod}).

This results in the following expression for (rigid) vegetative roughness:

$$n = \frac{y_n^{2/3}}{\sqrt{2g}} \sqrt{Md} \left[1.7 \frac{\left(1 - \sqrt{\frac{d}{X_{mod}}}\right)^{1.9}}{\left(1 - \frac{d}{Y}\right)^{1.8}} \right] \qquad (15.10)$$

which implies that the Manning coefficient for rigid vegetation increases with stem density, stem diameter and longitudinal spacing, and decreases with transverse spacing (in each case when all other parameters are fixed). Returning to the discussion in Section 15.2, this provides a basis for the estimation of the roughness of rigid floodplain vegetation, to which must be added component coefficients for bed roughness and other sources of energy loss.

15.6.8 An Alternative Means of Measuring Velocity in the Field

Given the uncertainties associated with estimating mean velocity from point measurements in the laboratory, the prospects of defining a reliable sampling protocol in the field are highly limited. In order to estimate mean velocities of flow through the willow stands on the floodplain in the Wienfluss large-scale flume, an alternative approach adopted involved tracer studies, in which injections of concentrated salt solution were tracked as the advecting and dispersing wave passed through a field of conductivity sensors. Using the initial time of rise of the conductivity as the salt wave passed a sensor, or the peak or centroid of the wave, it was possible to measure the time of travel, and thus the velocity. This provides a reach-scale average velocity, and in a series of experiments at constant discharge with a flow depth over the floodplain of 1–1.5 m, the velocities estimated by this method averaged 0.95–1.1 m s^{-1} for sensors in the lower half of the flow depth, and 1.0–1.2 m s^{-1} for sensors closer to the surface. During the same experiments, dye was added to the injected water, and was tracked using three video cameras; this surface indicator of velocity completed the velocity profile by giving a mean velocity for the leading edge of the dye of 1.20–1.30 m s^{-1}. Using evidence from the array of sensors and the videos, it was also possible to suggest that there was a pattern of secondary circulation over the floodplain similar to that illustrated in Figure 15.4, in which surface water flows away from the channel while the bed current is directed towards it.

One complication, of course, with flow through vegetation stems is that mechanical dispersion (Nepf *et al.*, 1997) takes place as the flow is deflected and forced to accelerate by the physical obstacles of the stems (see Figure 15.10). This may result in a somewhat different pattern of dispersion from that involving classic advection and dispersion, as parts of the advecting salt-laden flow pass on different sides of stems, and then follow different sinuous paths. Some evidence of this process is suggested in Figure 15.12, in which individual conductivity traces at particular sensors downstream from the injection point are seen to have multiple peaks, and to be potentially described by linear combinations of two or three log-normal distributions of travel times, each with different means and variances. This evidence of mechanical dispersion would repay further investigation.

15.6.9 Modelling the Wienflüss Flows

During the experiments conducted in the Wienflüss full-scale flume, the water surface was monitored. This means that to model the flow conditions during the experiment

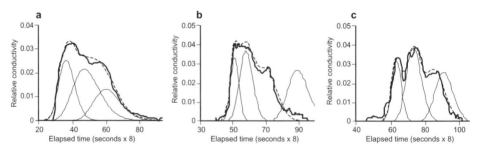

Figure 15.12 Conductivity traces following a salt solution injection recorded at three floodplain centre-line sensors at increasing distances from the injection point (from left to right, 4 m, 5.5 m and 7 m). Three log-normal curves have been optimised as a fit to the data in each case; note the distorted timescale. The three composite curves are displaced in time to reflect the increasing distance, but in each case their component curves peak at about 1.5 seconds apart

requires selection of roughness coefficients for the channel and the floodplain, in order to match this water surface and the known flow velocities at points where they were measured (such as in the stretch of floodplain where the conductivity tracing was undertaken). Because the water surface is known, hydraulically this is a 'fixed lid' model, and variations in the roughness coefficients alter the proportions of flow passing down the channel or over the floodplain. Of course, all estimates of roughness are uncertain, and the model set-up can approximate the available data with a variety of coefficient values.

An initial estimate of the Manning coefficient for the *channel*, which has large (>0.25 m) but flat bed material, and in which the velocity and slope were measured to enable back-calculation of roughness, is $n = 0.038$. The roughess coefficient for the floodplain at the conductivity tracing site, based on back-calculation from the velocity and slope measured here, appears to be 0.087, but forward derivation of this from measurable properties remains a question. Firstly, the floodplain bed surface was measured and (from Equation 15.2) a value of n_b was estimated as 0.025. Bed irregularity (n_1 in Equation 15.1) is likely to add to this, but is less well constrained. Given the density of willows and the diameter of their stems in the region of the conductivity experiments, Equation (15.10) predicts a vegetation roughness (n_4 in Equation 15.1) that is also 0.025. Because the willows are young and flexible, there was significant vibration and oscillation, which is likely to augment this value. In addition, there is curvature in the channel upstream from the experimental section; significant energy losses in turbulent boils and surface waves; and energy dissipation at the channel–floodplain interface. This leads to an approximate estimate of the total floodplain roughness coefficient of 0.07.

A one-dimensional step-backwater model (HEC-RAS), with separate roughness coefficients specified for channel and floodplain, was used to model the water surface profile. The water surface at the sections where conductivity measurements were undertaken was best matched with a floodplain roughness of 0.08 and a channel roughness of 0.03. This also gave the best match to the measured floodplain velocity. This joint manipulation of the channel and floodplain roughness results in changes in the proportions of flow occurring down the channel and over the floodplain.

15.7 Conclusions

The review and evidence reported in this chapter shows that the procedures for estimating the flow resistance of vegetation are gradually approaching the stage where properties of vegetation can be measured and used to determine the roughness coefficients required in hydraulic modelling; in short, equivalents to Equation (15.2) (for estimating bed roughness) are now becoming available for vegetation (e.g, equation 15.10). This may seem surprising, given the huge spatial and temporal variability of aquatic and riparian vegetation. However, the evidence in Section 15.6 reveals that, beginning with simple models of rigid vegetation, and progressively adding further complexity (mixed heights, flexibility, different spatial patterns), the interactions between flow and vegetation are explicable and potentially predictable. Nevertheless, it is also apparent that the vegetation component in Equation (15.1) is itself a composite of several influences, as evident from the form of Equation (15.10).

Progress in this area is of considerable significance. As noted in the introduction, there is increasing interest in restoration of 'natural' hydroecological conditions in rivers and on floodplains, but this trend runs counter to a history in which channels have been managed with a strong focus on flood control and stabilisation. This history has resulted in many land uses on floodplains that, inappropriate though they may be, cannot be altered in the short term because they confer a high value on the land, which pays for their flood defences. In addition, the complex pattern of land ownership of floodplains will not always permit significant alteration of land uses to more natural (and potentially less economically valued) functions. The implication is that restoration of channel and floodplain hydroecology will necessarily be part of integrated catchment management practices in which occasional opportunities for restoration arise, but must be tested first to ensure that their location and scale cannot increase risks in adjacent on nearby locations.

This will require that hydrological and hydraulic modelling are coupled at the catchment scale, with two- or three-dimensional hydraulic models simulating the consequences of restoration, requiring spatially distributed information on vegetation roughness, and temporally dynamic information on the evolving character of the vegetation, and its consequence for local flood levels. The capacity to simulate with this level of complexity and sophistication is not far away, and once the representation of vegetative roughness in such models is at the appropriate level of complexity, it will enable improved modelling of flood attenuation, floodplain drainage rates, nutrient fluxes, suspended sediment inputs and deposition rates, and a wide range of other processes that depend on accurate specification of flow velocities, depths and shear stresses.

References

Arcement, G.J. and Schneider, V.R. (1989) *Guide for Selecting Manning's Roughness Coefficients for Natural Channels and Flood Plains*. United States Geological Survey Water-Supply Paper, 2339.

Bölscher, J., Ergenzinger, P.-J. and Obenauf, P. (2005) Hydraulic, sedimentological and ecological problems of multi-functional riparian forest management. *Berliner Geographische Abhandlungen*, **65**, 1–145.

Chow, V.T. (1959) *Open Channel Hydraulics*. McGraw Hill, New York.

Costanza, R., d'Arge, R., de Groot, R., Farberk, S., Grasso, M., Hannon, B., Limburg, K., Naeem, S., O'Neill, R.V., Paruelo, J., Raskin, R.G., Sutton, P. and van den Belt, M. (1997) The value of the world's ecosystem services and natural capital. *Nature*, **387**, 253–260.

Cowan, W.L. (1956) Estimating hydraulic roughness coefficients. *Agricultural Engineering*, **37**, 473–475.

Darby, S.E. (1999) Effect of riparian vegetation on flow resistance and flood potential. *Journal of Hydraulic Engineering, ASCE*, **125**, 443–454.

Dawson, F.H. (1978) Aquatic plant management in semi-natural streams: the role of marginal vegetation. *Journal of Environmental Management*, **6**, 213–621.

Dawson, F.H. (1989) *Ecology and Management of Water Plants in Lowland Streams*. Freshwater Biological Association Annual Report, Ambleside, pp. 43–60.

Einstein, H.A. and Banks, R.B. (1950) Fluid resistance of composite roughness. *Transactions, American Geophysical Union*, **31**, 603–610.

Fathi-Moghadam, M. (1996) Momentum absorption in non-rigid, non-submerged, tall vegetation along rivers. PhD Thesis, University of Waterloo, Waterloo, Ontario, Canada.

Fathi-Moghadam, M. and Kouwen, N. (1997) Nonrigid, nonsubmerged, vegetative roughness on floodplains. *Journal of Hydraulic Engineering, ASCE*, **123**, 51–57.

Freeman, G.E., Rahmeyer, W.J. and Copeland, R.R. (2000) *Determination of resistance due to shrubs and woody vegetation*. ERDC/CHL TR-00-25, US ACE Engineer Research and Development Center, Coastal and Hydraulics Labaoratory, Washington DC.

Green, J.C. (2005) Modelling flow resistance in vegetated streams: review and development of new theory. *Hydrological Processes*, **19**, 1245–1259.

Hasegawa, K., Asai, S., Kanetaka, S. and Baba, H. (1999) Flow properties of a deep open experimental channel with a dense vegetation bank. *Journal of Hydroscience and Hydraulic Engineering, JSCE*, **17**, 59–70.

Henderson, F.M. (1966) *Open Channel Flow*. MacMillan, New York.

Herbich, J.B. and Shulits, S. (1964) Large-scale roughness in open channel flow. *Journal of the Hydraulics Division, ASCE*, **90**, 203–230.

Hughes, D. (1980) Floodplain inundation: processes and relationships with channel discharge. *Earth Surface Processes*, **5**, 297–304.

Hughes, F.M.R., Colston, A. and Mountford, J.O. (2005) Restoring riparian ecosystems: the challenge of accommodating variability and designing restoration trajectories. *Ecology and Society* **10**(1), article 12.

Järvelä, J. (2004) Determination of flow resistance caused by non-submerged woody vegetation. *International Journal of River Basin Management*, **2**, 61–70.

Johnson, J.W. (1944) Rectangular artificial roughness in open channels. *Transactions, American Geophysical Union, Papers, Hydrology*, 906–912.

Knight, D.W. and Shiono, K. (1996) River channel and floodplain hydraulics. In Anderson, M.G., Walling, D.E. and Bates, P.D. (eds) *Floodplain Processes*. John Wiley & Sons Ltd, Chichester, pp. 139–181.

Kouwen, N. (1992) Modern approach to design of grassed channels. *Journal of the Irrigation and Drainage Division, ASCE*, **118**, 733–743.

Kouwen, N. and Fathi-Moghadam, M. (2000) Friction factors for coniferous trees long rivers. *Journal of Hydraulic Engineering, ASCE*, **126**, 732–740.

Kouwen, N. and Li, R.-M. (1980) Biomechanics of vegetative channel linings. *Journal of the Hydraulics Division, ASCE*, **106**, 1085–1103.

Kouwen, N. and Unny, T.E. (1973) Flexible roughness in open channels. *Journal of the Hydraulics Division, ASCE*, **99**, 713–728.

Kouwen, N., Unny, T.E. and Hill, H.M. (1969) Flow retardance in vegetated channels. *Journal of The Irrigation and Drainage Division, ASCE*, **95**, 329–342.

Kutija, V. and Hong, H.T.M. (1996) A numerical model for assessing the additional resistance to flow introduced by flexible vegetation. *Journal of Hydraulic Research*, **34**, 99–114.

Larsen, T., Frier, J.O. and Vestergaard, K. (1990) Discharge/stage relations in vegetated Danish streams. In White, W.R. (ed) *River Flood Hydraulics*. John Wiley & Sons, Ltd, Chichester, pp. 187–195.

Li, R.M. and Shen, H.W. (1973) Effect of tall vegetation on flow and sediment. *Journal of the Hydraulics Division, ASCE*, **99**, 793–814.

McMahon, T.A. (1975) The mechanical design of trees. *Scientific American*, **233**, 92–102.

Naef, F., Kull, D. and Thoma, C. (2000) How do floodplains influence the discharge of extreme floods? In Bronstert, A., Bismuth, C. and Menzel, L. (eds) *European Conference on Advances in Flood Research*. Potsdam Institute for Climate Impact Research (PIK), Potsdam, pp. 644–652.

Naot, D., Nezu, I. and Nakagawa, H. (1996) Hydrodynamic behaviour of partly vegetated open channels. *Journal of Hydraulic Engineering, ASCE*, **122**, 625–633.

Nepf, H.M. (1999) Drag, turbulence and diffusion in flow through emergent vegetation. *Water Resources Research*, **35**, 479–489.

Nepf, H.M., Mugnier, C.G. and Zavistoski, R.A. (1997) The effects of vegetation on longitudinal dispersion. *Estuarine, Coastal and Shelf Science*, **44**, 675–684.

Petryk, S. and Bosmajian, G. (1975) Analysis of flow through vegetation. *Journal of the Hydraulics Division, ASCE*, **101**, 871–884.

Petts, G.E. (1990) The role of ecotones in aquatic landscape management. In Naiman, R.J. and Decamps, H. (eds), *The ecology and management of aquatic-terrestrial ecotones*. UNESCO/MAB Series 4, Parthenon, New Jersey, USA, 227–261.

Pitlo, H. and Dawson, F.H. (1990) Flow resistance of aquatic weeds. In Pieterse, A.H. and Murphy, K.J. (eds) *Aquatic Weeds: The Ecology and Management of Nuisance Aquatic Vegetation*. Oxford University Press, Oxford, pp. 74–84.

Ree, W.O. and Crow, F.R. (1977) *Friction factors for vegetated waterways of small slope*. ARS-S-151, Agricultural Research Service, US Dept of Agriculture.

Richards, K.S., Brasington, J. and Hughes, F. (2002) Geomorphic dynamics of floodplains: ecological implications and a potential modelling strategy. *Freshwater Biology*, 47, 1–22.

Schnauder, I. (2004) *Strömungsstruktur und Impulsasutausch in gegliederten Gerinnen mit Vorlandvegetation*. PhD Thesis, Institut für Wasserwirtschaft und Kulturtechnik Universität Karlsruhe (TH), Karlsruhe.

Smart, G.M. (1999) Coefficient of friction for flow resistance in alluvial channels. *Proceedings Institution of Civil Engineers Water, Maritime and Energy*, 136, 205–210.

Stillwater Outdoor Hydraulic Laboratory (1947, revised 1954). *Handbook of Channel Design for Soil and Water Conservation*. SCS-TP-61, United States Department of Agriculture, Washington DC.

Stone, B.M. and Shen, H.T. (2002) Hydraulic resistance of flow in channels with cylindrical roughness. *Journal of Hydraulic Engineering*, ASCE, 128, 500–506.

Trieste, D.J. and Jarrett, R.D. (1987) Roughness coefficients of large floods. In James, L.G. and English, M.J. (eds) *Irrigation and Drainage Division Speciality Conference 'Irrigation Systems for the 21st Century'*. Proceedings, New York, American Society of Civil Engineers, Portland, Oregon, pp. 32–40.

Tsujimoto, T., Shimizu, Y., Kitamura, T. and Okada, T. (1992) Turbulent open-channel flow over bed covered by rigid vegetation. *Journal of Hydroscience and Hydraulic Engineering, JSCE*, 10, 13–25.

Tsujimoto, T., Kitamura, T., Fujii, Y. and Nakagawa, H. (1996) Hydraulic resistance of flow with flexible vegetation in open channel. *Journal of Hydroscience and Hydraulic Engineering, JSCE*, 14, 47–56.

van Prooijen, B.C., Battjes, J.A. and Uijttewaal, W.S.J. (2005) Momentum exchange in straight uniform compound channel flow. *Journal of Hydraulic Engineering*, 131, 175–183.

Ward, J.V., Tockner, K., Uehlinger, U. and Malard, F. (2001) Understanding natural patterns and processes in river corridors as the basis for effective river restoration. *Regulated Rivers – Research and Management*, 17, 311–323.

Wooding, R.A., Bradley, E.F. and Marshall, J.K. (1973) Drag due to regular arrays of roughness elements of varying geometry. *Boundary-Layer Meteorology*, 5, 285–308.

16

Hydrogeomorphological and Ecological Interactions in Tropical Floodplains: The Significance of Confluence Zones in the Orinoco Basin, Venezuela

Judith Rosales, Ligia Blanco-Belmonte and Chris Bradley

16.1 Introduction

Rivers and their floodplains are characterised by lateral and longitudinal variations in hydrological and geomorphological processes that have significant implications for species composition and diversity. Traditionally, the tendency has been to regard these interrelationships as driven by progressive downstream changes in hydrological and geomorphological processes through the catchment (Vannote *et al.*, 1980). Increasingly though, the significance of local spatial and temporal variability in physical processes has been recognised with attention focusing on habitat patches (Frissell *et al.*, 1986), the nature of the disturbance regime such as flooding (Junk *et al.*, 1989), identification of areas where new habitats are being created (Hughes, 1997), and the extent to which vegetation succession can itself influence habitat heterogeneity within the floodplain (Gregory *et al.*, 1991). However, although the interdependence of physical and ecological processes is recognised, a suitable theoretical framework that can encompass the spatial and temporal variability of hydrological and geomorphological processes, and their ecological implications, has still to be established. Some advocate a landscape ecology approach (Wiens, 2002), whilst Benda *et al.* (2004) focus on the dynamic nature of the river network, and specifically the importance of tributary junctions or confluence zones. If the

Hydroecology and Ecohydrology: Past, Present and Future, Edited by Paul J. Wood, David M. Hannah and Jonathan P. Sadler
© 2007 John Wiley & Sons, Ltd.

confluence zone is considered to be the area within which the effects of the junction of two channels is evident, in some cases the zone may extend for considerable distances both below the point of confluence, and upstream along both the main channel and tributary river.

Confluence zones are characterised by complex variations in hydrological and geomorphological processes in which distinct flow zones may be identified. These include: flow acceleration downstream of the confluence, flow stagnation at the apex of the confluence and flow divergence at the tributary margin, with reduced velocity upstream (De Serres *et al.*, 1999). Within the confluence zone, morphological differences reflect a number of factors including the relative size of tributary and main-stem rivers, relative differences in stream power, as well as network pattern and junction geometry (Benda *et al.*, 2004), that will also influence local groundwater dynamics (Lambs, 2004). To date, most work has been on fluvial systems dominated by coarse sediment supply (Rice *et al.*, 2001), and in many cases complementary ecological data are lacking. Moreover, the confluence zones that have been studied are predominantly at high latitudes within Europe and North America, although the catchments of large tropical rivers in low latitudes are characterised by numerous tributary junctions, where rivers with markedly different hydrological and hydrochemical regimes join. Given the low gradient environment that typifies a significant proportion of many tropical basins, the confluence zone may extend a considerable distance upstream in a tributary as a result of a backwater effect from the main river. For example, Herrera and Rondón (1985) found that the Orinoco River exerted a discernible backwater effect more than 50 km up the Apure River, as far as San Fernando de Apure (Figure 16.1).

Large tropical rivers are characterised by seasonally variable flow dynamics dominated by a high and predictable flood pulse (Junk *et al.*, 1989). A diverse fauna and flora is associated with the highly dynamic variations in water, sediment and nutrient availability within river systems that in many cases remain largely pristine, are of considerable spatial extent, and with variable physiography. Research in these environments has been constrained to a considerable extent by the scale and diversity of the riparian ecosystem, and not least by the problems of data availability and logistical constraints in working in these environments. In this chapter we review recent hydrogeomorphological and ecological work in large tropical catchments, principally that of the Orinoco in South America, and illustrate how macroscale hydrological and geomorphological processes can influence floodplain diversity, before presenting a model of the confluence zone driven by a backwater effect as it might be applied in a tropical context.

16.2 Hydrogeomorphological Dynamics

The Orinoco has a mean annual discharge at Ciudad Guayana of $36\,000\,m^3\,s^{-1}$ and drains an area of $1\,100\,000\,km^2$ of Venezuela and Columbia, with tributary rivers extending east from the Andes mountains and across the sedimentary plain of the Llanos, and flowing west and north from the Guiana Shield (Figure 16.1; Table 16.1). The hydrological and hydrochemical characteristics of the Orinoco and its tributaries reflect differences in geology and topography of subcatchments across the basin, runoff pathways (whether at the surface or through alluvial sediments), and local variations in precipitation. Precipita-

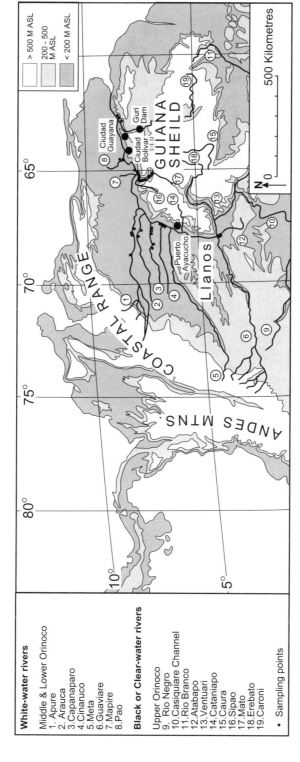

Figure 16.1 The catchment of the Orinoco river and its principal tributaries draining the Guayana shield, the Andes mountains, and the Llanos sedimentary plain

Table 16.1 Characteristics of the Orinoco river and selected tributary rivers

		Basin area (km^2)	Mean Discharge (m^3 s^{-1})	Specific discharge (mm yr^{-1})	pH	EC	Suspended solids (mg l^{-1})	Notes
Main stem	Lower Orinoco	108 000	36 000		6.8		80	Characterised by increasing seasonal variation in river level downstream (14 m at Puerto Ayachu; 16 m at Ciudad Bolívar)
White-water rivers	Apure	170 000	2 300	300	7.2		235	Flows from the Andes Mountains, with sandstone outcrops in the upper basin, and
	Mapire	282	6					Floodplain constricted by adjacent hills; Vegas-Vilarrúba & Herrera (1993) document a 'confluence effect' as Mapire waters are impounded by floodwaters in the Orinoco.
Black/clear-water rivers	Meta	110 000	5 600	1 600	7.4		362	
	Caroni	95 000	5 000	1 700	6.4		5	The Lower Caroni is impounded by the Guri Dam, above which is an artificial lake covering >4250 km^2.
	Caura	48 000	3 500	2 000	6.5		12	Brown colour, nutrient-poor, very low DOC, but high concentration of principal cations

tion variability is high and displays marked seasonality in response to the movement of the Bolivian high pressure cell and the inter-tropical convergence zone. In the Orinoco basin, annual precipitation averages 2000–3600 mm y^{-1} in the upper basin, 1000–2000 mm y^{-1} in the centre of the basin, and 1400–2000 mm y^{-1} in the Orinoco delta (Hamilton and Lewis, 1990), but with a strong seasonal component that produces a unimodal flood peak, with peak flows of ~47 000 m^3 s^{-1} in the middle Orinoco at Ciudad Bolívar in August or September.

The characteristics of the floodplain and alluvial sediments vary distinctly through the catchment. The upper reaches of rivers draining the Andes mountains actively meander by lateral erosion and point bar deposition, and are comparable to the Andes tributaries of the Amazon (Salo et al., 1986). In lower reaches, three types of tropical floodplain can be distinguished: fringing floodplain, internal delta and coastal deltaic floodplain (Welcomme, 1979). An internal delta is found between the Apure and Arauca Rivers, while the fringing floodplain is discontinuous with intermittent constriction points at bedrock outcrops. Along much of its course, the Orinoco has maintained a stable channel through the Holocene, reflecting the location of the river against the western margin of the Guiana shield, the low slope, and the predominantly fine sediment load. Consequently, floodplain construction has occurred mainly through overbank sedimentation, but with a marked difference between the floodplain on the west and east banks: the floodplain is less extensive to the south and east, near the Guiana shield, whilst most of the floodplain lakes are concentrated on the western floodplain.

In common with many tropical rivers, the Orinoco exhibits a considerable variety of channel forms, to the extent that in many cases it would be inappropriate to classify the river as straight, meandering or braided (Latrubesse et al., 2005). The floodplain of the Orinoco and its principal tributaries appears to have more affinity with complex multi-channel or anabranching systems, reflecting the low gradient environment and predominant supply of fine sediment, as described for other tropical rivers such as the Rio Negro (Latrubesse and Franzinelli, 2005). In these tropical catchments, the river and floodplain should be regarded as one system, which become completely integrated during seasonal high flows. As the discharge of the Orinoco and its tributary rivers increases, extensive areas are inundated, whether by the increase in local river flow or through a backwater effect where the flow of tributary rivers is impounded by higher water levels in the Orinoco. Vegas-Vilarrúbia and Herrera (1993) describe one example of the latter effect for the Mapire River and floodplain (Figure 16.1). In this case, the seasonal increase in flow of the Orinoco acts as an effective dam to the discharge of the Mapire River, which inundates the local floodplain to form a standing body of water that persists for between 5 and 6 months. At such times small creek systems provide the continuity either between individual floodplain lakes, or between the floodplain and the main channel. These creeks are important and have been little studied (Blake and Ollier, 1971); they facilitate water flow through the floodplain, providing nutrients, and preventing the development of anoxic conditions as floodwaters rise. For practical reasons, it has been very difficult to determine the significance of creek systems. Hydraulic studies of the Caura River by local hydrologists assumed negligible water flow through the floodplain (and with rating curves based on in-channel measurements), although Hamilton and Lewis (1990) observed flow velocities of ~10–20 cm s^{-1} in the fringing floodplain of the Orinoco. It is also difficult to transfer conventional expressions of the relationship between channel characteristics and discharge and sediment load from temperate to tropical rivers (see Pickup and Warner, 1984), which apparently reflects a number of very specific differences in geomorphic processes, environmental history, and in the role of floodplain vegetation in tropical compared with temperate floodplains.

Floodplain inundation arises both through the seasonal rise in river stage, as a result of high precipitation in the catchment headwaters, as well as through impeded drainage of local precipitation. Mertes (1997) has described a 'perirheic' zone characterised by the mixing of river and local water:

$$\Delta S_r = time \times (Q_{in} + Q_{tr} + Q_l + Q_p + Q_{gw}) - (Q_{out} + Q_e + Q_{gw})$$

where ΔS_r is change in storage within a floodplain reach with inputs including discharge (Q) from the river (in), major tributary (tr), local tributary (l), direct precipitation (p) and groundwater (gw), and outputs are river outflow (out), evapotranspiration (e) and groundwater seepage (gw). This equation indicates the variety of water types that are likely to be found in large continental floodplains, and whilst spatial and temporal trends in flood extent can be determined by synthetic aperture radar (Alsdorf, 2003) or satellite radar altimetry (Birkett et al., 2002), differences in water source and water-flow pathways have still to be considered in detail. The variability in groundwater flow, for example, has been neglected, although the dimensions of major channels may be sufficiently large to intercept regional groundwater flow. At a local scale, flow routes linking bodies of water can be described, or inferred, within a GIS (Alsdorf, 2003). Alternatively, mixing models may help to determine the contribution to total flow (and inundation) of individual subcatchments (Hamilton et al., 2004). Further quantification of the pattern of interaction of waters from diverse sources within tropical floodplains would be of use in resolving wider questions related to water movement through the floodplain, for example determining local water, sediment and nutrient budgets, (and the fate of organic carbon), and the implications for the floodplain biota, and floodplain development over time.

The significance of inundation dynamics and seasonal and inter-annual variations in discharge is implicit in the flood pulse concept (Junk et al., 1989). This highlights the importance of a regular and 'predictable' annual flood, which enables the floodplain to be reconnected to the river, and inundated to depths of several meters for an extended period. The concept has focused attention on cross-sectional (2-D) profiles across floodplains, characterised by differences in the intensity and duration of flooding, reflecting major floodplain features such as levees and backswamps. However, in some areas there may be a considerable variation in flood extent reflecting the inter-annual variability in peak flows, whilst low flows are also important, indicating trends in the intensity and duration of desiccation, and deoxygenation in floodplain water bodies. In general, there has been relatively little investigation of 'macroscale' processes that might explain hydrological variability in the Orinoco Basin. The El Nino – Southern Oscillation (ENSO) has been described as one of the main drivers of environmental variability in the tropics, and has been used with some success in the Amazon Basin (e.g., Foley et al., 2002; Eltahir et al., 2004). Although it seems that ENSO may correlate with periods of drought, by itself, it is a poor predictor of trends in the annual flow of the Orinoco (from 1923), as shown in Figure 16.2. However, inter-annual flow variations in the Orinoco catchment have been reproduced with some success using the Palmer Drought Severity Index (Dai

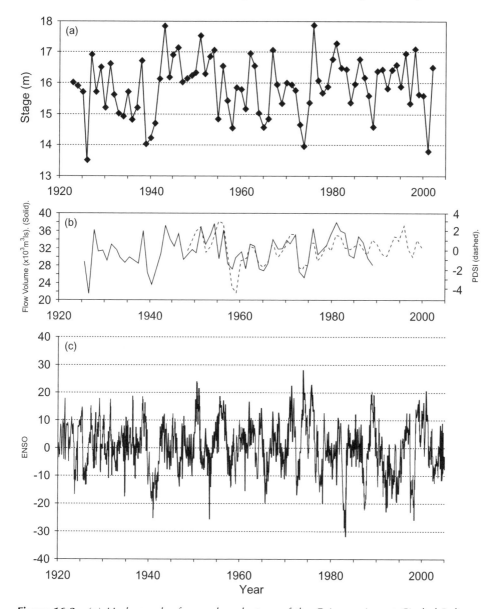

Figure 16.2 (a) Hydrograph of annual peak stage of the Orinoco river at Ciudad Bolivar, 1923–2002; (b) correlation between annual flow of the Orinoco river and the Palmer Drought Severity Index (from Dai et al., 2004); (c) variation in the El Nino Southern Oscillation (1923–2002)

et al., 2004). Dai *et al.* compared the annual discharge of a number of continental rivers, including the Amazon, the Congo and the Orinoco, with a basin-averaged measure of the PDSI. A correlation of 77% was obtained for the Amazon and Congo rivers, and 64% for the Orinoco (for the period 1947–1987). The PDSI is a function of precipitation and evapotranspiration over an extended period, and is moderated by soil-water storage. The success of a simple empirical index in representing discharge dynamics over such diverse catchment areas is significant, demonstrating the importance of specifying the hydrological budget appropriately and apportioning precipitation between local storage and runoff.

The heterogeneous character of large continental catchments is demonstrated by the differences in hydrochemistry of the principal tributary rivers of the Orinoco. These have traditionally been classified on the basis of their optical properties as black-water, clear-water, or white-water rivers (Wallace, 1853; Sioli, 1984). White-water rivers are eutrophic with high sediment and solute loads, and generally have headwaters in the western mountain chains. Black-water rivers, draining the resistant and highly weathered Shield area, have a low sediment and solute load; they are oligotrophic with a low pH and a dark colour reflecting a high concentrations of organic acids and dissolved organic carbon. In contrast, clear-water rivers have a low sediment and solute load, but as a result of their geology and vegetation cover, the pH of these rivers is higher and they are not discoloured by organic acids. The variability in pH and electrical conductivity (EC) of river waters measured at ~100 points (identified on Figure 16.1), on the main stem of the Orinoco river and a selection of tributary rivers in the catchment during a reconnaissance expedition in March 1999 is shown in Figure 16.3 (Rosales, Bradley and Gilvear, unpublished

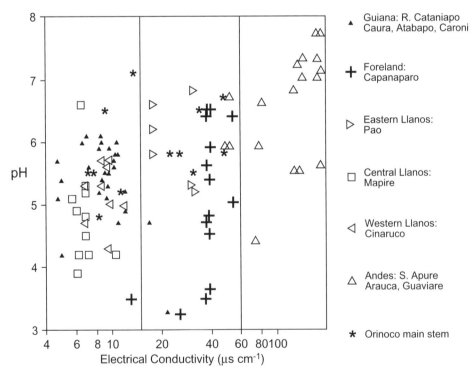

Figure 16.3 Geochemical variation in river-waters sampled at the point of confluence of the Orinoco and selected tributary rivers

data). Both pH and EC are characterised by considerable variability, however, the distinction in EC between rivers draining the Guiana shield (the Cataniapo, Caura, Atabapo and Caroni rivers), and the white-water rivers associated with drainage of the Andes mountains (the Apure, Arauca and Guaviare rivers) is particularly marked. These trends are matched by the geochemistry of fluvial sands within the Orinoco basin which are closely associated with the differences between black-water, clear-water and white-water rivers. The white-water rivers draining the Andes and the Llanos contribute weathered, orogenically derived sand, whilst rivers draining the Shield yield feldspars (Johnsson *et al.*, 1991). Thus, although tropical climates are generally associated with high rates of weathering, actual denudation rates are low: on the Guiana shield denudation rates are 15–20 mmy^{-1} on the highlands and 5 mmy^{-1} on lowlands.

16.3 The Riparian Ecosystem

The riparian flora and fauna reflects the variation in hydrological and geomorphological processes across tropical catchments, and are closely associated with the differences between white-, clear- and black-water rivers described above. Prance (1979) distinguished two major types of riparian forest: *várzea* (on white-water rivers) and *igapó* (on clear and black-water rivers), and with subdivisions of these communities reflecting hydrological and geomorphological dynamics. A number of studies have described these communities for the Amazon and its tributaries, but they have been less fully documented for the Orinoco (Irion *et al.*, 1997; Rosales *et al.*, 1999). For example, Wittman *et al.* (2004) described how community structure can be related to the height and duration of annual inundation and sedimentation in várzea communities of the Amazon: the density of trees varying from: 490 ha^{-1} where the flood height was 3–7 m, corresponding to an early successional stage; ~1000 h at a secondary successional stage; and ~430 trees ha^{-1} at a late successional stage, and with considerable variation in species richness between the sites (4, 45, and 91 tree species respectively). In such cases, distinctive vegetation communities can be associated with individual geomorphic units. This is illustrated by the floodplain cross section in Figure 16.4 which shows the variation in vegetation communities along one section across the floodplain of the Caura River ~20 km above the confluence with the Orinoco. The vegetation communities differ markedly along the transect, reflecting variability in the flood regime, the rate of sedimentation, nutrient availability and the stress posed by high water flows along the gradient in elevation.

Comparable vegetation cross sections have been derived for other tropical floodplains, such as the Tana River in Kenya (Hughes, 1990), and in many cases there is a clear association between individual vegetation communities and geomorphic units. For example, Rosales *et al.* (2003) distinguished four vegetation communities associated with nutrient-poor floodplains: herbaceous vegetation colonising exposed areas; marginal flooded forests; understorey herbaceous vegetation in backswamps and rheophylic (or 'flow-loving') aquatic vegetation. However, the composition and structure of each vegetation community varied as a result of local environmental factors, including flood frequency, organic matter content, lithology and topography.

Riparian forests, whether várzea or igapó, have a fundamental role in transferring energy and nutrients to adjacent aquatic and terrestrial ecosystems in the Orinoco basin;

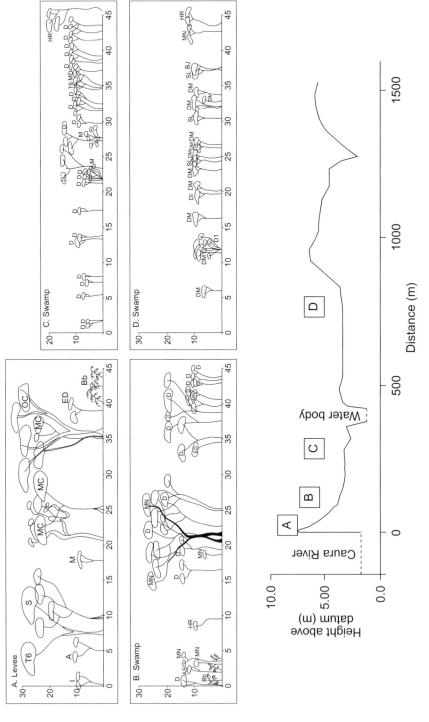

Figure 16.4 Topographic section across the floodplain of the Caura river, about 3 km above the Orinoco river indicating differences in vegetation communities between the levée, adjacent backswamps, and the swamp edge

however, when considering the interrelationships between hydrogeomorphological processes and the riparian fauna and flora, it is necessary to examine the wider spatial context beyond the two-dimensional transect across the floodplain (indicative of flood gradients). Thus, in the following sections, we examine physical and ecological processes within confluence zones of the Orinoco, focusing in particular at the intersection of eutrophic and oligotrophic rivers, characterised by várzea and igapó communities respectively.

16.4 Longitudinal Gradients at Confluence Zones

Data from several confluence zones in the Orinoco river basin (Rosales *et al.*, 1999; Rosales, 2000) suggest that their biological diversity may reflect a backwater effect characterised by distinct biogeochemical and hydrodynamics gradients in areas where oligotrophic, black-water, tributaries flow into the eutrophic Orinoco river. Estuarine hydrodynamic models can be used to explain the relationships between the environment and riparian biota (Levesque and Patino, 2001; Neill *et al.*, 2003). The hydraulic models suggest that sediment geochemistry, particle size, habitat, vegetation and benthic diversity are characterised by considerable and linked spatial variation through the confluence zone to the extent that conservation of these river network nodes should be actively encouraged. This is illustrated by data from the confluences of the clear- and black-water Caura river and Mapire rivers with the Orinoco river (Figure 16.1; Table 16.1). A survey of the confluence zones of both tributary rivers was undertaken in February and March 1999, during which a variety of physical and ecological data were collected:

16.4.1 Sediment Gradient

Sediment cores, integrated over a depth of 0–0.2 m, were collected from in-channel bars at regular distances relative to the tributary channel length at their confluence zones and with respect to the number and beginning and end of individual meander bends above the junctions of the Caura (14) and Mapire (10) rivers with the lower Orinoco river (Rosales *et al.*, 1999). Samples were collected at points where changes in vegetation community indicated a potential biogeochemical or hydrological control. The samples were analysed subsequently to determine USDA proportions of six grain sizes in the clastic fraction after clay separation, extractable major cations (Ca, Mg, Na, K), micronutrients (Fe, Mn, Cu) and P. At these sites also, the presence or absence of species indicators of nutrient-rich river waters was noted in addition to changes associated with a greater flooding depth and duration (e.g., open understorey, presence of sponges in forest trees, development of adventitious roots). The grain size and composition of the samples displayed variations that could only be explained in the context of the unique biogeochemical and hydraulic conditions of the confluence zone. Figure 16.5 shows how the mean Φ grain size changes through the confluence zones of both rivers, in a comparable fashion to the sum of cations and pH. These physical variables are important indicators that enable the várzea communities of the Orinoco to be distinguished from igapó and mixed igapó–várzea floodplain types associated with the Mapire and Caura confluence zones (Rosales, in preparation). A similar pattern has been observed in other confluence zones of the Orinoco river system that were studied by Rosales (2000). The clear-water rivers Capanaparo and Cinaruco are

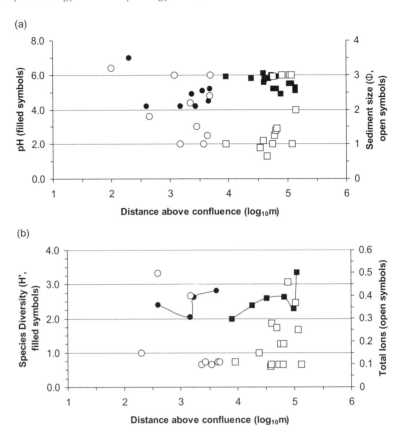

Figure 16.5 (a) Trends in pH and sediment size for the Mapire River (●, ○), and the Caura river (■, □) with distance above both the confluence of both rivers with the Orinoco; (b) variation in species diversity and total ions for the Mapire river (●, ○) and Caura river (■, □)

also characterised by changes in grain size and soil cations that appear to be related to the presence of plant species that are considered indicative of várzea vegetation, such as *Alchornea castaneifolia*. The significance of these results lies in their role in determining biological gradients along river confluences as illustrated below.

16.4.2 Biological Gradient

Forest vegetation Confluence zones in the Orinoco River system display clear gradients in species composition with floristic groups varying between Orinoco- and tributary-related types. This is demonstrated in Figure 16.5 by changes in parameters including the number of species, alpha diversity, the Shannon–Wiener index of diversity (H'), and Pielou's Species Evenness (J') (Greig-Smith, 1983), which had been determined at different locations along longitudinal gradients within the confluence zones of the Caura and

Mapire rivers with the Orinoco river, using data on individual trees (with diameter > 0.1 m at breast height) from floodplain levees. In the Caura floodplain these were derived from study plots of 0.1 ha area, and in the Mapire floodplain from four different locations within the confluence area (and by integrating three groups of ten 0.01 ha plots at different locations).

Marked changes in the vegetation species composition were evident with distinct vegetation patches at the confluence zones of the tributaries surveyed. For example in both the Mapire and Caura floodplains, species composition changed with distance upstream from the confluence of each river with the Orinoco from a várzea to igapó forest type (Rosales, 1990; Rosales et al., 2001). Key várzea species such as the Euphorbiaceae *Piranhea trifoliata*, the Lecythidaceae *Gustavia augusta*, the Flacourtiaceae *Homalium racemosum*, the Euphorbiaceae *Alchornea castaneifolia* provided useful indicators of the extent of influence of Orinoco backwaters on the confluence above the tributary–main river junction. The distance upstream of the junction where this change occurs is variable and seems to be related to the length, discharge, slope and stream-power of the tributary. For example, the influence of the Orinoco backwaters is relatively small for the large rivers, the Caroni and Cataniapo, and they have a short confluence zone. However, the Caura and Sipao rivers have a lower slope and stream power in downstream reaches, and have a longer confluence zone.

A conceptual model of the confluence zone that has been proposed by Rosales et al. (1999) is given in Figure 16.6. The combination of a longitudinal gradient along the tributary river, coupled with significant lateral trends in the physical environment accounts may explain why riparian forests within tributary confluence zones on large tropical rivers

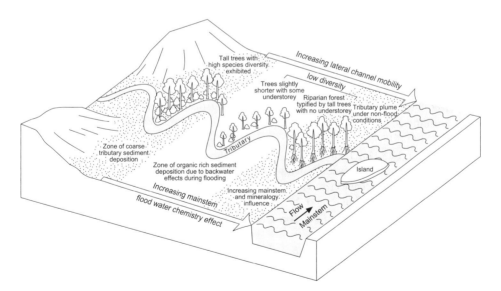

Figure 16.6 Schematic representation of the confluence zone of a large tropical river, and how the effects may vary along the tributary river with distance upstream from the confluence point

with a significant flood pulse have higher beta diversity (in relation to the number of habitats), than other riparian forests on the main stem of the river. Lateral movement of the river creates new riparian areas for colonisation and induces vegetation succession sequences (Salo et al., 1986), whilst there is marked hydraulic variability along the lower reaches of the tributary near the actual point of junction with the main river. The conceptual model envisages that during the flood pulse a substantial backwater effect is exhibited along the lower reaches of the tributary as the floodplain is inundated. As a result, there is a substantial reduction of in-channel and floodplain flow velocities as the tributary waters mix with the floodwaters of the main river. Typically, close to the upstream limit of the backwater effect, deposition of the coarser fraction of the tributary sediment load occurs within the channel but also in the floodplain. It includes both inorganic (mainly sands) and organic materials (i.e., trunks and woody debris). At the downstream limit of the confluence zone, close to the junction of the tributary and the main river, finer tributary-derived sediments settle out on the margins of the channel and on the floodplain. However, a substantial volume of overbank deposition associated with the main river occurs due to significant areas of 'stagnant' flow near the junction. Field evidence was found comparing the sediment size distribution with other reaches. This indicated areas with low flow and low sediment-transport capacity. This sediment typically has fine particles that are organic rich and with a different mineralogy from upstream, resulting from different sediment sources of the main river.

Benthic invertebrates The conceptual model, summarised in Figure 16.6, was devised to explain trends in vegetation communities, and geomorphic units in the confluence zones of the Caura and Mapire rivers. However, the importance of confluence zones is also demonstrated by studies of the distribution of aquatic invertebrates in a confluence zone where two small tributaries, the Caño Mato and Caño Sipao, join the lower Caura river (Blanco-Belmonte, 2006). In each case, at the point of the tributary confluence with the Caura river, the Caura is itself affected by a backwater effect from the Orinoco river at high flow. There were distinctive changes in species density (Shannon–Wiener H'), biomass and taxa richness at specific locations through the Caño Mato and Caura river confluence zone. Changes in species diversity were associated with different substrate types (sediment and green leaves) identified along lateral transects extending, in each case, from the centre of a floodplain lake to the tributary river. Nine transects (30 sites) along the lower 100 km of the Caura across Caño Mato, Caño Sipao and Caura river were sampled from mid-March (low water) to August 1999 (high water) (Blanco-Belmonte, 2006). Four replicates of sediment samples and three replicates of macroinvertebrates associated with green leaves from the adjacent riparian flooded forest were collected using a BK core sampler and a quadrat (625 cm^2, 250 micrometre mesh) at each site respectively. All samples were preserved in 4% formaldehyde. A selection of the results of this study is presented below.

Species diversity (Shannon–Wiener, H'), biomass (g m^{-2}) and taxa richness of macroinvertebrates found in sediment at three sites on the Caño Mato (Patiquin, Cejal, and Pozo Rico) and at Naparaico on the Caura river are shown in Figure 16.7 for periods of high- and low-water. At each site there were three sampling points: (i) in the centre of the floodplain lake (LC); (ii) at the margin of the lake (LS); and (iii) in the adjacent riparian forest (FOR). The results suggest a similar lateral increase in species diversity, biomass

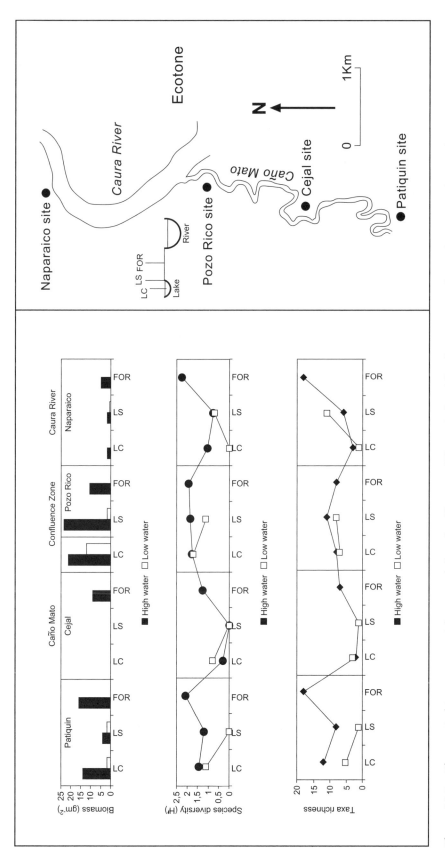

Figure 16.7 Changes in biomass, taxa richness and diversity (Shannon–Wiener index, H') of macroinvertebrates living in sediment along four sections of the Caño Mato and Caura river during periods of high and low water (LC: lake centre, LS: lake side, FOR: flooded forest)

and taxa richness for almost all sampling points within the confluence zone in the two periods. There was also a higher species diversity and taxa richness in the adjacent flooded riparian forest (shown by arrows in Figure 16.7).

Taxa richness, species diversity (H') and biomass (g.m^{-2}) results from macroinvertebrates associated with green leaves (which represent a temporary substrate during high water periods) from the adjacent flooded riparian forest are presented in Figure 16.8. These represent the first empirical data on taxa richness, diversity and biomass of invertebrates in the Caura flooded forest. The results indicate that relative abundance of species are often characterized by relatively few dominant species and a larger number of rare species: 66% of the total abundance was represented by three families, Chironomidae, Staphylinidae and Naididae, while 24 families (comprising 44% of total richness) occurred only once in the confluence zone. In the other sites (Patiquin, Naparaico and Brava) 70–82% of the total abundance was represented by two families, Chironomidae and Naididae, whilst 13–20 families (18–30% of the total richness) occurred only once. These data come from poorly studied habitats in the tropics, and complement the results of other published studies. For example, Vinson and Hawkins (1998) found that the richness of aquatic invertebrates can vary markedly in relation to habitat structure, and that a physically complex substrate type, as in our example, generally supports more taxa than structurally simple substrates such as sand and bedrock. Likewise, when examining data from tropical South America, Central America and West Africa, Claro-Jr. et al. (2004) and Tedesco et al. (2005) found that the highest fish species richness was associated with channels that were directly connected to the rain-forest refuges. There also appeared to be a strong association between higher richness and forested landscapes.

Variations in the species composition of the forest may also explain the differences that were observed in biomass, species diversity and species richness of macroinvertebrates at the flooded forest site in the confluence zone (Richardson et al., 2005), and the hydraulic properties of the confluence zone are clearly also important in influencing the quantity and quality of available habitats (Statzner and Higler, 1986).

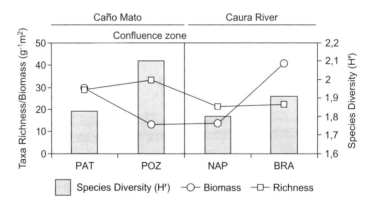

Figure 16.8 Changes in biomass, taxa richness and diversity (Shannon–Wiener index, H') of macroinvertebrates associated with green leaves from the flooded forest of the Caño Mato and Caura river during high water

The distribution of taxa at different points across the confluence zone can be compared between sites, substrates and individual points (Blanco-Belmonte, 2006) and could be divided in a similar manner to that described by Lamb and Mallik (2003) for vegetation traits across a riparian-zone / forest ecotone:

(1) species that are characteristic of specific habitats (temporary substrates such as coarse woody debris, green leaves and litter; permanent substrates such as alluvial sediments, and bedrock) found in low abundance at the confluence zone: Nematoda in sediment, Tanypodinae in green leaves, *Asthenopus curtus* (Ephemeroptera) and *Pristina* spp. (Annelida) in coarse woody debris;
(2) species that do not change in abundance across the confluence zone: *Cyrnellus* spp. (Trichoptera) in woody debris and *Cypridae* spp. (Ostracoda) in sediment and *Ceratopogonidae* spp. in green leaves;
(3) specialist species, which are rare in specific substrates but which may occur at relatively high abundance in the more dynamic confluence zone: *Stenochironomus* spp., in sediment, *Staphilinidae* spp., *Cyclestheria hislopi* and *Ancylidae* in green leaves and *Tenagobia* spp., *Hydrochus* spp. and *Helobdella stagnalis* in coarse woody debris.

The shift in the distribution and richness of stream invertebrates at the confluence of Mato tributary probably represents a response to changing environmental conditions found within the confluence zone (Lamb and Mallik, 2003). However, the macroinvertebrate data suggest that the confluence zones of selected tributaries of the Caura river present comparable gradients in respect of macroinvertebrate diversity, as the conceptual model for riparian vegetation summarised in Figure 16.6.

16.5 Synthesis and Conclusions

Tropical floodplains are highly productive and diverse environments, which in common with their temperate analogues, reflect the interactions between physico chemical, geomorphological and hydrological processes at a variety of scales (e.g., Richards *et al.*, 2002). At the macroscale, regional differences in geology and sediment and solute fluxes determine nutrient availability. This produces distinct riparian plant communities that are closely associated with river type, with the floodplains of white-water rivers contrasting with those of clear-water or black-water rivers. However, the implications of the work described above are that there is a discernible 'confluence effect' in which trends in water and sediment chemistry, particle size, vegetation species and habitat diversity vary in a consistent fashion along the floodplain of tributary rivers in the vicinity of their confluence. The resulting patterns confirm the interdependence between physical and ecological processes that have been described in a number of studies, and are suggested by initial work on confluence zones in temperate regions (e.g., Rice *et al.*, 2001) but they remain to be fully documented and quantified in tropical regions. The trends are suggested by the patterns of vegetation and invertebrate communities and, although not described here, extend also to fish populations. For example, Hoeinghaus *et al.* (2003) describe variations in fish species in floodplain creeks of the Cinaruco river, an eastern flowing tributary of the Orinoco (Figure 16.1). They recorded a total of 60 species, with a clear relationship between species assemblage, the depth of sampling sites and location along a longitudinal gradient from the Orinoco. In such studies, it is clearly important to consider fully how the relationship between ecology and hydrology may vary through the annual flood cycle as the riparian forest becomes reintegrated with increasing river stage. The inundated floodplain acts as a refuge and food source for fishers, as described by Conrad Vispo in Rosales *et al.* (2001). He analysed the stomach contents of *Piaractus brachypomus*, a large member of the piranha family, which appears to take full advantage of the annual flood cycle: its strong teeth and jaws enable the fish to feed on a wide variety of fruits and seeds in the flooded forest, and reproduction is timed to coincide with the period of rising water levels.

In general, it is noticeable how few genuinely interdisciplinary studies have sought to investigate the interrelationships between physical and biological processes in large tropical floodplains and there is enormous potential for future work. For understandable reasons, for many years, the emphasis has been on monitoring and documentation to categorise the variety of habitats and understand the dynamism of the tropical floodplain. We still do not understand fully the ways in which key formative processes, such as nutrient recycling (Newbold, 1992), are different in tropical as opposed to temperate floodplains, and how some of these differences might be explained. In a stimulating review, Klemeš (1993) argued that the characteristics of tropical areas were so distinct as to require an alternative methodological approach. This realisation is centred upon three principal artefacts of tropical vis-à-vis temperate biomes. Firstly, the substantially greater scale of approach required in tropical areas, in which macrohydrological land surface–atmosphere process interactions are highly significant. Secondly, the greater importance of ecohydrological processes in the tropics as illustrated by the proportionately greater contribution of vegetation (and evapotranspiration) to hydrological fluxes in tropical regions. Thirdly, the likelihood that a nonstationary process-response will limit the application of deterministic hydrological models. Here, for example, Klemeš (1993) mentions the possibility of complex changes in atmospheric CO_2 concentrations, related to the essentially unpredictable interactions between vegetation, nutrient retention and decomposition rates, and land-use change, and advocates an approach to humid tropical studies rooted in biophysiological models rather than the 'energy' based approach that characterises temperate models.

Tropical floodplains differ in other ways from those of temperate regions. Aside from their different hydrological and sedimentological regimes, they have also experienced very different histories through the Holocene with contrasting patterns of alluvial architecture and sedimentary composition, which influences rates of subsurface flow through the floodplain. However, few process-based studies have sought to quantify the implications of these differences in determining water flow pathways. Hydrological dynamics have been inferred from relatively sparse data (i.e., spot surface-water measurements) with insufficient quantification of seasonal variability or the variation in the source of water with depth in floodplain soils. This may be particularly important within confluence zones where it is conceivable that water from four distinct sources may mix (precipitation, nutrient-rich river water, nutrient-poor river water and groundwater), to produce significant seasonal differences in nutrient availability with depth, particularly in the rooting zone. As root water uptake from deep soil layers in tropical floodplains appears to be important in sustaining evapotranspiration during dry periods (see Nepstad *et al.*, 1994), seasonal differences in root water sources may be expected as a result of the flood pulse in tropical floodplains.

There also appears to be considerable scope for identifying and quantifying the importance of 'boundaries' at a number of scales: between different vegetation communities within the floodplain, and at the floodplain margins, between riparian communities and those of the *terra firma* rainforest, or savanna. This may enable better identification of vegetation units within the floodplain and their association with sediment type and hydrological regime, which has a number of implications given the scale of these units in tropical floodplains and the degree to which they may influence global biogeochemical cycles. Smith *et al.* (2000), for example, measured methane emissions at 21 sites in the lower Orinoco and in the floodplains of the Caura and Caroni rivers (Figure 16.1). They estimated that methane emissions were of the order of $0.17 \, \text{Tg} \, \text{y}^{-1} \, \text{m}^{-2}$ of water surface, with rates that were five times higher in the floodplain forest, and which occurred primarily as the forest drained at the end of the seasonal flood pulse when the carbon content of sediment was high and the soil-water content exceeded 25%. However, the ways in which emissions of methane, carbon dioxide, and nitrous oxide may vary as a result of environmental processes have yet to be fully determined. For example, Davidson *et al.* (2004) undertook an experiment on 1 ha of tropical forest near Santarem, in Brazil, in which throughfall was excluded to simulate the effects of drought. They found that emissions of nitrous oxide fell by over 40% whilst methane emissions were reduced by a factor of four. The work suggests that gaseous emissions from the *terra firma* rainforest are sensitive to changes in precipitation, results that should apply in part to tropical floodplains.

The implications of the work discussed in this paper, are that tropical floodplains are characterised by the complex interaction of physical and biological processes, with marked spatial and temporal variability. They are unique in a number of respects, not least as a result of their diversity and extent, but in many areas they are increasingly vulnerable to environmental change that may have unforeseen implications (Latrubesse *et al.*, 2005). In most cases, it would be incorrect to regard the tropical floodplains as pristine habitat, given the number of indigenous communities, and the scale of current or past activities such as gold mining, rice cultivation and rubber extraction. However, the scale of present activities (including deforestation), changes in precipitation and pressure for river regulation for navigation and hydropower development suggest that these tropical environments face an uncertain future. For these reasons, quantifying the importance of the interrelationships between physical and biological processes, and focusing conservation on areas with higher species diversity, such as confluence zones, should appear in any programme to conserve large tropical rivers and floodplains.

Acknowledgements

We would like to thank Anne Ankcorn and Kevin Burkhill for cartography, and we are very grateful for the helpful comments of David Hannah and Etienne Muller on an earlier version of this chapter.

References

Alsdorf DE. 2003. Water storage of the central Amazon floodplain measured with GIS and remote sensing imagery. *Annals of the Association of American Geographers*, **93**: 55–66.

Benda L, Poff NL, Miller D, Dunne T, Reeves G, Pess G, Pollock M. 2004. The network dynamics hypothesis: how channel networks structure riverine habitats. *Bioscience*, **54**: 413–427.

Birkett CM, Mertes LAK, Dunne T, Costa MH, Jasinski MJ. 2002. Surface-water dynamics in the Amazon Basin: application of satellite radar altimetry. *Journal of Geophysical Research – Atmosphere*, **107, D20**, art no. 8059.

Blake DH, Ollier CD. 1971. Alluvial plains of the Fly River, Papua. *Zeitschrift für Geomorphologie, Suppl.*, **12**: 1–17.

Blanco-Belmonte L. 2006. Variación espacial de la comunidad de macro-invertebrados en tres fases hidrológicas en el Bajo Caura (Venezuela). Magíster Thesis, Universidad Nacional Experimental de Guayana, Ciudad Guayana.

Claro-Jr L, Ferreira E, Zuanon J, Araujo-Lima C. 2004. O efeito da floresta alagada na alimentação de três espécies de peixes onívoros em lagos de várzea da Amazônia Central, Brasil. *Acta Amazonica*, **34**: 133–137.

Dai A, Trenberth KE, Qian T. 2004. A global data set of Palmer Drought Severity Index for 1870–2002: relationship with soil moisture and effects of surface warming. *Journal of Hydrometeorology*, **5**: 1117–1130.

Davidson EA, Ishida FY, Nepstad DC. 2004. Effects of an experimental drought on soil emissions of carbon dioxide, methane, nitrous oxide, and nitric oxide in a moist tropical forest. *Global Change Biology*, **10**: 718–730.

De Serres B, Roy AG, Biron PM, Best JL. 1999. Three-dimensional structure of flow at a confluence of river channels with discordant beds. *Geomorphology*, **26**: 313–325.

Eltahir EAB, Loux B, Yamana TK, Bomblies A. 2004. A see-saw oscillation between the Amazon and Congo basins. *Geophysical Research Letters*, **31**: 23, art no. L23201.

Foley JA, Botta A, Cole MT, Costa MH. 2002. El Nino-Southern Oscillation and the climate, ecosystems and rivers of Amazonia. *Global Biogeochemical Cycles*, **16**: 4, art no. 1132.

Frissel C, Wiss W, Warren C, Huxley M. 1986. A hierarchical framework for stream classification: viewing streams in a watershed context. *Environmental Management*, **10**: 199–214.

Gregory SV, Swanson FJ, McKee WA, Cummins KW. 1991. An ecosystem perspective of riparian zones. *Bioscience*, **41**: 540–551.

Greig-Smith P. 1983. *Quantitative Plant Ecology*. Blackwell Science: Oxford.

Hamilton SK, Lewis Jr. WM. 1990. Physical characteristics of the fringing floodplain of the Orinoco River, Venezuela. *Interciencia*, **15**: 491–500.

Hamilton SK, Sippel SJ, Melack JM. 2004. Seasonal inundation patterns in two large savanna floodplains of South America: the Llanos de Moxos (Bolivia) and the Llanos del Orinoco (Venezuela and Columbia). *Hydrological Processes*, **18**: 2103–2116.

Herrera JK, Rondón J. 1985. Estudio fotogeomorfológico del Rió Orinoco y confluencia con el Rió Apure. Sector Puerto Aycacucho, Rió Caura. In I *Simposio Amazónico*. M.I. Muñoz (ed), Boletín de Geología (Venezuela), Publicación Especial, **10**: 361–372.

Hoeinghaus DJ, Layman CA, Arrington DA, Winemiller KO. 2003. Spatiotemporal variation in fish assemblage structure in tropical floodplain creeks. *Environmental Biology of Fishes*, **67**: 379–387.

Hughes FMR. 1990. The influence of flooding regimes on forest distribution and composition in the Tana River Floodplain, Kenya. *Journal of Applied Ecology*, **27**: 475–491.

Hughes FMR. 1997. Floodplain biogeomorphology, *Progress in Physical Geography*, **21**: 501–529.

Irion G, Junk WJ, de Mello JASN. 1997. The large Central Amazonian River floodplains near Manaus: geological, climatological, hydrological and geomorphological aspects. In *The Central Amazon Floodplain*. W.J. Junk (ed.), Springer, Ecological Studies, **126**: 23–46.

Johnsson MJ, Stallard RF, Lundberg N. 1991. Controls on the composition of fluvial sands from a tropical weathering environment: sands of the Orinoco River drainage basin, Venezuela and Columbia. *Geological Society of America Bulletin*, **103**: 1622–1647.

Junk WJ, Bayley PB, Sparks RE. 1989. The flood pulse concept in river-floodplain systems. *Canadian Special Publication of Fisheries and Aquatic Sciences*, **106**: 110–127.

Klemeš V. 1993. The problems of the humid tropics – opportunities for reassessment of hydrological methodology. In *Hydrology and Water Management in the Humid Tropics. Hydrological Research Issues and Strategies for Water Management*. M. Bonnell, M.M. Hufschmidt, J.S. Gladwell (eds), Cambridge University Press: Cambridge, pp. 45–52.

Lamb EG, Mallik AU. 2003. Plant species traits across a riparian-zone/forest ecotone. *Journal of Vegetation Science*, **14**: 853–858.

Lambs L. 2004. Interactions between groundwater and surface water at river banks and the confluences of rivers. *Journal of Hydrology*, **288**: 312–326.

Latrubesse EM, Franzinelli E. 2005. The late Quaternary evolution of the Negro River, Brazil: implications for island and floodplain formation in large anabranching tropical systems. *Geomorphology*, **70**: 372–397.

Latrubesse EM, Stevauz JC, Sinha R. 2005. Tropical Rivers. *Geomorphology*, **70**: 187–206.

Levesque VA, Patino E. 2001. Hydrodynamic characteristics of estuarine rivers along the South western coast of Everglades National Park. Abstract, Florida Bay Science Conference, Key Largo, FL, April 24–26.

Mertes LAK. 1997. Documentation and significance of the perirheic zone on inundated floodplains. *Water Resources Research*, **33**: 1749–1762.

Neill S, Copeland S, Ferrier G, Folkard A. 2003. Observations and numerical modelling of a non-buoyant front in the Tay Estuary, Scotland. *Estuarine, Coastal and Shelf Science*, **59**: 173–184.

Nepstad DC, de Carvalho CR, Davidson EA, Jipp PH, Lefebvre PA, Negreiros GH, da Silva ED, Stone TA, Trumbore SE, Vierira S. 1994. The role of deep roots in the hydrological and carbon cycles of Amazonian forests and pastures. *Nature*, **372**: 666–669.

Newbold JD. 1992. Cycles and spirals of nutrients. In *Rivers Handbook*, Volume 1. P. Calow, G.E. Petts (eds), Blackwell Science: Oxford, pp. 379–408.

Pickup G, Warner RF. 1984. Geomorphology of tropical rivers. II. Channel adjustment to sediment load and discharge in the Fly and Lower Purari, Papua New Guinea. *Catena Supplement*, **5**: 19–41.

Prance GT. 1979. Notes on the vegetation of Amazonia. III. The terminology of Amazonian forest types subject to inundation. *Britonnia*, **31**: 26–38.

Rice SP, Greenwood MT, Joyce CB. 2001. Tributaries, sediment sources, and the longitudinal organisation of macroinvertebrate fauna along river systems. *Canadian Journal of Fisheries and Aquatic Sciences*, **58**: 828–840.

Richards K, Brassington J, Hughes F. 2002. Geomorphic dynamics of floodplains: ecological implications and a potential modelling strategy. *Freshwater Biology*, **47**: 559–579.

Richardson BA, Richardson MJ, Soto-Adames FN. 2005. Separating the effects of forest type and elevation on the diversity of litter invertebrate communities in a humid tropical forest in Puerto Rico. *Journal of Animal Ecology*, **74**: 926–936.

Rosales J. 1990. Análisis florístico-estructural y algunas relaciones ecológicas en un bosque estacionalmente inundable en la boca del Río Mapire, Edo. Anzoategui. MSc Thesis. Instituto Venezolano de Investigaciones Científicas, Venezuela.

Rosales J. 2000. An ecohydrological approach for riparian forest biodiversity conservation in large tropical rivers. Unpublished PhD Thesis, University of Birmingham.

Rosales J, Petts G, Salo J. 1999. Riparian flooded forests of the Orinoco and Amazon basins: a comparative review. *Biodiversity and Conservation*, **8**: 551–586.

Rosales J, Bradley C, Petts G, Gilvear D. 2000. A conceptual model of hydrogeomorphological-vegetation interactions within confluence zones of large tropical rivers; Orinoco, Venezuela. In *Proceedings of the Symposium of Hydrological and Geochemical processes in Large Scale River Basins with Special Emphasis on the Amazon and other Tropical Basins*. Manaus, Brazil, 15–19 November.

Rosales J, Petts G, Knab-Vispo C. 2001. Ecological gradients in riparian forests of the lower Caura River, Venezuela. *Plant Ecology*, **152**: 101–118.

Rosales J, Bevilacqua M, Diaz W, Pérez R, Rivas D, Caura S. 2003. Riparian vegetation communities of the Caura River basin, Bolívar State, Venezuela. In *A Biological Assessment of the Aquatic*

Ecosystems of the Caura River Basin, Bolívar State, Venezuela. B. Chernoff, A. Machado-Allison, K. Riseng, and J.R. Montambault (eds) RAP Bulletin of Biological Assessment, **28**. Conservation International, Washington, DC.

Salo J, Kallioloa R, Hakkinen I, Makinen Y, Niñéemela P, Puhakkam M, Coley P. 1986. River dynamics and the diversity of the Amazon lowland forest. *Nature*, **322**: 254–258.

Sioli H. 1984. The Amazon and its main affluents: hydrography, morphology of the river courses and river types. In *The Amazon: Limnology and Landscape Ecology of a Mighty Tropical River and its Basin*. H. Sioli (ed.), W. Junk Publishers: Dordrecht, pp. 125–165.

Smith LK, Lewis Jr. WM, Chanton JP, Cronin G, Hamilton SK. 2000. Methane emissions from the Orinoco River floodplain, Venezuela. *Biogeochemistry*, **51**: 113–140.

Statzner B, Higler B. 1986. Stream hydraulics as a major determinant of benthic invertebrate zonation patterns. *Freshwater Biology*, **16**: 127–139.

Tedesco PA, Oberdorff T, Lasso CA, Zapata M, Hugueny B. 2005. Evidence of history in explaining diversity patterns in tropical riverine fish. *Journal of Biogeography*, **32**: 1899–1907.

Vannote RL, Minshall GW, Cummins KW, Sedell JR, Cushing CC. 1980. The river continuum concept. *Canadian Journal of Fisheries and Aquatic Sciences*, **37**: 130–137.

Vegas-Vilarrúbia T, Herrera R. 1993. Seasonal alternation of lentic / lotic conditions in the Mapire system, a tropical floodplain lake in Venezuela. *Hydrobiologia*, **262**: 43–55.

Wallace AR. 1853. *Palm Trees of the Amazon and Their Uses*. John van Voorst: London.

Welcomme RL. 1979. *Fisheries Ecology of Floodplain Rivers*. Longman: London.

Wiens JA. 2002. Riverine landscapes: taking landscape ecology into the water. *Freshwater Biology*, **47**: 517–539.

Wittman F, Junk WJ, Piedade MTF. 2004. The varzea forests in Amazonia: flooding and the highly dynamic geomorphology interact with natural forest succession. *Forest Ecology and Management*, **196**: 199–212.

17

Hydroecological Patterns of Change in Riverine Plant Communities

Birgitta M. Renöfält and Christer Nilsson

17.1 Introduction

Hydroecology and ecohydrology have been presented as a new science in which hydrologists and ecologists interact to understand the dynamics of ecosystems influenced by hydrological processes (e.g., Bond, 2003; Hannah *et al.*, 2004). Ecologists may already consider the abiotic environment as an integral part of their research and have therefore unconsciously worked in the field of hydroecology (Bond, 2003). For example, a plant ecologist focusing on rivers cannot make progress without recognising the importance of the hydrological cycle. Given the overarching significance of hydrological processes in governing, directly or indirectly, riparian and aquatic vegetation in running waters, and the influence of vegetation on hydrology, bilateral collaboration between riverine (i.e., riparian and river) plant ecologists on the one hand, and hydrologists on the other is vital for deepening the understanding of rivers and for formulating relevant management policies. Examples of such fruitful collaborations include Ward *et al.* (1999, 2002), Tockner *et al.* (2000), Gurnell *et al.* (2001), Gurnell and Petts (2002), Guilloy-Froget *et al.* (2002), Tockner *et al.* (2003), Jiang *et al.* (2004), Zhao *et al.* (2004), Hudon *et al.* (2006), and Stromberg *et al.* (2006).

Hannah *et al.* (2004) reviewed some of the definitions of hydroecology (and ecohydrology) put forward in the scientific literature. The simplest definition comprises the scientific overlap between hydrology and ecology, and the mutual impacts between disciplines (Kundzewicz, 2002; Zalewski, 2002). Dunbar and Acreman (2001) provided an even wider definition by presenting hydroecology as 'the linking of knowledge from hydrological, hydraulic, geomorphological and biological/ecological sciences to predict the response of freshwater biota and ecosystems to variation of abiotic factors over a range of spatial and temporal scales'. Their definition is unilateral, in the way that it looks only at the biotic response to hydrology. Hannah *et al.* (2004) proposed a more two-sided definition, and presented their own list of criteria, or study foci that should be met for useful definition of the term. Their list included hydrological–ecological interactions, process understanding, the focus system's full range of habitats and their species, various scales, and interdisciplinarity. Although this latter definition might sound like a viable approach for a truly new and dynamic discipline, much work in riparian ecology already fits the definition of Dunbar and Acreman (2001). The advancement of ecohydrology as a science would probably be favoured if more work focused not only on investigating hydrology's role on biota, but also took into account the reciprocal effect of biota on hydrological processes. This is a major area for further investigation, and can only be achieved by a proper understanding of both disciplines.

The aim of this chapter is to provide an idea of the spatial and temporal patterns of vegetation change in riverine landscapes, and of how such changes relate to hydrology. We define 'hydroecological patterns of change' as the way hydrological processes affect and shape ecological processes of riverine vegetation, and vice versa. There is a multitude of examples in the literature on how hydrology affects plant ecology. Examples of reverse effects are fewer, although Tabacchi *et al.* (2000) reviewed some important ones. We have adopted a broad definition of 'patterns of change' in plant communities, encompassing changes that occur under natural conditions through space and time, but also changes that take place when the hydrology of a river is altered anthropogenically. We begin with a brief description of the characteristics of riverine vegetation, and continue by summarising interactions between hydrological and plant ecological processes. This general overview is followed by some examples of changes, natural as well as anthropogenic, presented in a four-dimensional landscape context (Ward, 1998).

17.2 Vegetation in Riverine Habitats

The characteristics of riverine vegetation vary with the location, size, slope, and geomorphology of the river and with the dynamics of its flow. Small, sometimes steep, headwater streams in mountains are shaped by short, frequent flooding events, and may be regularly impacted by ground frost, and events of channel ice formation and break-up. Such streams may have coarse riparian substrates dominated by cryptogams and small herbs. In contrast, the lower reaches of major rivers may show wide heterogeneous floodplains and deltas of fine-grade sediments with dense forests governed by long-lasting flooding events. The riparian zone is generally the best vegetated riverine habitat. Naiman and Décamps (1997) defined it as the area that 'encompasses the stream channel between the low and high water marks and that portion of the terrestrial landscape from the high water mark toward the uplands where vegetation may be influenced by elevated water tables or flooding and by the ability of the soils to hold water'. Since flooding varies greatly over time, the exact delineation of this area is somewhat arbitrary.

A globally common feature of riparian vegetation is that it is often considerably more species rich than the surrounding uplands (e.g., Nilsson *et al.*, 1989; Junk and Piedade,

1993; Goebel *et al.*, 2003), although some areas, such as parts of Australia and South Africa, provide exceptions (Hancock *et al.*, 1996). The riparian system can be very apparent in dry regions where riparian zones form lush green arteries through a dry matrix of desert or savannah. In contrast, moist regions such as rainforests do not show the same contrast between riparian and upland vegetation. Aquatic areas in streams are also usually inhabited by plants, stopping at depths where light is too low to support plant growth. The community type depends on substrate characteristics, nutrient availability, water temperature, wave exposure and flow velocity (Nilsson, 1999). Aquatic plant communities may vary from formations of scattered, tiny plants on channel bottoms, to dense, several metres tall stands of reeds, especially in tranquil river reaches (e.g., Murphy *et al.*, 2003).

17.3 Hydrological–Ecological Interactions

17.3.1 Hydrological Drivers

Hydrological processes such as inundation, and wave and flow action, create environmental gradients for several plant-related variables and thus have a major role in governing riparian vegetation (Table 17.1). Moving water sorts substrates through erosion and sedimentation, it redistributes nutrients and organic matter throughout the riverscape, and it serves as a dispersal vector for many plants (Goodson *et al.*, 2001). Environmental

Table 17.1 Plant-related processes and services maintained by different flow levels (modified from Postel and Richter, 2003)

Flow level		Processes and services
Low flow	Drought	• Recruit certain floodplain plants • Eliminate some invasive alien species • Control distribution and abundance of riparian and instream plants
	Base flow	• Provide habitat for aquatic flora • Support the hyporheic zone
Flood flow	Normal	• Connect channel and riparian zone • Disperse propagules • Deposit nutrients in riparian zone • Disturb riparian vegetation • Perform small-scale sorting of substrates • Create environmental gradients
	Extreme	• Rejuvenate system • Control distribution and abundance of floodplain and instream plants • Perform large-scale sorting of substrates and lateral movement of river channel (shape physical habitat) • Open up gaps for colonising plants • Deposit nutrients • Eliminate some invasive alien species • Recharge floodplain water table

gradients created by water movements shift in space and time, hence opening up niches to allow communities to restructure and species to coexist, resulting in dynamic, nonequilibrium plant communities. An important hydrology-based concept that is useful in riparian ecology is that of the natural flow regime (Poff *et al.*, 1997), suggesting that natural, unaltered flows sustain ecosystem integrity and biodiversity of rivers and organise ecosystems. It also proposes that each river has a characteristic combination of various hydrological components (i.e., a hydrological 'fingerprint'; Puckridge *et al.*, 1998). A previous, related but less general, model is the flood pulse concept (Junk *et al.*, 1989), which states that the periodic recurrence of flood pulses is the principal driving force for biodiversity in river floodplain ecosystems. It also emphasises the lateral connectivity between the channel and the floodplain. This concept is based mainly on large tropical lowland rivers. Tockner *et al.* (2000) further developed it to incorporate temperate areas by including information derived from near-natural proglacial, headwater and lowland floodplains. They emphasised the importance of expansion–contraction cycles, flow pulses, occurring well below bank full stage, and highlighted the complexity of expansion–contraction events and their consequences on habitat heterogeneity and functional processes.

We will briefly present some examples of plant responses to three of the main hydrological drivers of riverine vegetation.

Inundation Fluctuating water levels govern the structure and dynamics of riparian plant communities (e.g., Nilsson and Keddy 1988). The most important components of such fluctuations are their timing, duration, magnitude, frequency and rate of change. Rivers with large natural fluctuations in water-level provide opportunities for well-developed plant zonation, often with high species diversity. Several studies have shown that plant species richness in riparian zones peaks at intermediate levels of flood frequency (e.g., Pollock *et al.*, 1998; Chapin *et al.*, 2002) in concordance with the intermediate disturbance hypothesis (Connell, 1978). Interannual differences in inundation regimes create a dynamism that further contributes to this diversity, on both habitat and species levels (Naiman *et al.*, 1993; Naiman and Décamps, 1997).

Wave action Wave action varies with fetch and bottom characteristics. It sorts substrates and creates gradients of substrate fineness. The composition and vertical extent of plant communities vary along this gradient (Keddy, 1985). Strongly wave-exposed riparian areas may have sparse vegetation, often restricted to their uppermost parts, whereas sheltered areas usually have well-developed vegetation across the riparian zone to the lower limits for aquatic plants. Although conditions for plant growth tend to be worst at strongly wave-exposed areas where substrate is coarsest, the dispersal of floating propagules is most effective where waves are strong (Nilsson *et al.*, 2002).

Flow action Variation in flow action is a basic characteristic of rivers, expressed by the variation between turbulent and tranquil reaches. Turbulent reaches are characterised by fast and turbulent flow, low sedimentation rate, and channels of coarse (eroded) material, such as bedrock, boulders, stones and gravel. Tranquil reaches have slower and more laminar flow, high sedimentation rate, and channels of fine material with much sand and silt (Nilsson, 1999). The interaction between erosion and deposition is important for propagule dispersal, and the degree to which river reaches lose and receive propagules varies with the level of flow and wave action combined (Nilsson *et al.*, 2002).

During baseflow conditions, the effects of flow action are reflected in the distribution of aquatic plants, which are usually absent or sparse in turbulent reaches but may be abundant in tranquil reaches. Floods affect the vegetation by the force of the water itself and by objects transported by flow, such as large wood and ice (Filip *et al.*, 1989; Nilsson *et al.*, 1993). Flow action mechanically injures plants, but also shapes dispersal (Merritt and Wohl, 2002; Nilsson *et al.*, 2002) and often creates a shifting mosaic of plant community patches.

17.3.2 Ecological Drivers

There are at least two major aspects of vegetation control on hydrology, i.e. run-off control and water processing (Tabacchi *et al.*, 2000).

Run-off (in rivers) Riverine vegetation affects flow velocity in the stream, thereby also affecting the deposition of sediments (Gurnell *et al.*, 2006). Dense vegetation, especially riparian scrubs or forest but to some extent also in-channel vegetation, reduces current velocities by providing 'roughness'. It also affects run-off through evapotranspiration. The best example of how vegetation controls run-off is probably when large woody debris jams up, dams water and reroutes river channels (O'Connor *et al.*, 2003).

Water processing The plant water cycle, driven by water uptake through roots and transpiration through pores from leaves, is a basic ecosystem process that varies with vegetation type. Its importance becomes apparent when catchments are clear-cut and overland flow increases (see Post and Jakeman, 1996; Jones and Post, 2004). Jones and Grant (1996) concluded that tree harvesting in the Cascades Range, western Oregon increased peak discharges by 50% in small basins and by 100% in large river basins, over the past 50 years. They attributed these increases to changes in flow routing by roads, and to changes in water balance following harvesting and vegetation succession. Guillemette *et al.* (2005) compared peak flow studies in 50 basins and concluded that globally, the risk for large peak flows increases with the intensity of tree harvesting.

17.4 Natural Patterns of Change

The landscape scale is useful for examining hydroecological patterns of change in riverine environments because it is sufficiently broad to encompass the spatially dynamic nature of these systems (Ward *et al.*, 2002). Ward (1998) used a four-dimensional framework to describe linkages within river basins. His framework includes (i) a lateral dimension describing interactions between the channel, the riparian zone and the surrounding uplands; (ii) a vertical dimension covering interactions between groundwater and surface water; (iii) a longitudinal dimension encompassing upstream–downstream interactions; and (iv) a temporal dimension that describes changes with time in all three spatial dimensions. Free-flowing rivers have hydrological connectivity throughout these dimensions. Here we use this four-dimensional framework as a basis for describing spatial and

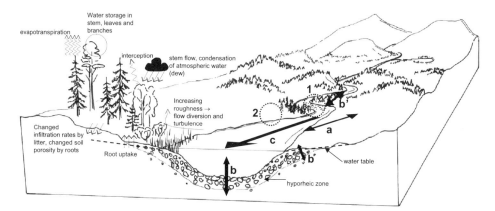

Figure 17.1 Hydroecology in a riparian landscape. The example illustrates a boreal free-flowing river. The right side of the picture illustrates the three spatial dimensions within the riverine corridor. (a) This lateral dimension is primarily governed by environmental gradients created by seasonally fluctuating water levels. Large natural fluctuations give rise to a well-developed plant zonation. Species richness often peaks at intermediate levels of flooding. (b) The vertical dimension is primarily affected by the availability of subsurface (hyporheic) water, differing chemical composition between down- and upwelling water and the presence of groundwater discharge. (b') Hyporheic water is also an integral part of the longitudinal dimension (the hyporheic corridor) with complex flow paths. (c) The longitudinal dimension is primarily affected by geomorphological and hydrological interactions. Differences in flow velocity, flooding dynamics and wave action between (1) turbulent and (2) tranquil sections of a river give rise to differences in floristic composition and resilience.

The left side of the picture illustrates some examples of how vegetation affects the hydrological cycle. Rain water and condensed atmospheric water (dew) is either evaporated back to the atmosphere (interception), or the flow is concentrated on leaves, branches and stems to the ground. Infiltration rates are affected by litter on the ground and soil porosity is affected by roots. Water (of river, hyporheic and groundwater origin) is taken up by roots and stored in stems, leaves and branches, or released to the atmosphere through evapotranspiration. Vegetation also influences flow by increasing turbulence and diversion, for example by means of log jams

temporal patterns of vegetation change in relation to hydrology and hydraulics (Figure 17.1).

17.4.1 Lateral Dimension

Disturbance, productivity and spatial heterogeneity in available resources are main factors regulating species richness in plant communities (Ricklefs and Schluter, 1993; Huston, 1994). Natural riparian zones possess a number of qualities that support high species richness. Even when the riparian zone is a narrow band along the river, it is species rich compared with adjacent upland communities. Silvertown *et al.* (1999) found that species richness in English floodplain meadows can be explained by species' tolerances to drought and aeration. Pollock *et al.* (1998) demonstrated bell-shaped relationships both between flood frequency and species richness, and between productivity and species richness in Alaskan riparian vegetation (cf. Grime, 1973; Connell, 1978) with intermediate levels of flood frequency and productivity resulting in highest species richness. They also showed that species richness increased linearly with the spatial variation of flood frequencies within a site. This spatial variation is caused by microtopographical differences. A nonlinear regression model relating species richness to flood frequency and spatial variation of flood frequencies explained much of the variation in species richness between communities. Species-rich sites had intermediate flood frequencies and high spatial variation. Species-poor sites, on the other hand, had low or abundant flooding and low spatial variation. These data suggest that small-scale spatial variation can dramatically alter the impact of disturbances.

The gradient in environmental factors is reflected in the vegetation. Species distributions are governed by the capacity of species to cope with inundation (e.g., Blom and Voesenek, 1996; Blom, 1999). Therefore, plant zonation varies with hydrology. This is a good example of how the relationship between hydrology and plant biology structures plant communities. For example, boreal rivers are characterized by a major peak in flooding as a result of snowmelt in late spring (Nilsson, 1999). They usually have a relatively confined and geomorphologically stable channel, and often gentle riparian (i.e., lateral) slopes. This fosters a distinct lateral zonation with vegetation communities characterised by forest communities at the top, followed by shrubs, sedges, horse-tails, and amphibious plants at the bottom (Figure 17.2; Nilsson, 1999), reflecting species-specific tolerance to different frequencies and severities of disturbance.

The factors controlling the riparian flora may differ between different types of stream and their interactions between fluvial processes and landforms. Hupp and Osterkamp (1996) compared streams across the United States, and found that riparian vegetation along high-gradient streams in humid environments is mainly controlled by frequency, duration and intensity of floods. Vegetation along low-gradient, floodplain streams, with a more meandering or braided channel may be more controlled by fluxes in sediment deposition and erosion. Vegetation along streams in arid environments is more controlled by water availability. Specific riverbed variables such as soil texture, land form, soil moisture, soil chemistry and seepage are important within specific sites or rivers (e.g., Higgins *et al.*, 1997; Williams and Wiser, 2004).

17.4.2 Vertical Dimension

Most studies on subsurface ('hyporheic') flow deal with the importance of depth to the groundwater for growth of riparian woodlands of obligate phreatophytic trees, such as cottonwood/poplar (*Populus* spp.) and willows (*Salix* spp.) (e.g., Rood *et al.*, 2003). The complex interactions between surface and hyporheic flows have received far less attention. Hyporheic zones are subsurface ecotones where surface water and groundwater mix. Boulton *et al.* (1998) defined the hyporheic zone as 'a spatially fluctuating ecotone between the surface stream and the deep groundwater where ecological processes and their requirements and products are influenced at a number of scales by water movement, permeability, substrate, particle size, resident biota and the physiochemical features of the overlying stream and adjacent aquifers'. Areas with downwelling water are described as losing and areas with upwelling water as gaining reaches. Downwelling zones are often located in the upstream end of a floodplain, at the head of turbulent reaches and bases of

324 *Hydroecology and Ecohydrology: Past, Present and Future*

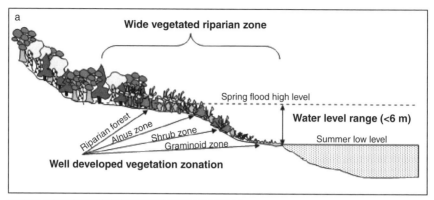
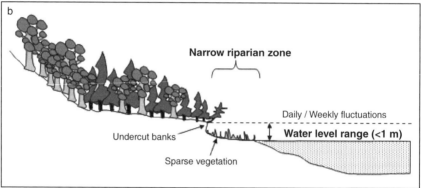
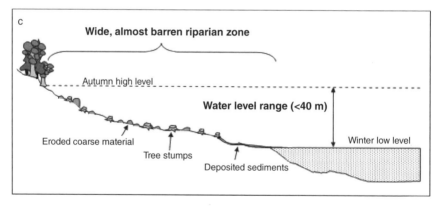

Figure 17.2 Riparian zones along free-flowing and regulated rivers in northern Sweden. (a) Free-flowing river: spring flooding (water-level range <6m) results in well-developed zonation, species-rich, productive and diverse vegetation. (b) Run-of-river impoundment: daily or weekly fluctuation (water-level range <1m) produces narrow riparian zones, a sparse and species-poor vegetation similar to that of lakes, and undercut banks. (c) Storage reservoir: seasonal water-level fluctuations (water level range <40m) with raised water levels during summer and a maximum during late summer and autumn, resulting in extensive, almost barren land. In the latter case, the upper end of the riparian zone consists of coarse, eroded material and the lower parts of finer, deposited material

tranquil reaches, whereas upwelling zones are at the base of turbulent reaches and heads of tranquil reaches (Stanford and Ward, 1988; Valett *et al.*, 1994; Lambs, 2004). Upwelling waters are often richer in nutrients compared with surface waters, as a result of microbiological activity in the interstitial spaces, and may also have more algae (e.g., Valett *et al.*, 1994; Dahm *et al.*, 1998).

The hyporheic zone has been recognised for a diverse array of aquatic invertebrates (e.g., Boulton *et al.*, 1998), but until recently it has received little attention from plant ecologists. Harner and Stanford (2003) compared cottonwood growth between a gaining and a losing reach on the Nyack Floodplain, Flathead River, Montana, US, and found that trees in the gaining reach grew fastest. They associated this difference with a greater availability of growth-promoting resources in the gaining reach, such as improved hydration and fertilisation. The greater availability of nutrients was reflected in lower C:N ratios in the leaves of the trees in the gaining reach. Hyporheic flow is also an integral part of the longitudinal dimension, a so-called hyporheic corridor (Figure 17.1). Its alternation between downwelling and upwelling areas probably exerts important control on biodiversity and production of riparian vegetation (Stanford and Ward, 1993).

17.4.3 Longitudinal Dimension

The river continuum concept was an early attempt to model the distribution of biota along a river (Vannote *et al.*, 1980). This model predicts that the abundance and diversity of organisms maximise at intermediate reaches of a river. In general, most analyses of riverine species diversity have dealt with animal diversity (e.g., Naiman *et al.*, 1987; Reyjol *et al.*, 2003); much less attention has been paid to plant diversity. Nilsson *et al.* (1989, 1991) found a peak in species richness of riparian plants in the middle reaches of free-flowing rivers in northern Sweden. The peak coincided with the location of the former highest coastline, which corresponds to a shift from morainic to sedimentary deposits. Similar bell-shaped patterns in species richness have also been found in French and North American rivers (Planty-Tabacchi *et al.*, 1996), suggesting generality in structuring factors. Renöfält *et al.* (2005a) found that germination of experimentally sown seeds was most effective in the middle reaches of the free-flowing Vindel River. These findings suggest that the middle reaches have an intrinsic potential to accumulate more species than do other reaches.

Nilsson and Jansson (1995) proposed several hydrology-related mechanisms behind the quadratic, longitudinal, pattern of plant species richness. They hypothesised that maximum species richness resulted from intermediate disturbance by floods, and/or maximum substrate heterogeneity at the transition from more stable, morainic deposits in the upper half of the river to easily eroded sediments further downstream. Another hypothesis is that plant dispersal by water would become more important downstream, but that potential species richness would not be realised because of too strong a substrate disturbance in the lower reaches of the rivers (Nilsson and Jansson, 1995). Rivers that are fragmented by dams exhibit an opposite pattern, with lowest diversity in their middle reaches, indicating that effective downstream dispersal of plant propagules is hindered (Nilsson and Jansson, 1995). Impounded rivers also present larger discontinuities in the flora, as compared with free-flowing rivers (Jansson *et al.*, 2000b).

While the river continuum concept considers river systems as continua of physical variables and associated biotic structures, Naiman et al. (1988) suggested that the longitudinal dimension of a river could be viewed as a series of discrete patches with relatively distinct boundaries. This view should be regarded as a complement to the river continuum concept, in which the relative importance of longitudinal versus lateral linkages can be evaluated. For example, tranquil and turbulent reaches may have radically different types of plant communities. In northern Sweden, turbulent reaches have more species and more alpine species, whereas tranquil reaches are floristically most similar to plant communities closer to the coast (Nilsson, 1999). Renöfält et al. (2007) found that plant communities in tranquil reaches are more dynamic over time than those in turbulent ones. A survey conducted over a period of three decades, including three repeated surveys of the riparian vegetation along a free-flowing, boreal river in northern Sweden showed that extreme floods reduced riparian plant species richness in tranquil reaches, but that a subsequent period of less extreme flood events facilitated recovery. Species richness in turbulent river reaches remained high and stable through-out the period. Tranquil river reaches were also more prone to invasion by ruderal species following major floods. One possible explanation for the sensitivity of tranquil reaches to floods could be that their soils become more anaerobic (lower redox potential) during floods, thus resulting in higher plant mortality and more space for colonisation. Also, lateral plant zonation is often less defined in turbulent reaches and flood-intolerant species are sometimes found at low elevations (Nilsson, 1979), probably depending on a lower risk for anaerobic conditions.

Nilsson et al. (2002) demonstrated that differences in floristic composition between tranquil and turbulent reaches relate to floating time of seeds. They distinguished between short-floating and long-floating seeds, based on whether they could float for less or more than 2 days. Turbulent reaches had a higher proportion of short-floaters, whilst the opposite was true for tranquil reaches. Many short-floating seeds sink in tranquil reaches without reaching the riparian zone but are kept in motion in turbulent reaches until washed ashore.

17.4.4 Temporal Dimension

River flow varies within and between years, and this variation affects patterns of plant distribution on various scales. Van der Nat et al. (2003) mapped habitat changes in active channels of the Tagliamento river in north-eastern Italy during 2.5 years. They found that aquatic floodplain habitats were highly dynamic over this short time scale, and that even small flow pulses can impose major habitat change affecting fauna and flora. However, the relative abundance of habitats within the active corridor remained almost constant, suggesting a state of shifting mosaics (e.g., Callaway and Davis, 1993).

Salo et al. (1986) suggested that the diversity of the Amazon rain forest is driven and maintained by large-scale contemporary and historical disturbances of meandering rivers. The lateral movements of the river channel across wide areas give rise to successional rainforest stages. They used satellite images and found that, contrary to the traditional view of a relatively stable Amazonian rainforest driven mainly by gap dynamics, all types of the western Amazon lowlands have been affected by river dynamics. They proposed that the disturbance is partly responsible for the high diversity of the system, and for several reasons. First, the erosional and depositional nature of river dynamics gives rise to high habitat diversity. Second, these habitat types are relatively stable in species composition and relatively short-lived, which would reduce the effect of competitive exclusion. Third, the floodplains may differ profoundly in water and soil chemistry, mode of alluvial sedimentation and case-historical biogeographical events. Salo *et al.* (1986) also suggested river dynamics as a mode for allopatric speciation with the high habitat heterogeneity leading to isolation of populations.

Large floods are often viewed as rejuvenators of riverine ecosystems. For example, a large flood caused a fivefold increase in riverine marshland along the Hassayampa river in south-western US (Stromberg *et al.*, 1997). The flood eroded terraces, widened the channel and recharged the floodplain aquifer. Furthermore, lowering of the floodplain surface changed riparian plant composition. For two seasons following the flood, obligate wetland species were abundant in areas with saturated surface soils or shallow water tables. The flood also rejuvenated the riparian forests of cottonwood and willow. The expansion of riverine marsh and young stands of cottonwood and willow took place at the expense of mature forests of cottonwood, willow and deep-rooted velvet mesquite. Further development of riparian marshland was obstructed by sedimentation following subsequent small floods, elevating the floodplain relative to the water table. These results illustrate the dynamic nature of floodplains in alluvial, arid-land rivers and highlight the importance of natural flows to maintain riverine diversity.

The effect of large floods is less apparent in high-gradient rivers with more stable channels. However, Renöfält *et al.* (2005b) found that species richness per site in a Swedish boreal river was lower in years following large floods, compared to years preceded by a period of low to moderate flooding. They also found that the downstream pattern of species richness differed between these years. After large floods, species richness decreased linearly towards the coast and resembled the longitudinal pattern of the surrounding upland vegetation. Following a period with low to moderate flooding, species richness peaked in the middle reaches of the river.

17.5 Human Impacts

The temporal aspect of hydroecological change also includes the human dimension. River hydrology provides services that humans depend upon for economic growth and prosperity. Unfortunately, human use of rivers has degraded riverine ecosystems (Nilsson *et al.*, 2005b; Syvitski *et al.*, 2005). In the mid-1990s humans expropriated more than half of the world's run-off (Postel *et al.*, 1996) for services such as irrigation, hydropower production, transportation and potable water. This figure has most likely risen since then, due to the increasing demand for water. The fact that humans wield a major control of hydrological drivers implies that the natural patterns of riverine vegetation change. Here we provide examples of how human activities affect natural vegetation and describe how patterns are altered in all the dimensions mentioned above.

In dry regions, human activities have substantially reduced flow in rivers. Human demands for water in these areas have led to groundwater depletion and stream dewatering up to the point where perennial rivers have become ephemeral (Stromberg *et al.*, 1996). The damming of rivers in the arid south-western US has radically altered the magnitude, timing, and frequency of floods that promote regeneration and maintenance of extensive riparian cottonwood (*Populus deltoides*) forests, leading to a decline of these forests and a 'desertification' of the riparian flora (Stromberg *et al.*, 1996; Scott *et al.*, 1999; Stromberg, 2001).

Flow regulation seriously alters riverine vegetation (Nilsson and Svedmark, 2002). The individuality of flow regimes (e.g., Poff *et al.*, 1997) implies that there are river-specific as well as more general regulation effects. Interactions with other factors, such as fluvial geomorphological settings, further complicate the ecological response (Petts, 1979; Johnson, 1998). In rivers with reduced flow variation and no major floods, upland communities usually expand towards the damming waterline (Grelsson and Nilsson, 1980; Johnson, 1994; Toner and Keddy, 1997). In lake-like sections of impounded rivers (i.e., run-of-river impoundments) riparian vegetation is usually confined to a narrow strip along the waterline and its species composition reflects traits of lakeshore vegetation (Figure 17.2). Storage reservoirs may be subjected to huge water-level fluctuations. In boreal reservoirs water levels are kept high during most of the growing season but are reduced during winter and spring. This artificial rhythm is inhospitable for plant growth, and unless there is abundant (moist) fine-grade material that may support a few plants, reservoir shorelines can become barren land (Figure 17.2). In warmer, frost-free areas such as the tropics, terrestrial grasses may grow on reservoir shorelines during drawdown periods or the entire water surface might be covered by floating vegetation, such as *Eichhornia crassipes* (Baxter, 1977; Bini *et al.*, 1999).

Although the history of regulated river ecology goes some 25 years back (Stanford and Ward, 2001), many studies of the effect of river regulation on riparian vegetation have been limited to smaller scales, such as effects of individual dams. In reality, many rivers have been transformed into series of lake-like reservoirs, disrupted by dams and underground passages. Few studies have dealt with the cumulative effects of multiple dams. Jansson *et al.* (2000a) compared regulated and free-flowing rivers in northern Sweden to evaluate the effects of river regulation on riparian vegetation. Even though storage reservoirs and run-of-river impoundments differ in hydrology, both had fewer species per site and a sparser plant cover than did free-flowing rivers. In a later comparison between Sweden and Canada, similar differences were found (Dynesius *et al.*, 2004). Johansson and Nilsson (2002) tested whether the responses of riparian plants to flooding differed between regulated and free-flowing rivers. They found that for a range of species, plants performed worse in regulated rivers. They concluded that a substantial reduction in flood duration and frequency is needed to improve plant performance in the regulated systems.

Between-year water-level fluctuations in reservoirs can result in a considerable variation in shoreline vegetation (Ekzertzev, 1979; Nilsson and Keddy, 1988). Reservoirs with low water-level fluctuations can develop shoreline vegetation similar to that of lakes. For example, reservoirs on the Volga River, filled and held at a constant level during the growing season, developed stable plant communities within a decade. Shallow parts fostered lush growth (Ekzertzev, 1979). However, the long-term and large-scale effects of river regulation on riverine vegetation are difficult to evaluate since long-term data are scarce. Nilsson *et al.* (1997) investigated the community development in 45 run-of-river impoundments and 43 storage reservoirs ranging between 1 and 70 years old. Shortly after the onset of regulation, species richness was low and plant cover sparse in both run-of-river impoundments and storage reservoirs regardless of whether raised water levels intersected former upland soils or overlapped with pristine water levels. Storage reservoirs remained sparsely vegetated regardless of age. Species richness increased during the first 34 years but thereafter decreased to levels similar to those found immediately after regulation. One reason for this could be that suitable habitats are gradually eroded, but also that the species pool becomes deprived because of constraints on dispersal. The fate of run-of-river impoundments differed slightly depending on whether sites were pre-upland or pre-riparian, but both followed the pattern of storage reservoirs, with increasing species richness during the first years followed by a decrease in the older impoundments. Pre-riparian sites had significantly higher species richness than pre-upland sites. They also found evidence for recovery of species composition in older pre-riparian sites when comparing regulated and free-flowing sites, but pre-upland sites remained floristically different compared with the free-flowing situation.

Invasion by alien species is one of the most severe threats to biodiversity (Vitousek *et al.*, 1997; Wilcove *et al.*, 1998; Mack *et al.*, 2000). Riparian corridors include a range of different landscape elements and form dispersal networks connecting different landscapes (Forman and Godron, 1986; Ward *et al.*, 2002). This makes them highly sensitive to plant invasion (DeFerrari and Naiman, 1994; Stohlgren *et al.*, 1999; Brown and Peet, 2003). The hydrological connectivity is a major cause for the high species diversity of river corridors (Ward *et al.*, 2002), but also serves as a dispersal vector for alien invaders. The spread of exotic plant species is an almost worldwide phenomenon, but only few studies have explored landscape-level patterns of invasiveness that go beyond noting that ecosystems are differently invaded. Planty-Tabacchi *et al.* (1996) investigated longitudinal patterns of plant invasiveness in the riparian corridors of the French Adour river and the North American McKenzie river and found that the proportion of alien plant species increased downstream. Nilsson *et al.* (1989) found the same pattern for ruderal plants in the riparian corridors of the free-flowing Torne and Kalix rivers in northern Sweden. They suggested this pattern to be dependent on downstream areas being more affected by agriculture and having a higher population density than upstream areas.

There are several examples of alien infestations that have caused severe problems in riparian corridors, and humans try to control these invaders at heavy expense (Pimentel *et al.*, 2000; Le Maitre *et al.*, 2002). One of the most well-known is the invasion of salt cedar (*Tamarix* spp.) in North American watercourses. Salt cedar spreads at rates up to 20 km/year (Di Tomaso, 1998), and is now the dominant woody species in many perennial rivers in the arid south-western US and north-western Mexico (Glenn and Nagler, 2005). The spread of salt cedar may relate to the fact that all large rivers in this region have been channelised and regulated. Eliminated flood pulses and lowered base flows have led to drier and more saline riparian soils. This seems to favour salt cedar while native willows and cottonwoods decline. In rivers that still have a flood-pulse regime, or in rivers where prescribed flooding has been used, cottonwoods and willows establish despite the presence of salt cedar (Glenn and Nagler, 2005). It has been reported that *Tamarix* trees use more water compared with native trees, leading to reduced groundwater levels, elimination of natural vegetation, and reduced water flows in river channels (Di Tomaso, 1998). Recent research, however, has shown no unusually high evapotranspiration rates in this species that would make it capable of emptying watercourses (Glenn and Nagler, 2005). Regardless, allowing for a more dynamic and natural hydrological regime might dampen the problem of this pest species.

17.6 Ways Forward

We have discussed the importance of hydrological processes for governing plant communities in rivers at different spatial and temporal scales. We have also provided examples of how plants can affect the hydrology of rivers, and summarised human impacts showing that humans have appropriated a fair-sized proportion of the governance of hydrological processes. This human stewardship has implied major drawbacks for the environment (e.g., Naiman *et al.*, 2002) but may also provide opportunities for the future. Following the increasing awareness of how important it is for rivers to have as natural a hydrology as possible, we now see an increasing number of restoration efforts. These include re-operation of dams (i.e., environmental flow prescriptions), implying that more water is let through the sluice gates – and at times when river ecosystems downstream need water – removal of dikes, restoration of channelised or dry river reaches, and also complete removal of dams and reservoirs.

Re-operation of dams Parts of the Colorado river ecosystem were greatly altered following closure of Glen Canyon Dam in 1963, as flood control and daily fluctuating releases from the dam caused large ecological changes. In the 1990s dam releases were modified both for scientific purposes and protection of the river ecosystem. For example, high flows, which carry sediment and build habitats, were introduced as part of initial efforts to restore parts of the river. In 1996, a 7-day experimental controlled flood, lower than pre-dam spring floods, was released downstream of the dam site in order to study closely the effects of a high flow event equivalent to those proposed for future dam management (Patten *et al.*, 2001). It is an example of modification of operations of a large dam to balance economic gains with ecological protection. This flood successively restored sandbars throughout the river corridor and was specifically timed to avoid dispersal and germination of alien plants. Other concerns were also taken into consideration. For example, more than half of the individuals of the endangered Kanab ambersnail were removed from the riparian zone before the flood (Stevens *et al.*, 2001). The test flood did not appear to have any substantial effect on fish (Valdez *et al.*, 2001).

Restoration of channelised sites In northern Sweden, where the majority of rivers were used for timber-floating for about 150 years, restoration is now underway in a few of the rivers that remain unimpounded. This restoration means that boulders are returned to the river channel, stone piers and wing dams are removed, and cut-off side channels are reconnected to the main channel (Figure 17.3). Nilsson *et al.* (2005a) summarised the predicted outcomes of these alterations, both in a short-term and in a more long-term perspective. They hypothesised a number of geomorphological, hydraulic/hydrological and ecological responses. For example, following increases in channel area, sinuosity and roughness, flow resistance and water depth are expected to increase and water velocity to decrease. This is predicted to lead to increases towards predisturbance levels in variables such as habitat complexity, dispersal and migration of organisms, retention of detritus, production, and species diversity. Several timber-floating constructions are kept for culture-historical reasons, but in general, the rivers are likely to attain a better hydro-ecological status than the one that has prevailed during the last century.

Before restoration

- Simplified homogenous watercourse

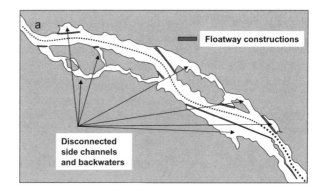

After restoration

- More channels
- Increased sinuosity
- Slower and more variable flow
- Increased substrate heterogeneity
- Increased retention capacity
- Increased frequency and magnitude of inundation

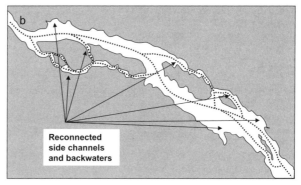

Figure 17.3 (a) A river reach in northern Sweden converted to fit the demands of timber-floating: side channels and back waters have been cut off by stone piers and the river bed has been cleared of large obstacles, such as boulders and large woody debris, resulting in a simplified and homogenous watercourse. (b) Recently restored river reach: floatway constructions have been removed resulting in a heterogeneous river reach with diverse processes

Dam removal   The majority of dam removals have occurred in the US. So far, there are over 500 documented dam removals there, involving obstructions less than 12 m high. Since reservoirs upstream of dams may contain large amounts of sediment, dam removal will be accompanied by intense sediment redistribution, implying that head-cuts will move upstream and eroded sediment be transported and deposited further downstream (e.g., Doyle *et al.*, 2003). This type of sediment pulse stresses existing communities of plants and macroinvertebrates but also creates new ground for colonisation (e.g., Shafroth *et al.*, 2002).

Restoration projects need to be reviewed constantly and appraised by a wide range of specialists. We have focused on projects where hydroecological processes dominate. Such restoration efforts will not reach their full potential unless the underlying hydroecological science is developed. To reach this goal, ecologists and hydrologists need to integrate their approaches, identify common scales, and develop common tools, a common language and common results (Janauer, 2000). Progress within this field will foster a better understanding of ecosystem processes and increase the opportunities of elaborating sustainable management for an increasing number of important water bodies.

Acknowledgements

We thank Catherine Reidy and two anonymous reviewers for valuable comments on the manuscript. Funding was provided by the Swedish Research Council and the Swedish Research Council Formas.

References

Baxter, R. M. 1977. Environmental effects of dams and impoundments, *Annual Review of Ecology and Systematics*, **8**, 255–283.

Bini, L. M., Thomaz, S. M., Murphy, K. J. and Camargo, A. F. M. 1999. Aquatic macrophyte distribution in relation to water and sediment conditions in the Itaipu Reservoir, Brazil, *Hydrobiologia*, **415**, 147–154.

Blom, C. W. P. M. 1999. Adaptations to flooding stress: from plant community to molecule, *Plant Biology*, **1**, 261–273.

Blom, C. W. P. M. and Voesenek, L. A. C. J. 1996. Flooding: the survival strategies of plants, *Trends in Ecology and Evolution*, **11**, 290–295.

Bond, B. 2003. Hydrology and ecology meet and the meeting is good, *Hydrological Processes*, **17**, 2087–2089.

Boulton, A. J., Findlay, S., Marmonier, P., Stanley, E. H. and Valett, H. M. 1998. The functional significance of the hyporheic zone in streams and rivers, *Annual Review of Ecology and Systematics*, **29**, 59–81.

Brown, R. L. and Peet, R. K. 2003. Diversity and invasibility of Southern Appalachian plant communities, *Ecology*, **84**, 32–39.

Callaway, R. M. and Davis, F. W. 1993. Vegetation dynamics, fire, and the physical environment in coastal Central California, *Ecology*, **74**, 1567–1578.

Chapin, D. M., Beschta, R. L. and Shen, H. W. 2002. Relationships between flood frequencies and riparian plant communities in the upper Klamath Basin, Oregon, *Journal of the American Water Resources Association*, **38**, 603–617.

Connell, J. H. 1978. Diversity in tropical rain forests and coral reefs, *Science*, **199**, 1302–1310.

Dahm, C. N., Grimm, N. B., Marmonier, P., Valett, H. M. and Vervier, P. 1998. Nutrient dynamics at the interface between surface waters and groundwaters, *Freshwater Biology*, **40**, 427–451.

DeFerrari, C. M. and Naiman, R. J. 1994. A multiscale assessment of the occurrence of exotic plants on the Olympic peninsula, Washington, *Journal of Vegetation Science*, **5**, 247–258.

Di Tomaso, J. M. 1998. Impact, biology, and ecology of saltcedar (*Tamarix* spp.) in the southwestern United States, *Weed Technology*, **12**, 326–336.

Doyle, M. W., Stanley, E. H. and Harbor, J. M. 2003. Channel adjustments following two dam removals in Wisconsin, *Water Resources Research*, **39**, no. 1011.

Dunbar, M. J. and Acreman, M. C. 2001. Applied hydro-ecological sciences for the twenty-first century, in Acreman, M. C. (ed.) *Hydro-Ecology: Linking Hydro and Aquatic Ecology*, International Association of Hydrological Sciences Publication No. 266. Wallingford (UK): IAHS Press, pp. 1–17.

Dynesius, M., Jansson, R., Johansson, M. E. and Nilsson, C. 2004. Intercontinental similarities in riparian-plant diversity and sensitivity to river regulation, *Ecological Applications*, **14**, 173–191.

Ekzertsev, A. 1979. The higher aquatic vegetation of the Volga, in Mordukhai-Boltovskoi, P. D. (ed.) *The River Volga and Its Life*. The Hague (Netherlands): Junk, pp. 271–294.

Filip, G. M., Bryant, L. D. and Parks, C. A. 1989. Mass movement of river ice causes severe tree wounds along the Grande-Ronde River in northeastern Oregon, *Northwest Science*, **63**, 211–213.

Forman, R. T. T. and Godron, M. 1986. *Landscape Ecology*, New York (USA): John Wiley & Sons, Inc.

Glenn, E. P. and Nagler, P. L. 2005. Comparative ecophysiology of *Tamarix ramosissima* and native trees in western US riparian zones, *Journal of Arid Environments*, **61**, 419–446.

Goebel, P. C., Palik, B. J. and Pregitzer, K. S. 2003. Plant diversity contributions of riparian areas in watersheds of the Northern Lake States, USA, *Ecological Applications*, **13**, 1595–1609.

Goodson, J. M., Gurnell, A. M., Angold, P. G. and Morrissey, I. P. 2001. Riparian seed banks: structure, process and implications for riparian management, *Progress in Physical Geography*, **25**, 301–325.

Grelsson, G. and Nilsson, C. 1980. Colonization by *Pinus sylvestris* of a former middle-geolittoral habitat, on the Umeälven river in northern Sweden, following river regulation for hydroelectric power, *Holarctic Ecology*, **3**, 124–128.

Grime, J. P. 1973. Competitive exlusion in herbaceous vegetation, *Nature*, **242**, 344–347.

Guillemette, F., Plamondon, A. P., Prevost, M. and Levesque, D. 2005. Rainfall generated stormflow response to clearcutting an boreal forest: peak flow comparison with 50 world-wide basin studies, *Journal of Hydrology*, **302**, 137–153.

Guilloy-Froget, H., Muller, E., Barsoum, N. and Hughes, F. M. R. 2002. Dispersal, germination, and survival of *Populus nigra* L. (Salicaceae) in changing hydrologic conditions, *Wetlands*, **22**, 478–488.

Gurnell, A. M. and Petts, G. E. 2002. Island-dominated landscapes of large floodplain rivers, a European perspective, *Freshwater Biology*, **47**, 581–600.

Gurnell, A. M., Petts, G. E., Hannah, D. M., Smith, B. P. G., Edwards, P. J., Kollmann, J., Ward, J. V. and Tockner, K. 2001. Riparian vegetation and island formation along the gravel-bed Fiume Tagliamento, Italy, *Earth Surface Processes and Landforms*, **26**, 31–62.

Gurnell, A. M., van Oosterhout, M. P., de Vlieger, B. and Goodson, J. M. 2006. Reach-scale interactions between aquatic plants and physical habitat: River Frome, Dorset, *River Research and Applications*, **22**, 667–680.

Hancock, C. N., Ladd, P. G. and Froend, R. H. 1996. Biodiversity and management of riparian vegetation in western Australia, *Forest Ecology and Management*, **85**, 239–250.

Hannah, D. M., Wood, P. J. and Sadler, J. P. 2004. Ecohydrology and hydroecology: a new paradigm? *Hydrological Processes*, **18**, 3439–3445.

Harner, M. J. and Stanford, J. A. 2003. Differences in cottonwood growth between a losing and a gaining reach of an alluvial floodplain, *Ecology*, **84**, 1453–1458.

Higgins, S. I., Rogers, K. H. and Kemper, J. 1997. A description of the functional vegetation pattern of a semi-arid floodplain, South Africa, *Plant Ecology*, **129**, 95–101.

Hudon, C., Wilcox, D. and Ingram, J. 2006. Modeling wetland plant community response to assess water-level regulation scenarios in the Lake Ontario–St. Lawrence River basin, *Environmental Monitoring and Assessment*, **113**, 303–328.

Hupp, C. R. and Osterkamp, W. R. 1996. Riparian vegetation and fluvial geomorphic features, *Geomorphology*, **14**, 277–295.

Huston, M. A. 1994. *Biological Diversity: The Coexistence of Species on Changing Landscapes.* Cambridge (UK): Cambridge University Press.

Janauer, G. A. 2000. Ecohydrology: fusing concepts and scales, *Ecological Engineering*, **16**, 9–16.

Jansson, R., Nilsson, C., Dynesius, M. and Andersson, E. 2000a. Effects of river regulation on river-margin vegetation: a comparison of eight boreal rivers, *Ecological Applications*, **10**, 203–224.

Jansson, R., Nilsson, C. and Renöfält, B. 2000b. Fragmentation of riparian floras in rivers with multiple dams, *Ecology*, **81**, 899–901.

Jiang, H., Liu, S. R., Sun, P. S., An, S. Q., Zhou, G. Y., Li, C. Y., Wang, J. X., Yu, H. and Tian, X. J. 2004. The influence of vegetation type on the hydrological process at the landscape scale, *Canadian Journal of Remote Sensing*, **30**, 743–763.

Johansson, M. E. and Nilsson, C. 2002. Responses of riparian plants to water-level variation in free-flowing and regulated boreal rivers: an experimental study, *Journal of Applied Ecology*, **39**, 971–986.

Johnson, W. C. 1994. Woodland expansion in the Platte River, Nebraska: patterns and causes, *Ecological Monographs*, **64**, 45–84.

Johnson, W. C. 1998. Adjustment of riparian vegetation to river regulation in the great plains, USA, *Wetlands*, **18**, 608–618.

Jones, J. A. and Grant, G. E. 1996. Peak flow responses to clear-cutting and roads in small and large basins, western Cascades, Oregon, *Water Resources Research*, **32**, 959–974.

Jones, J. A. and Post, D. A. 2004. Seasonal and successional streamflow response to forest cutting and regrowth in the northwest and eastern United States, *Water Resources Research*, **40**, No. W05203.

Junk, W. J. and Piedade, M. T. F. 1993. Herbaceous plants of the Amazon floodplain near Manaus: species diversity and adaptations to the flood pulse, *Amazonia*, **12**, 467–484.

Junk, W. J., Bayley, P. B. and Sparks, R. E. 1989. The flood pulse concept in river-floodplain systems, *Canadian Special Publication in Fisheries and Aquatic Sciences*, **106**, 110–127.

Keddy, P. A. 1985. Wave disturbance on lakeshores and the within-lake distribution of Ontario's Atlantic coastal flora, *Canadian Journal of Botany*, **63**, 656–660.

Kundzewicz, Z. W. 2002. Ecohydrology: seeking consensus on interpretation of the notion, *Hydrological Sciences Journal*, **47**, 799–804.

Lambs, L. 2004. Interactions between groundwater and surface water at river banks and the confluence of rivers, *Journal of Hydrology*, **288**, 312–326.

Le Maitre, D. C., van Wilgen, B. W., Gelderblom, C. M., Bailey, C., Chapman, R. A. and Nel, J. A. 2002. Invasive alien trees and water resources in South Africa: case studies of the cost and benefits of management, *Forest Ecology and Management*, **160**, 143–159.

Mack, R. N., Simberloff, D., Lonsdale, W. M., Evans, H., Clout, M. and Bazzaz, F. A. 2000. Biotic invasions: causes, epidemiology, global consequences and control, *Ecological Applications*, **10**, 689–710.

Merritt, D. M. and Wohl, E. E. 2002. Processes governing hydrochory along rivers: hydraulics, hydrology, and dispersal phenology, *Ecological Applications*, **12**, 1071–1087.

Murphy, K. J., Dickinson, B., Thomaz, S. M., Bini, L. M., Dick, K., Greaves, K., Kennedy, M. P., Livingstone, S., McFerran, H., Milne, J. M., Oldroyd, J. and Wingfield, R. A. 2003. Aquatic plant communities and predictors of diversity in a sub-tropical river floodplain: the upper Rio Paraná, Brazil, *Aquatic Botany*, **77**, 257–276.

Naiman, R. J. and Décamps, H. 1997. The ecology of interfaces: riparian zones, *Annual Review of Ecology and Systematics*, **28**, 621–658.

Naiman, R. J., Melillo, J. M., Lock, M. A., Ford, T. E. and Reice, S. R. 1987. Longitudinal patterns of ecosystem processes and community structure in a sub-arctic river continuum, *Ecology*, **68**, 1139–1156.

Naiman, R. J., Décamps, H., Pastor, J. and Johnston, C. A. 1988. The potential importance of boundaries to fluvial systems, *Journal of the North American Benthological Society*, **7**, 289–306.

Naiman, R. J., Décamps, H. and Pollock, M. 1993. The role of riparian corridors in maintaining regional biodiversity, *Ecological Applications*, **3**, 209–212.

Naiman, R. J., Bunn, S. E., Nilsson, C., Petts, G. E., Pinay, G. and Thompson, L. C. 2002. Legitimizing fluvial ecosystems as users of water: an overview, *Environmental Management*, **30**, 455–467.

Nilsson, C. 1979. Geolittoral extent of *Vaccinium myrtillus* and vegetational differentiation on a bank of the Muddusjokk river, North Sweden, *Holarctic Ecology*, **2**, 137–143.

Nilsson, C. 1999. Rivers and streams, *Acta Phytogeographica Suedica*, **84**, 135–148.

Nilsson, C. and Jansson, R. 1995. Floristic differences between riparian corridors of regulated and free-flowing boreal rivers, *Regulated Rivers: Research and Management*, **11**, 55–66.

Nilsson, C. and Keddy, P. A. 1988. Predictability of change in shoreline vegetation in a hydro-electric reservoir, northern Sweden, *Canadian Journal of Fisheries and Aquatic Sciences*, **45**, 1896–1904.

Nilsson, C. and Svedmark, M. 2002. Basic principles and ecological consequences of changing water regimes: riparian plant communities, *Environmental Management*, **30**, 468–480.

Nilsson, C., Grelsson, G., Johansson, M. and Sperens, U. 1989. Patterns of plant species richness along riverbanks, *Ecology*, **70**, 77–84.

Nilsson, C., Grelsson, G., Dynesius, M., Johansson, M. E. and Sperens, U. 1991. Small rivers behave like large rivers: effects of postglacial history on plant species richness along riverbanks, *Journal of Biogeography*, **18**, 533–541.
Nilsson, C., Nilsson, E., Johansson, M. E., Dynesius, M., Grelsson, G., Xiong, S., Jansson, R. and Danvind, M. 1993. Processes structuring riparian vegetation, in Menon, J. (ed.) *Current Topics in Botanical Research*, Trivandrum (India), pp. 419–431.
Nilsson, C., Jansson, R. and Zinko, U. 1997. Long-term responses of river-margin vegetation to water-level regulation, *Science*, **276**, 798–800.
Nilsson, C., Andersson, E., Merritt, D. M. and Johansson, M. E. 2002. Differences in riparian flora between river banks and river lakeshores explained by dispersal traits, *Ecology*, **83**, 2878–2887.
Nilsson, C., Lepori, F., Malmqvist, B., Törnlund, E., Hjerdt, N., Helfield, J. M., Palm, D., Östergren, J., Jansson, R., Brännäs, E. and Lundqvist, H. 2005a. Forecasting environmental responses to restoration of rivers used as log floatways: an interdisciplinary challenge, *Ecosystems*, **8**, 779–800.
Nilsson, C., Reidy, C. A., Dynesius, M. and Revenga, C. 2005b. Fragmentation and flow regulation of the world's large river systems, *Science*, **308**, 405–408.
O'Connor, J. E., Jones, M. A. and Haluska, T. L. 2003. Flood plain and channel dynamics of the Quinault and Queets Rivers, Washington, USA, *Geomorphology*, **51**, 31–59.
Patten, D. T., Harpman, D. A., Voita, M. I. and Randle, T. J. 2001. A managed flood on the Colorado River: background, objectives, design, and implementation, *Ecological Applications*, **11**, 635–643.
Petts, G. E. 1979. Complex response of river channel morphology subsequent to reservoir construction, *Progress in Physical Geography*, **3**, 329–362.
Pimentel, D., Lach, L., Zuniga, R. and Morrison, D. 2000. Environmental and economic costs of non-indigenous species in the United States, *BioScience*, **50**, 53–65.
Planty-Tabacchi, A.-M., Tabacchi, E., Naiman, R. J., DeFerrari, C. and Décamps, H. 1996. Invasibility of species rich communities in riparian zones, *Conservation Biology*, **10**, 598–607.
Poff, N. L., Allan, J. D., Bain, M. B., Karr, J. R., Prestergaard, K. L., Richter, B. D., Sparks, R. E. and Stromberg, J. 1997. The natural flow regime: a paradigm for river conservation and restoration, *BioScience*, **47**, 769–784.
Pollock, M. M., Naiman, R. J. and Hanley, T. A. 1998. Plant species richness in riparian wetlands: a test of the biodiversity theory, *Ecology*, **79**, 94–105.
Post, D. A. and Jakeman, A. J. 1996. Relationships between catchment attributes and hydrological response characteristics in small Australian mountain ash catchments, *Hydrological Processes*, **10**, 877–892.
Postel, S. L., Daily, G. C. and Ehrlich, P. R. 1996. Human appropriation of available fresh water, *Science*, **271**, 785–788.
Postel, S. and Richter, B. 2003. *Rivers for Life: Managing Water for People and Nature*, Washington, DC (USA): Island Press.
Puckridge, J. T., Sheldon, F., Walker, K. F. and Boulton, A. J. 1998. Flow variability and the ecology of large rivers, *Marine and Freshwater Research*, **49**, 55–72.
Renöfält, B. M., Jansson, R. and Nilsson, C. 2005a. Spatial patterns of plant invasiveness in a riparian corridor, *Landscape Ecology*, **20**, 165–176.
Renöfält, B. M., Nilsson, C. and Jansson, R. 2005b. Spatial and temporal patterns of species richness in a riparian landscape, *Journal of Biogeography*, **32**, 2025–2037.
Renöfält, B. M., Merritt, D. M. and Nilsson, C. 2007. Connecting variation in vegetation and stream flow: the role of geomorphic context in vegetation response to large floods along boreal rivers, *Journal of Applied Ecology*, **44**, 147–157.
Reyjol, Y., Compin, A., A-Ibarra, A. and Lim, P. 2003. Longitudinal diversity patterns in streams: comparing invertebrates and fish communities, *Archiv für Hydrobiologie*, **157**, 525–533.
Ricklefs, R. E. and Schluter, D. 1993. *Species Diversity in Ecological Communities: Historical and Geographical Perspectives*. Chicago (USA): University of Chicago Press.
Rood, S. B., Gourley, C. R., Ammon, E. M., Heki, L. G., Klotz, J. R., Morrison, M. L., Mosley, D., Scoppettone, G. G., Swanson, S. and Wagner, P. L. 2003. Flows for floodplain forests: a successful riparian restoration, *BioScience*, **53**, 647–656.

Salo, J., Kalliola, R., Häkkinen, I., Mäkinen, Y., Niemelä, P., Puhakka, M. and Coly, P. D. 1986. River dynamics and the diversity of Amazon lowland forest, *Nature*, **322**, 254–258.

Scott, M. L., Shafroth, P. B. and Auble, G. T. 1999. Responses of riparian cottonwoods to alluvial water table declines, *Environmental Management*, **23**, 347–358.

Shafroth, P. B., Friedman, J. M., Auble, G. T., Scott, M. L. and Braatne, J. H. 2002. Potential responses of riparian vegetation to dam removal, *BioScience*, **52**, 703–712.

Silvertown, J., Dodd, M. E., Gowning, D. J. G. and Mountford, J. O. 1999. Hydrologically defined niches reveal a basis for species richness in plant communities, *Nature*, **400**, 61–63.

Stanford, J. A. and Ward, J. V. 1988. The hyporheic habitat of river ecosystems, *Nature*, **335**, 64–66.

Stanford, J. A. and Ward, J. V. 1993. An ecosystem perspective of alluvial rivers: connectivity and the hyporheic corridor, *Journal of the North American Benthological Society*, **12**, 48–60.

Stanford, J. A. and Ward, J. V. 2001. Revisiting the serial discontinuity concept, *Regulated Rivers: Research and Management*, **17**, 303–310.

Stevens, L. E., Ayers, T. J., Bennett, J. B., Christensen, K., Kearsley, M. J. C., Meretsky, V. J., Phillips, A. M., Parnell, R. A., Spence, J., Sogge, M. K., Springer, A. E. and Wegner, D. L. 2001. Planned flooding and Colorado River riparian trade-offs downstream from Glen Canyon Dam, Arizona, *Ecological Applications*, **11**, 701–710.

Stohlgren, T. J., Binkley, D., Chong, G. W., Kalkhan, M. A., Schell, L. D., Bull, K. A., Otsuki, Y., Newman, G., Bashkin, M. and Son, Y. 1999. Exotic plant species invade hot spots of native plant diversity, *Ecological Monographs*, **69**, 25–46.

Stromberg, J. C. 2001. Restoration of riparian vegetation in the south-western United States: importance of flow regimes and fluvial dynamism, *Journal of Arid Environments*, **49**, 17–34.

Stromberg, J. C., Tiller, R. and Richter, B. 1996. Effects of groundwater decline on riparian vegetation of semiarid regions: the San Pedro, Arizona, *Ecological Applications*, **6**, 113–131.

Stromberg, J. C., Fry, J. and Patten, D. T. 1997. Marsh development after large floods in an alluvial, arid-land river, *Wetlands*, **17**, 292–300.

Stromberg, J. C., Lite, S. J., Rychener, T. J., Levick, L. R., Dixon, M. D. and Watts, J. M. 2006. Status of the riparian ecosystem in the upper San Pedro River, Arizona: Application of an assessment model, *Environmental Monitoring and Assessment*, **115**, 145–173.

Syvitski, J. P. M., Vörösmarty, C. J., Kettner, A. J. and Green, P. 2005. Impact of humans on the flux of terrestrial sediment to the global coastal ocean, *Science*, **308**, 376–380.

Tabacchi, E., Lambs, L., Guilloy, H., Planty-Tabacchi, A.-M., Muller, E. and Décamps, H. 2000. Impacts of riparian vegetation on hydrological processes, *Hydrological Processes*, **14**, 2959–2976.

Tockner, K., Malard, F. and Ward, J. V. 2000. An extension of the flood pulse concept, *Hydrological Processes*, **14**, 2861–2883.

Tockner, K., Ward, J. V., Arscott, D. B., Edwards, P. J., Kollmann, J., Gurnell, A. M., Petts, G. E. and Maiolini, B. 2003. The Tagliamento River: a model ecosystem of European importance, *Aquatic Sciences*, **65**, 239–253.

Toner, M. and Keddy, P. A. 1997. River hydrology and riparian wetlands: a predictive model for ecological assembly, *Ecological Applications*, **7**, 236–246.

Valdez, R. A., Hoffnagle, T. L., McIvor, C. C., McKinney, T. and Leibfried, W. C. 2001. Effects of a test flood on fishes of the Colorado River in Grand Canyon, Arizona, *Ecological Applications*, **11**, 686–700.

Valett, H. M., Fisher, S. G., Grimm, N. B. and Camill, P. 1994. Vertical hydrologic exchange and ecological stability of a desert stream, *Ecology*, **75**, 548–560.

van der Nat, D., Tockner, K., Edwards, P. J., Ward, J. V. and Gurnell, A. M. 2003. Habitat change in braided flood plains (Tagliamento, NE-Italy), *Freshwater Biology*, **48**, 1799–1812.

Vannote, R. L., Minshall, G. W., Cummings, K. W., Sedell, J. S. and Cushing, C. E. 1980. The river continuum concept, *Canadian Journal of Fisheries and Aquatic Sciences*, **37**, 130–137.

Vitousek, P. M., D'Antonio, C. M., Loope, L. L., Rejmánek, M. and Westbrooks, R. 1997. Introduced species: a significant component of human-caused global change, *New Zealand Journal of Ecology*, **21**, 1–16.

Ward, J. V. 1998. The four dimensional nature of lotic ecosystems, *Journal of the North American Benthological Society*, **8**, 2–8.

Ward, J. V., Tockner, K. and Schiemer, F. 1999. Biodiversity of floodplain river ecosystems: ecotones and connectivity, *Regulated Rivers: Research and Management*, **15**, 125–139.

Ward, J. V., Malard, F. and Tockner, K. 2002. Landscape ecology: a framework for integrating pattern and processes in river corridors, *Landscape Ecology*, **17**, 35–45.

Wilcove, D. S., Rothstein, D., Dubow, J., Phillips, A. and Losos, E. 1998. Quantifying threats to imperilled species in the United States, *BioScience*, **48**, 607–615.

Williams, P. A. and Wiser, S. 2004. Determinants of regional and local patterns in the floras of braided riverbeds in New Zealand, *Journal of Biogeography*, **31**, 1355–1372.

Zalewski, M. 2002. Ecohydrology: the use of ecological and hydrological processes for sustainable management of water resources, *Hydrological Sciences Journal*, **47**, 823–832.

Zhao, W. Z., Chang, X. L. and He, Z. B. 2004. Responses of distribution pattern of desert riparian forests to hydrologic process in Ejina oasis, *Science in China Series D-Earth Sciences*, Suppl. I, **47**, 21–31.

18
Hydroecology of Alpine Rivers

Lee E. Brown, Alexander M. Milner and David M. Hannah

18.1 Introduction

Alpine zones are found in mountainous regions above the permanent treeline and are widely distributed throughout both hemispheres (Figure 18.1; Körner, 2003). In undisturbed areas, the treeline approximates the 10 °C July (northern hemisphere) or January (southern hemisphere) isotherm (Remmert, 1980). Above this altitude, the terrestrial environment is generally characterised by bare rock surfaces, poorly developed soils and prostrate vegetation, except close to river margins, where riparian vegetation can sometimes be extensive. Snowpacks and glaciers dominate at higher altitudes. In Scandinavia, birch usually dominates forest below the treeline, whereas in the Rocky Mountains of North America various species of pine, fir and spruce are typical. The elevation of alpine river systems can vary markedly according to location. For example, alpine rivers can occur near sea level at high latitudes or >4000 m a.s.l (metres above sea level) in tropical mountains (Figure 18.1).

Alpine river systems are fed by glacial icemelt, snowmelt, and groundwater (Brown *et al.*, 2003), individually and jointly producing characteristic discharge regimes and a distinctive suite of physicochemical properties (Brittain *et al.*, 2001). These properties (e.g., water temperature, stream discharge, suspended sediment concentration, hydrochemistry) change over temporal scales ranging from inter-annual to diurnal (Smith *et al.*, 2001) and are a major determinant of biotic community structure. Being extreme environments, alpine streams typically support many specialised taxa that are adapted for harsh conditions. Therefore, they occupy the 'declining limb' of habitat harshness – ecosystem diversity curves (Tockner *et al.*, 1997).

Hydroecology and Ecohydrology: Past, Present and Future, Edited by Paul J. Wood, David M. Hannah and Jonathan P. Sadler
© 2007 John Wiley & Sons, Ltd.

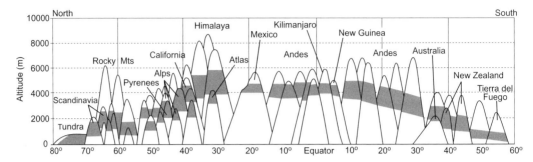

Figure 18.1 Occurrence of the alpine zone (grey band) as a function of latitude (modified from Körner, 2003). Permanent snowfields and glaciers are often found above the alpine zone

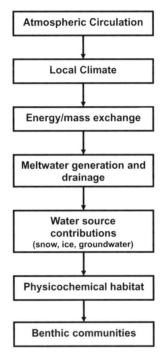

Figure 18.2 Flow diagram of hypothesised links in the climate-hydrology-ecology process cascade within an alpine, glacierized river basin (modified from Hannah et al., 2007)

Many glacierized alpine river basins are located at their climatic limits (see Oerlemans, 2005) making them extremely sensitive to climate change (Krajick, 2004). This sensitivity is enhanced by strong coupling between climate, snowpacks/glacier mass-balance, stream flow, water quality (physicochemical habitat), and river ecology (Figure 18.2; Hannah et al., 2007). Identifying early signals of, and subtle shifts in, hydrological and ecological response to climate change/variability is particularly important in these environments because scenarios of future climate change suggest that alpine ecosystems will show greater impacts than those at lower altitude (Beniston, 2000).

This chapter reviews current understanding of water source–physicochemical properties–biota relationships for rivers in alpine glacierized basins. The chapter begins with an overview of alpine water source dynamics, then considers recent advances in understanding relationships between water source and river physicochemical characteristics, and ends by examining the ecological characteristics of alpine river systems. The chapter concludes by discussing the utility of a novel method seeking to integrate alpine river system understanding within a single hydroecological framework, before identifying significant emergent research gaps. This synthesis aims to inform and encourage interdisciplinary research in the field of alpine river system hydroecology.

18.2 Water Sources Dynamics in Alpine River Systems

Variability in the magnitude, timing and duration of peak snowmelt, glacial icemelt and groundwater contributions plays an important role in influencing alpine basin streamflow (Malard *et al.*, 1999). A considerable amount of research during the 1980s and 1990s greatly improved understanding of glacial hydrology (Fountain and Walder, 1998). More recently, the role of water source dynamics in a wider alpine basin context has received attention with the hydrological importance of groundwater systems being identified (Ward *et al.*, 1999; Smith *et al.*, 2001; Brown *et al.*, 2006c).

The surface of a glacier consists of an ablation zone comprising ice and seasonal snowpacks at lower elevations, and snow (above the firn line) at higher elevations in the accumulation zone. Large-scale atmospheric circulation influences glaciers by supplying and removing mass (i.e., precipitation; Figure 18.2) and energy at the glacier and snowpack surface (Hannah *et al.*, 1999a). Different air masses, with contrasting thermodynamic properties, significantly affect snow- and icemelt. Typically, cyclonic weather patterns are associated with low ablation, whereas high melt is associated with anticyclonic systems (Hannah and McGregor, 1997). At smaller spatial scales, shading, altitude and transient snowline position (surface albedo) influence the distribution of energy receipt and, in turn, ablation patterns across glaciers (Hodson *et al.*, 1998; Hannah *et al.*, 2000a).

In the ablation zone, snowpacks retain meltwaters during the early melt season, moderating peak stream discharge (Fountain, 1996). As snowpacks recede up-glacier over the melt season, ice is revealed and supraglacial meltwater channels develop. Exposure of ice allows faster drainage of meltwater compared with slower percolation through snowpacks (Hannah *et al.*, 1999b). Some meltwater drains over the glacier surface into proglacial streams but a large proportion enters the glacier interior (englacial) drainage system through crevasses and moulins. Water reaching the glacier bed enters the subglacial drainage system and may be routed to the glacier snout in major arterial conduits, smaller conduits linking cavities and/or thin water films (Fountain and Walder, 1998; Fountain *et al.*, 2005). As the englacial and subglacial drainage systems expand and become more connected over the melt season, diurnal fluctuations in proglacial stream discharge become more pronounced (Hannah *et al.*, 1999b, 2000b; Swift *et al.*, 2005). As water moves through the glacier drainage system, it is physicochemically altered due to entrainment of fine sediment and geochemical weathering (Gurnell and Clark, 1987; Tranter *et al.*, 2002).

With distance from the glacier margin, glacial influence decreases and snow- and ice-melt runoff mixes with groundwater to change the hydrology of alpine streams (Malard *et al.*, 1999; Brown *et al.*, 2006c). Alpine basins have traditionally been regarded as predominantly surface water fed (i.e., snowpacks and glaciers) but a study of water quantities from different sources in the Taillon-Gabiétous basin, French Pyrénées, revealed that groundwater can contribute ~25% of streamflow at only 1.5 km from the Taillon Glacier (Brown *et al.*, 2006c). Although groundwater sources are strongly dependent on snowmelt and precipitation recharge (Flerchinger *et al.*, 1992), water becomes physicochemically modified due to the long residence times of water in subsurface aquifers.

18.3 Physicochemical Properties of Alpine Rivers

Spatio-temporal variability of water source contributions makes rivers in alpine glacierized basins distinct from other lotic systems (Smith *et al.*, 2001). The timing and volume of bulk meltwater production (Hannah and Gurnell, 2001), along with inputs of groundwater from springs, seeps and upwellings (Ward *et al.*, 1999) generate distinct patterns of stream discharge, water temperature, hydrochemistry, electrical conductivity, turbidity and suspended sediment concentration over annual, seasonal and diurnal time scales (Hannah *et al.*, 2007). Heterogeneity of the physicochemical environment creates a spatial and temporal mosaic of stream habitats related to differences in water source contributions, and hydrological connectivity (Malard *et al.*, 2000; Burgherr *et al.*, 2002; Malard *et al.*, 2006). The dynamics of these ecologically relevant physicochemical variables are outlined below.

18.3.1 Stream Discharge

Glacier-fed rivers have a characteristic annual flow regime with peak flow in mid-summer. However, this seasonal 'glacial flow-regime' can be punctuated by episodic precipitation events to yield a more variable discharge regime (Hannah *et al.*, 2000b; Smith *et al.*, 2001). In contrast, many alpine groundwater streams are characterised by relatively constant discharge over inter-annual time scales (Ward *et al.*, 1999), although Brown *et al.* (2006b) suggest that some alpine groundwater streams can freeze for short periods during winter.

Seasonal changes in the diurnal hydrographs of proglacial rivers are influenced by transient snowline retreat, seasonal patterns in atmospheric heating and shifts from snow- to icemelt dominated runoff. Hannah *et al.* (1999b, 2000b) inferred changes in glacier drainage system hydrology by classifying diurnal outflow hydrographs based on their form (shape) and magnitude (size) using data from the Taillon glacier stream, French Pyrénées. The approach has also been applied successfully at Haut Glacier d'Arolla, Switzerland (Swift *et al.*, 2005). During the early melt season when snowpacks cover large areas of alpine basins, diurnal hydrographs are characterised by flows that build over the day, peaking in the late evening as a result of snowpacks delaying meltwater movement to the proglacial stream. During the later melt season (July–August) mid-afternoon peak flows (linked to energy receipt at the glacier surface; Hannah *et al.*, 2000a) are characteristic of glacier-fed rivers (Figure 18.3).

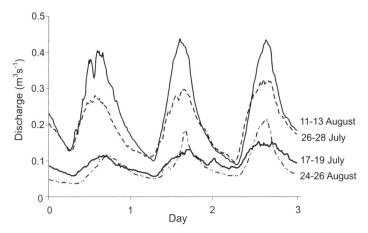

Figure 18.3 *Diurnal discharge fluctuations approximately 100 m downstream of the Taillon glacier margin during the 1995 melt season (see Hannah et al., 2000b for methodological details)*

18.3.2 Stream Temperature

Alpine glacier-fed rivers are characterised by large annual temperature fluctuations, with water temperatures close to zero or sub-zero during winter, increasing temperatures during spring–early summer, then decreasing during late summer–autumn (Uehlinger et al., 2003; Brown et al., 2006b). In contrast, many alpine groundwater streams are more thermally stable (Ward et al., 1999). However, during the summer melt season, some hillslope groundwater streams have temperature ranges of up to 14 °C (Brown et al., 2005). In summer, diurnal temperature fluctuations of up to 2 °C occur for meltwater streams near the glacier margin but, following snowpack melting and uncovering of streams, further downstream diurnal variability can rise to >10 °C (Brown et al., 2004a, 2005, 2006a).

Downstream increases in water temperature are typical of glacier-fed rivers (Uehlinger et al., 2003; Brown et al., 2005) as streams become more exposed to atmospheric heating and less influenced by initial source temperatures. Over a distance of about 1 km, mean water column temperatures increased by 7.0–8.6 °C, in the Taillon glacier stream, French Pyrénées (Brown et al., 2004a, 2006a). In the much larger Roseg river, Swiss Alps, mid-summer temperature increases of 7–8 °C were observed over a distance of >10 km (Uehlinger et al., 2003). However, groundwater tributary inflows can modify longitudinal temperature profiles in glacier-fed rivers. For example, Taillon Glacier stream temperatures decreased (average 0.4 °C) due to karstic groundwater stream inflows, whereas warmer water predominantly sourced from hillslope groundwater streams resulted in higher temperatures (average 1.7–1.8 °C; Brown et al., 2006a).

Vertical thermal gradients (from the water column into the riverbed) have received very little attention in alpine streams (Malard et al., 2001b; Brown et al., 2005, 2006a). Streambed temperatures have been found to be colder and more variable in glacier-fed rivers compared with groundwater streams (Figure 18.4) due to the magnitude of temperature fluctuations in the water column (Brown et al., 2006a). In addition, differences

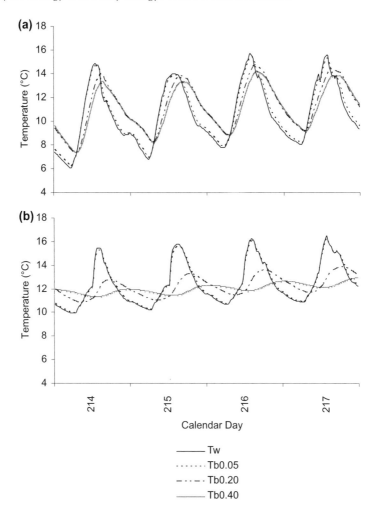

Figure 18.4 Water column (Tw) and streambed (Tb) temperatures (0.05–0.40 m depth) measured in the (a) Taillon glacier stream, and (b) the Tourettes predominantly groundwater-fed tributary between days 214–217, 2003 (modified from Hannah et al., 2007)

in channel sediments and groundwater–surface water interactions may also influence streambed thermal profiles (Malard *et al.*, 2001b). Overall, the thermal heterogeneity of alpine rivers is influenced by the relative volume of different water source contributions, proximity to source (i.e., potential for exposure to atmospheric heating/cooling), and prevailing weather and river flow conditions (Malard *et al.*, 2001b; Uehlinger *et al.*, 2003; Brown *et al.*, 2005; Brown and Hannah, 2007).

18.3.3 Suspended Sediment Concentration

Suspended sediment concentration in alpine glacier-fed streams has been studied since the early 1980s with the aim of determining sediment budgets and quantifying erosion rates to inform geomorphological and applied (e.g., hydroelectric power generation) studies (Gurnell, 1987). Glaciers are the major source of fluvial-transported fine sediment in alpine rivers and, during summer melt periods when subglacial channelised drainage systems are most efficient, glacial rivers often have high suspended sediment concentrations (>500 mg l^{-1}; Gurnell and Clark, 1987; Smith *et al.*, 2001). Spatial and temporal suspended sediment concentration dynamics result from interactions between sediment supply at the glacier bed and connectivity of the stream channel with sediment sources. Additionally, suspended sediment can derive from re-entrainment of transiently stored sediment along the margins of stream channels during peak flows and heavy precipitation events (Richards and Moore, 2003).

In contrast to glacier-fed streams, those sourced predominantly from melting snowpacks transport little suspended sediment (Ward, 1994). Any sediment in rivers sourced from snowmelt is likely to come from entrainment of river bank sediments (Richards and Moore, 2003). Groundwater streams have suspended sediment concentrations close to zero (Ward *et al.*, 1999) in stark contrast to glacier-fed streams (Figure 18.5). However,

Figure 18.5 (See also colour plates) *The confluence of a small karstic groundwater tributary (bottom; width approximately 1 m) and the Taillon glacier stream, French Pyrénées (flow right to left), illustrating the marked contrast in suspended sediment concentrations*

Tockner et al. (1997) describe a predominantly groundwater-fed stream carrying relatively high suspended sediment concentration due to sediment-laden glacial water infiltrating through floodplain alluvial sediments.

18.3.4 Hydrochemistry

In glacierized basins, meltwaters acquire solute from two principal sources: (i) the atmosphere (precipitation and dry deposition) that provides sea salts, acidic aerosols of N and S, plus CO_2 and O_2 to drive chemical weathering reactions; (ii) the chemical weathering of rock in subglacial and ice-marginal environments (Tranter et al., 1996). Hydrochemical characteristics are strongly influenced by seasonal snowmelt, and the development of glacier drainage systems (Hodson et al., 2002; Tranter et al., 2002). Therefore, variability in alpine stream solute concentrations over time reflects changes in water source contributions to streams.

A distinctive early melt season peak in some solutes occurs during snowmelt with preferential elution of SO_4^{2-}, NO_3^- and Cl^- from snowpacks (Johannessen and Henriksen, 1978). The initial release of acid anions (SO_4^{2-} and NO_3^-) often results in a marked decline in proglacial stream pH (e.g., Lepori et al., 2003). Thereafter, melting snowpacks produce waters that dilute more solute-rich subglacial runoff and so result in a characteristic inverse relationship between stream discharge and electrical conductivity (Fenn, 1987). Meltwater solute acquisition is enhanced from glacier comminuted rock flour and an abundance of reactive minerals, resulting in waters enriched in carbonates and sulphates (Tranter et al., 2002). For groundwater streams in alpine river basins, higher solute concentrations are indicative of subterranean water's long residence time (Cooper et al., 2002). Consequently, alpine groundwater streams are often characterized by relatively high electrical conductivity and enrichment of major ions and silica (Ward et al., 1999; Malard et al., 2000; Brown et al., 2006c).

18.4 Biota of Alpine Rivers

Ecological research in alpine river systems has largely focused on longitudinal patterns of biotic communities with respect to distance from the glacier margin, in particular the conceptual model of benthic macroinvertebrate community structure proposed by Milner and Petts (1994). However, following the Arctic and Alpine Stream Ecosystem Research (AASER) initiative (Brittain and Milner, 2001), the biota of alpine streams fed predominantly by snow and groundwater has begun to be studied in more detail. Although physiochemical habitats in alpine rivers exert a major constraint on biotic communities, spatial and temporal habitat heterogeneity resulting from different water sources enables a diverse flora and fauna to colonize and persist (Milner et al., 2001a).

18.4.1 Biota of Alpine Glacier-fed Rivers

Similar longitudinal patterns in benthic macroinvertebrate fauna downstream from glaciers have been documented in Europe, Greenland and New Zealand (Castella et al., 2001; Friberg et al., 2001; Milner et al., 2001a,b) as a function of channel stability, water temperature and inputs of allochthonous organic matter (Figure 18.6). The chironomid genus

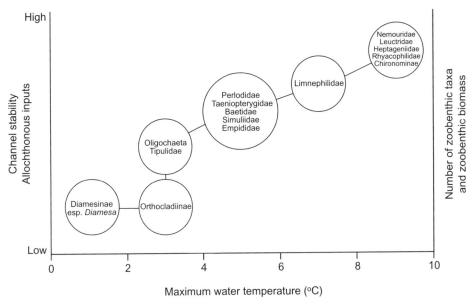

Figure 18.6 *Conceptual model describing the likely first appearance of macroinvertebrate taxa along a longitudinal continuum from the glacier margin for European glacier-fed rivers (modified from Milner et al., 2001a)*

Diamesa typically dominates near the glacier margin where maximum water temperature is <2 °C and river channel stability is low, although Füreder *et al.* (2001) discuss the presence of some Plecoptera, Ephemeroptera and Coleoptera close to the glacier snout in Austrian alpine streams. Some Orthocladiinae (Chironomidae) can tolerate low stability habitats where maximum water temperature reaches 4 °C, with Oligochaeta and Tipulidae colonizing higher stability channels at these temperatures. In New Zealand, *Diamesa* spp. are absent and leptophlebid mayflies (*Delatidium* spp.) are found near to glacier margins at low temperatures (Milner *et al.*, 2001b).

Further from the glacier snout, channel stability and maximum water temperature increase and Ephemeroptera, Plecoptera and Trichoptera taxa become increasingly abundant along with Simuliidae and other chironomids (e.g., Chironominae). Malard *et al.* (2001a, 2003) described longitudinal increases in invertebrate richness with distance from the glacier margin in the Roseg river, Switzerland, closely linked to increased water temperature, particulate organic matter, and upwelling groundwater. However, glacier-fed stream longitudinal (downstream) faunal patterns are influenced by changing physico-chemical habitat conditions, particularly at tributary inflows where biotic diversity may increase but also decrease downstream (Saltveit *et al.*, 2001; Knispel and Castella, 2003). Clear longitudinal changes in species traits are also evident along glacial rivers, with small, crawling, deposit feeders and scrapers dominating invertebrate communities close to the glacier margin before trait diversity increases downstream (Snook and Milner, 2002; Ilg and Castella, 2006).

Alpine glacier-fed streams above the treeline typically receive minimal allochthonous inputs, and retention is low even where inputs are high (Zah and Uehlinger, 2001). Thus, secondary production is dependent on autochthonous energy sources (Bürgi *et al.*, 2003). Primary production is dominated by the chrysophyte *Hydrurus foetidus*, and diatoms belonging to the genera *Fragilaria, Achnanthes, Cymbella* and *Pinnularia* (Uehlinger *et al.*, 1998; Hieber *et al.*, 2001; Brown *et al.*, 2004b; Robinson and Kawecka, 2005; Rott *et al.*, 2006). Clear longitudinal patterns in algal biomass (Uehlinger *et al.*, 1998), chlorophyll *a* (Lods-Crozet *et al.*, 2001) and diatom taxonomic richness (Brown *et al.*, 2004b) are evident, predominantly related to changes in water temperature and channel stability.

18.4.2 Biota of Other (nonglacier-fed) Alpine Rivers

The difference in habitat conditions between streams fed by different water sources plays an important role influencing stream biota. In the Austrian Alps, Chironomidae abundance and total taxonomic richness were found to be 2–8×, and 2–3× higher, respectively, in the groundwater-fed Königsbach river compared with the nearby glacier-fed Rotmoosache river (Füreder *et al.*, 2001). Of the total 126 taxa, 30 were found only in the spring-fed system whereas only six taxa were specific to the glacial river. Brown *et al.* (2006c; 2007a) found higher total macroinvertebrate abundance, diversity and persistence (the relative constancy of taxa presence/absence over time) in the predominantly groundwater-fed Tourettes stream, French Pyrénées, compared with the nearby Taillon glacier stream. However, stream community stability (the similarity of community composition over time with respect to relative abundance) was typically highest in the glacier-fed stream. This suggests that sites with more stable physicochemical habitat variables (e.g., groundwater-fed streams) may not necessarily support the most stable communities, perhaps because species abundance patterns are influenced more by changes in resource availability and biotic interactions (Zah *et al.*, 2001).

Klein and Tockner (2000) identified 15 major groups of macroinvertebrates from different stream types in the Val Roseg floodplain, Switzerland. Heteroptera, Coleoptera, Gastropoda and Bivalvia were restricted to spring-fed streams, and Ostracoda was rarely found, apart from in groundwater channels. Similarly, of five endemic Pyrénéan species found in the Taillon-Gabiétous basin, French Pyrénées, four were strongly associated with groundwater streams (Snook and Milner, 2001). The Plecopteran *Chloroperla breviata* was found exclusively in the groundwater-fed Tourettes stream, upstream of its confluence with the Taillon glacier stream. *Habroleptoides berthelemyi* was mostly found in the Tourettes stream but also occasionally downstream of its confluence with the Taillon glacier stream during periods of limited meltwater production. Together, these studies suggest alpine groundwater springs and streams are important contributors to biodiversity. Furthermore, groundwater-fed streams may be potential sources of macroinvertebrates for the more physically disturbed meltwater-fed streams (Robinson *et al.*, 2004).

Between-stream differences in habitat also appear to be important in determining algal communities in alpine rivers (Rott *et al.*, 2006). Uehlinger *et al.* (1998) suggested that high algal biomass in groundwater streams compared with the adjacent glacier-fed Roseg river, Switzerland, was due to lower suspended sediment concentrations (thus, increased light penetration) and more stable flows reducing stress on benthic organisms. Towards the end of the summer melt season, main channel sites sourced from the Tschierva glacier, Val Roseg, had high relative abundances of *H. foetidus*, whereas groundwater channels were dominated by diatoms (Uehlinger et al., 1998). Groundwater streams on average supported higher total algal biomass and were major sources of autochthonous energy inputs to the Val Roseg glacial river during the summer melt season.

18.4.3 Temporal Variability of Biota in Alpine Rivers

Inter-annual Scale   A comparison of benthic macroinvertebrates in streams of the Macun basin, Swiss National Park, between 2002 and 2003 highlighted clear inter-annual differences in community composition (Rüegg and Robinson, 2004). In a permanent flowing meltwater stream, the macroinvertebrate community was dominated by chironomids in 2002. However, in 2003 the flatworm *Crenobia alpina* (Turbellaria) accounted for ~50% of the community. Overall, assemblage structure appeared to shift forward by approximately three weeks in 2003 due to relatively high water temperature and lower precipitation (i.e., reduced disturbance). Additionally, changes in the life cycle of *Prosimuliium latimucro* (Simuliidae) were observed, with faster developmental rates in 2003.

The negative mass-balance of the Taillon glacier, French Pyrénées (retreat averaging $5.5\,m\,y^{-1}$; Association Moraine Pyrénéenne de Glaciologie, 2003) and earlier snowmelt (inferred from higher snowline altitudes; Hannah et al., 1999b; Brown et al., 2004a) resulted in lower melt season discharge and higher mean water temperature in the Taillon glacier stream in June 2002–03 compared with June 1996–97. Brown et al. (2006c) show that benthic macroinvertebrates responded to these changes, with lower persistence of late-melt season glacier-fed stream communities compared with a nearby groundwater-fed stream. However, the specific reasons for changes in community structure over intermediate years of the study were unclear as stream habitat conditions were not monitored between 1997 and 2002, and winter conditions were unknown due to poor accessibility and logistical problems of reliable winter data collection in mountainous terrain.

Seasonal Scale   The few studies that have considered benthic communities outside the summer melt season have typically adopted a quarterly sampling strategy to examine seasonal variations (Lods-Crozet et al., 2001; Burgherr et al., 2002). Seasonal shifts in benthic community richness, abundance, dispersal and colonization are influenced by the snowmelt and glacier-melt 'flood-pulse' and associated stream habitat variability (Robinson et al., 2001; Schütz et al., 2001; Malard et al., 2006). In major proglacial channels, algal and macroinvertebrate communities may be strongly influenced by changes in water sources, with reduced densities during summer periods of peak glacier-melt (Burgherr et al., 2002; Malard et al., 2006) and higher abundances at other times of the year due to changes in hydrochemistry, suspended sediment concentration and stream discharge/disturbance. Ephemeroptera and Plecoptera only became abundant in the glacier-fed Roseg river, Swiss Alps, during late autumn/early winter (Burgherr et al., 2002). Environmental conditions were suggested to be more favourable for these taxa at this time due to a greater contribution of groundwater.

Seasonal variability in physicochemical variables can open 'windows of opportunity' (Uehlinger et al., 2002) for biotic communities in alpine rivers. During early spring, melting snow provides a relatively constant source of water to streams, suspended sediment concentration is low and elution of nutrients may stimulate primary production (Rott *et al.*, 2006). Harsher conditions occur during peak summer melt (i.e., higher suspended sediment concentration, more variable flows) before relatively benign conditions occur as glacial melt decreases into autumn. More harsh conditions are typical of winter, with low temperature and potentially frozen streams. Biotic abundance and diversity vary seasonally but typically increase when physicochemical habitat variables are more favourable (Milner *et al.*, 2001a). For example, the growth of algae is often greatest following snowmelt in spring, leading to increased abundances of grazers such as Diamesinae and Orthocladiinae (Chironomidae) (Lavandier and Décamps, 1984). More stable year-round patterns in habitat and benthic communities are found in glacial streams when groundwater contributions are high (Füreder *et al.*, 2001; Burgherr *et al.*, 2002).

The strength of seasonal fluctuations in stream physicochemical habitat can be envisaged in the context of the harsh–benign concept of Peckarsky (1983) and along an additional gradient of water sources from meltwater to groundwater (Figure 18.7). In predominantly meltwater-fed streams, 'windows of opportunity' occur between the harsh summer and winter periods. In these 'windows', habitat conditions become more benign, biodiversity and production increases and biotic interactions may be more important. However, habitat conditions are typically more benign as meltwater contributions decline at the expense of groundwater even during the summer and winter, facilitating increased diversity and abundance year round.

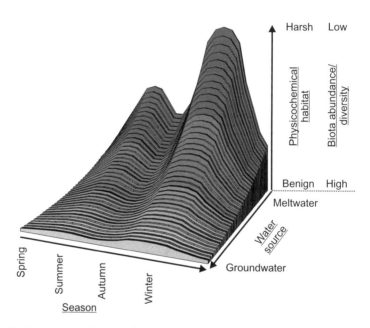

Figure 18.7 Conceptual diagram illustrating stream biotic diversity and abundance along a harsh–benign physicochemical habitat gradient, as a function of time (season) and water source contributions

18.5 Towards an Integrated Hydroecological Understanding of Alpine River Systems

Distinguishing water source contributions to alpine rivers clearly has implications for understanding physicochemical habitat variability and the structure and function of stream flora and fauna. Early approaches linking water sources and stream ecology classified streams into three categories, based upon the perceived primary water source and water temperature (Ward, 1994). Streams fed by glacial meltwaters were termed 'kryal', by snowmelt named 'rhithral' and groundwater streams classified as 'krenal'. This approach has recently been refined because streams are rarely fed from one distinct source (Brown et al., 2003). For example, large quantities of what is considered 'glacial' runoff by stream ecologists is derived hydrologically from both snow- and glacial ice-melt during the summer melt season (Fountain and Walder, 1998). Furthermore, with increasing distance from glaciers, tributary and/or upwelling inputs of groundwater often supplement melt-water contributions to traditional 'kryal' streams (Brown et al., 2006c), changing the physicochemical habitat.

This modified approach to understanding hydroecological relationships in alpine streams is based on quantification of relative water source contributions to streamflow (Brown et al., 2003; 2006c). Streams are classified into nine categories based upon their proportional water source contributions (Figure 18.8). These classification categories relate water source contributions with physicochemical habitat variables such as stream discharge, suspended sediment concentration and hydrochemistry (Table 18.1). By providing quantitative explanations of the stream 'type' as determined by proportions of water sources, this approach seeks to facilitate more accurate comparison between (i) stream reaches and (ii) river basins (in space), and also (iii) changes over time.

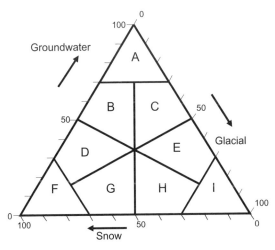

Figure 18.8 Alternative alpine stream habitat classification based upon water source contributions (modified from Brown et al., 2003). See Table 18.1 for description of classes A–I

Table 18.1 Characteristic physicochemical variables predicted for the alpine stream classification categories shown in Figure 18.8 (modified from Brown et al., 2003)

Category	Suspended sediment concentration	Hydrochemistry	Flow variation
A: Krenal	Constantly very low <10 mg L^{-1}	High silica, bicarbonate and sulfate	Constant flow with little diurnal variation
B: Kreno-nival	Low but >krenal as snowmelt may contribute a small amount	High silica (<krenal) and relatively high chloride (<nival)	Relatively constant flow, small diurnal variation from snowmelt input
C: Kreno-kryal	Low concentration due to groundwater dilution. Small diurnal variation likely	High silica (<krenal) and relatively high calcium and sulfate (<kryal)	Baseflow relatively constant but glacial influence may cause diurnal variation
D: Nivo-krenal	Low but slightly > kreno-nival due to greater proportion of snowmelt	High chloride (<nival) and relatively high silica (<krenal)	Diurnal variations due to snow input (>kreno-nival)
E: Kryo-krenal	Relatively high as glacial source dominant. <kryal	High sulfate (<kryal) and silica (<krenal)	Glacial source dominant so fairly pronounced diurnal variation
F: Nival	Low but > krenal. Very low diurnal variation. Possible increase when high flows	High chloride and nitrate from snowpack elution, low silica	Diurnal cycle <kryal. Building towards late afternoon/early evening.
G: Nivo-kryal	Elevated levels compared with nival due to glacial influence. Diurnal variations	High chloride as snow dominant (<nival) Relatively high sulfate. Low silica.	Diurnal peak intermediate between nival and kryal. <kryo-nival as snow dominant
H: Kryo-nival	High but < kryal. Pronounced diurnal variations	High sulfate as glacial meltwater source dominates Relatively high chloride and nitrate	Pronounced diurnal peak intermediate between nivo-kryal and kryal
I: Kryal	Very high during peak flows with large diurnal range	High sulfate and calcium, low chloride and silica.	Large diurnal cycle related to solar radiation input. Peak late afternoon

The approach has successfully illustrated temporal and longitudinal changes in water source contributions in the Taillon basin, French Pyrénées in 1996, 1997, 2002 and 2003 (Brown *et al.*, 2003; 2006c; 2007b) but requires validation in other alpine basins. Ongoing studies comparing water source contributions with stream biota in the Taillon-Gabiétous basin indicate significant increases in invertebrate richness associated with greater proportions of groundwater contribution to streamflow (Figure 18.9). The increase in taxonomic

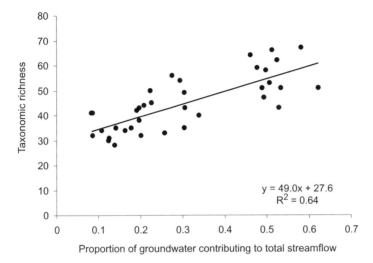

Figure 18.9 *Relationship between proportion of groundwater contributing to total streamflow and taxonomic richness (P < 0.01). Data were collected from the predominantly groundwater-fed Tourettes stream and two sites on the Taillon glacier stream, French Pyrénées (see Brown et al., 2006c for site details)*

richness is related to more favourable stream environments for invertebrates as groundwater contributions increase. By accurately quantifying water source contributions of alpine streams and linking these with physiochemical and ecological data, this approach should allow more accurate prediction of the effects of any future changes in alpine river water sources.

18.6 Conclusions and Future Research Directions

Many advances in understanding the hydrology and ecology of glacierized alpine river systems have been made during the last two decades. It is only recently that hydrologists and ecologists have started working together to develop a more complete understanding of how these rivers function. From the synthesis of current hydroecological research in alpine streams, several important research gaps emerge:

(i) Most studies of alpine rivers have focused on the summer melt season due to logistical difficulties of accessing remote alpine basins, and maintaining reliable data collection, particularly during winter. Relationships between alpine stream water sources, physicochemical habitat and benthic macroinvertebrate communities based only on this short time-frame clearly do not provide a complete picture of how these rivers function over longer time-scales (Brown *et al.*, 2006c). Given the potential for significant changes, in alpine basin hydrological functioning under scenarios of future climate change, and despite the logistics and additional costs, year-round studies are required to understand fully how stream community composition and diversity are related to seasonal shifts in water sourcing and physicochemical habitat conditions. A potential approach for integrating water sourcing – stream habitat – stream ecology in alpine glacierized basins has been outlined but requires more widespread testing (Brown et al., 2003).

(ii) In terrestrial alpine ecosystems, plant and animal distributions are rapidly changing in response to climate change (Krajick, 2004). To better understand the implications of these changes, a worldwide network (GLobal Observation Research Initiative in Alpine environments; GLORIA) has been established to track long-term changes in terrestrial plant community biodiversity and distribution (Grabherr et al., 2000). The initiative provides intensive data collection at a network of 'master' monitoring sites that are supplemented by baseline data from a worldwide multi-site network. The data sets from this initiative should facilitate large-scale comparisons of vegetation change in alpine environments, and will form a basis for modelling climate change effects on alpine vegetation. We suggest that a similar coordinated network of monitoring sites is essential for alpine river systems, because continuous estimates for mountain glaciers predominantly show decreases in mass balance over the last century and suggest that glacier retreat may even be accelerating in some regions (IAHS (ICSI)–UNEP–UNESCO, 2005). This network would provide continuous, long-term data sets (>decadal with continuous year-to-year hydroecological monitoring) to identify the importance of any changes in the timing and magnitude of meltwater runoff and associated stream habitat conditions for stream communities (Brown et al., 2007b).

(iii) Despite the hypothesised importance of the streambed (hyporheic zone) as a refuge from harsh, variable habitat conditions during peak summer melt, and for over-wintering to avoid frozen conditions (e.g., Lods-Crozet et al., 2001; Milner et al., 2001a; Brown et al., 2006b), virtually no data exists to our knowledge regarding the faunal composition and dynamics of alpine hyporheic zones. Only limited reference has been made to the presence of permanent aquatic taxa, including Cyclopoida, Harpacticoida and Ostracoda (Malard et al., 2001a, 2003). Furthermore, as there have been few concurrent studies of benthic and streambed faunal abundance and diversity except those in the Val Roseg, Switzerland (Malard et al., 2001a, 2003), very limited information exists to assess comprehensively how streambed communities contribute to alpine river biodiversity. The streambed typically has warmer and less variable water temperatures than the water column (Brown et al., 2005), sediments are subject to less frequent disturbance, and particulate organic matter (an important food source for invertebrates) may accumulate in interstices (Malard et al., 2003). We hypothesize that the streambed acts as an important habitat in alpine rivers for meiofauna, which are unable to maintain large populations in the benthos, and invertebrates as a refuge from the harsher surface habitat conditions.

(iv) Our present understanding of alpine stream ecosystems relates mainly to the influence of habitat on benthic macroinvertebrates (Figure 18.10a) or primary producers (Figure 18.10b). The role of biotic interactions in alpine river systems remains unclear. By hypothesizing how feeding interactions fit into alpine river ecosystems, more complex scenarios of habitat influencing producers with subsequent influence on macroinvertebrate communities can be envisaged (Figure 18.10c), or habitat influencing macroinvertebrates with subsequent influences on producers (Figure 18.10d). Nevertheless, a more likely hypothesis is that habitat influences both pro-

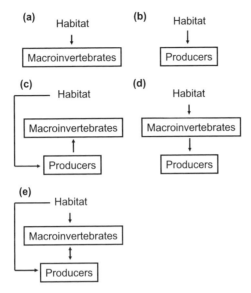

Figure 18.10 Potential habitat–producer–macroinvertebrate relationships: (a) stream habitat-macroinvertebrate (Milner et al., 2001a), and (b) stream habitat–producer relationships (Lods-Crozet et al., 2001; Brown et al., 2004b); and more complex hypothesized: (c) habitat–producer–macroinvertebrate interactions; (d) habitat–macroinvertebrate–producer interactions, and (e) bidirectional interactions

ducers and consumers and biotic interactions are variable (even bidirectional) in space and time (Figure 18.10e). Given the important influence of temporal changes in glacierized catchment water sourcing and related habitat characteristics throughout the year (Malard *et al.*, 2006), this is likely to result in shifts between bottom-up (producer-driven) and top-down (consumption-driven) interactions. For example, primary producers often increase in abundance when early season snowpack nutrient elution occurs, or during winter when disturbance is less frequent (Rott *et al.*, 2006), potentially providing more resources for consumers (Zah *et al.*, 2001; Füreder *et al.*, 2003). Outside of these periods (i.e., harsh periods; Figure 18.7), primary producers may be limited by nutrients and controlled more by invertebrate consumption. Because we aspire to use alpine stream communities as indicators of glacier fluctuations, and shifts in species distributions can be complicated by biotic interactions (Davis *et al.*, 1998), data are clearly necessary to test these hypotheses. There is also a pressing need to understand the role of fungi and bacteria in alpine stream food webs (Gessner *et al.*, 1998; Battin *et al.*, 2004; Robinson and Jolidon, 2005).

This chapter has summarized some of the important linkages between hydrology and ecology for alpine river systems. As a consequence of the strong coupling between climate and hydrology in these systems, and in most cases relatively low anthropogenic influence, these rivers provide ideal environments for studying climate change/variability effects on river ecosystems. Although the structure of benthic macroinvertebrate communities is now relatively well known in alpine glacier-fed streams, their function and response to changing dynamic water source contributions requires further investigation. Given the potential for significant changes in alpine catchment hydrology under future scenarios of climate change, it is important that ecological response can be predicted. We anticipate that this synthesis of current alpine river system understanding will enable further bridging of gaps between hydrological and ecological sciences, and potentially stimulate further research on alpine river systems.

Acknowledgements

We are grateful to Leopold Füreder and Christopher Robinson for providing helpful reviews that improved the final chapter.

References

Association Moraine Pyrénéenne de Glaciologie. 2003. *Reconstitution des variations frontales de trois glaciers Pyrénéens depuis la fin du Petit Age Glaciaire (1850)*. Association Moraine Pyrénéenne de Glaciologie.

Battin, T. J., Wille, A., Psenner, R., and Richter, A. 2004. Large-scale environmental controls on microbial biofilms in high-alpine streams. *Biogeosciences*, **1**: 159–171.

Beniston, M. 2000. *Environmental Change in Mountains and Uplands*. London: Arnold.

Brittain, J. E. and Milner, A. M. 2001. Ecology of glacier-fed rivers: current status and concepts. *Freshwater Biology*, **46**: 1571–1578.

Brittain, J. E., Adalsteinsson, H., Castella, E., Mar Gislason, G., Lencioni, V., Lods-Crozet, B., Maiolini, B., Milner, A. M., Petts, G. E., and Saltveit, S. J. 2001. Towards a conceptual understanding of arctic and alpine streams. *Verhandlungen International Vereinigung Limnologie*, **27**: 740–743.

Brown, L. E. and Hannah, D. M. 2007. Alpine stream temperature response to storm events. *Journal of Hydrometeordogy*, **8**: 952–967.

Brown, L. E., Hannah, D. M., and Milner, A. M. 2003. Alpine stream habitat classification: An alternative approach incorporating the role of dynamic water source contributions. *Arctic Antarctic and Alpine Research*, **35**: 313–322.

Brown, L. E., Hannah, D. M., and Milner, A. M. 2004a. Alpine stream temperature variability and potential implications for benthic communities. In Webb, B. W., Acreman, M. C., Maksimovic, C., Smithers, H., and Kirby, C. (eds) *Hydrology: Science and Practice for the 21st Century*. London: British Hydrological Society, pp. 28–38.

Brown, L. E., Sherlock, C., Milner, A. M., Hannah, D. M., and Ledger, M. E. 2004b. Longitudinal distribution of diatom communities along a proglacial stream in the Taillon-Gabiétous catchment, French Pyrénées. *Pirineos*, **158/159**: 73–86.

Brown, L. E., Hannah, D. M., and Milner, A. M. 2005. Spatial and temporal water column and streambed temperature dynamics within an alpine catchment: Implications for benthic communities. *Hydrological Processes*, **19**: 1585–1610.

Brown, L. E., Hannah, D. M., and Milner, A. M. 2006a. Hydroclimatological influences upon water column and streambed temperature dynamics in an alpine river system. *Journal of Hydrology*, **325**: 1–20.

Brown, L. E., Hannah, D. M., and Milner, A. M. 2006b. Thermal variability and streamflow permanency in an alpine river system. *River Research and Applications*, **22**: 493–501.

Brown, L. E., Milner, A. M., and Hannah, D. M. 2006c. Stability and persistence of alpine stream macroinvertebrate communities and the role of physicochemical habitat variables. *Hydrobiologia*, **560**: 159–173.

Brown, L. E., Hannah, D. M., Milner, A. M., Soulsby, C., Hodson, A., and Brewer, M. J. 2006c. Water source dynamics in an alpine glacierized river basin (Taillon-Gabiétous, French Pyrénées). *Water Resources Research*, **42**: W08404.

Brown, L. E., Milner, A. M., and Hannah, D. M. 2007a. Groundwater influence on alpine stream ecosystems. *Freshwater Biology*, **52**: 878–890.

Brown, L. E., Hannah, D. M., and Milner, A. M. 2007b. Vulnerability of alpine stream biodiversity to shrinking glaciers and snowpacks. *Global Change Biology*, **13**: 958–966.

Burgherr, P., Ward, J. V., and Robinson, C. T. 2002. Seasonal variation in zoobenthos across habitat gradients in an alpine glacial floodplain (Val Roseg, Swiss Alps). *Journal of the North American Benthological Society*, **21**: 561–575.

Bürgi, H. R., Burgherr, P., and Uehlinger, U. 2003. Aquatic Flora. In Ward, J. V. and Uehlinger, U. (eds) *Ecology of a Glacial Flood Plain*. Dordrecht, Netherlands: Kluwer Academic Publishers, pp. 139–151.

Castella, E., Adalsteinsson, H., Brittain, J. E., Gislason, G. M., Lehmann, A., Lencioni, V., Lods-Crozet, B., Maiolini, B., Milner, A. M., Olafsson, J. S., Saltveit, S. J., and Snook, D. L. 2001. Macrobenthic invertebrate richness and composition along a latitudinal gradient of European glacier-fed streams. *Freshwater Biology*, **46**: 1811–1831.

Cooper, R. J., Wadham, J. L., Tranter, M., Hodgkins, R., and Peters, N. E. 2002. Groundwater hydrochemistry in the active layer of the proglacial zone, Finsterwalderbreen, Svalbard. *Journal of Hydrology*, **269**: 208–223.

Davis, A. J., Jenkinson, L. S., Lawton, J. H., Shorrocks, B., and Wood, S. 1998. Making mistakes when predicting shifts in species range in response to global warming. *Nature*, **391**: 783–786.

Fenn, C. R. 1987. Electrical Conductivity. In Gurnell, A. M. and Clark, M. J. (eds) *Glacio-fluvial Sediment Transfer*. Chichester, UK: John Wiley & Sons, Ltd, pp. 377–414.

Flerchinger, G. N., Cooley, K. R., and Ralston, D. R. 1992. Groundwater response to snowmelt in a mountainous watershed. *Journal of Hydrology*, **133**: 293–311.

Fountain, A. G. 1996. Effect of snow and firn hydrology on the physical and chemical characteristics of glacial runoff. *Hydrological Processes*, **10**: 509–521.

Fountain, A. G. and Walder, J. S. 1998. Water flow through temperate glaciers. *Reviews of Geophysics*, **36**: 299–328.

Fountain, A. G., Jacobel, R. W., Schlichting, R., and Jansson, P. 2005. Fractures as the main pathways of water flow in temperate glaciers. *Nature*, **433**: 618–621.

Friberg, N., Milner, A. M., Svendsen, L. M., Lindegaard, C., and Larsen, S. E. 2001. Macroinvertebrate stream communities along regional and physico-chemical gradients in Western Greenland. *Freshwater Biology*, **46**: 1753–1764.

Füreder, L., Schütz, C., Wallinger, M., and Burger, R. 2001. Physico-chemistry and aquatic insects of a glacier-fed and a spring-fed alpine stream. *Freshwater Biology*, **46**: 1673–1690.

Füreder, L., Welter, C., and Jackson, J. K. 2003. Dietary and stable isotope ($d^{13}C$, $d^{15}N$) analyses in alpine stream insects. *International Review of Hydrobiology*, **88**: 314–331.

Gessner, M. O., Robinson, C. T., and Ward, J. V. 1998. Leaf breakdown in streams of an alpine glacial floodplain: dynamics of fungi and nutrients. *Journal of the North American Benthological Society*, **17**: 403–419.

Grabherr, G., Gottfried, M., and Pauli, H. 2000. GLORIA: A global observation research initiative in Alpine environments. *Mountain Research and Development*, **20/2**: 190–192.

Gurnell, A. M. 1987. Suspended sediment. In Gurnell, A. M. and Clark, M. J. (eds) *Glacio-fluvial Sediment Transfer*. Chichester UK: John Wiley & Sons, Ltd, pp. 305–354.

Gurnell, A. M. and Clark, M. J. 1987. *Glacio-fluvial Sediment Transfer*. Chichester: John Wiley & Sons, Ltd.

Hannah, D. M. and Gurnell, A. M. 2001. A conceptual linear reservoir runoff model to investigate melt season changes in cirque glacier hydrology. *Journal of Hydrology*, **246**: 123–141.

Hannah, D. M. and McGregor, G. 1997. Evaluating the impact of climate on snow and ice melt in the Taillon basin, French Pyrénées. *Journal of Glaciology*, **43**: 563–568.

Hannah, D. M., Gurnell, A. M., and McGregor, G. R. 1999a. Identifying links between large-scale atmospheric circulation and local glacier ablation climates in the French Pyrénées. *IAHS Publication 256*, pp. 155–164.

Hannah, D. M., Gurnell, A. M., and McGregor, G. R. 1999b. A methodology for investigation of the seasonal evolution in proglacial hydrograph form. *Hydrological Processes*, **13**: 2603–2622.

Hannah, D. M., Gurnell, A. M., and McGregor, G. R. 2000a. Spatio-temporal variation in microclimate, the surface energy balance and ablation over a cirque glacier. *International Journal of Climatology*, **20**: 733–758.

Hannah, D. M., Smith, B. P., Gurnell, A. M., and McGregor, G. R. 2000b. An approach to hydrograph classification. *Hydrological Processes*, **14**: 317–338.

Hannah, D. M., Brown, L. E., Milner, A. M., Gurnell, A. M., McGregor, G. R., Petts, G. E., Smith, B. P. G., and Snook, D. L. 2007. Integrating climate–hydrology–ecology for alpine river systems. *Aquatic Conservation: Marine and Freshwater Ecosystems*. **17**: 636–656.

Hieber, M., Robinson, C. T., Rushforth, S. R., and Uehlinger, U. 2001. Algal communities associated with different alpine stream types. *Arctic Antarctic and Alpine Research*, **33**: 447–456.

Hodson, A. J., Gurnell, A. M., Washington, R., Tranter, M., Clark, M. J., and Hagen, J. O. 1998. Meteorological and runoff time series characteristics in a small, high-Arctic glaciated basin, Svalbard. *Hydrological Processes*, **12**: 509–526.

Hodson, A., Tranter, M., Gurnell, A., Clark, M., and Hagen, J. O. 2002. The hydrochemistry of Bayelva, a high Arctic proglacial stream in Svalbard. *Journal of Hydrology*, **257**: 91–114.

IAHS (ICSI)–UNEP–UNESCO. 2005. *Fluctuations of Glaciers VIII.* Paris: IAHS (ICSI)–UNEP–UNESCO.

Ilg, C. and Castella, E. 2006. Patterns of macroinvertebrate traits along three glacial stream continuums. *Freshwater Biology*, **51**: 840–853.

Johannessen, M. and Henriksen, A. 1978. Chemistry of snow meltwater: changes in concentration during melting. *Water Resources Research*, **14**: 615–619.

Klein, P. and Tockner, K. 2000. Biodiversity in alpine springs of a glacial floodplain (Val Roseg, Switzerland). *Verhandlungen International Vereinigung Limnologie*, **27**: 704–710.

Knispel, S. and Castella, E. 2003. Disruption of a longitudinal pattern in environmental factors and benthic fauna by a glacial tributary. *Freshwater Biology*, **48**: 604–618.

Körner, C. 2003. *Alpine Plant Life*. Berlin: Springer-Verlag.

Krajick, K. 2004. All downhill from here? *Science*, **303**: 1601–1602.

Lavandier, P. and Décamps, H. 1984. Estaragne. In Whitton, B. A. (ed.) *Ecology of European Rivers*. Oxford, UK: Blackwell Scientific, pp. 237–264.

Lepori, F., Barbieri, A., and Ormerod, S. J. 2003. Effects of episodic acidification on macroinvertebrate assemblages in Swiss Alpine streams. *Freshwater Biology*, **48**: 1873–1885.

Lods-Crozet, B., Castella, E., Cambin, D., Ilg, C., Knispel, S., and Mayor-Simeant, H. 2001. Macroinvertebrate community structure in relation to environmental variables in a Swiss glacial stream. *Freshwater Biology*, **46**: 1641–1662.

Malard, F., Tockner, K., and Ward, J. V. 1999. Shifting dominance of subcatchment water sources and flow paths in a glacial floodplain, Val Roseg, Switzerland. *Arctic Antarctic and Alpine Research*, **31**: 135–150.

Malard, F., Tockner, K., and Ward, J. V. 2000. Physico-chemical heterogeneity in a glacial riverscape. *Landscape Ecology*, **15**: 679–695.

Malard, F., Lafont, M., Burgherr, P., and Ward, J. V. 2001a. A comparison of longitudinal patterns in hyporheic and benthic Oligochaete assemblages in a glacial river. *Arctic Antarctic and Alpine Research*, **33**: 457–466.

Malard, F., Mangin, A., Uehlinger, U., and Ward, J. V. 2001b. Thermal heterogeneity in the hyporheic zone of a glacial floodplain. *Canadian Journal of Fisheries and Aquatic Sciences*, **58**: 1319–1335.

Malard, F., Galassi, D., Lafont, M., Doledec, S., and Ward, J. V. 2003. Longitudinal patterns of invertebrates in the hyporheic zone of a glacial river. *Freshwater Biology*, **48**: 1709–1725.

Malard, F., Uehlinger, U., Zah, R., and Tockner, K. 2006. Flood-pulse and riverscape dynamics in a braided glacial river. *Ecology*, **87**: 704–716.

Milner, A. M. and Petts, G. E. 1994. Glacial rivers: physical habitat and ecology. *Freshwater Biology*, **32**: 295–307.

Milner, A. M., Brittain, J. E., Castella, E., and Petts, G. E. 2001a. Trends of macroinvertebrate community structure in glacier-fed rivers in relation to environmental conditions: a synthesis. *Freshwater Biology*, **46**: 1833–1848.

Milner, A. M., Taylor, R. C., and Winterbourn, M. J. 2001b. Longitudinal distribution of macroinvertebrates in two glacier-fed New Zealand rivers. *Freshwater Biology*, **46**: 1765–1776.

Oerlemans, J. 2005. Extracting a climate signal from 169 glacier records. *Science*, **308**: 675–677.

Peckarsky, B. L. 1983. Biotic interactions or abiotic limitations? A model of lotic community structure. In Fontaine, T. D. and Bartell, S. M. (eds) *Dynamics of Lotic Ecosystems* Michigan: Ann Arbor Science, pp. 303–323.

Remmert, H. 1980. *Arctic Animal Ecology*. Berlin: Springer-Verlag.

Richards, G. and Moore, R. D. 2003. Suspended sediment dynamics in a steep, glacier-fed mountain stream, Place Creek, Canada. *Hydrological Processes*, **17**: 1733–1753.

Robinson, C. T. and Jolidon, C. 2005. Leaf breakdown and the ecosystem functioning of alpine streams. *Journal of the North American Benthological Society*, **24**: 495–507.

Robinson, C. T. and Kawecka, B. 2005. Benthic diatoms of an Alpine stream/lake network in Switzerland. *Aquatic Sciences*, **67**: 492–506.

Robinson, C. T., Uehlinger, U., and Hieber, M. 2001. Spatio-temporal variation in macroinvertebrate assemblages of glacial streams in the Swiss Alps. *Freshwater Biology*, **46**: 1663–1672.

Robinson, C. T., Tockner, K., and Burgherr, P. 2004. Drift benthos relationships in the seasonal colonization dynamics of alpine streams. *Archiv für Hydrobiologie*, **160**: 447–470.

Rott, E., Cantonati, M., Füreder, L., and Pfister, P. 2006. Benthic algae in high altitude streams of the Alps – a neglected component of the aquatic biota. *Hydrobiologia*, **562**: 195–216.

Rüegg, J. and Robinson, C. T. 2004. Comparison of macroinvertebrate assemblages of permanent and temporary streams in an Alpine flood plain, Switzerland. *Archiv für Hydrobiologie*, **161**: 489–510.

Saltveit, S. J., Haug, L., and Brittain, J. E. 2001. Invertebrate drift in a glacial river and its non-glacial tributary. *Freshwater Biology*, **46**: 1777–1789.

Schütz, C., Wallinger, M., Burger, R., and Füreder, L. 2001. Effects of snow cover on the benthic fauna in a glacier-fed stream. *Freshwater Biology*, **46**: 1691–1704.

Smith, B. P. G., Hannah, D. M., Gurnell, A. M., and Petts, G. E. 2001. A hydrogeomorphological context for ecological research on alpine glacial rivers. *Freshwater Biology*, **46**: 1579–1596.

Snook, D. L. and Milner, A. M. 2001. The influence of glacial runoff on stream macroinvertebrate communities in the Taillon catchment, French Pyrénées. *Freshwater Biology*, **46**: 1609–1623.

Snook, D. L. and Milner, A. M. 2002. Biological traits of macroinvertebrates and hydraulic conditions in a glacier-fed catchment (French Pyrénées). *Archiv für Hydrobiologie*, **153**: 245–271.

Swift, D. A., Nienow, P. W., Hoey, T. B., and Mair, D. W. F. 2005. Seasonal evolution of runoff from Haut Glacier d'Arolla, Switzerland and implications for glacial geomorphic processes. *Journal of Hydrology*, **309**: 133–148.

Tockner, K., Malard, F., Burgherr, P., Robinson, C. T., Uehlinger, U., Zah, R., and Ward, J. V. 1997. Physico-chemical characterization of channel types in a glacial floodplain ecosystem (Val Roseg, Switzerland). *Archiv für Hydrobiologie*, **140**: 433–463.

Tranter, M., Brown, G. H., Hodson, A. J., and Gurnell, A. M. 1996. Hydrochemistry as an indicator of subglacial drainage system structure: a comparison of alpine and sub-polar environments. *Hydrological Processes*, **10**: 541–556.

Tranter, M., Sharp, M. J., Lamb, H. R., Brown, G. H., Hubbard, B. P., and Willis, I. C. 2002. Geochemical weathering at the bed of Haut Glacier d'Arolla, Switzerland – a new model. *Hydrological Processes*, **16**: 959–994.

Uehlinger, U., Zah, R., and Bürgi, H. 1998. The Val Roseg project: temporal and spatial patterns of benthic algae in an Alpine stream ecosystem influenced by glacier runoff. *IAHS Publication* 248. pp. 419–424.

Uehlinger, U., Tockner, K., and Malard, F. 2002. Ecological windows in glacial stream ecosystems. *EAWAG News*, **54a**: 20–21.

Uehlinger, U., Malard, F., and Ward, J. V. 2003. Thermal patterns in the surface waters of a glacial river corridor (Val Roseg, Switzerland). *Freshwater Biology*, **48**: 284–300.

Ward, J. V. 1994. Ecology of alpine streams. *Freshwater Biology*, **32**: 277–294.

Ward, J. V., Malard, F., Tockner, K., and Uehlinger, U. 1999. Influence of ground water on surface water conditions in a glacial floodplain of the Swiss Alps. *Hydrological Processes*, **13**: 277–294.

Zah, R., Burgherr, P., Bernasconi, S. M., and Uehlinger, U. 2001. Stable isotope analysis of macroinvertebrates and their food sources in a glacier stream. *Freshwater Biology*, **46**: 871–882.

Zah, R. and Uehlinger, U. 2001. Particulate organic matter inputs to a glacial stream ecosystem in the Swiss Alps. *Freshwater Biology*, **46**: 1597–1608.

19
Fluvial Sedimentology: Implications for Riverine Ecosystems

Gregory H. Sambrook Smith

19.1 Introduction

Much work investigating links between river properties and ecology has taken a predominantly reach-scale perspective, such as the River Continuum Concept (Vanotte *et al.*, 1980), which describes how river ecology varies from upstream to downstream. Linked with this has been the view that it is the broad characteristics of the flow regime (Poff *et al.*, 1997; Bunn and Arthington, 2002) that form the key physical control on riverine ecology. Another influential idea in riverine ecology, the Flood Pulse Concept (Junk *et al.*, 1989; see also Tockner *et al.*, 2000), also has the flow regime as its key driver. These ideas, and subsequent modifications of them, have undoubtedly led to a greatly enhanced understanding of how river ecology both influences and is influenced by its physical environment. However, there has been relatively little research conducted at the bar scale (i.e., exposed riverine sediments) that investigates specific links of ecology with bar geomorphology and, in particular, sedimentology (i.e., physical habitat dynamics). Even where the influence of sedimentology is addressed (e.g., Claret *et al.*, 1997), this is often dealt with in a qualitative sense (e.g., fine or coarse sediment, presence or absence of fines) that relates overwhelmingly to just surface grain size. There is little appreciation of the 4-D heterogeneity of barforms; thus, from a landscape ecology perspective, only certain parts of the patch hierarchy have been adequately investigated. For example, Poole (2002) describes how river systems can be viewed as a hierarchy of patches of different spatial scale linked by trans-scale processes (physical and biological). The patches within the hierarchy can thus be influenced by either top-down or bottom-up processes. It is generally only the top-down processes that are synthesised in current ecological models, as Poole (2002, p. 649) states '... bottom-up influences in the physical hierarchy and biological feedback that affect stream structure are not addressed by the river continuum concept, serial discontinuity concept, flood pulse concept or hyporheic corridor concept'. Additionally, as reviewed by Sadler *et al.* (2004), given that bars have high productivity, species richness, fidelity and rarity, there is thus an urgent need for a more detailed understanding of how the interactions between ecology (primarily riparian plants and insects), geomorphology and sedimentology operate.

Bars (defined here as having lengths proportional to channel width and heights similar to the flow depth; ASCE, 1966) represent a fundamental element of rivers that are both controlled by external factors (e.g., sediment supply, discharge regime, valley topography) as well as themselves controlling within channel processes (e.g., water and sediment routing). This chapter will discuss the origins of bar formation, migration and modification. The discussion relates primarily to those barforms that are exposed for most of the time but suffer periodic submersion during flood events, rather than those that are submerged for most of the time and only exposed during droughts. Particular emphasis will be given to explaining the processes of change and the causes of change at a range of spatial and temporal scales from minor modifications to bar surfaces during an event to major changes to bar complexes caused by avulsions. Additionally, the sedimentology of bars and their relationship to bar formation, migration and modification will be addressed. The detailed characteristics of bars, including their sedimentology is relevant to ecologists for the following reasons; (i) grain size variability will dictate hydraulic conductivity and hence moisture content; (ii) small spatial and temporal changes in grain size across a bar surface during a flood event can lead to significant changes in shear stress on macroinvertebrates; (iii) bar evolution and modification can result in significant changes in bar elevation and hence in depth to water table and bar edge length, which will also influence water and nutrient availability, and (iv) patterns of bar erosion, deposition and channel avulsion will control the frequency and extent of habitat disturbance and habitat creation.

19.2 The Sedimentology of Barforms

19.2.1 Ecological Implications

Studies of the relationship between grain size variability and ecological functioning have been dominated by within-channel considerations, such as the assessment of bed grain size distribution for salmonid spawning (e.g., Kondolf, 2000) or general downstream trends of how grain size may influence macroinvertebrate fauna (e.g., Rice *et al.*, 2001a, b). The role of barforms has received much less attention, as emphasised by Ward *et al.* (1999, p. 71) who stated that 'The role of islands has been almost totally ignored by stream ecologists'. However, grain size variability within and on the surface of barforms is also important. Firstly, changes in grain size and sorting dictate the porosity, permeability and hence hydraulic conductivity within the bar sediments. This will have an important impact upon plant ecology by determining moisture retention and thus seed germination (Hughes *et al.*, 2000) as well as the subsequent levels of stress the plant will experience.

For example, Francis *et al.* (2006) highlight the importance of grain size and how this varies between species in their study on the braided Tagliamento river, Italy. They found that *Salix elaeagnos* seedlings had a preference for coarse (>2 mm) bar surfaces and *Populus nigra* for fine (<2 mm). Such trends are missed when research is carried out at a larger scale, an important oversight given the importance of both species for plant communities along the Tagliamento. Hughes *et al.* (2000) highlight how even the response of the different sexes of the same plant species may be influenced by grain size. For example, they found that the females tended to favour wetter sites than the males, which were more tolerant of free draining gravel-rich substrates. Likewise, the influence of grain size on hydraulic conductivity is important for insect populations as it will determine the likelihood of desiccation of invertebrates. For example, Bates *et al.* (in press), in their study of the microspatial distribution of beetles on bars, found that *Bembidion punctulatum* was significantly positively associated with finer sediment.

Secondly, during flood events bar surfaces will be inundated. Surface roughness across a bar will impact on the flow structure and hence the shear stress acting on small-scale fauna (e.g., Silvester and Sleigh, 1985; Davis and Barmuta, 1989; Carling, 1992). The precise grain size distribution and structure across the bar surface thus becomes important in terms of dictating sites of flow refugia (e.g., Lancaster, 1999), in a similar way as for within the channel.

19.2.2 Grain Size and Sorting

According to one of the most often cited models of bar initiation (Bridge, 1993) all bars, whether point or braid bars, have a common origin. Bar initiation is based on the evolution of lobate fronted (generally submerged at most flows) barforms arranged on alternating sides of the channel, either as a single row for evolution into point bars or a double row for evolution into braid bars. These barforms themselves will be composed of stacked ripples/dunes (sand bed) or bedload sheets (gravel bed). Flume experiments, based on a gravel bed river (Ashmore, 1991) corroborated the presence of alternating bars formed by bedload sheets and their subsequent evolution into fully emergent bar forms. As the bar grows it triggers bank erosion, causing the channel to increase in width and the bar form to become more emergent. These bars are called unit bars and are defined (after Smith, 1978) as having a shape that remains relatively unmodified during migration, and being simple forms that are not amalgamated/superimposed upon other barforms. Given the common origin of unit bars, in the initial stages of growth the similarity between bars is probably greater than their differences. At this stage the barforms will also be at their most homogenous in terms of grain size variation within the bar. There will be a general fining up of grain size and potentially some variability within the bar form. For example, Ashmore's (1991) study found that the bedload sheets were coarser at their downstream end and finer near the centre. Likewise, Lunt and Bridge (2004) report that individual unit bars coarsen downstream towards the bar crest. Based on a quantification of ground penetrating radar surveys, Sambrook Smith *et al.* (2006) also found that facies distributions within a unit bar are quite homogenous with >60% of the deposits classified as of a single facies type related to the migration of dune bedforms. However, as the bar evolves over time and experiences a wider range of discharge events, migration rates and periods of erosion and deposition, its morphology and sedimentology will become more complex;

such bars are usually referred to as compound bars. Compound bars are defined as forms that comprise more than one unit bar and evolve through several erosion and deposition events (Bridge, 2003). Any reach of river will normally comprise a range of unit and compound bars (Figure 19.1). Thus the overall sedimentology of any river reach will be diverse and not well represented by average measures of surface grain size that do not fully account for the 3-D variability within and between different bar types.

Once a bar has become emergent and has a greater overall range of topography, particles moving across and around the barform will be increasingly subject to both fluid forces and gravitational forces (due to the slope associated with bar margins for example). These two forces act to produce the characteristic grain-size sorting found on most bar surfaces. The fluid force is normally expressed by the shear stress, which is primarily a function of depth and slope; thus, as flow goes from channel to bartop the depth decreases. This means shear stress reduces, resulting in the preferential deposition of coarser particles at the barhead. Since the gravitational force acting on a slope is proportional to particle mass while the drag force of the flow is proportional to particle area, coarse particles will be transported downslope more readily than are fine particles (see review in Powell, 1998). This has the effect of concentrating the coarser particles at the base of bars. The combined effect of these processes is the characteristic barhead to bartail fining and cross-bar fining commonly seen in point bars. Braid bars also exhibit barhead to bartail fining for the same reasons as outlined above (Ashworth *et al.*, 1992). However, given that flow can top a braid bar laterally from either side cross-bar fining may either be absent or very pro-

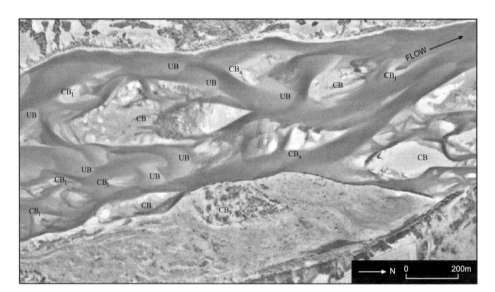

Figure 19.1 Aerial photograph of a section of the study reach taken in April 2000 illustrating the key morphological features of the South Saskatchewan River. CB = compound bars; CB_v = vegetated compound bars; CB_l = compound bars with limbs; CB_a = asymmetric compound bars; UB = unit bars. From Sambrook Smith et al. (2006), copyright (2006) with permission from Blackwell Publishing

nounced (if there is a predominant direction of flow over the bartop). For flora and fauna on the bar surface it is thus likely that the barhead area will provide a greatly contrasting habitat as compared with that in the bartail. For example, in the braided Sagavanirktok river in Alaska, Lunt and Bridge (2004) found that average grain size was commonly >8 mm at the bar head reducing to nearer 2 mm at the bar tail, often associated with decimetre-thick deposits of sand. Both point and braid bars also exhibit grain size variability in the vertical. Given the more predictable migration history of point bars (see below), where the bar migrates laterally and downstream, the vertical succession of sediments will generally display an overall fining up character (e.g., Bridge et al. 1995). Given their greater complexity of evolution, the vertical sorting in braid bars may be less distinct and punctuated by several fining up cycles relating to different cross-bar channels or episodes of unit bar accretion. For both point and braid compound bars Lunt and Bridge (2004) reported that the most pronounced fining-up will be in bartail areas with little variation in upstream and medial sections of the bar.

However, these broad generalisations belie a great deal of variability that is common to most bars, and occurs at scales of direct relevance to their ecology. Bluck (1971) provides one of the classic and most comprehensive accounts of size segregation on a point bar (Figure 19.2). He splits a bar into platform and supra-platform, the former being

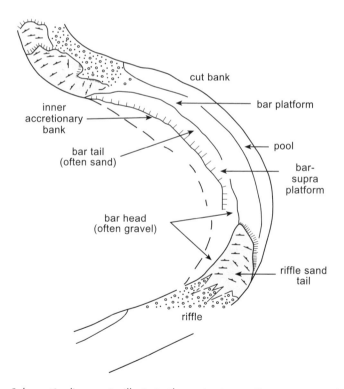

Figure 19.2 Schematic diagram to illustrate the grain size sorting across a typical point bar from the River Endrick, Scotland. From Bluck (1971), reproduced with permission from the Scottish Journal of Geology

largely submerged for most of the time while the latter is generally only inundated during higher stage. The supra-platform is further split into bar head, bar lee, bartail and inner accretionary bank sections; grain size generally goes from coarse to fine in that order. However, as Bluck (1971) describes, there is great variability within each of these areas, no more so than in the barhead area. These areas may contain open framework gravels, framework gravels, interbedded sands and gravels or gravel infilled with clay that has been deposited as flow stage falls. All these different types of sedimentary structure will have very different hydraulic conductivites over a very small area. Permeability is largely a function of D_{10} (the grain diameter that 10% of all the particles in a sample are finer than) and sorting, thus the infilling of pore spaces between gravel with even modest amounts of fines can significantly affect hydraulic conductivity at the bar surface. Hazen (1893) provides a simple equation for estimating hydraulic conductivity (K);

$$K = AD_{10}^2$$

A is equal to 1 when K is in $cm\,s^{-1}$ and D in mm. Here and subsequently D_i refers to the particle diameter for which i % of particles are finer than. While developed for soils, Anderson et al. (1999) and Moreton et al. (2002) have found it can also be applied to alluvial fine-sands and gravels. Although probably secondary to grain size, sorting also needs to be quantified, Masch and Denny (1966) suggest using the following measure of standard deviation (σ_1) to assess this;

$$\sigma_1 = ((D_{16} - D_{84})/4) + ((D_5 - D_{95})/6.6)$$

They found that with a constant D_{50} as σ_1 increases then hydraulic conductivity decreases. Moreton et al. (2002) conducted some flume experiments based on the braided gravel bed Ashburton river, New Zealand. Their analysis of the resulting sediments indicated that permeability could vary by three orders of magnitude between different deposits. Open framework gravels, for example, will have the highest permeabilities as they are devoid of sediment in the pore spaces between gravel particles, which would otherwise impede water movement. Lunt et al. (2004) report that the permeability of open framework gravels is in the range 10^3–10^4 darcys. This is one or two orders of magnitude greater than for sandy gravel and up to four orders of magnitude greater than for sand (Table 19.1).

Perhaps the most pronounced permeability variations in meandering systems occur in the tight meanders of confined systems where concave bank benches (e.g., Hickin, 1979; Page and Nanson, 1982) form. These features form on the concave bank of meanders due to flow separation and the subsequent deposition of fine material (sand, silt and clay) as a thin bar, which may eventually become attached as sedimentation progresses (Figure 19.3). Deposits such as this will firstly increase the total bar edge length within the system, and secondly provide an area of fine-grained deposits that is contemporaneous but very different to adjacent gravely point bar deposits. There is thus a great variation in permeability over a short distance at such sites.

Variability in the sedimentology of braid bars is also common as reported by Sambrook Smith (2000), based on work in the gravel bed braided Sunwapta river. From a distal reach of the river (D_{50} = 15 mm) there is a general downbar fining and also cross-bar

Table 19.1 Summary of permeability values for a range of fluvial sands and gravels, from Lunt et al. (2004). OFG = open framework gravel

Literature sources	Method	K (m/day)	K (darcys)	Porosity (%)	D_{50} (mm)	Sediment type
Anderson et al. (1999)	Grain size (Hazen)	1040	1718		9.13	OFG
		0.588	1		0.93	Very fine sand?
		25.9	43		0.57	Medium sand
		46–147	76–243		0.55–12.4	Sandy gravel
Klingbeil et al. (1999)	Pneumatic tests and tracers	342	568	27.0		Planar/trough gravel
		10.0	17	30.0		Bimodal gravel
		21 800	36 159	36.0		OFG
		95.0	158	42.0		Sand
Titzel (1997), Welch outcrop	Air permeameter	9.7–32.8	16–54			Fine-medium sand
		54.3	90			Massive gravel
		46.7	77			Stratified gravel
Hamilton outcrop		71.5	118			Medium sand
		137	228			Stratified gravel
Welch outcrop mean		27.7	46			
Hamilton outcrop mean		123	204			
Huggenberger (1993)	Permeameter and grain size (Kozeny)	12.1	20	20.0		Grey gravel
		2.59	4			Brown gravel
		25.9	43	43.0		Sand
		43 200	71 743	35.0		OFG
		8.64–51.8	14–86	14.0		Bimodal gravel
		0.0864	0			Silt
Jussel et al. (1994)	Permeameter (disturbed)	12.1–19	20–31	17.5–20.1	12.9–13.1	Grey/brown gravel
		13 000	21 475	34.9	9.7	OFG
		147	243	43.4	0.47	Sand
Sagavanirktok River, corelab data	Permeameter	12.1	>20	20.7	9.02	OFG
		8.03	13	18.7	4.9	Sand and gravel
		1.51	3	21.3	0.17	Sand
Lab permeameter	(disturbed)	280	464		11.31	OFG
		64.8	107		0.32	Sand

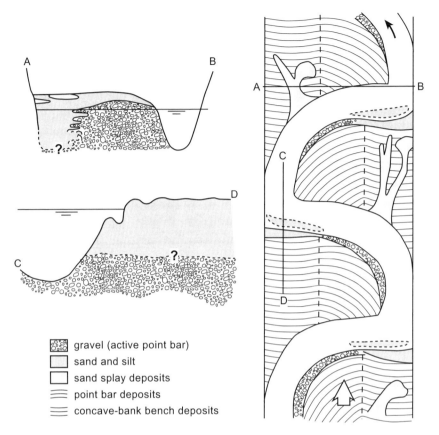

Figure 19.3 Schematic diagram to illustrate the relationship between meander planform and associated point bar and concave-bank bench deposits. From Hickin (1986), copyright (1986) with permission from Blackwell Publishing

fining, especially pronounced across the bartail due to the diversion of flow over the bar in this area (Figure 19.4). Also apparent is the presence of thin clay layers within the profile associated with deposition at low stage. This can have an important impact on the hydraulic conductivity acting as a barrier to water movement. It should be noted that this will not necessarily be well related to the surface sediment, thus emphasising the importance of quantifying the subsurface sedimentology of surface patches. Sambrook Smith (2000) found that more proximal braid bars displayed less variability in their sedimentology, although evidence of clay drapes could be found within the profile. Recent studies using ground penetrating radar have enabled the development of much more quantitative 3-D representations of braid bars (e.g., Bridge and Lunt, 2006; Sambrook Smith *et al.*, 2006; Mumpy *et al.*, in press), which highlight their spatially variable internal sedimentology. For example, Sambrook Smith *et al.* (2006) found that one side of their study bar comprised 39% of a facies composed of high-angle strata, yet this was largely absent on the other side of the bar. They concluded that the facies variability within and between bars made it difficult to derive a generalised model of braid bar sedimentology. This

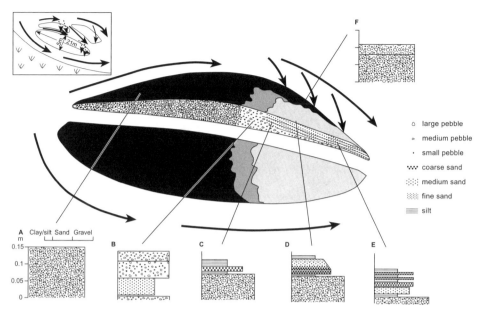

Figure 19.4 Pseudo-3D diagram showing stratigraphic logs and spatial sorting of sediments in a typical braid bar of the central reach of the Sunwapta river, Canada. The shading indicates the broad pattern of surface grain size variation with darker tones representing coarser sizes. From Sambrook Smith (2000), copyright (2000) with permission from Elsevier

variability has been noted previously and has led to attempts to describe bar surfaces based on morphostratigraphic units (defined on the basis of morphology and sedimentology). For example, Brierley (1991) split bars into four units; bar-platform, chute channel, ridge, and remnant floodplain. Using this system he found that the different units were more randomly organized for compound braid bars but displayed more predictable downstream and arcuate patterns for compound point bars (Figure 19.5). Whether the pattern of units may be a useful tool for understanding ecological distributions is unclear, as each unit comprises different facies and there is much overlap of facies between units (Table 19.2).

Perhaps of more direct relevance is the detailed consideration of permeability changes within a unit braid bar as presented by Lunt *et al.* (2004). They found that sandy gravel was the dominant deposit (by volume), with open framework gravel and sands making up 15% and 4% of the sediments respectively. In terms of the spatial arrangement of these facies they were able to determine some general patterns. Thick sets of open framework gravel were most likely to occur in steeply dipping strata at bar margins, in which case it could be spatially extensive; widths similar to the bar width, lengths up to 30% of the bar length and thicknesses up to 66% of the bar height. Thus permeability, and by implication water availability, of bar edges can be highly spatially variable (Figure 19.6). This has implications as to how these sites may function as habitats; the larvae of Carabidae (ground beetles) and Staphylinidae (rove beetles) live in the interstices of gravel near the water line and are very sensitive to desiccation (Tockner *et al.*, 2006). Likewise,

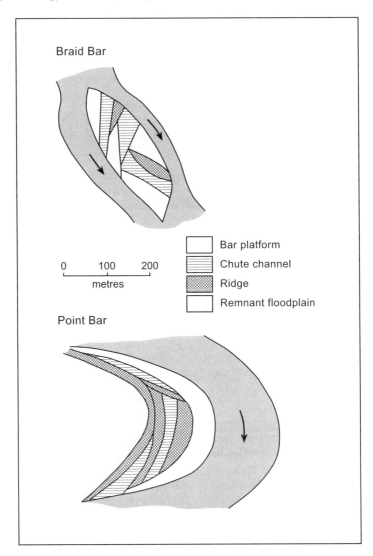

Figure 19.5 *Schematic diagram to illustrate the spatial arrangement of morphostratigraphic units on a typical braid and point bar of the Squamish river, Canada. Note that in the point bar the units are orientated in an arcuate pattern whereas in the braid bar they are more random. From Brierley (1991), copyright (1991) with permission of Blackwell Publishing*

Tockner *et al.* (2006) also suggest that measures of permeability can be useful for assessing the potential of a site as a refugia. However, it should be noted that any influence of grain size is likely to be secondary to that of elevation, as discussed by Bates *et al.* (in press).

A further layer of complexity is added to bar surface sedimentology when the bars become dissected by cross-bar channels. These can occur on braid and point bars, but are

Table 19.2 Facies composition of morphostratigraphic units in the Squamish river, Canada (Brierley, 1991). Only those facies which make up >= 10% of each unit are included. Sw = wavy-bedded, medium-fine sands; Sr = internally graded medium or fine sands; Ss = massive, medium-coarse sands; St = cross-bedded troughs, medium-coarse sand, occasionally pebbly, internally graded; Sp = planar–tabular cross-beds, medium-coarse sand, occasionally pebbly, internally graded; Fm = massive or finely laminated fine sands and silts

Morphostratigraphic unit	Facies Composition		
	Contemporary bar surfaces	Lateral sequences	Longitudinal sequences
Remnant floodplain	24% Sw	40% Sr	41% Sw
	20% Sp	34% Sw	35% Sr
	16% Sr	24% St	12% St
	12% St	33% Sw	45% Sw
	11% Ss	32% Sr	34% Sr
	10% Fm	24% Fm	12% Fm
Ridge	30% Sr	61% Sw	43% Sw
	22% Sw	18% Sr	43% Sr
	19% Sp		
	10% St		
Chute channel	27% Sr	30% Sw	56% Sr
	25% Sw	26% St	26% Sw
	16% Sp	14% Sp	
	13% St		
Bar platform	34% Sp	32% St	32% St
	18% St	25% Sp	23% Sr
	18% Sw	19% Sw	14% Sw
	18% Sr		

more likely to be across the inner bank on point bars (sometimes referred to as chute channels), with a more random orientation on braid bars. The presence of cross-bar channels may be more likely on larger compound bars that have a greater elevation and are thus more likely to be dissected on falling stage. Moreton *et al.* (2002) found that these cross-bar channels initially had high permeabilities within them as a result of stalling coarse-grained lobes of sediment. However, once the channel was abandoned and filled with finer grained deposits, permeability was reduced. An additional point to note about cross-bar channels is that they will be the foci for standing water on the bar surfaces. Thus, in effect, the bar edge length is increased although this extra edge length is not connected to the surrounding channel (except during high flow events). Another standing water feature is a bar-top hollow (Best *et al.*, 2006). These form largely by bar-top deposition due to the convergence of two inwardly accreting bartail limbs, together with the occasional cutoff/infill of cross-bar channels.

19.2.3 Bar Surface Structure and Hydraulics

At the smallest spatial and temporal scales, grain size sorting will vary with the individual bedforms, e.g., the foresets of gravel bedload sheets have high porosity and permeability

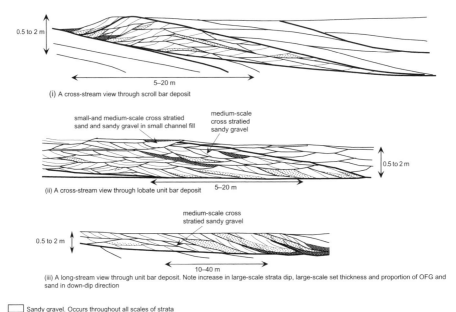

Figure 19.6 *Sections through unit bar deposits of the Sagavanirktok river showing the typical spatial arrangement of sediments. Note that the permeability of open-framework gravels is generally an order of magnitude greater than for sandy gravel and up to four orders of magnitude greater than for sand. From Lunt* et al. *(2004), copyright (2004) with permission of SEPM (Society for Sedimentary Geology)*

(Brayshaw *et al.*, 1996). However, at these scales it may be the impact of the bar surface structure on the hydraulics that becomes more ecologically relevant. The full range of bedforms associated with the channel bed such as, ripples, dunes, bed-load sheets, pebble clusters, sand ribbons, and transverse ribs, are also found on bar surfaces. A common feature of bar surfaces in gravel rivers is the cluster bedform. They usually occupy 5–10 % of a gravel bed and can be between 0.1 to 1.2 m long, with widths approximately half that distance (Brayshaw, 1984). They are characterised by a central large clast (often $>D_{95}$) with imbricated gravels on the stoss side (sizes between D_{74} and D_{94}) and finer particles in the lee (sizes between D_8 and D_{46}). Studies have shown how cluster bedforms have a significant impact on the turbulent structure around them (e.g., Buffin-Belanger and Roy, 1998; Lawless and Robert, 2001). Briefly this comprises: (i) flow acceleration up the stoss-side; (ii) flow separation in the wake; (iii) vortex shedding from the crest; (iv) flow reattachment downstream; (v) upwelling of flow downstream of the reattachment point, and (vi) recovery of flow. Furthermore, Buffin-Belanger and Roy (1998) suggest that the impact of a pebble cluster on the flow structure will be felt for a distance 9–15 times the height of the central largest clast. The complexity of flow structures around bedforms is of direct relevance to a number ecological applications. For example, studies have begun to relate near-bed flow properties to biofilm metabolism (Cardinale *et al.*, 2002), macroinvertebrate distributions (Bouckaert and Davis, 1998), flow refugia (Lancaster, 2000; Lancaster *et al.*, 2006) and river algae (Matthaei *et al.*, 2003).

All the studies mentioned above, however, have looked at within-channel bedforms rather than those at the bar surface, and results may not be directly comparable. Firstly, variability in structure and patch size can be more pronounced on bar surfaces than within-channel environments where finer grained sediments may be in transport more often and so less subject to sorting processes. Sambrook Smith and Nicholas (2005) demonstrate that even modest amounts of sand deposition around gravel particles can significantly change the turbulent structure, such that reverse flow in the lee of gravel particles is eliminated, near-bed shear stress and turbulent kinetic energy decrease, and burst and sweep events become less frequent around gravel particles. Secondly, as bar surfaces become emergent they modify flow patterns, resulting in a more variable orientation of bedforms than would be found within channel (Bridge, 2003). This may lead to differences in the spacing between bedforms on bar surfaces as compared within channel with associated changes in turbulent structure. Where bedforms are closely spaced it gives rise to 'skimming flow', as opposed to 'wake interaction' when bedforms are more widely spaced (Nowell and Church, 1979).

A final point to note is that recent studies (e.g., Best, 2005; Sambrook Smith and Nicholas, 2005) using particle imaging velocimetry (PIV) can now measure mean and turbulent properties of the flow to within 1 mm of the bed. These have revealed even more detail about the complex relationship between grain size and flow structure and at a scale more appropriate to that of macroinvertebrates. For example, Sambrook Smith and Nicholas (2005) show that data collected at a height of 2 cm above the bed differed greatly to that obtained in the bottom 3 mm of flow. Thus the pattern of hydraulic patches and edge types is significantly simplified when based on mean velocity data obtained above the bed as compared with turbulence characteristics taken at the bed.

19.3 The Evolution of Barforms

19.3.1 Ecological Implications

The way in which bars evolve are important ecologically for two main reasons. Firstly, and most importantly, the spatial extent and timing of erosional and depositional events of barforms controls habitat destruction and construction or turnover, the rate of change between habitat types. For braid bars in particular, turnover of habitat patches is the most important factor in determining biodiversity through its control on the processes of disturbance and succession (Sadler, 2005). Turnover will vary between different bar types within the same river and similar bar types but at different stages of their evolution. The frequency of avulsions (the movement of a channel to a new location on the floodplain) will determine how often these habitats are completely reset. Of less importance, but still relevant, are modifications to existing habitat. Hughes *et al.* (2001) demonstrate the importance of such events to seedlings of *Populus nigra* and *Salix alba*. They found that seedlings up to 2 weeks old were very intolerant of burial by sediment. Conversely seedlings at 4–6 weeks of age could actually be stimulated to increased growth rates by partial burial. Secondly, the manner in which bars either grow (laterally or vertically) or diminish in size will affect their overall area, bar edge length and vertical range in topography.

Tockner *et al.* (2006) discuss how bar edges are a key feature in maintaining high species diversity within braided rivers. They state that a high ratio of bar edge length to stream area will provide maximum efficiency of the transfer rate of energy across the aquatic–terrestrial boundary. For example, Sadler *et al.* (2004) found that ground beetles are often more abundant on bar edges, which they ascribe to the fact that aquatic drift is a major source of food for such organisms. This has been quantified by recent experimental work (Paetzold *et al.*, 2005, 2006), which demonstrated that carabid beetles of the genus *Bembidion* and *Nebria picicornis* fed entirely on aquatic insects. Additionally, as bars vary in height this will affect the distance to the water table that will impact both the flora and fauna. For example, Francis *et al.* (2006) found that cuttings of *Salix elaeagnos* and *Populus nigra* performed best on lower elevations of bars whilst the opposite was true for seedlings of the same species.

19.3.2 Bar Migration

In broad terms the migration rate of all bars will increase with discharge, water surface slope and sediment supply and will decrease with increasing bank silt-clay content, bank vegetation and bank height (Hickin and Nanson, 1975). At the individual bar scale it is perhaps simplest to view bar stability as being related to the overall bar morphology, thus is it either bank attached, i.e. point bars (relatively more stable) or unattached, i.e., braid bars (relatively less stable). This general approach has ecological significance as reflected in the work of Arscott *et al.* (2002) on the Tagliamento river, Italy. They found that over a year the aquatic habitat turnover was 62% for a braided (i.e., predominantly unattached bars) headwater section of the river, decreasing to 20% for a meandering (i.e., predominantly attached bars) reach further downstream. For bank-attached unit bars the rate of downstream migration rapidly decreases as channel sinuosity increases. Thus the flow diversion and bank erosion associated with the growth of unit bars means that they rapidly stabilise and grow into compound point bars. Conversely, it is not unusual for unattached unit bars to migrate a distance downstream equivalent to their length (i.e., 100% turnover) over an annual time period (e.g., Werritty and Leys, 2001; Sambrook Smith *et al.*, 2006). Generally, however, unit bars of either type are often short lived features and can be destroyed within a single flow event or accrete and become part of a much larger compound braid bar.

The migration of compound point bars in meandering channels has been extensively studied and is well understood (classic studies include Hickin and Nanson, 1975; Hooke, 1977; Ferguson, 1981). Starting from an initially straight channel, migration rates are low, as the bend increases in sinuosity (ratio of the radius of channel curvature to channel width approximately 3) migration rates increase and the bend may display extension as well as translation (Figure 19.7). This will increase the bar edge length so providing important additional habitat. Maximum annual migration rates will be equivalent to the channel width of the system, although rates will vary especially in relation to bank material type. For example, Mount *et al.* (2005) found that straighter reaches of the River Trannon, with cohesive silt–clay banks, had mean migration rates $<0.07 \, \text{m y}^{-1}$. Conversely, areas of higher sinuosity, with gravel–sand banks, had average migration rates of $0.39 \, \text{m y}^{-1}$, with maximum rates up to $5.53 \, \text{m y}^{-1}$. Once a bend becomes tight, however,

Fluvial Sedimentology: Implications for Riverine Ecosystems 375

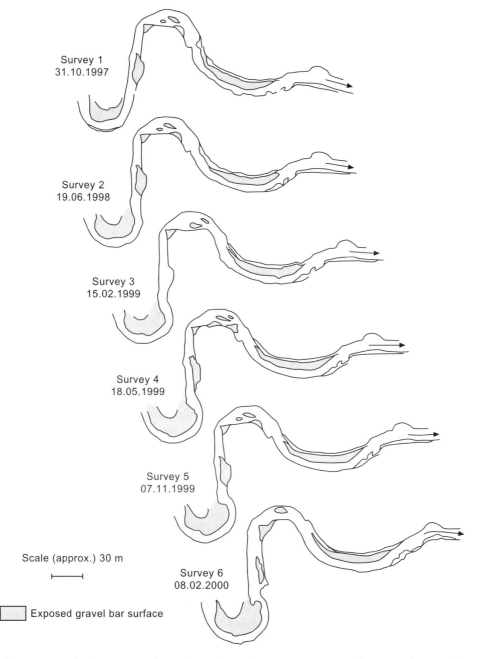

Figure 19.7 Planform maps of a study reach on the River Trannon, Wales. Note the growth in size of the most upstream point bar, also associated with a rotational development to the meander. Repeat surveys between 1997 and 1999 of the study reach shows that the volume of the point bar in the reach grew by 287.8 m^3. Contrast this with the very limited change to the downstream point bar in a less sinuous section of the river. Adapted from Mount (2000)

there is an increasing likelihood that a neck cutoff will occur reducing the sinuosity and so starting a new cycle of migration with a new point bar greatly reduced in size. The abandoned channel next to the old point bar will infill with sediment ranging in size from sands to clay and will provide a marked contrast in habitat to that of the adjacent old point bar.

Compound braid bars commonly grow episodically as unit bars accrete onto them. For example, Lunt and Bridge (2004), working in the Sagavanirktok river, Alaska, measured the average growth of a compound bar they studied as $1.9\,\mathrm{m\,y^{-1}}$ with a maximum of $8.5\,\mathrm{m\,y^{-1}}$ due to unit bar accretion in the bar tail region. This was similar to the rate of bank erosion which averaged $0.5–1.5\,\mathrm{m\,y^{-1}}$ with a maximum of $6.31\,\mathrm{m\,y^{-1}}$, so highlighting the relationship between these two processes. The frequency of occurrence of these processes can be greater in braided rivers, resulting in compound braid bars with a more complex sedimentology than that of point bars in meandering rivers. As each unit bar accretes onto a compound bar, its inherent grain size variability is added to that of the compound bar as a whole, resulting in a mosaic of grain size (and facies) variability superimposed on the general fining trends outlined above. When bars accrete onto each other there will also be a reduction in the total bar edge length of the system; however, this is likely to be relatively shortlived as new bars grow and evolve from upstream. Of longer term significance is where a compound bar accretes onto the channel edge; in such cases bank edge length will be reduced for a significantly longer period of time. This can occur when unit bars migrating onto the head of a compound bar gradually block the adjacent channel as the bar accretes upstream. The channel subsequently fills over time until the bar becomes attached to the bank. During this period an isolated area of standing water may be present in the topographic low of the pre-existing channel, until this too also fills. The sediments of the channel fill usually fine downstream, with the scale of bedforms also decreasing in the same direction (Bridge, 2003). Substantial areas of clay will be restricted to the most downstream sections of the channel fill. Channel fills thus provide an additional diversity of sedimentology as compared with those of the adjacent bar. Conversely point bars can evolve such that they become detached from the bank and function as a braid bar. In such instances bar edge length increases significantly and the bar loses its connection to the floodplain. Such a transformation typically occurs when the chute channel cutting across the inside of the point bar enlarges to such a degree that it becomes a permanent feature. For example Lunt and Bridge (2004) document an initially meandering reach of the Saganvirktok River, Alaska, in 1949 that had become braided by 1983 after a chute channel had initially formed across the point bar in 1968.

19.3.3 Avulsion

Avulsion is where the river moves to a new location at a lower elevation. This will create totally new barforms and will in effect reset the ecological system. The frequency of avulsion is thus of major significance to a fuller understanding of ecological communities. As the intermediate disturbance hypothesis (Connell, 1978) suggests, it will be those systems that experience moderate levels of avulsion that will have the greatest biodiversity. Bridge (2003) reports average inter-avulsion periods as ranging from 28 years (Kosi, India) to 1380 years (lower Mississippi, USA). These differences being related to such factors as tectonics, climate change, base level change and the ratio of cross-valley to down-valley slope. However, these factors relate to avulsions of the entire channel belt. Of more relevance to ecology are the smaller scale within channel belt avulsions that are common within braided river systems. Within braidplain avulsions fit into two basic categories that involve either the creation of new channels and bars (primary anastomosis; Church, 1972) or the reoccupation of previously abandoned channels (secondary anastomosis; Church, 1972), this latter often being associated with rising stage. Primary anastomosis will involve turnover of sediments and habitat while the latter will only increase bar-edge length, and often for only fairly limited periods of time over the course of a single flood hydrograph. The mechanism of primary anastomosis is usually described as relating to sediment lobes, which may stall and deflect flow (see Leddy *et al.*, 1993), often at bifurcations (Warburton *et al.*, 1993), causing partial blockage that diverts water to a new route. Federici and Paola (2003) suggest that bifurcations typically display an inherent instability, with flow switching from just one channel to two and then back again, which may also drive the small scale avulsions commonly seen in braided rivers. Studies have shown (Laronne and Duncan, 1992; Warburton *et al.*, 2002) that relatively modest amounts of sediment are required to achieve this. Thus avulsions need not be particularly related to flow stage; it is the sediment supply rate that is the major driving variable of avulsion frequency, as demonstrated by the physical modelling experiments of Ashworth *et al.* (2004). They found that an increase in the ratio of sediment to water discharge led to a greater channel mobility, bifurcation and avulsion. As sediment supply was increased avulsion frequency increased by a factor of up to 5. Thus systems with high rates of sediment supply (e.g., proglacial systems for example) are likely to experience frequent avulsions that may lead to restricted biodiversity. For example, in the proglacial gravel bed braided Sunwapta River, Canada, Goff and Ashmore (1994) describe how within their study reach, a cycle of medial bar destruction and then evolution took place twice over periods of 5 and 2 days.

19.4 Discussion and Conclusion

Having presented detail at the habitat (bar) and microhabitat (grain) scale it is important, as suggested by Poole (2002), to place this work in the context of the higher patch scales, i.e. the reach, segment or network. A simple example can be used to illustrate the importance of this approach. In their study of habitat change in a braided river, van der Nat *et al.* (2003) classified the braidplain into channels, alluvial channels, backwaters, ponds, bare alluvium, vegetated islands and riparian forest. The relative proportions of these habitats were then mapped from September 1999–January 2002. They concluded that the relative habitat composition within the active corridor remained almost constant. However, by categorising all surfaces together ignores the variety of forms in which bars can manifest themselves. As Poole (2002, p. 643) states, 'from a catchment hierarchy view, it seems logical that rearranging the patches along a river's course will change the ecological dynamics of the system even if the relative proportion of patch-types in the system remains the same'. Thus as has been presented, different bar types will have different properties of dynamics and evolution that may be reflected in different ecological populations. If a reach has the same proportion of bars but in one it is all mid-channel medial bars and the other it is all bank attached compound bars, will these two reaches function in the same way? Or at a finer scale if two bars are the same size and shape and one is composed of a single grain size and the other is more heterogeneous, will these systems have the same ecological populations? Only by asking questions about the role of complexity at all scales can such questions be answered. Bates *et al.* (in press) illustrate this point when they conclude that large bars have longer lateral gradients, which supports a large range of species. In comparison, the lateral gradient of smaller less elevated bars contracts quickly as stage rises and so supports only a restricted range of species.

The requirement to consider how the small-scale fits with the larger scales within the overall hierarchy can be considered by looking at the broader temporal and spatial controls on individual bars. For example, numerous studies have shown how bar evolution can change in response to the system switching between meandering and braided states (Werritty and Ferguson, 1980; Macklin, 1997; Macklin *et al.*, 1992; Passmore and Macklin 2000; Warburton *et al.*, 2002). Some of these cycles may be related to climate change or landuse change whereas others may be simply a function of the normal pattern of endogenously driven change. For example, Werritty and Leys (2001) demonstrate how some Scottish rivers increase in sinuosity during periods of low incidence of flooding and then decrease in sinuosity and become more braided during periods with a high incidence of floods. Active areas can increase in size by factors ranging between 5.59 and 1.53, with changes in number of bars linked with this. Harvey (1991, 1994, 2001) reports similar switching between different states but triggered more by variability in sediment supply rather than discharge. In this study in the Howgill Fells, slope channel coupling is strong, thus sediment from alluvial fans or landslides readily discharges into the river system. Where high rates of sediment input are common, rates of change can remain high. Conversely where the slope–channel coupling is not well developed or is interrupted, change can diminish. A significant issue is that this temporal variability needs to be understood and taken into account if long term ecological populations are to be managed appropriately. A related point is that as a first approximation slope–channel coupling, and hence sediment supply as a ratio of water discharge, will decrease downstream from the headwaters. This may partly explain the longitudinal gradient in community structure (Framenau *et al.*, 2002)) and biodiversity, which tends to peak in the mid to lower reaches (Sadler *et al.*, 2004), i.e., in relation to moderate levels of disturbance.

Sediment supply and transport rarely show simple downstream patterns; thus, controls on bar evolution can be complicated, as expressed in the concept of storage and transfer zones within a river system (Church, 1983). Hence a small section of a river may be laterally confined such that sediment moves through rapidly (transfer zone) and bar growth is restricted. The next reach, if it can spread laterally, will allow aggradation of sediment with the braid bars and high rates of change that go with it (storage zones). Church (1983) found that storage zones were generally located at tributary junctions. This is because sediment from upstream accumulated behind the small-scale alluvial fans that formed in the confluence zone, so restricting sediment movement. Likewise, others have shown how this variability in sediment supply can lead to significant changes in bar area between adjacent sites (Madej, 1982; Meade, 1985; Jacobson and Bobbitt Gran, 1999). These concepts apply equally to the bar scale as demonstrated by Wathen *et al.* (1997), who showed that the way sediment interacts between channel and bar can heavily influence how the bar evolves. Thus sediment pulses can be just as important as flow pulses.

In its broadest sense this notion of variability in bar response relates to traditional ideas of landscape sensitivity (Downs and Gregory, 1995), albeit applied at a smaller scale for individual bars. However, the idea of assessing how readily individual bars can absorb change with either modest or extreme adjustments is still valid. It should also be noted that with recent advances in remote sensing techniques, identifying and quantifying change within rivers (including submerged areas) is now possible at much higher resolutions and over much broader areas than has hitherto been possible (e.g., Westaway et al., 2000, 2003; Lane, 2006).

Historically, there has been a focus on within channel processes and the role of water discharge as the key controlling variable for explaining riverine ecosystems. The main conclusion of this paper is that ecologically, sediment also matters through its influence on permeability, hydraulics, bar form and disturbance. However, relationships between sedimentology and ecology have been fairly simplified to date and a proper quantification of links between the two is still in its infancy. Some potential areas of collaboration between sedimentologists and ecologists to address this might include the following:

(i) To what extent is the variability in sedimentology of different bars reflected in their ecological communities? For example, would similarly sized point and braid bars in the same river display significant differences in their ecology based on the differences in sedimentology and dynamics discussed previously? How might these relationships vary with both the scale of the barform and the organism under consideration? To begin to assess these questions information on population structures needs to be mapped onto detailed 3-D images of bar sedimentology quantified from ground penetrating radar (Naegeli et al., 1996) surveys linked with cores (from which permeability can be quantified). In such a way the important (ecologically) scales of sediment patch size and type can be quantified.

(ii) How do sediment structures influence exchange processes operating across the bar surface? Sediments are primary sites for the storage of organic matter and contaminants. The quality of within sediment habitat will thus be highly dependent on stream–subsurface exchange dynamics. However, the dynamics of the stream–subsurface exchange system and how this is mediated by sedimentological bedforms is poorly understood, being largely based on inference rather than direct measurement (Packman et al., 2004). New understanding would be made possible by using new technologies such as particle imaging velocimetry (PIV). This would allow, for example, coupled measurements of flow and colloid dynamics to be taken across the gravel surface–subsurface interface from which relationships between the free flow, pore flow and contaminant dynamics could be quantified.

(iii) How does bar sedimentology and evolution affect exchanges across the aquatic–terrestrial boundary? Permeability of bar edges and the ratio of bar edge to stream area both dictate the efficiency of transfer across the aquatic–terrestrial boundary (Tockner et al., 2006). These factors are both influenced by the local sediment supply rate, in driving the frequency of avulsion, for example. However, as a bar evolves from a point to braid type, following a local avulsion, the ecological response remains poorly quantified. Additionally, recent research has highlighted the importance of floating organic matter that gets stranded on bar edges as an important source of food for riparian arthropods (Paetzold et al., 2005, 2006). This organic matter becomes particularly concentrated in slow moving areas. Work using computational fluid dynamics (CFD) can provide simulations of flow velocities within a braided river (e.g., Lane and Richards, 1998; Nicholas and Sambrook Smith, 1999), from which it may be possible to highlight areas of likely concentration of floating organic matter and so assess the overall energy transfer from the river to the riparian zone. Perhaps more importantly, models can be run to determine how the flow patterns around the bars of a reach would be affected by river regulation, and hence to what degree the aquatic–terrestrial linkages might be changed (also see discussion in Leclerc, 2005, for applications of CFD to ecology).

The overall goal must be to quantify properly sedimentological and ecological interactions so that they can form the basis for prediction. Some progress has begun in this area, for example, it is now known that as plants grow they begin to influence the processes of sediment deposition and bar evolution (e.g., see Gurnell and Petts, 2002; Tooth and Nanson, 2000) rather than just being a function of bar form and structure. Building on this, models are now beginning to be developed of river evolution that take account of vegetation parameters (e.g., Richards *et al.*, 2002; Murray and Paola, 2003; Nicholas *et al.*, 2006). In such a way models of bar dynamics can be improved and predictions of population community dynamics developed. This would be a powerful tool in river restoration projects that often have to rely on qualitative descriptions of idealised systems as a basis for management (Paola *et al.*, 2006).

References

Anderson, M. P., Aiken, J. S., Webb, E. K. and Mickelson, D. M. (1999) Sedimentology and hydrogeology of two braided stream systems. *Sedimentary Geology*, **129**, 187–200.

Arscott, D. B., Tockner, K., van der Nat, D. and Ward, J. V. (2002) Aquatic habitat dynamics along a braided alpine river ecosystem (Tagliamento River, Northeast Italy). *Ecosystems*, **5**, 802–814.

ASCE (1966) Nomenclature for bed forms in alluvial channels. Task force on Bedforms in Alluvial Channels. *American Society of Civil Engineers Proceedings, Journal of the Hydraulics Division*, **92** (HY3), 51–64.

Ashmore, P. E. (1991) How do gravel-bed rivers braid? *Canadian Journal of Earth Sciences*, **28**, 326–341.

Ashworth, P. J., Ferguson, R. I. and Powell, M. D. (1992) Bedload transport and sorting in braided channels. In *Dynamics of Gravel-bed Rivers*. Billi, P., Hey, C. R., Thorne, C. R. and Tacconi, P. (eds). John Wiley & Sons, Ltd, Chichester, pp. 497–515.

Ashworth, P. J., Best, J. L. and Jones, M. (2004) The relationship between sediment supply and avulsion frequency in braided rivers. *Geology*, **32**, 21–24.

Bates, A. J., Sadler, J. P. and Fowles, A. P. (in press) The microspatial distribution of beetles (Coleoptera) on exposed riverine sediments (ERS) during stable flow levels and weather conditions. *Entomologia Experimentalis et Applicata*.

Best, J. L. (2005) The kinematics, topology and significance of dune-related macroturbulence: Some observations from the laboratory and field. In *Fluvial Sedimentology VII*. Blum, M. D., Marriott, S. B. and Leclair, S. F. (eds). *Special Publication of the IAS*, **35**, 41–60.

Best, J. L., Woodward, J., Ashworth, P. J., Sambrook Smith, G. H. and Simpson, C. (2006) Bar-top hollows: A new element in the architecture of sandy braided rivers – Morphology, origin and subsurface structure. *Sedimentary Geology*, **190**, 241–255.

Bluck, B. J. (1971) Sedimentation in the meandering River Endrick. *Scottish Journal of Geology*, **7**, 93–138.

Bouckaert, F. W. and Davis, J. (1998) Microflow regimes and the distribution of macroinvertebrates around stream boulders. *Freshwater Biology*, **40**, 77–86.

Brayshaw, A. C. (1984) Characteristics and origin of cluster bedforms in coarse-grained alluvial channels. In *Sedimentology of Gravels and Conglomerates*. Koster, E. H. and Steel, R. J. (eds). Canadian Society of Petroleum Geologists, Memoir 10, pp. 77–85.

Brayshaw, A. C., Davies, G. W. and Corbett, P. W. M. (1996) Depositional controls on primary permeability and porosity at the bedform scale in fluvial reservoir sandstones. In *Advances in Fluvial Dynamics and Stratigraphy*. Carling, P. A. and Dawson, M. R. (eds). John Wiley & Sons Ltd, Chichester, pp. 374–394.

Bridge, J. S. (1993) The interaction between channel geometry, water flow, sediment transport and deposition in braided rivers. In *Braided Rivers*. Best, J. B. and Bristow, C. S. (eds). Geological Society Special Publication No. 75, pp. 13–71.

Bridge, J. S. (2003) *River and Floodplains: Forms, Processes, and Sedimentary Record*. Blackwell Publishing, Oxford.

Bridge, J. S. and Lunt, I. A. (2006) Depositional models of braided rivers. In *Braided Rivers*. Sambrook Smith, G. H., Best, J. L., Bristow, C and Petts, G. E. (eds). Special Publication of the IAS, **36**, 11–50.

Bridge, J. S., Alexander, J., Collier, R. E. L., Gawthorpe, R. L. and Jarvis, J. (1995) Ground penetrating radar and coring used to document the large-scale structure of point-bar deposits in 3-D. *Sedimentology*, **42**, 839–852.

Brierley, G. J. (1991) Floodplain sedimentology of the Squamish River, British Columbia: relevance of element analysis. *Sedimentology*, **38**, 735–750.

Buffin-Belanger, T. and Roy, A. G. (1998) Effects of a pebble cluster on the turbulent structure of a depth-limited flow in a gravel-bed river, *Geomorphology*, **25**, 249–267.

Bunn, S. E. and Arthington, A. H. (2002) Basic principles and ecological consequences of altered flow regimes for aquatic biodiversity. *Environmental Management*, **30**, 492–507.

Cardinale B. J., Palmer, M. A., Swan, C. M., Brooks, S. and Poff, N. L. (2002) The influence of substrate heterogeneity on biofilm metabolism in a stream ecosystem. *Ecology*, **83**, 412–422.

Carling, P. A. (1992) The nature of the fluid boundary layer and the selection of parameters for benthic ecology. *Freshwater Ecology*, **28**, 273–284.

Church, M. (1972) Baffin Island Sandurs: A study of arctic fluvial processes. *Geological Survey of Canada Bulletin*, vol. 216.

Church, M. A. (1983) Pattern of instability in a wandering gravel bed channel. *Special Publication of the International Association of Sedimentologists*, **6**, 169–180.

Claret, C., Marmonier, P., Boissier, J.-M., Fontvieille, D. and Blanc, P. (1997) Nutrient transfer between parafluvial interstitial water and river water: influence of gravel bar heterogeneity. *Freshwater Biology*, **37**, 657–670.

Connell, J. H. (1978) Diversity in tropical rain forests and coral reefs. *Science*, **199**, 1302–1310.

Davis, J. A. and Barmuta, L. A. (1989) An ecologically useful classification of mean and near-bed flows in streams and rivers. *Freshwater Biology*, **21**, 271–282.

Downs, P. W. and Gregory, K. J. (1995) Approaches to river channel sensitivity. *Professional Geographer*, **47**, 168–175.

Federici, B. and Paola C. (2003) Dynamics of channel bifurcations in noncohesive sediments. *Water Resources Research*, **39**, 1162.

Ferguson, R. I. (1981) Channel form and channel changes. In *British Rivers*. Lewin, J. (ed). Allen and Unwin, London, pp. 90–125.

Framenau, V. W., Manderbach, R. and Baehr, M. (2002) Riparian gravel banks of upland and lowland rivers in Victoria (south-east Australia): arthropod community structure and life-history patterns along a longitudinal gradient. *Australian Journal of Zoology*, **50**, 103–123.

Francis, R. A., Gurnell, A. M., Petts, G. E. and Edwards, P. J. (2006) Riparian tree establishment on gravel bars: interactions between plant growth strategy and the physical environment. In *Braided Rivers*. Sambrook Smith, G. H., Best, J. L., Bristow, C and Petts, G. E. (eds). Special Publication of the IAS, **36**, 361–380.

Goff, J. R. and Ashmore, P. (1994) Gravel transport and morphological change in braided Sunwapta River, Alberta, Canada. *Earth Surface Processes and Landforms*, **19**, 195–212.

Gurnell, A. M. and Petts, G. E. (2002) Island-dominated landscapes of large floodplain rivers, a European perspective. *Freshwater Biology*, **47**, 581–600.

Harvey, A. M. (1991) The influence of sediment supply on the channel morphology of upland streams: Howgill Fells, Northwest England. *Earth Surface Processes and Landforms*, **16**, 675–684.

Harvey, A. M. (1994) Influence of slope/stream coupling on process interactions on eroding gully slopes, Howgill Fells, Northwest England. In *Process Models and Theoretical Geomorphology*. Kirkby, M. J. (ed). John Wiley & Sons, Ltd, Chichester, pp. 247–270.

Harvey, A. M. (2001) Coupling between hillslopes and channels in upland fluvial systems: implications for landscape sensitivity, illustrated from the Howgill Fells, northwest England. *Catena*, **42**, 225–250.

Hazen, A. (1893) Some physical properties of sands and gravels with special reference to their filtration. Lawrence, Massachusetts, *24th Annual Report of the State Board of Health of Massachusetts for 1893*.

Hickin, E. J. (1979) Concave bank benches in the Squamish River, British Columbia. *Canadian Journal of Earth Science*, **16**, 200–203.

Hickin, E. J. (1986) Concave-bank benches in the floodplains of Muskwa and Fort Nelson Rivers, British Columbia. *The Canadian Geographer*, **30**, 111–122.

Hickin, E. J. and Nanson, G. C. (1975) The character of channel migration on the Beatton River, Northeast British Columbia, Canada. *Geological Society of America Bulletin*, **86**, 487–494.

Hooke, J. M. (1977) The distribution and nature of changes in river channel patterns: the example of Devon. In *River Channel Changes*. Gregory, K. J. (ed). John Wiley & Sons, Ltd, Chichester, pp. 265–280.

Huggenberger, P. (1993) Radar facies: recognition of characteristic braided river structures of the Pleistocene Rhine gravel (NE part of Switzerland). In *Braided Rivers*. Best, J. L. and Bristow, C. S. (eds). Geological Society of London, Special Publication No. 75, pp. 163–176.

Hughes, F. M. R., Barsoum, N., Richards, K. S., Winfield, M. and Hayes, A. (2000) The response of male and female black poplar (*Populus nigra* L. subspecies *betulifolia* (Pursh) W. Wettst) cuttings to different water table depths and sediment types: implications for flow management and river corridor biodiversity. *Hydrological Processes*, **14**, 3075–3098.

Hughes, F. M. R., Adams, W. M., Muller, E., Nilsson, C., Richards, K. S., Barsoum, N., Decamps, H., Foussadier, R., Girel, J., Guilloy, H., Hayes, A., Johansson, M., Lambs, L., Pautou, G., Peiry, J.-L., Perrow, M., Vautier, F. and Winfield, M. (2001) The importance of different scale processes for the restoration of floodplain woodlands. *Regulated Rivers: Research and Management*, **17**, 325–345.

Jacobson, R. B. and Bobbitt Gran, K. (1999) Gravel sediment routing from widespread low-intensity landscape disturbance, Current River basin, Missouri. *Earth Surface Processes and Landforms*, **24**, 897–917.

Junk, W. J., Bayley, P. B. and Spinks, R. E. (1989) The flood-pulse concept in river–floodplain systems. *Canadian Special Publication of Fisheries and Aquatic Sciences*, **106**, 110–127.

Jussel, P., Stauffer, F. and Dracos, T. (1994) Transport modelling in heterogeneous aquifers. 1. Statistical description and numerical generation of gravel deposits. *Water Resources Research*, **30**, 1803–1817.

Klingbeil, R., Kleinedam, S., Asprion, U., Aigner, T. and Teutsch, G. (1999) Relating lithofacies to hydrofacies, outcrop-based hydrogeological characterisation of Quaternary gravel deposits. *Sedimentary Geology*, **129**, 299–310.

Kondolf, G. M. (2000) Assessing salmonid spawning gravel quality. *Transactions of the American Fisheries Society*, **129**, 262–281.

Lancaster, J. (1999) Small-scale movements of lotic macroinvertebrates with variations in flow. *Freshwater Biology*, **41**, 605–619.

Lancaster J. (2000) Geometric scaling of microhabitat patches and their efficacy as refugia during disturbance. *Journal of Animal Ecology*, **69**, 442–457.

Lancaster, J., Buffin-Belanger, T, Reid, I. and Rice, S. (2006) Flow- and substratum-mediated movement by a stream insect. *Freshwater Biology*, **51**, 1053–1069.

Lane, S. N. (2006) Approaching the system-scale understanding of braided river behaviour. In *Braided Rivers*. Sambrook Smith, G. H., Best, J. L., Bristow, C. and Petts, G. E. (eds). Special Publication of the IAS, **36**, 107–135.

Lane, S. N. and Richards, K. S. (1998) Two-dimensional modelling of flow processes in a multi-thread channel. *Hydrological Processes*, **12**, 1279–1298.

Laronne, J. B. and Duncan, M. J. (1992) Bedload transport paths and gravel bar formation. In *Dynamics of Gravel-bed Rivers*. Billi, P., Hey, C. R., Thorne, C. R. and Tacconi, P. (eds). John Wiley & Sons, Ltd, Chichester, pp. 177–200.

Lawless, M. and A. Robert (2001) Three-dimensional flow structure around small-scale bedforms in a simulated gravel-bed environment, *Earth Surface Processes and Landforms*, **26**, 507–522.

Leclerc, M. (2005) Ecohydraulics: a new interdisciplinary frontier for CFD. In *Computational Fluid Dynamics: Applications in Environmental Hydraulics*. Bates, P. D., Lane, S. N. and Ferguson, R. I. (eds). John Wiley & Sons, Ltd, Chichester, pp. 429–460.

Leddy, J. O., Ashworth, P. J. and Best, J. L. (1993) Mechanisms of anabranch avulsion within gravel-bed braided rivers: observations from a scaled physical model. In *Braided Rivers*. Best, J. B. and Bristow, C. S. (eds). Geological Society Special Publication, No. 75, pp. 119–127.

Lunt, I. A. and Bridge, J. S. (2004) Evolution and deposits of a gravely braid bar Sagavanirktok River, Alaska. *Sedimentology*, **51**, 1–18.

Lunt, I. A., Bridge, J. S. and Tye, R. S. (2004) Development of a 3-D depositional model of braided-river gravels and sands to improve aquifer characterization. In *Aquifer Characterization*. Bridge, J. S. and Hyndman, D. W. (eds). SEPM Special Publication No. 80, pp. 139–169.

Macklin, M. G. (1997) Fluvial geomorphology of north-east England (Black Burn, Cumbria). In *Fluvial Geomorphology of Great Britain*. Gregory, K. J. (ed). Chapman and Hall, London, pp. 205–209.

Macklin, M. G., Rumsby, B. T. and Heap, T. (1992) Flood alluviation and entrenchment–Holocene valley floor development and transformation in the British uplands. *Geological Society of America Bulletin*, **104**, 631–643.

Madej, M. A. (1982) Sediment transport and channel changes in an aggrading stream in the Puget Lowland, Washington. In *Sediment Budgets and Routing in Forested Drainage Basins*. Swanson, F. J., Janda, R. J., Dunne, T. and Swanston, D. N. (eds). USDA Forest Service Technical Report, PNW-141, pp. 97–109.

Masch, F. D. and Denny, K. J. (1966) Grain-size distribution and its effect on the permeability of unconsolidated sands. *Water Resources Research*, **2**, 665–677.

Matthaei, C. D., Guggelberger, C. and Huber, H. (2003) Local disturbance history affects patchiness of benthic river algae. *Freshwater Biology*, **48**, 1514–1526.

Meade, R. H. (1985) Wavelike movement of bed load sediment, East Fork River, Wyoming. *Environmental Geology and Water Sciences*, **7**, 215–225.

Moreton, D. J., Ashworth, P. J. and Best, J. B. (2002) The physical scale modelling of braided alluvial architecture and estimation of subsurface permeability. *Basin Research*, **14**, 265–285.

Mount, N. J. (2000) Medium-term response of lowland river reaches to changes in upland land use. Unpublished PhD thesis, Liverpool John Moores University.

Mount, N. J., Sambrook Smith, G. H. and Stott, T. A. (2005) An assessment of the impact of upland afforestation on lowland river reaches: the Afon Trannon, Mid-Wales. *Geomorphology*, **64**, 255–269.

Mumpy, A. J., Jol, H. M., Kean, W. F. and Isbell, J. L. (in press). Architecture and sedimentology of an active braid bar in the Wisconsin River based on 3D ground penetrating radar. In *Stratigraphic Analysis using Ground Penetrating Radar*. Baker, G. S., Jol, H. M. (eds). GSA Special Paper.

Murray, A. B. and Paola, C. (2003) Modelling the effect of vegetation on channel pattern in braided rivers. *Earth Surface Processes and Landforms*, **28**, 131–143.

Naegeli, M. W., Huggenburger, P. and Uehlinger, U. (1996) Ground penetrating radar for assessing sediment structures in the hyporheic zone of a prealpine river. *Journal of the North American Benthological Society*, **15**, 353–366.

Nicholas, A. P. and Sambrook Smith, G. H. (1999) Numerical simulation of three-dimensional flow hydraulics in a braided channel. *Hydrological Processes*, **13**, 913–929.

Nicholas, A. P., Thomas, R. and Quine, T. A. (2006) Cellular modelling of braided river form and process. In *Braided Rivers*. Sambrook Smith, G. H., Best, J. L., Bristow, C. and Petts, G. E. (eds). Special Publication of the IAS, **36**, 137–151.

Nowell, A. R. M. and Church, M. (1979) Turbulent flow in a depth-limited boundary layer. *Journal of Geophysical Research*, **84**, 4816–4824.

Packman, A. I., Salehin, M. and Zaramella, M. (2004) Hyporheic exchange with gravel beds: Basic hydrodynamic interactions and bedform-induced advective flows. *Journal of Hydraulic Engineering*, **130**, 647–656.

Paetzold, A., Schubert, C. J. and Tockner, K. (2005) Aquatic terrestrial linkages along a braided-river: Riparian arthropods feeding on aquatic insects. *Ecosystems*, **8**, 748–759.

Paetzold, A., Bernet, J. F. and Tockner, K. (2006) Consumer-specific responses to riverine subsidy pulses in a riparian arthropod assemblage. *Freshwater Biology*, doi:10.1111/j.1365-2427.2006.01559.x.

Page, K. and Nanson, G. C. (1982) Concave bank formation and associated floodplain formation. *Earth Surface Processes and Landforms*, **7**, 529–543.

Paola, C., Foufoula-Georgiou, E., Dietrich, W. E., Hondzo, M., Mohrig, D., Parker, G., Power, M. E., Rodriguez-Iturbe, I., Voller, V. and Wilcock, P. (2006) Toward a unified science of the Earth's surface: Opportunities for synthesis among hydrology, geomorphology, geochemistry and ecology. *Water Resources Research*, **42**, W03S10, doi:10.1029/2005WR004336.

Passmore, D. G. and Macklin, M. G. (2000) Late Holocene channel and floodplain development in a wandering gravel-bed river: the River South Tyne at Lambley, Northern England. *Earth Surface Processes and Landforms*, **25**, 1237–1256.

Poff, N. L., Allan, J. D., Bain, M. B., Karr, J. R., Prestegaard, K. L., Richter, B. D., Sparks, R. E. and Stromberg, J. C. (1997) The natural flow regime. *Bioscience*, **47**, 769–784.

Poole, G. C. (2002) Fluvial landscape ecology: addressing uniqueness within the river discontinuum. *Freshwater Biology*, **47**, 641–660.

Powell, D. M. (1998) Patterns and processes of sediment sorting in gravel-bed rivers. *Progress in Physical Geography*, **22**, 1–32.

Richards, K., Brasington, J. and Hughes, F. (2002) Geomorphic dynamics of floodplains: ecological implications and a potential modelling strategy. *Freshwater Biology*, **47**, 559–579.

Rice, S. P., Greenwood, M. T. and Joyce, C. B. (2001a) Macroinvertebrate community changes at coarse sediment recruitment points along two gravel bed rivers. *Water Resources Research*, **37**, 2793–2803.

Rice, S. P., Greenwood, M. T. and Joyce, C. B. (2001b) Tributaries, sediment sources, and the longitudinal organisation of macroinvertebrate fauna along river systems. *Canadian Journal of Fisheries and Aquatic Sciences*, **58**, 824–840.

Sadler, J. P. (2005) The ecological functioning of exposed riverine sediments. In *Emerging Concepts for Integrating Human and Environmental Water Needs in River Basin Catchment Management*. Petts, G. E. and Kennedy, R. (eds). US Army Corps of Engineers, ERDC/EL TR-05-13.

Sadler, J. P., Bell, D. and Fowles, A. P. (2004) The hydroecological controls and conservation value of beetles on exposed riverine sediments in England and Wales. *Biological Conservation*, **118**, 41–56.

Sambrook Smith, G. H. (2000) Small-scale cyclicity in alpine proglacial fluvial sedimentation. *Sedimentary Geology*, **132**, 217–231.

Sambrook Smith, G. H. and Nicholas, A. P. (2005) Effect on flow structure of sand deposition on a gravel bed: Results from a two-dimensional flume experiment. *Water Resources Research*, **41**, W10405, doi:10.1029/2004WR003817.

Sambrook Smith, G. H., Ashworth, P. J., Best, J. L., Woodward, J. and Simpson, C. J. (2006) The sedimentology and alluvial architecture of the sandy braided South Saskatchewan River, Canada. *Sedimentology*, **53**, 413–434.

Silvester, N. R. and Sleigh, M. A. (1985) The forces on microorganisms at surfaces in flowing water. *Freshwater Biology*, **15**, 433–448.

Smith, N. D. (1978) Some comments on terminology for bars in shallow rivers. In *Fluvial Sedimentology*. Miall, A. D. (ed). Canadian Society of Petroleum Geologists Memoir, No. 5, pp. 85–88.

Titzel, C. S. (1997) Quantification of the permeability distribution within sand and gravel lithofacies in a southern portion of the Miami Valley aquifer. Unpublished MSc thesis, Wright State University, Dayton, Ohio.

Tockner, K., Malard, F. and Ward, J. V. (2000) An extension of the flood pulse concept. *Hydrological Processes*, 14, 2861–2883.

Tockner, K., Paetzold, A., Karaus, U., Claret, C. and Zettel, J. (2006) Ecology of braided rivers. In *Braided Rivers*. Sambrook Smith, G. H., Best, J. L., Bristow, C. and Petts, G. E. (eds). Special Publication of the IAS, 36, 339–359.

Tooth, S. and Nanson, G. C. (2000) Anabranching rivers on the Northern Plains of arid central Australia. *Geomorphology*, 29, 211–233.

Van der Nat, D., Tockner, K., Edwards, P. J., Ward, J. V. and Gurnell, A. M. (2003) Habitat change in braided flood plains (Tagliamento, NE-Italy). *Freshwater Biology*, 48, 1799–1812.

Vanotte, R. L., Minshall, G. W., Cummins, K. W., Sedell, J. R. and Cushing, C. E. (1980) The river continuum concept. *Canadian Journal of Fisheries and Aquatic Sciences*, 37, 130–137.

Warburton, J., Davies, T. R. H. and Mandl, M. G. (1993) A meso-scale field investigation of channel change and floodplain characteristics in an upland braided gravel-bed river, New Zealand In *Braided Rivers*. Best, J. B. and Bristow, C. S. (eds). Geological Society Special Publication, No. 75, pp. 241–255.

Warburton, J., Danks, M. and Wishart, D. (2002). Stability of an upland gravel-bed stream, Swinhope Burn, Northern England. *Catena*, 49, 309–329.

Ward, J. V., Tockner, K., Edwards, P. J., Kollmann, J., Bretschko, G., Gurnell, A. M., Petts, G. E. and Rossaro, B. (1999) A reference system for the Alps: the 'Fiume Tagliamento'. *Regulated Rivers: Research and Management*, 15, 63–75.

Wathen, S. J., Hoey, T. B. and Werritty, A. (1997) Quantitative determination of the activity of within reach sediment storage in a small gravel bed river using transit time and response time. *Geomorphology*, 20, 113–134.

Werritty, A. and Ferguson, R. I. (1980) Pattern changes in a Scottish braided river over 1, 30, and 200 years. In *Timescales in Geomorphology*. Cullingford, R. A., Davidson, D. A. and Lewin, J. (eds). John Wiley & Sons, Ltd, Chichester, pp. 53–68.

Werritty, A. and Leys, K. F. (2001) The sensitivity of Scottish rivers and upland valley floors to recent environmental change. *Catena*, 42, 251–273.

Westaway, R. M., Lane, S. N. and Hicks, D. M. (2000) The development of an automated correction procedure for digital photogrammetry for the study of wide, shallow, gravel-bed rivers. *Earth Surface Processes and Landforms*, 25, 209–226.

Westaway, R. M., Lane, S. N. and Hicks, D. M. (2003) Remote survey of large-scale braided rivers using digital photogrammetry and image analysis. *International Journal of Remote Sensing*, 24, 795–816.

20

Physical–Ecological Interactions in a Lowland River System: Large Wood, Hydraulic Complexity and Native Fish Associations in the River Murray, Australia

Victor Hughes, Martin C. Thoms, Simon J. Nicol and John D. Koehn

20.1 Introduction

Interdisciplinary science involves the 'explicit joining of two or more areas of understanding into a single conceptual–empirical structure' (Pickett *et al.*, 1999). Integration of disciplines in this way can be done along additive or extractive lines. The additive case is where two areas of study are combined, more or less intact, into a new composite understanding; in the extractive case, by contrast, different areas of study provide components that are fused to yield a new understanding. Both processes can be used in river science, depending on the nature of the problem at hand and the state of knowledge in the different disciplines.

In the science of ecohydrology–hydroecology, where the investigator seeks to unravel mutual interactions between the hydrological cycle and ecosystems at different scales (Porporato and Rodriguez-Iturbe, 2002), additive studies have dominated (e.g., Statzner and Higler, 1986; Newson and Newson, 2000; Young, 1993). Such a science can be regarded as a subset of the broader field of ecogeomorphology in which three well-advanced disciplines are integrated: river ecology, hydrology, and fluvial geomorphology.

Hydroecology and Ecohydrology: Past, Present and Future, Edited by Paul J. Wood, David M. Hannah and Jonathan P. Sadler
© 2007 John Wiley & Sons, Ltd.

Despite an acceleration in the number of research publications in ecohydrology–hydroecology since the 1980s, few have been extractive in nature (but see Poff and Alan, 1995; Parsons and Thoms, 2007, for exceptions). Thus, the case can be made that development of new paradigms within this emerging discipline has been restricted (Nuttle, 2002).

With increasing pressures on the environment, there is a strong incentive to manage rivers as ecosystems (Palmer and Bernhardt, 2006). This provides a basis for extractive studies in ecohydrology–hydroecology, in this way potentially bridging the gap between the traditional subject boundaries of hydrology and ecology (Hannah et al., 2004).

In this study we bring these general principles to bear and describe the hydraulic complexity of large wood in the main channel of the River Murray, a large lowland system in southeastern Australia. In particular, we examine hydraulic complexity by observing the use of these habitats by fish. We also provide hypotheses on which to base further investigations of large wood reinstatement in these types of river system.

The importance of large wood in riverine ecosystems has been well documented in the literature (Angermeier and Karr, 1984; Tillma et al., 1998; Roni and Quinn, 2001; Hughes and Thoms, 2002). Indeed the reintroduction of large wood is a popular river-management tool (Reich, 1999). However, efforts to date have been crude, being aimed at placing individual pieces of large wood in high-energy river channels so as to create habitat with physical features resembling riffles and pools (Till, 2000). Physically, large wood affects the hydraulic (Gippel et al., 1996; Daniels and Rhoads, 2003), morphological (Piegay and Gurnell, 1997; Montgomery and Piegay, 2003) and ecological (Crook and Robertson, 1999; Pettit et al., 2005) character of rivers, although no simple relationship exists between the abundance or complexity of large wood and the response of riverine ecosystems (Hupp et al., 1995; Gurnell et al., 2000).

The effect of large wood on stream flow, at a range of scales, is a popular area of research. Large wood has been demonstrated to increase in-channel flow resistance (Shields and Gippel, 1995), modify flow structures (Gippel et al., 1996) and influence the range of velocities in the downstream direction (Mutz, 2002). Much of this research is limited to one-dimensional velocity measurements, with the effects in three dimensions being inferred from visual observations. Recent advances in technology and the use of acoustic Dopplers – such as acoustic Doppler current profilers (ADCP; Gaumeuman and Jacobson, 2005) and velocimeters (ADV; Daniels and Rhoads, 2003) – have allowed three-dimensional flow structures and the influence of large wood on them to be directly measured in the field.

These acoustic Doppler studies have focused on small streams (e.g., Daniels and Rhoads, 2003, 2004) and higher-energy systems (Jacobson and Heuser, 2002; Gaueuman and Jacobson, 2005). Associations between large wood and in-channel form and process differ between higher-energy upland rivers and lower-energy lowland rivers (Piegay and Gurnell, 1997; Hughes and Thoms, 2002). In high-energy rivers, large wood can be highly mobile and be transported considerable distances; it can significantly alter flow patterns by forming in-channel geomorphic units (Gurnell et al., 2000) and debris jams (Assani and Petit, 1995). In lower-energy rivers, large wood is relatively immobile, essentially remaining where it falls for extended periods (e.g., greater than 100 years; Hughes and Thoms, 2002) and has strong effects on flow structures at channel margins (Gippel et al., 1996). Interactions between large wood and in-channel flow structures are an area of focus for ecohydrology–hydroecology, albeit limited to higher-energy environments; however associations between these interactions and ecological phenomena are also limited.

Large wood is an important ecological component of many riverine environments, providing physical habitat for aquatic organisms by acting as a substratum for plants (Biggs, 1996), invertebrates (Hax and Golloday, 1998) and fish eggs (Harris and Rowland, 1996). Its effect on flow provides hydraulic relief for organisms such as plants (Biggs and Stokseth, 1996) and fish (Aadland, 1993). Fish often cluster around patches of suitable habitat (Inoue and Nakano, 1998), influenced by water depth, water velocity, instream cover, riverbed substratum (Bain *et al.*, 1988) and the presence of predators and prey (Kramer *et al.*, 1997). Associations occur between the choice of preferred habitat and bioenergetics, with a trade-off between the energy expended in swimming and the energy gained from capturing prey (Hughes and Dill, 1990). River-dwelling fish have been shown to select hydraulic habitats that maximise advantage and minimise disadvantage in order to attain net energy gains (Fausch, 1984; Rosenfeld and Boss, 2001; Grossman *et al.*, 2002). Juvenile salmonids, for example, occupy stream positions of low water velocity to minimise the energy expended on swimming, yet close enough to swift currents to maximise access to invertebrate drift (Fausch, 1984). In addition, fish use large wood as foraging sites, spawning substrates, refuges from predators or prey, and as protection from strong river currents (Lehtinen *et al.*, 1997).

In Australian lowland river systems, Murray cod, *Maccullochella peelii peelii*, and river blackfish, *Gadopsis marmoratus*, have both been reported to use large wood as spawning substrates (Cadwallader and Backhouse, 1983; Jackson, 1978). Similarly, in upland rivers, blackfish use stream positions where water velocities are low (Koehn *et al.*, 1994). Large wood is also a major provider of hydraulic resistance (Gippel *et al.*, 1996), but it is unknown whether in Australian lowland rivers this resistance creates water velocity refuges suitable for fish. Relationships between hydraulic conditions and fish habitat preference have been demonstrated in other types of river systems, notably high-energy gravel-bed rivers (Lamouroux and Souchon, 2002), but the relationships for large low-energy lowland systems remain unknown.

Large wood is a primary source of structural habitat for fish in Australian lowland rivers (Koehn and Nicol, 1998, Growns *et al.*, 2004). Surface water velocity can be used as a discriminator of native fish habitat (Koehn and Nicol, 1998), but it is unclear whether this result is related to river flow patterns or to the velocity refuges created by the hydraulic complexity associated with large wood. The widespread removal of large wood from Australian lowland rivers (Treadwell *et al.*, 1999) is thought to have reduced hydraulic complexity and fish habitat (Hughes and Thoms, 2002); the reinstatement of such habitat is now a popular rehabilitation measure to increase fish populations (MDBC, 2004). As foreshadowed, here we describe the hydraulic complexity of large wood in the main channel of the River Murray and examine hydraulic complexity by observing the use of these habitats by fish; some relevant hypotheses are formulated for the management of large wood in rivers.

20.2 Study Area

This study was conducted in a 95-km reach of the River Murray in SE Australia, between Yarrawonga and Tocumwal (Figure 20.1). The study reach, located within the Riverine Plains Tract of the River Murray, is typical of lowland Australian rivers: low bed slopes, extensive floodplains and a highly meandering river channel (Thoms and Sheldon, 2000). There are four main tracts along the River Murray: the Headwaters Tract or upland area

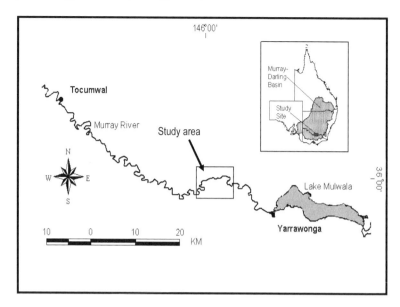

Figure 20.1 The study reach, located along the River Murray in SE Australia

between the source of the river and Albury; the Riverine Plains Tract, which extends westward to the junction of the River Murray and Murrumbidgee River; the Mallee Tract between there and the junction with the Darling River; and the South Australian Tract between the Darling River and Lake Alexandrina (Walker, 1992). Generally, the lowland Riverine Plains, Mallee and South Australian tracts cut into Tertiary alluvium of relatively fine sand (Twidale *et al.*, 1978). These lowland tracts have well-developed floodplains (up to 30 km wide) that are dissected by meandering and anastomosing channels. A feature of the River Murray is an extended long profile (Thoms and Walker, 1993), so that 89% of its length (2560 km) has a channel gradient of less than $0.0017\,\mathrm{m\,km^{-1}}$.

In general, dimensions of the study reach's bankfull channel are: widths up to 150 m, maximum depths of 8 m, width:depth ratios of 20 and cross-sectional areas from 390 to 500 m^2. High flow variability is a feature of the River Murray and derives from erratic rainfall, low relief and high evaporation potential. The coefficient of variation for the long-term annual flows (1900–1998) at Tocumwal is 125%, compared with an average of 33% for rivers of similar size elsewhere in the world (Finlayson and McMahon, 1988). The average annual discharge of the River Murray is 150 m^3 s^{-1} at Tocumwal (range 20–1564 m^3 s^{-1} over the same period).

Large wood is a major physical in-channel structure in this lowland river ecosystem (Thoms and Sheldon, 2000). In the River Murray, large wood is mostly immobile; after being recruited by bank erosion (largely from riparian forests of river red gum, *Eucalyptus camaldulensis*), trees generally remain where they fall. They are passively realigned rather than actively transported (Hughes and Thoms, 2002). Unlike many other reaches of the River Murray, the study reach between Yarrawonga and Tocumwal has not been subjected to extensive removal of large wood – although upstream, between Yarrawonga and Albury, over 10 000 pieces of large wood have been removed from the river channel resulting in major changes to stage–discharge relationships. Removal of large wood from the in-channel environment lowered bankfull stages by over 0.5 m at some sites between Albury and Yarrawonga. Along the study reach, pieces of large wood vary in length (up to 35 m), are structurally complex (from single unbranched logs to complete trees), have angles to the main flow varying from 0 to 90°, and range in proximity to the bank from 0 to 45 m (Hughes and Thoms, 2002).

The study reach is known to contain populations of the native Murray cod, *Maccullochella peelii peelii*; trout cod, *M. macquariensis*; golden perch, *Macquaria ambigua*; and introduced carp, *Cyprinus carpio* (Koehn and Nicol, 1998). These species utilise large wood as primary habitat, and hydraulic complexity is thought to be a factor that may explain observed differences in habitat use between these species (Koehn and Nicol, 1998).

20.3 Methods

Sampling was undertaken when flows in the River Murray were $23.1 \, \text{m}^3 \, \text{s}^{-1}$ at Yarrawonga, just upstream of the study reach and are equal to the 98th percentile on the flow–duration curve. In total, 27 independent large pieces of wood (at least 200 m apart) were surveyed. We considered the sites independent from each other because they were either in different meander bends or in different quarters of the same meander. The following physical variables were recorded for each large wood piece: angle to flow, as one of three classes (class 1 = 0° to 30°, class 2 = >30° to 60°, class 3 = >60° to 90°); distance from bank (measured from the part of the large wood closest to the bank); length of large wood; structural complexity as one of four classes (see Figure 20.2); and bend position (inner or outer bend, 1st, 2nd, 3rd or 4th quarter – Figure 20.3).

To measure hydraulic complexity and fish occupancy for each piece of large wood, the following sampling protocol was used. Initially each piece of large wood was electrofished using a boat-mounted 7.5 GPP electrofishing unit (Smith-Root Inc., Portland, OR, USA). The boat, with a crew consisting of a driver, dip-netter and recorder, approached each piece of large wood from its downstream end. Once the anodes of the boat were

Class	Description	Examples	
1	A single trunk or branch – a root mass or small part of a branch may be present.		
2	Double trunk or branch; or single trunk with one level of branching for most of the length.		
3	Trunk with multiple branches for most of length, with second level of branching present		
4	Complete tree with extensive branching – OR an accumulation of snags in which individual pieces could not be resolved		

Figure 20.2 Structural complexity classes used to describe pieces of large wood

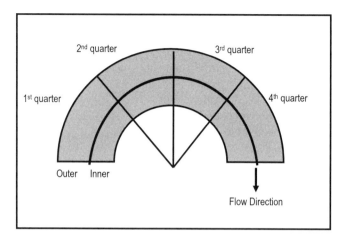

Figure 20.3 Planform bend positions

positioned above the large wood, voltage was applied for a maximum of 8 seconds. The electrofishing unit was set to 1000 V, 120 pulses per sec and 60% duty cycle. Stunned fish were then dip-netted and records made of observed capture position and fish length. Water clarity at the time of sampling was high and fish position was not affected by drift from the point of stunning. In the case of the exotic fish species (carp) only one specimen was captured for measurement; all others were observed and informally considered to be of similar size. To minimise human disturbance to fish occupancy, electrofishing at each piece of large wood was undertaken prior to hydraulic measurements.

Immediately after electrofishing, a second boat fitted with a 1200 kHz Acoustic Doppler Current Profiler (ADCP; RD Instruments, San Diego, CA, USA) recorded a series of velocity profiles using RDI 'Transect' software. In-stream flows can be expressed as three variables: a direction relative to the Earth (given as a compass bearing), a horizontal component (the speed at which the water is flowing in that direction), and a vertical component (the speed at which the water is flowing relative to the horizontal) (Lemmin and Rolland, 1997). The ADCP and Transect software record these components of water velocity averaged over an operator-selected time interval (in this study 5 seconds). The ADCP was mounted at the front of the boat with its transducer faces submerged to a depth of 100 mm and pointing down into the water column. The RDI instrument measured the water velocity in depth cells of 25 cm (RDI, 1994) and produced a vertical velocity profile for each 5-second interval (Table 20.1). The boat was driven around each large wood at $<1\,\mathrm{m\,s^{-1}}$ and approximately 50 velocity profiles were recorded around each piece of large wood. The Transect files were post-processed to remove unreliable data (RDI, 1994). Then, at each piece of large wood, the velocity profiles were combined and summarised to give the following hydraulic variables for each large wood: horizontal velocity (mean, minima, maxima, standard deviation and distribution kurtosis and skewness); vertical velocity (mean, minima, maxima, range and standard deviation); and velocity direction (bearing and standard deviation).

Dissimilarity matrices were prepared for the physical character, hydraulic character, combined physical and hydraulic characteristics and number of fish by species. Each

Table 20.1 *Example of a velocity profile*

Depth (m)	Horizontal velocity (cm s^{-1})	Flow direction (degrees)	Vertical velocity (cm/s^{-1})
0.5 to <0.75	9.2	191.3	−7.9
0.75 to <1.0	47.9	342.4	3.6
1.0 to <1.5	45.7	234.7	−25.2
1.5 to <1.75	47.8	265	5.3
1.75 to <2.0	29.2	181.8	−9.8

dissimilarity matrix was derived using the Gower association measure, as recommended for nonbiological data (Belbin, 1993). Mantel tests were performed to compare the following matrix pairs: physical compared to hydraulic, physical to fish, hydraulic to fish, and combined physical–hydraulic to fish (as recommended by Quinn and Keough, 2002).

Graphic summaries of the snag hydraulic data were visually examined to see if particular hydraulic variables were associated with an increased abundance of fish (e.g., 77% of fish were caught at a piece of large wood with a mean velocity below the median). This was followed by a multivariate classification of the 27 individual large wood pieces using seven hydraulic variables: range of vertical velocity; mean, standard deviation, minimum and skewness of horizontal velocity; bearing; and standard deviation of direction. Data were classified using the flexible 'unweighted pair-groups using arithmetic averages' (UPGMA) fusion strategy (Belbin and McDonald, 1993). Similar groups were independently selected from the dendrogram by using a dissimilarity threshold of 0.334, as recommended by Quinn and Keough (2002). A multidimensional scaling (MDS) ordination was also undertaken to further investigate the association of large wood groups in the study reach. To assess those physical variables that may influence hydraulic conditions at individual pieces of large wood, a principal axis correlation (PCC; Belbin, 1993) was conducted, with the resulting vectors plotted on the ordination graph.

To test if the presence or absence of fish could be explained by the occurrence of regions of low flow (i.e., velocity shelters); the proportion of flow records measured at each piece of large wood that were within low, medium or high categories were calculated for both horizontal and vertical velocity profiles. In the horizontal plane, low velocities were considered to be those below 30 cm s^{-1}, medium velocities between 30 and 50 cm s^{-1} and high greater than 50 cm s^{-1}. Vertical velocities were considered high if they were greater than 8 cm s^{-1} upwards or 11 cm s^{-1} downwards. Correspondingly, medium velocities were those between 5 and 8 cm s^{-1} (upwards) and −5 and −11 cm s^{-1} (downwards); low velocities were those between 0 and 5 cm s^{-1} (upwards) and 0 and −5 cm s^{-1} (downwards). These categories emerged by dividing the range of values in the data into thirds. All data were root-transformed in order to ensure normality. The mean and 95% confidence intervals for each category were computed for two groups of large wood groups: fish present and fish absent. These statistics were then visually examined to decide whether any probabilistic tests were required. Formal analysis of the data for each species was not considered because of the low sample size of each group (seven large wood pieces with Murray cod, ten with trout cod, five with carp, four with golden perch and 11 with no fish).

20.4 Results

20.4.1 Physical and Hydraulic Characteristics of Large Wood

Mantel test comparisons of dissimilarity matrices showed no statistically significant correlations between any of the variable pairs physical to hydraulic, physical to fish, hydraulic to fish, or combined physical–hydraulic to fish. The classification dendrogram did reveal four groups of large wood (plus two outliers) and this was confirmed by the MDS ordination (Figure 20.4). Only five physical variables had an $r^2 > 0.75$ in the PCC analysis (angle, bend position, distance, length, complexity) and these were associated with the different large wood groups (as shown by their vector position in the MDS ordination, Figure 20.4). The character of the different large wood groups, as given in Table 20.2, are as below.

Group A The 11 pieces of large wood in this group were the most perpendicular to the flow, the closest to the bank, longest and most structurally complex. Compared with the other groups, these pieces had the lowest mean horizontal velocity and were more hydraulically complex, with the largest range of vertical velocities, coefficient of variances for

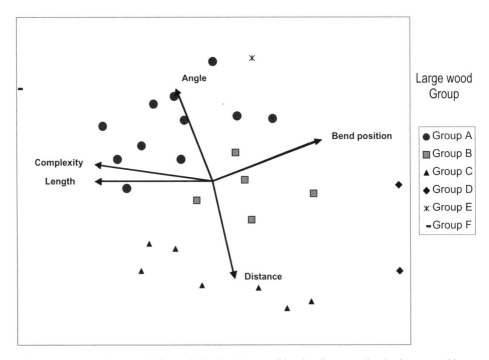

Figure 20.4 *Ordination of the eight hydraulic variables for the 27 individual pieces of large wood. PCC vectors for the physical variables of large wood in hydraulic space are also displayed. The stress level for this ordination was 0.12 in two dimensions. Distance = distance from bank. Large wood groups set* a priori *from the classification dendrogram*

Table 20.2 Summary of the physical and hydraulic variables of the four groups elicited from the classification of selected large wood hydraulic variables and associated fish numbers

Variable (mean of)	Group (no. of large wood in group)			
	A (11)	B (5)	C (7)	D (2)
Physical character				
Angle class	2.4	2.2	1.6	1.5
Distance to bank (m)	10.5	22.0	20.0	15.0
Length (m)	18.6	16.0	16.4	15.0
Complexity class	1.9	1.4	1.3	1.0
Hydraulic character				
Vertical velocity range (cm s^{-1})	21.2	20.8	15.7	28.8
Horizontal velocity (cm s^{-1})	23.2	31.1	31.4	41.2
Horizontal velocity standard deviation (cm s^{-1})	11.7	13.6	11.5	15.8
Horizontal velocity coefficient of variance	50.3	43.5	36.6	38.3
Horizontal velocity minimum (cm/s^{-1})	1.5	2.9	3.8	3.2
Horizontal velocity distribution skewness	0.48	0.07	−0.10	0.17
Direction standard deviation (°)	53.3	44.7	26.3	27.6
Direction range (°)	349.6	311.9	224.5	161.2

horizontal velocities, and ranges and standard deviations of velocity directions. Rather than standard deviation, coefficient of variance was used to provide a measure of between-group comparison of the variability of horizontal velocity. All large wood in this group had strongly positively skewed velocity distributions (range 0.45 to 0.88), indicating a predominance of velocities below the median. The PCC vectors indicated that this group of large wood was associated with the physical characteristics of being close to the bank and perpendicular to the flow (Figure 20.4).

Group B This group contained five individual pieces of large wood and collectively they were less perpendicular to flow, relatively more distant from the bank, shorter, and simpler in structure compared with Group A. Large wood in this group also had higher mean horizontal velocities than those in Group A and were less hydraulically complex, with a lower range of vertical velocities and directions. This group of large wood also had less variability in horizontal velocities and directions of flow. The skewness of velocity distributions was approximately normal, with values ranging from −0.12 to +0.16.

Group C The seven large wood pieces in this group were more aligned to the flow than large wood in Group B, but were similar in other physical characteristics. However, this group had a smaller range of vertical velocities than either Group A or B. Mean horizontal velocities were similar to Group B, but the coefficient of variance was lower. They also had a lower range, and less variability, of flow directions. Large wood in this group had horizontal velocity distributions that varied between +0.33 and −0.45. The PCC vectors indicated that this group of snags was associated with the physical characteristics of being distant from the bank and parallel to the flow (Figure 20.4).

Group D The two pieces of large wood had a similar alignment to flow as Group C, but were closer to the bank, shorter and structurally simple (single, unbranched logs). This group of large wood had the highest range of vertical velocities and the highest mean horizontal velocities. Variability of horizontal velocities and of flow directions were similar to Group C, but the range of flow directions was the lowest of all groups. The skewness of velocity distributions ranged between −0.05 and +0.38. The PCC vectors indicated that this group of snags was associated with the physical characteristics of being structurally simple and short (Figure 20.4).

Group E This group consisted of a single piece of large wood that was perpendicular to the flow, close to the bank, of medium length and structurally simple. Its mean horizontal velocity was similar to the average values of groups A–D and the distribution of horizontal velocities was negatively skewed. It had a high range of velocity directions and no fish were caught near this piece.

Group F This too was a single piece of large wood aligned perpendicular to the flow, close to the bank and of medium length, but had a structural complexity of 2 (Figure. 20.2). It had a low mean horizontal velocity and the distribution of horizontal velocities was negatively skewed. It had a high range of velocity directions and no fish were caught nearby.

20.4.2 Fish Capture

A total of 47 fish from four species were captured from the 27 pieces of large wood, with trout cod and carp the most abundant (Tables 20.3 and 20.4). The size range of fish differed considerably (181–980 mm) between species. The Murray cod varied between 288

Table 20.3 Number of fish caught at each large wood group

Fish species (totals for all large wood)	Group A	Group B	Group C	Group D
Murray Cod	7	1	0	1
Trout Cod	7	4	5	0
Carp	14	2	2	0
Golden Perch	4	0	0	0
Total number of fish per group	32	7	7	1
Mean number of fish per large wood	2.9	1.4	1.0	0.5

Table 20.4 Number, mean length and length range of fish species used in analysis

Fish species	Total no.	Mean length	Length range
Murray Cod	9	496	288–980
Trout Cod	17	301	181–660
Golden Perch	5	450	404–591
Carp	16	535	

and 980 mm and the trout cod between 181 and 660 mm. Golden perch were of medium size, with a more restricted range of sizes (404–591 mm). Only one carp was measured (535 mm), although others were informally observed to be of similar size. Large wood Group A had the highest number of fish per piece: all species were well represented and it was the only group where golden perch were recorded. Seven fish (including four trout cod) were caught at large wood Group B. Seven fish (five trout cod and two carp) were recorded at large wood Group C, and one fish (a Murray cod) was caught at Group D.

20.4.3 Analysis of Fish Abundance in Areas of Low, Medium and High Velocities

Visual examination of the mean and 95% confidence intervals of the hydraulic velocity categories around individual pieces of large wood with fish presence and absence revealed no differences in the means within each category (Fig. 5a,b). Consequently, no formal probabilistic test was undertaken.

20.5 Discussion

Pieces of large wood with different physical characteristics were associated with different hydraulic conditions. Lower mean velocities and relatively more diverse hydraulic conditions (greatest diversity of horizontal velocity and flow direction) were associated with physically more complex large wood (Group A). Decreases in the physical complexity of large wood were associated with an increase in mean velocities and decreased hydraulic complexity (Groups B, C and D, progressively).

A comparison of the physical and hydraulic characteristics of different large wood groups suggests that particular large wood physical properties may be associated with certain hydraulic characteristics. The greatest physical difference between large wood of Groups A and B was the proximity to bank; the greatest hydraulic differences were in the mean and variability of horizontal velocity. By contrast, the greatest physical difference between Groups B and C was in angle to the flow; the greatest hydraulic differences were in the range and variability of flow direction. It is pertinent to note that no single physical variable is the sole predictor for hydraulic complexity – each of the four physical variables appears to influence hydraulic complexity and the variables act together to increase hydraulic complexity. Because the technology to measure three-dimensional flows in the field is a relatively recent development, detailed investigations of the effect of large wood on the fluvial dynamics of meander bends are limited. Daniels and Rhoads (2003, 2004) demonstrated that large wood strongly influenced flow direction and near-bank velocities in small streams, but this study did not compare large wood of varying complexity. In addition, Gippel *et al.* (1996) demonstrated that large wood is a source of hydraulic resistance and this increases when large wood presents greater angles to the flow. However, the study of Gippel *et al.* (1996) was limited to investigating flow in one dimension – the longitudinal direction.

In the present study, fish were found to be more abundant at large wood that was hydraulically complex and had lower horizontal velocities. This suggests that areas of lower velocities (velocity shelters) may be preferred by fish in Australian lowland rivers, a result similar to that seen in higher-energy rivers. Fausch (1984) found that trout utilised boulders to maintain favourable (for predation) stream positions with minimal energy expenditure. Similarly, Rosenfeld and Boss (2001) demonstrated that the preference of juvenile cutthroat trout for pools (slow-flowing) over riffles (faster flowing) could be explained by energetics – fish lost weight in riffles and gained weight in pools.

In the present study, the species composition of the fish caught was associated with the position of large wood in the river channel. All four fish species caught in this study were found in association with large wood close to the bank (Group A); however trout cod were also found in association with large wood more distant from the bank (Groups B and C). In general, these pieces of large wood were found to be hydraulically less complex and had higher mean horizontal velocities (e.g., Group A). Murray cod and trout cod have been previously reported to favour both locations in other reaches of the River Murray (Koehn and Nicol, 1998); however, this earlier report did not indicate whether the size of large wood and its physical complexity were associated with fish occupancy or species composition. The results of this study (Figure 20.5a,b) showed no obvious difference in the distribution of low, medium and high velocity classes for vertical and horizontal velocities around large wood – regardless of whether fish were present or absent. However, areas of low velocity in both the horizontal and vertical profiles were the dominant category observed, supporting the hypothesis that fish use velocity shelters constructed by large wood. Fish presence at some pieces of large wood and not others may be explained by other factors such as position in the channel (as discussed above). In an extensive study of the habitat used by the species in this study, there was no reported association between these species and open water habitats in this reach (Nicol, 2005). However, the inclusion of an open water control would be advisable if considering other fish species.

Preference for certain hydraulic habitats is affected by the behavioural and ecological traits of each species and their life stage. Murray cod and trout cod are largely benthic species that prefer shelter near the bottom, whereas golden perch and carp are midwater schooling species that may require wood higher in the water column (Cadwallader and Backhouse, 1983; Koehn and Nicol, 1998). This level of detail has not been explored in this paper. Optimal foraging has been used to predict microhabitat selection in drift-feeding stream salmonids (Fausch, 1984; Hill and Grossman, 1993; Hughes, 1998) and focal-point velocities occupied by several species of stream minnow (Grossman et al., 2002). There is an opportunity for investigation of the microhabitat choice of hydraulic areas for foraging by Australian lowland river species. An immediate hypothesis to explore would be to test the association between velocity refuges and fish occupancy for a range of species and life stages. Furthermore, altered flows associated with river regulation and irrigation typically create long periods of constant flows. Understanding the consequent changes in hydraulic diversity, and hence changes in availability of suitable fish habitat, provides insight into the importance of large wood as a source of hydraulic diversity in the River Murray. This is important, given the widespread removal of large wood in the past and the recognition of large wood reinstatement as a fish habitat rehabilitation measure (MDBC, 2004).

Although the presence or absence of fish was not associated with differences in the relative distribution of low, medium and high vertical and horizontal velocities around large wood, areas of low velocity in both the horizontal and vertical directions were the dominant category observed. This result suggests that while areas of low velocity are a

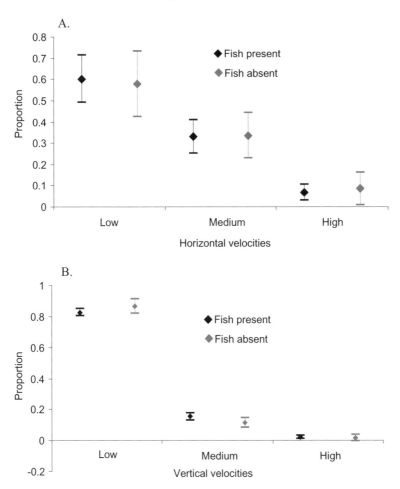

Figure 20.5 Distribution of low, medium and high horizontal (A) and vertical (B) velocities – fish capture points compared to large wood pieces where fish absent. Velocities were root transformed to normalise data. Velocity classes are: for horizontal velocities, low is $<30\,cm/s^{-1}$, medium is from 30 to $50\,cm/s^{-1}$, high is $>50\,cm/s^{-1}$; for vertical velocities, low is from >-5 to $<5\,cm/s^{-1}$, medium is from -11 to $-5\,cm/s^{-1}$ and from 5 to $8\,cm/s^{-1}$, high is $<-11\,cm/s^{-1}$ and $>8\,cm/s^{-1}$. Error bar = 2 × standard error

dominant hydraulic characteristic of large wood, these may not be the only habitat characteristic associated with fish.

20.6 Conclusions

Hydraulic complexity is an important factor in the choice of habitat by riverine fish species. In lowland rivers, large wood modifies flow patterns, influences hydraulic complexity and creates habitat refuges. Associations between hydraulic complexity and native fish abundance and species composition have been demonstrated in this study for the River Murray, a large lowland river system in SE Australia. Four distinct groups of large wood hydraulic character associations were found, ranging from high to low hydraulic complexity. Hydraulic complexities increased and mean flow velocities declined as large wood became more perpendicular to the flow, structurally more complex, and larger and closer to the bank. Fish sought out this hydraulic complexity, with mean velocity and the position of large wood within the river channel being identified as important factors associated with its use. Trout cod were associated with large wood that had higher mean velocities and was located in central river channel positions, whereas Murray cod and golden perch were associated with sites that had lower velocities and were closer to the river-bank. The observations from this study provide a starting point for further investigation into hydraulic complexity and habitat used by these lowland riverine fish species, as well as assisting with habitat restoration for these species.

The reintroduction of large wood is a popular river management tool (Reich, 1999). Efforts to date have been relatively crude and are generally aimed at placing individual pieces of large wood in high-energy river channels to create features such as riffles and pools (Till, 2000). In low-energy rivers, management of large wood is generally undertaken at the habitat scale because the aim of restoration projects is often linked to single-species conservation (e.g., Murray cod: Koehn *et al.*, 2004). There is no consideration of the interactions between channel morphology, large wood character, hydraulic habitat and fish use. The results of this study suggest that, at least in low-energy systems similar to the River Murray, reintroduction of large wood should focus on the position of individual large wood within meander bends and the character of the large wood. The establishment of physical–biological associations, such as those demonstrated in this study between large wood, hydraulic complexity and fish use, will help to restore the ecosystem integrity of large lowland rivers.

Many disciplines are often brought together to solve environmental problems. Interdisciplinary research in river ecosystems is a relatively young endeavour and one that is fraught with problems – linking across scales and integrating different disciplinary approaches and conceptual tools. Frameworks are useful tools for achieving this, helping to define the bounds for the selection and solution of problems; they indicate the role of empirical assumptions, carry the structural assumptions, show how facts, hypotheses, models and expectations are linked, and indicate the scope to which a generalisation or model applies. Interdisciplinary river science lacks such an integrative framework. A framework is neither a model nor a theory: models describe how things work and theories explain phenomena, whereas conceptual frameworks help to order phenomena and materials, thereby revealing patterns. In order to advance interdisciplinary arenas like ecogeomorphology or ecohydrology–hydroecology, we require the development and articulation of a framework to unify the field of study and ensure interdisciplinary interaction at appropriate scales.

Acknowledgements

Jason Lieschke, John Mahoney, Jarod Lyon and Adam Scott helped with fieldwork. This work was undertaken with the financial assistance of a grant from the Murray–Darling

Basin Commission (MT and SN) and a PhD scholarship from Land & Water Australia (VH).

References

Aadland, L.P. 1993. Stream habitat types: their fish assemblages and relationship to flow. *North American Journal of Fisheries Management*, **13**, 790–806.
Angermeier, P.L., Karr, J.R. 1984. Relationships between woody debris and fish habitat in a small warmwater stream. *Transactions of the American Fisheries Society*, **113**, 716–726.
Assani, A.A., Petit, F. 1995. Log-jam effects on bed-load mobility from experiments in Pacific coastal streams and rivers. In Naiman, R.J., Bilby, R.E. (eds), *River Ecology and Management: Lessons from the Pacific Coastal Ecoregion*. Springer, New York, pp. 324–346.
Bain, M.B., Finn, J.T., Booke, H.E. 1988. Streamflow regulation and fish community structure. *Ecology*, **69**, 382–392.
Belbin, L. 1993. *PATN – Pattern Analysis Package*. CSIRO Division of Wildlife and Ecology, Canberra.
Belbin, L., McDonald, C. 1993. Comparing three classification strategies for use in ecology. *Journal of Vegetation Science*, **4**, 341–348.
Biggs, B.J.F. 1996. Hydraulic habitat of plants in streams. *Regulated Rivers: Research and Management*, **12**, 131–144.
Biggs, B.J.F., Stokseth, S. 1996. Hydraulic habitat suitability for periphyton in rivers. *Regulated Rivers: Research and Management*, **12**, 251–261.
Cadwallader, P.L., Backhouse, G.N. 1983. *A Guide to the Freshwater Fish of Victoria*. Government Printer, Melbourne.
Crook D.A., Robertson, I.A. 1999. Relationships between riverine fish and woody debris: implications for lowland rivers. *Marine and Freshwater Research*, **50**, 941–953.
Daniels, M.D., Rhoads, B.L. 2003. Influence of a large woody debris obstruction on three-dimensional flow structure in a meander bend. *Geomorphology*, **51**, 159–173.
Daniels, M.D., Rhoads, B.L. 2004. Effect of large woody debris configuration on three-dimensional flow structure in two low-energy meander bends at varying stages. *Water Resources Research*, **10**, W11302.
Fausch, K.D. 1984. Profitable stream positions for salmonids: relating specific growth rate to net energy gain. *Canadian Journal of Zoology*, **62**, 441–451.
Finlayson, B.L., McMahon, T.A. 1988. Australia v the world: A comparative analysis of streamflow characteristics. In Warner, R.F. (ed.), *Fluvial Geomorphology of Australia*. Academic Press, Sydney, pp. 17–40.
Gaueuman, D., Jacobson, R.B. 2005. Aquatic habitat mapping with an acoustic Doppler current profiler: Considerations for data quality. *USGS Report* 2005-1163.
Gippel, C.J., Finlayson, B.L., O'Neill, I.C. 1996. Distribution and hydraulic significance of large woody debris in a lowland Australian river. *Hydrobiologia*, **318**, 179–194.
Grossman, G.D., Rincon, P.A., Farr, M.D., Ratajcak Jr., R.E. 2002. A new optimal foraging model predicts habitat use by drift-feeding stream minnows. *Ecology of Freshwater Fish*, **11**, 2–10.
Growns, I., Wooden, I., Schiller, C. 2004. Use of instream wood habitat by trout cod *Maccullochella macquariensis* (Cuvier) in the Murrumbidgee River. *Pacific Conservation Biology*, **10**, 261–265.
Gurnell, A.M., Petts, G.E., Hannah, D.M., Smith, B.P.G., Edwards, P.J., Kollman, J., Ward, J.V., Tockner, K. 2000. Wood storage within the active zone of a large European gravel-bed river. *Geomorphology*, **34**, 55–72.
Hannah, D.M., Wood, P.J., Sadler, J.P. 2004. Ecohydrology and hydroecology: A 'new paradigm'? *Hydrological Processes*, **18**, 3439–3445.
Harris, J.H., Rowland, S.J. 1996. Australian freshwater cods and basses. In McDowall, R.M. (ed.), *Freshwater Fishes of South-eastern Australia*. Reed Books, Sydney, pp. 150–167.

Hax, C.L., Golloday, S.W. 1998. Flow disturbance of macroinvertebrates inhabiting sediments and woody debris in a prairie stream. *American Midland Naturalist*, **139**, 210–223.

Hill, J., Grossman, G.D. 1993. An energetic model of microhabitat use for rainbow trout and rosyside dace. *Ecology*, **74**, 685–698.

Hughes, N.F. 1998. A model of habitat selection by drift-feeding stream salmonids at different scales. *Ecology*, **79**, 281–294.

Hughes, N.F., Dill, L.M. 1990. Position choice by drift-feeding salmonids: model and test for arctic grayling (*Thymallus arcticus*) in subarctic mountain streams, interior Alaska. *Canadian Journal of Fisheries and Aquatic Sciences*, **47**, 2039–2048.

Hughes, V., Thoms, M.C. 2002. Associations between channel morphology and large woody debris in a lowland river. *International Association of Hydrological Sciences*, **276**, 11–18.

Hupp, C.R., Osterkamp, W.R., Howard, A.D. (eds). 1995. *Biogeomorphology, Terrestrial and Freshwater systems*. Elsevier, Amsterdam.

Inoue, M., Nakano S. 1998. Effects of woody debris on the habitat of juvenile masu salmon (*Oncorhynchus masou*) in northern Japanese streams. *Freshwater Biology*, **40**, 1–16.

Jackson, P.D. 1978. Spawning and early development of the river blackfish, *Gadopsis marmoratus* Richardson (Gadopsiformes: Gadopsidae) in the McKenzie River, Victoria. *Australian Journal of Marine and Freshwater Research*, **29**, 293–298.

Jacobson, R.B., Heuser, J.L. 2002. Visualization of flow alternatives, Lower Missouri River, US Geological Survey Open-file Report 02-122.

Kemp, J.L., Harper, D.M. Crosa, G. 1999. Use of 'functional habitats' to link ecology with morphology and hydrology in river rehabilitation. *Aquatic Conservation: Marine and Freshwater Ecosystems*, **9**, 159–178.

Koehn, J., Nicol, S. 1998. Habitat and movement requirements of fish. In Banens, R.J., Lehane, R. (eds), *Riverine Environment Research Forum*. Proceedings of the inaugural Riverine Environment Research Forum. Murray–Darling Basin Commission, Canberra, pp. 1–6.

Koehn, J.D., Nicol, S.J., Fairbrother, P.S. 2004. Spatial arrangement and physical characteristics of structural woody habitat in a lowland river in south-eastern Australia. *Aquatic Conservation: Marine and Freshwater Ecosystems*, **14**, 457–464.

Koehn, J.D., O'Connor, N.A., Jackson, P.D. 1994. Seasonal and size-related variation in microhabitat use of a small Victorian stream assemblage. *Australian Journal of Marine and Freshwater Research*, **45**, 1353–1366.

Kramer D.L., Rangeley R.W., Chapman L.J. 1997. Habitat selection: patterns of spatial distribution from behavioural decisions. In Godin, J.J. (ed.), *Behavioural Ecology of Teleost Fishes*. Oxford University Press, Oxford, pp. 37–80.

Lamouroux, N., Souchon, Y. 2002. Simple predictions of instream habitat model outputs for fish habitat guilds in large streams. *Freshwater Biology*, **47**, 1531–1542.

Lehtinen, R.M., Mundahl, N.M., Madejczyk, J.C. 1997. Autumn use of woody large wood by fishes in backwater and channel border habitats of a large river. *Experimental Biology of Fishes*, **49**, 7–19.

Lemmin, U., Rolland, T. 1997. A monostatic acoustic velocity profiler for laboratory and field studies of turbulent flow. *Journal of Hydraulic Engineering*, **123**, 1089–1098.

MDBC. 2004. *Native Fish Strategy for the Murray–Darling Basin 2003–2013*. Murray–Darling Basin Commission, Canberra.

Montgomery, D.R., Piegay, H. 2003. Wood in rivers: interactions with channel morphology and processes. *Geomorphology*, **51**, 1–5.

Mutz, M. 2002. Influences of woody debris on flow patterns and channel morphology in a low energy, sand-bed stream reach. *International Review of Hydrobiology*, **85**, 107–121.

Newson, M.D., Newson, C.L. 2000. Geomorphology, ecology and river channel habitat: mesoscale approaches to basin-scale challenges. *Progress in Physical Geography*, **24**, 195–217.

Nicol, S.J. 2005. *Large Woody Debris in Freshwater Ecosystems*. PhD Thesis, The University of Melbourne, Australia.

Nuttle, W.K. 2002. Is ecohydrology one idea or many? *Hydrological Sciences*, **47**, 805–807.

Palmer, M.A., Bernhardt, E.S. 2006. Hydroecology and river restoration: Ripe for research and synthesis. *Water Resources and Research*, **42**, W03S07.

Parsons, M.E., Thoms, M.C. 2007. Hierarchical patterns of physical-biological associations in river ecosystems. *Geomorphology*, **89**, 127–146.

Pettit, N.E., Naiman, R.J., Rogers, K.H., Little, J.H. 2005. Post-flooding distribution and characteristics of large woody debris piles along the semi-arid Sabie River, South Africa. *River Research and Applications*, **21**, 27–38.

Piegay, H., Gurnell, A.M. 1997. Large woody debris and river geomorphological pattern: examples from S.E. France and S. England. *Geomorphology*, **19**, 99–116.

Pickett, S.T.A., Burch, W.R., Grove, J.M. 1999. Interdisciplinary research: maintaining the constructive impulse in a culture of criticism. *Ecosystems*, **2**, 302–307.

Porporato, A., Rodriguez-Iturbe, I. 2002. Ecohydrology – a challenging multidisciplinary research perspective. *Hydrological Sciences*, **47**, 811–821.

Quinn, G. P., Keough, M. J. 2002. *Experimental Design and Data Analysis for Biologists.* Cambridge University Press, Cambridge.

RDI. 1994. *Broadband Acoustic Doppler Current Profiler Technical Manual.* RD Instruments, San Diego.

Reich, M. 1999. Ecological, technical and economical aspects of stream restoration with large wood. *Zeitschrift für Okologie und Naturschutz*, **8**, 251–253.

Roni, P., Quinn, T.P. 2001. Effects of wood placement on movements of trout and juvenile Coho salmon in natural and artificial stream channels. *Transactions of the American Fisheries Society*, **130**, 675–685.

Rosenfeld, J.S., Boss, S. 2001. Fitness consequences of habitat use for juvenile cutthroat trout: energetic costs and benefits in pools and riffles. *Canadian Journal of Fisheries and Aquatic Sciences*, **58**, 585–593.

Shields, F.D., Gippel, C.J. 1995. Prediction of effects of woody debris removal on flow resistance. *Journal of Hydraulic Engineering*, **121**, 341–354.

Statzner, B., Higler, B. 1986. Stream hydraulics as a major determinant of benthic invertebrate zonation patterns. *Freshwater Biology*, **16**, 127–139.

Thoms, M.C., Sheldon, F. 2000. Australian lowland river systems: An introduction. *Regulated Rivers: Research and Management*, **16**, 375–383.

Thoms, M.C., Parsons, M.E. 2002. Ecogeomorphology: an interdisciplinary approach to river science. *International Association of Hydrological Sciences*, **276**, 113–120.

Thoms, M.C., Walker, K.F. 1993. A case history of the environmental effects of flow regulation on a semi-arid lowland river: The River Murray, South Australia. *Regulated Rivers: Research and Management*, **8**, 103–119.

Till, B., 2000. Large woody debris demonstration site on the Dandalup River, WA. *RipRap*, **16**, 26–28.

Tillma, J.S., Guy, C.S., Mammoliti, C.S. 1998. Relations among habitat and population characteristics of spotted bass in Kansas streams. *North American Journal of Fisheries Management*, **18**, 886–893.

Treadwell, S., Koehn, J., Bunn, S. 1999. Large woody debris and other aquatic habitat. In Lovett, S., Price, P (eds). *Riparian Land Management Technical Guidelines. Volume One, Part A: Principles of Sound Management.* Land and Water Resources Research and Development Corporation, Canberra, pp. 79–97.

Twidale, C.R., Lindsay, J.M., Bourne, J. 1978. Age and origin of the Murray River and gorge in South Australia. *Proceedings of the Royal Society of Victoria*, **90**, 27–41.

Walker, K.F. 1992. The River Murray, Australia: A semiarid lowland river. In Calow, P., Petts, G.E. (eds.), *The Rivers Handbook.* Vol. 1. Blackwell, London, pp. 472–488.

Young, W.J. 1993. Field techniques for the classification of near-bed flow regimes. *Freshwater Biology*, **29**, 377–383.

21
The Ecological Significance of Hydraulic Retention Zones

Friedrich Schiemer and Thomas Hein

21.1 Introduction

High species diversity and intensified ecological processes in rivers are related to zones of higher retention, also known as slackwater, transient storage or inshore retention zones. In the hydrological literature the term 'deadwater' is often used, which is misleading from an ecological perspective.

In low-order rivers, bed sediments and hyporheic zones form important transient storage compartments (Findlay, 1995; Brunke and Gonser, 1997; Battin, 2000). In large rivers, lateral exchange processes between in-stream, ecotonal zones and lentic water bodies in the floodplains by surface and subsurface flow paths are crucial (Hein *et al.*, 2005). Understanding the large-scale biogeochemical linkages requires a detailed mechanistic knowledge of functional processes in relation to retentiveness and hydrodynamic exchange in river corridors (Kaplan and Newbold, 2003). With regard to biogeochemical processes (e.g., retention, transformation and production of matter), storage zones have been identified as 'hot spots' in the river corridor network (McClain *et al.*, 2003; Thorp *et al.*, 2006).

They are equally important as habitats for a diverse biota. Storage relates to structural properties, physiographic habitat conditions and biotic interactions. Inshore retention zones and habitat connectivity between the river and floodplains are central in maintaining biodiversity patterns (Tockner *et al.*, 2000).

This chapter concentrates on retention areas in river corridors, including inshore and floodplain zones of large river systems. It describes the physiographic conditions defined by geomorphology and hydrology. The significance of such areas for the river system as a whole is analysed by reviewing their functions at small scales – both as habitats for selected species and in the framework of limnological processes. The small-scale characteristics must be 'scaled-up' in order to understand their significance for the river system. This has become a key topic in river management.

21.2 Geomorphology and Patch Dynamics Creating Retention Zones

The structural properties of river corridors reflect their particular landscape setting; they are shaped by the dynamic interaction between hydrology and geomorphology (Rosgen, 1994). The emerging concept of river ecology considers the geomorphology and morphodynamics of the channels and their riparian zones as being hierarchically scaled (Frissell *et al.*, 1986; Habersack, 2000; Naiman and Decamps, 1997; Pringle *et al.*, 1988; Townsend, 1996; Thoms and Parsons, 2002). Spatial heterogeneity is represented as nested patch mosaics at different scales from local substrate–water interfaces to the river reach level, and river–floodplain complexes. Under natural conditions the fluvial morphology at a particular landscape setting represents a dynamic equilibrium between hydrology and bed-load transport; this is characterized by a statistical predictability of habitat composition and characteristic habitat turnover times (Schiemer, 1999; Hohensinner *et al.*, 2004; Thorp *et al.*, 2006; Wu, 1999).

The geomorphic and hydrological properties of a river reach lead to characteristic retention conditions. An important task in river ecology is to formulate parameters that help analyse patch structure and patch dynamics, and their linkage to ecological processes at different spatio-temporal scales. At the macro-scale (10s–100s km), long-term changes in landscape structure can be analysed based on historical maps (e.g., Hohensinner *et al.*, 2004). At the meso-scale (km), a wide range of parameters identifying morphological structures have been proposed; they still need to be screened and tested for their comparability and ecological relevance (e.g., Middelkoop *et al.*, 2005). At a micro-scale (10s–100s m), the relative length of the shoreline flowage line, inshore sinuosity and the extent of shallow areas in dependence of river stage define indices for structural conditions and heterogeneity.

In the context of microhabitat conditions for 0+ fish (Schiemer *et al.*, 2001b) and source areas for potamoplankton development (Reckendorfer *et al.*, 1999), for example, a micro-scale index has been successfully applied (see below). Inshore sinuosity and retentiveness at the micro-scale change with river stage, i.e. they are dynamically controlled by hydrology. Figure 21.1 illustrates the relationship between sinuosity and river stage for the Austrian Danube downstream of Vienna. Highest values of shoreline structure are, under these specific conditions, characteristic for mean water levels. The high stochasticity and wide amplitude of water level fluctuations (average diel change in the water level 30 cm; Schiemer and Zalewski, 1992) causes considerable dynamics in the local conditions and promotes hydrodynamic exchange processes between inshore storage zones and the river. The interaction of inshore relief and hydrology creates the physiographic habitat conditions for the biota.

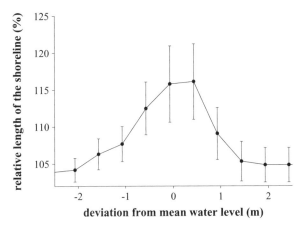

Figure 21.1 *Relative shoreline length (flowage line) of the free-flowing Austrian Danube downstream of Vienna. Mean and standard error are given for different water levels. Reproduced from Reckendorfer et al., Freshwater Biology,* **41**, *'Zooplankton abundance in the River Danube, Austria: the significance of inshore retention', Reproduced from Blackwell Publishing*

21.3 Retention, Hydraulics and Physiographic Conditions

Channel geometry and roughness create strong transversal gradients of river flow and zones of higher retentiveness. In large rivers these features are generally associated with natural river bank formations such as gravel bars, and with engineering constructions such as groynes. The ecological significance of such retention zones is determined by the specific hydraulic characteristics, the flow velocity pattern, residence time and the exchange rates with the main channel. Their geometry determines the flow pattern, the formation of gyres and the form and size of the interfacial mixing zones between retention area and main channel. All these parameters depend on river stage. Consequently, the physical environment is in permanent change, with pulses of nutrients and organic matter. Hydraulic models of exchange processes between retention zones and the main stem of rivers are in an early stage of development. They are mainly based on artificial, geometric engineering structures such as groynes and harbours (e.g., Czernuszenko *et al.*, 1998; Uijttewaal *et al.*, 2001; Mazijk, 2002; Cheong and Seo, 2003).

Slackwater zones are, by definition, areas of deposition and storage (Carling, 1992), and are defined by residence time, recirculation flow patterns and turbulence profiles. Reduced velocity and turbulent mixing lead to lower substrate dynamics and sedimentation of seston. Scour and sedimentation patterns depend on flow velocity vectors, resulting in substrate grain size sorting. Accumulation of fine sediment can compact the substrate and decrease exchange processes between the river, the hyporheic zone and groundwater (Brunke and Gonser, 1997).

The studies by Sukhodolov *et al.* (2002) and Engelhardt *et al.* (2004) on the correspondence between geohydrodynamic patterns and sedimentation in groyne fields illustrate the complexity of flow velocity and the small-scale pattern of sedimentation and erosion processes. The authors combined flume experiments and field studies on river bed topography, velocity measurements, tracer experiments to calculate the main hydrodynamic characteristics and the residence times within such areas. Figure 21.2 illustrates the flow pattern and the location of scouring and deposition zones within a groyne field. Engelhardt *et al.* (2004) showed the change in seston quantity and composition (from mineral fast-sinking particles to a higher organic content) along the flow path. Fine particle deposition is associated with the formation of recirculation patterns. Sediment layers are thickest in the central part of the gyres. Since patterns of current velocity change with river stage, substrate composition will be an integration of the flow regime over time.

The physiographic condition (local gradients of current velocity, temperature and substrate) in retention areas provides the habitat characteristics for specific biotic assemblages – potamoplankton, benthos and fish (see below).

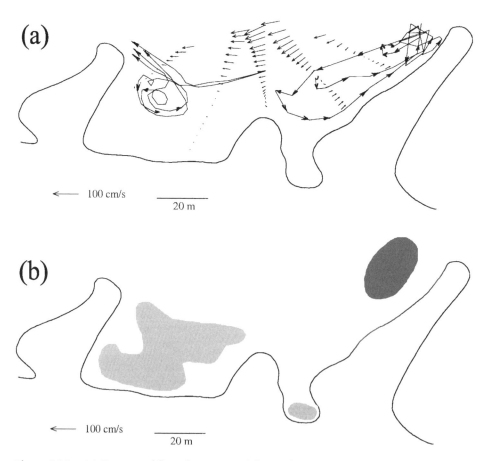

Figure 21.2 (a) Patterns of flow direction and flow velocity in groyne fields. (b) Long-term cumulative effects expressed as fine sediment deposition zones (light shading) and erosion zones (dark shading). Modified after Sukhodolov et al. (2002). Reproduced by permission of John Wiley & Sons, Ltd

21.4 Habitat Conditions for Characteristic Biota

Retention zones in rivers show a high species diversity of benthos, periphyton, plankton and fish. This reflects the specific characteristic conditions of reduced flow, increased substrate stability, sediment accumulation, light and temperature conditions, high nutrient availability, etc. Figure 21.3 illustrates, as an example, the specific temperature conditions in two types of inshore areas of the Austrian Danube compared with the official daily temperature recordings from the main river channel. The temperature regime of the inshore areas becomes decoupled from main channel conditions depending on water retention and exchange. Local temperature conditions are crucial for temperature-dependent processes of species bound to the littoral zones (Keckeis et al., 1997).

For fish, retention zones are important as feeding and over-wintering areas and as refuges during floods (Schiemer and Waidbacher, 1992). These zones are particularly significant as nurseries for most riverine species. Such nurseries are almost exclusively restricted to richly structured littoral zones of the river. Species number and diversity of fish fry are positively correlated with the structural diversity and retentiveness of a particular nursery area (Figure 21.4a). Requirements with regard to water currents, substrate type and food change dramatically during the early life history. The larvae are bound to sheltered bays. With increasing age and size, early juveniles migrate to adjoining shallow gravel banks. The individual success during this period is largely determined by structural shelter, food availability, temperature and current velocity pattern. Critical velocities are in the order of a few cm s^{-1} at the onset of exogenous feeding and progressively increase with fish size (Schiemer et al., 2001b). The form and sinuosity of inshore zones also

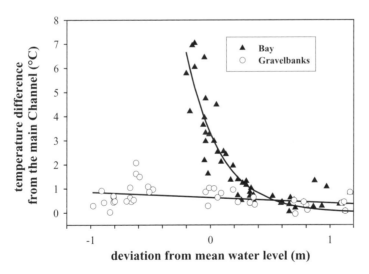

Figure 21.3 *Relationship between the water level of the Danube and the temperature in two types of inshore areas: a bay, and a gravel bank with reduced flow. The temperature is characterized as deviation from daily (07.00 h) recordings made along the main channel. From Keckeis* et al. *(1997). Reproduced from CNRS*

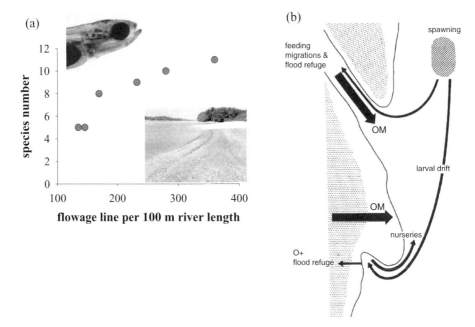

Figure 21.4 (a) Relationship between number of fish species and inshore sinuosity of the Danube River in Austria. Reproduced from Schiemer et al. (1999) and Keckeis and Schiemer (2002). (b) Schematic representation of inshore sinuosity with respect to their value as fish nurseries. The scheme indicates the river shoreline with a connected backwater and gravel bar. Stippled area = terrestrial vegetation, OM = organic material. For further explanation see text

determine the gradient of complementary microhabitat types necessary to meet the changing requirements of 0+ fish; they also create refugia during spates (Sedell et al., 1990; Hildrew, 1996). High hydrologic retention fosters recruitment and year class strength. The inshore retention zones therefore represent seeding areas for the whole river.

Under conditions of continuously and stochastically changing water levels, which are characteristic for large temperate rivers, the geohydromorphic settings at a range of scales are essential for fish recruitment (Figure 21.4b). On a larger spatial scale, the connectedness between spawning and nursery areas (in-channel structures, e.g., riffle–pool sequences) and the retentivity of a river reach determine the refuge capacity at strong water level changes (Schlosser, 1995). Fluctuating water levels will lead to wash-out effects, causing high mortality and an unidirectional, downstream shift of the fish fry population. Retention in this case is defined both by the hydraulic pattern and the capability of early juvenile fish to actively escape wash-out effects. On a mesoscale, microhabitat availability is determined by the form and sinuosity of inshore structure (e.g., structure of gravelbanks) and dependant of the river stage. Microhabitat gradients are necessary to cover the ontogenetic niche shifts of the growing young-of-the-year fish. On a microscale the physical shoreline structure and concomitant retention determine the microhabitat conditions (temperature, food availability, pattern of current velocity, current velocity).

Macrozoobenthos assemblages and their distribution pattern in large rivers are strongly governed by current conditions and substrate composition (Palmer and Ricciardi, 2004). In river transects, the abundance and diversity of benthos generally increase from the main channel towards the littoral banks (Humpesch and Fesl, 2005). Inshore retention zones are characterized by spatial heterogeneity and strong gradients, which in terms of macrozoobenthos is expressed in a mosaic pattern of species distribution (Fesl and Humpesch, 2006). Storage areas in which organically enriched soft sediments have accumulated exhibit a reduced species diversity but very high densities of endopelic forms (e.g., tubificids, spaeriids). Such fine sediment deposition zones are important feeding grounds for fish.

21.5 Retention and Water Column Processes

The significance of retention zones for 'water column' processes has been addressed in a number of publications showing that aquatic productivity is shaped by hydraulic exchange (e.g., Margalef, 1960; Reynolds et al., 1991; Carling, 1992; Reynolds and Descy, 1996; Basu and Pick, 1996, 1997; Lair and Reyes-Marchant, 1997; Reckendorfer et al., 1999; Schiemer et al., 2001a). Riverine phytoplankton production is clearly highest in these areas (Reynolds and Descy, 1996). Here, reduced flow velocity and turbulent mixing is coupled with seston sedimentation and enhanced light penetration at conditions of high nutrient availability, characteristic for large regulated rivers. This enhances phytoplankton production. A case study in the River Severn, UK (Reynolds et al., 1991), reported concentrations of algal chlorophyll up to 40-times higher in retention zones than in the transit river flow.

Despite the accumulated knowledge on the role of hydrology in plankton development in rivers, hydraulic controls of biotic processes and food chain interactions within the plankton community remain poorly understood. This is a major theme for future research. Detailed studies on pelagic processes and carbon dynamics in a river-sidearm system of the Danube have confirmed the significance of hydrological connectivity (Heiler et al., 1995; Hein et al., 1999; Tockner et al., 1999). These studies underlined the significance of water retention and exchange rates for plankton processes and ecological succession following flood pulses. They also pinpointed the necessity of selecting and defining appropriate ecohydrological parameters (see Monsen et al., 2002). A main parameter for understanding limnological processes is the 'age' of river water in the retention zones. The sequence from low to high water age after flood pulses correlates with successions of the planktonic communities and their interactions (Baranyi et al., 2002; Keckeis et al., 2003). These results on connected sidearm systems can serve as models for shorter-term processes in inshore retention zones within the main channel itself. For the river-sidearm systems the relevant time scale is on the order of days. For retention zones within the river, the time scale has to be extended to include water age on the order of minutes to hours (Sukhodolov et al., 2002). Figure 21.5 illustrates the relationship between nutrient availability and phytoplankton biomass following flood pulses; it also addresses phases of surface connectivity between the main channel and the storage zone. Clearly, the nutrient load coming from the river is rapidly used up while phytoplankton biomass increases. The resource utilization, i.e. the nutrient stripping by phytoplankton – the rapid decrease

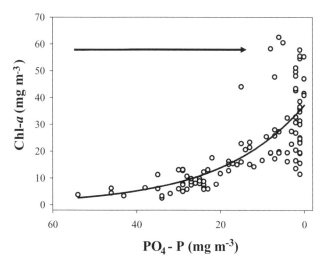

Figure 21.5 Relationship of inorganic phosphate to chlorophyll a concentration in a Danube side arm following a flood pulse. Note the reversed x-axis. The arrow indicates increasing water age from 0 to 7 days. Modified after Hein et al. (2005)

of inorganic phosphorous – becomes visible within a time scale of days (Hein *et al.*, 1999, 2005). This can lead to nutrient limitation of algal growth during periods of longer retention.

The change in biomass of different functional groups of river plankton with water age is illustrated in Figure 21.6. The sequence also implies an increasing complexity of biotic interactions. Increasing algal biomass in the initial phase is linked to conditions of high nutrient availability, improving light conditions and low initial grazing rates (see Baranyi *et al.*, 2002; Keckeis *et al.*, 2003). The phytoplankton biomass tends to peak at intermediate water age, while at higher age the algal biomass declines due to biotic control. The phytoplankton community composition will also change over the course of this succession.

An important aspect for the ecological functioning of large rivers is the coupling between phytoplankton and bacterial production in the water column of storage zones. Because of the high loads of dissolved (DOM) and particulate organic material (POM) transported by the river (e.g., Hein *et al.*, 2003), bacterial activities in the main flow are comparatively high (Berger *et al.*, 1995). Bacterial secondary production depends on the type and origin of DOM. Stable isotope techniques (e.g., Thorp *et al.*, 1998) and a fluorescence method that differentiates between the allochthonous and autochthonous DOM (McKnight *et al.*, 2001) were applied to screen for the terrestrial or aquatic origin of the material. The allochthonous organic material transported by the river supports a significantly lower bacterial productivity than the organic material derived from aquatic production. Bacterial activity is apparently strongly enhanced by phytoplankton primary production at sites of reduced flow and increased retention. Figure 21.6 indicates the initial parallel increase in both components, suggesting a stimulating effect. At higher residence times, the phytoplankton and bacterial biomass are independent of each other and the system changes from autotrophy to heterotrophy.

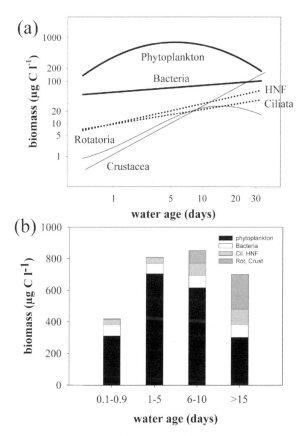

Figure 21.6 Biomass accumulation of plankton in side-arms of the Danube following a flood pulse. (a) The x-axis gives water age in days in a logarithmic scale. The biomass of all functional groups of plankton is expressed as µg carbon per litre. (b) Biomass composition of plankton at different stages of retention after a flood pulse, grouped in phytoplankton, bacteria, unicellular grazers (Cil = ciliates, HNF = heterotrophic nannoflagellates) and metazoan plankton (Rot = rotifers, Crust = crustaceans). The graph is based on a large set of partially unpublished data (see Hein et al., 2003, 2005)

Plankton communities of large rivers exhibit complex food web structures. Processes are intensified and specific successional interaction patterns develop dependant on retentiveness. Among the grazers the unicellular forms, especially heterotrophic flagellates, dominate (Weitere and Arntd, 2002; Weitere *et al.*, 2005). They apparently exert a moderate grazing pressure on phytoplankton communities. The results from the river-sidearm system of the Danube (Figure 21.6) demonstrate that, with water age, the biomass of heterotrophic flagellates (HNF) and ciliates increases at a moderate rate. Both groups are less stimulated by hydrological exchange (connectivity) than the phytoplankton, which exhibited a fast initial bloom. In general, microbial processes and accompanying unicellular grazers appear to be less controlled by hydrology. In contrast to the heterotrophic flagellates and ciliates, the micrometazoan plankton, especially rotifers, show strong population growth at an early stage after a flood pulse, followed by a strong increase of small cladocerans (*Bosmina*). Under these conditions the microzooplankton exerts a strong predation pressure on the phytoplankton communities (Keckeis *et al.*, 2003). This sequence can be explained by differences in the respective generation times (Reckendorfer *et al.*, 1999; Baranyi *et al.*, 2002). In summary, our present state of knowledge indicates that the pelagic food chain is primarily under hydrologic control. With increasing residence time, e.g. water age, a shift takes place from primary production to secondary production, i.e. a change from autotrophic to heterotrophic conditions, and from physical to biotic control. The zooplankton shows a distinct sequence from unicellular to metazoan grazers. Higher biomass of crustaceans, as efficient grazers, reduces the phytoplankton at longer retention phases.

There is also growing evidence that, within large rivers, coupling between the plankton community and benthic filter feeders (e.g., bivalves and amphipods) is important in slackwater areas, this coupling can influence the structure and function of 'potamoplankton' (Ietswaart *et al.*, 1999; Thorp and Caspers, 2002).

21.6 The Significance of Retention Zones for the River Network

Holistic concepts in river ecology emphasize that functional networks of river systems strongly depend on lateral interaction (e.g., Benda *et al.*, 2004). Local processes in storage areas, as well as their habitat and refuge value for biota, clearly have a cumulative effect on the whole system. Their overall significance depends on their areal extent.

Upscaling the effects of local processes in storage zones for the whole river system requires identifying the specific significance of these zones as sink and source areas. They form sinks in terms of sediment deposition, nutrient assimilation and absorption (Hein *et al.*, 2005). The nutrient recycling is locally concentrated and the nutrient spiral length decreases with hydraulic retention (see Thorp *et al.*, 2006). Hydrological routing and 'dead zone' models have been developed for the longitudinal solute transport in rivers. They identify the cumulative effect of transient storage zones along the river course (Cheong and Seo, 2003; Czernuszenko *et al.*, 1998; Mazijk, 2002).

The 'river productivity model' emphasizes that autochthonous production in the channel and connected backwaters is significant for the riverine food web (Thorp and Delong, 1994, 2002). Accordingly, autochthonous carbon production depends on the spatial and temporal availability of such areas. The 'paradox of potamoplankton' comprising algae, rotifers and crustaceans under critical growth conditions with regard to light attenuation, turbidity and flow velocity in large rivers has been frequently discussed (e.g., Reynolds and Descy, 1996). Hydraulic properties necessary to account for algal dynamics were discussed by Reynolds and Descy (1996); they identified the cumulative storage capacity of a river reach in relation to growth performances of algal species as a major prerequisite for potamoplankton development and dynamics. This capacity can explain the nature of river phytoplankton associations and their productivity (Reynolds *et al.*, 1991; Reynolds and Glaister, 1993). The explanation is linked to the fact that the productive storage zones in the main channel and its floodplains function as source areas.

The enhanced primary productivity in retention zones stimulates microbial processes by producing nonrefractory dissolved organic carbon. This in a cumulative form is essential for the decomposition and natural water purification processes of the overall river system.

Lair *et al.* (1997) concluded from the longitudinal and seasonal distribution pattern of zooplankton that the 'inoculum points' are significant in explaining the population dynamics of both algae and micrometazoan plankton in large rivers. Even though zooplankton recruitment and production is limited to current velocities below $10\,\mathrm{cm.s}^{-1}$ (Vranovsky, 1995), the drift of large rivers like the Danube or Rhine always contains zooplankton. There is good evidence that this riverine zooplankton is produced in inshore areas with high retention, i.e. in bays forming and being isolated along the shoreline at certain water levels. Large populations of rotifers develop under such conditions and are released into the channel when water exchange is enhanced. Zooplankton densities in the river correlate with the index values of shoreline configuration (Reckendorfer *et al.*, 1999). In low-water conditions, coinciding with shrinking boundary areas, the concentration of zooplankton is low. Conversely, zooplankton densities in the drift increase with increasing inshore structure at higher water levels. The interpretation is that the extent of inshore configuration is an immediate correlate to dead zone conditions, which enhance zooplankton production. The time lag between inshore sinuosity and zooplankton abundance depends on the population development rate of plankton taxa. For instance, the highest correlations for rotifers were found by relating the sinuosity index values 6–10 days prior to sampling – the time required to build up sufficient population sizes.

Besides their significance for functional processes, retention zones may contribute substantially to the overall biodiversity of a river system. The significance of ecotonal and riparian greenways, corridors, and stepping stones for the distribution of species, for population exchange processes, and as a refuge zone for specific biota has been repeatedly emphasized (Poff, 1997; Naiman and Decamps, 1997; Decamps *et al.*, 2004). The ecotonal areas are crucial source areas for riverine fish and form the key fish nursery zones. Finally, they form important feeding and winter resting places for waterbirds.

Both for limnological processes and biotic diversity, the role of storage zones for the overall riverine landscape will depend on the cumulative retention capacity. A better understanding of retention characteristics at different spatio-temporal scales is needed. We propose a 'patch scale retention hypothesis' that deals with specific retention characteristics in terms of processes and refuge values of differently sized geohydrodynamic patches. The present contribution suggests that inshore storage capacity is hydraulically scaled with respect to specific biological functions. Small patches will have different retention and, over time, will develop qualitatively different conditions than patches with long-term residence characteristics. For example, spatially small storage zones with short retention times allow the production of riverine phytoplankton, whereas structural properties leading to longer storage are necessary to allow zooplankton development; structural properties on several scales will be required as microhabitats and refugia for larval fish. This is represented in Figure 21.7, which distinguishes three spatial and hydrological scales:

(i) Large-scale retention areas: e.g. input, output and internal process in concert with flood pulses extending into terrestrial parts of wetlands. Large-scale retention zones are lakes or reservoirs in the longitudinal course of the river; they function as discontinuities (Ward and Stanford, 1995). The significance of such large storage zones

416 Hydroecology and Ecohydrology: Past, Present and Future

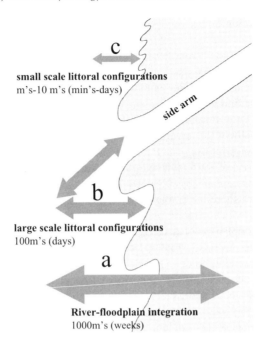

Figure 21.7 *Schematic presentation of the spatio-temporal scaling of retention characteristics in river–floodplain systems. The lightly shaded areas represent the semiterrestrial and terrestrial floodplain zones. The dark-shaded arrows represent retention and exchange characteristics. (a) Large-scale exchange with the floodplain at flood events. (b) Mesoscale 'within channel' exchange at larger shoreline configurations and with side arms. (c) Exchange at small scale shoreline configurations*

for nutrient export from rivers into the sea has been addressed by Ittekot *et al.* (2000). The retention time will be of the order of weeks.

(ii) Mesoscale retention areas within the main stem of the river or river-sidearm connections exposed to water level fluctuations of the river (Tockner *et al.*, 2000), with retention times of days.

(iii) Small-scale and short-term retention caused by littoral configuration. The different effects will depend on relative exchange volumes and on their temporal pattern. The retention time will be of the order of minutes to hours.

21.7 Implications for River Management

River engineering, regulation and damming have changed the natural geohydrodynamics of most rivers and have considerably degraded their structural and functional properties (e.g., Stanford *et al.*, 1996; Buijse *et al.*, 2005). The loss in geomorphic dynamics due to channel straightening, shoreline fortifications and confinement of natural inundation areas has strongly compromised the overall retention capacities of most river systems (Ward and Stanford, 1995). This capacity has decreased over a range of spatial scales: from a loss of large-scale connectivities and storage capacities of floodplains and formerly multibranched river systems, to a drop in small-scale slackwater areas of structured inshore zones. As discussed, retention is a main element for understanding ecosystem processes and biodiversity. The decline in patch dynamics has reduced successional processes. The decline in habitat quality in the ecotonal zones is reflected in a large proportion of endangered species in regulated rivers in which both the in-channel retention and the connectivity between the river and its side arms has markedly decreased (Schiemer, 1999). Reduced retention capacities have also impacted ecosystem functions, the 'green services' of rivers like self-purification processes along with nutrient deposition and storage.

Modern restoration schemes emphasize the necessity of restoring the geohydromorphological conditions and characteristic patch dynamics for specific river types. Improving retention capacities is a main management issue and restoration target. Enhancing patch dynamics, spatial heterogeneity and related retention capacities has become a primary objective from a conservation point of view but also from the standpoint of water quality control. In a European context, this represents a major challenge with respect to the EU Water Framework Directive (Buijse *et al*., 2005). The mitigation and restoration concepts currently under development therefore address riverbank rehabilitation and sidearm reconnections as primary ecological objectives (Reckendorfer *et al*., 2005).

The development of management and restoration strategies for large regulated rivers requires a detailed understanding of the functioning of the various subunits and their integration in river–floodplain systems. This refers to their general functions, as well as to the sensitivity of target species to hydraulic retentiveness. This approach will provide a main framework and should become a main research focus in river ecology. In order to derive clear management practices, we have to develop a better mechanistic understanding of the ecohydrology of these retention zones. Developing mechanistic models that relate hydraulic retention to biological functions requires an integrative approach – a challenge for ecologists and hydrologists alike.

References

Baranyi C, Hein T, Holarek C, Keckeis S, Schiemer F. 2002. Zooplankton biomass and community structure in a Danube River floodplain system: effects of hydrology. *Freshwater Biology*, **47**: 473–482.
Basu BK, Pick FR. 1996. Factors regulating phytoplankton and zooplankton biomass in temperate rivers. *Limnology and Oceanography*, **41**: 1572–1577.
Basu BK, Pick FR. 1997. Phytoplankton and zooplankton development in a lowland river. *Journal of Plankton Research*, **19**: 237–253.
Battin TJ. 2000. Hydrodynamics is a major determinant of streambed biofilm activity: From the sediment to the reach scale. *Limnology and Oceanography*, **45**: 1308–1319.
Benda L, Poff LR, Miller D, Dunne T, Reeves G, Pollock M, Pess G. 2004. The network dynamics hypothesis: How channel networks structure riverine. *BioScience*, **54**: 413–427.
Berger B, Hoch B, Kavka G, Herndl GJ. 1995. Bacterial metabolism in the River Danube: parameters influencing bacterial production. *Freshwater Biology*, **34**: 601–616.
Brunke M, Gonser T. 1997. The ecological significance of exchange processes between rivers and groundwater. *Freshwater Biology*, **37**: 1–33.

Buijse AD, Klijn F, Leuven RSEW, Middelkoop H, Schiemer F, Thorp JH, Wolfert HP. 2005. Rehabilitation of large rivers: references, achievements and integration into river management. *Archiv für Hydrobiologie Supplement*, **155**: 715–738.

Carling PA. 1992. In-stream hydraulics and sediment transport. *The River Handbook*. P. Calow and G.E. Petts (eds), Blackwell Scientific Publications, Oxford, pp. 101–125.

Cheong TS, Seo IW. 2003. Parameter estimation of the transient storage model by a routing method for river mixing processes. *Water Resources Research*, **39**: Art. No. 1074.

Czernuszenko W, Rowinski PM, Sukhodolov A. 1998. Experimental and numerical validation of the dead-zone model for longitudinal dispersion in rivers. *Journal of Hydraulic Research*, **36**: 269–280.

Decamps H, Pinay G, Naiman RJ, Petts GE, McClain ME, Hillbricht-Ilkowska A, Hanley TA, Holmes RM, Quinn J, Gibert J, Planty Tabacchi A-M, Schiemer F, Tabacchi E, Zalewski M. 2004. Riparian Zones: Where biochemistry meets biodiversity in management practice. *Polish Journal of Ecology*, **52**: 3–18.

Engelhardt C, Kruger A, Sukhodolov A, Nicklisch A. 2004. A study of phytoplankton spatial distributions, flow structure and characteristics of mixing in a river reach with groynes. *Journal of Plankton Research*, **26**: 1351–1366.

Fesl C, Humpesch UH. 2006. Spatio-temporal variability of benthic macroinvertebrate community attributes and their relationship to environmental factors in a large river (Danube, Austria). – *Archiv für Hydrobiologie Supplement*, **158**: 329–350.

Findlay S. 1995. Importance of surface–subsurface exchange in stream ecosystems: The hyporheic zone. *Limnology and Oceanography*, **40**: 159–164.

Frissell CA, Liss WJ, Warren CE, Hurley MC. 1986. A hierarchical framework for stream habitat classification: viewing streams in a watershed context. *Journal of Environmental Management*, **10**: 199–214.

Habersack HM. 2000. The river-scaling concept (RSC): a basis for ecological assessments. *Hydrobiologia*, **422–423**: 49–60.

Heiler G, Hein T, Schiemer F, Bornette G. 1995. Hydrological connectivity and flood pulses as the central aspect for the integrity of a river-floodplain system. *Regulated Rivers: Research & Management*, **11**: 351–361.

Hein T, Heiler G, Pennetzdorfer D, Riedler P, Schagerl M, Schiemer F. 1999. The Danube restoration project: functional aspects and planktonic productivity in the floodplain system. *Regulated Rivers: Research & Management*, **15**: 259–270.

Hein T, Baranyi C, Herndl GJ, Wanek W, Schiemer F. 2003. Allochthonous and autochthonous particulate organic matter in floodplains of the River Danube: the importance of hydrological connectivity. *Freshwater Biology*, **48**: 220–232.

Hein T, Reckendorfer W, Thorp J, Schiemer F. 2005. The role of slackwater areas and the hydrologic exchange for biogeochemical processes in river corridors: examples from the Austrian Danube. *Archive für Hydrobiologie Supplement*, **155**: 425–442.

Hildrew AG. 1996. Whole river ecology: spatial scale and heterogeneity in the ecology or running waters. *Archiv für Hydrobiologie Supplement*, **113**: 25–43.

Hohensinner S, Habersack H, Jungwirth M, Zauner G. 2004. Reconstruction of the characterstics of a natural alluvial river–floodplain system and hydromorphological changes following human modifications: The Danube River (1812–1991). *River Research and Applications*, **20**: 25–41.

Humpesch UH, Fesl C. 2005. Biodiversity of macrozoobenthos in a large river, the Austrian Danube, including quantitative studies in a free-flowing stretch below Vienna: a short review. *Freshwater Forum*, **24**: 3–23.

Ietswaart T, Breebaart L, van Zanten B, Bijerk R. 1999. Plankton dynamics in the river Rhine during downstream transport as influenced by biotic interactions and hydrological conditions. *Hydrobiologia*, **410**: 1–10.

Ittekkot V, Humborg C, Schaefer P. 2000. Hydrological alteration on land and marine biogeochemistry: a silicate issue? *BioScience*, **50**: 776–782.

Kaplan LA, Newbold JD. 2003. The role of monomers in stream ecosystem metabolism. In Findlay SEG, Sinsabaugh RL (eds) *Aquatic Ecosystems Interactivity of Dissolved Organic Matter*. Academic Press, San Diego, London, pp. 97–119.

Keckeis H, Winkler G, Flore L, Reckendorfer W, Schiemer F. 1997. Spatial and seasonal characteristics of 0+ fish nursery habitats of Nase, *Chondrostoma nasus* in the River Danube, Austria. *Folia Zoologica*, **46**: 133–150.

Keckeis S, Baranyi C, Hein T, Holarek C, Riedler P, Schiemer F. 2003. The significance of zooplankton grazing in a floodplain system of the River Danube. *Journal of Plankton Research*, **25**: 243–253.

Lair N, Reyes-Marchant P. 1997. The potamoplankton of the Middle Loire and the role of the 'moving littoral' in downstream transfer of algae and rotifers. *Hydrobiologia*, **356**: 33–52.

Margalef R. 1960. Ideas for a synthetic approach to the ecology of running waters. *Internationale Revue der gesamten Hydrobiologie*, **45**: 133–153.

Mazijk A. 2002. Modelling the effects of groyne fields on the transport of dissolved matter within the Rhine Alarm-Model. *Journal of Hydrology*, **264**: 213–229.

McClain ME, Boyer EW, Dent CL, Gergel SE, Grimm NB, Grooman PM, Hart SC, Harvey JW, Johnston CA, Mayorga E, McDowell WH, Pinay G. 2003. Biochemical hot spots and hot moments at the interface of terrestrial and aquatic ecosystems. *Ecosystems*, **6**: 301–312.

McKnight DM, Boyer EW, Westerhoff PK, Doran PT, Kulbe T, Anderson DT. 2001. Spectrofluorometric characterization of dissolved organic matter for indication of precursor organic material and aromaticity. *Limnology and Oceanography*, **46**: 38–48.

Middelkoop H, Schoor MM, Wolfert HP, Maas GJ, Stouthamer E. 2005. Targets of ecological rehabilitation of the lower Rhine and Meuse based on a historic-geomorphic reference. *Archiv für Hydrobiologie Supplement*, **155**: 36–88.

Monsen NE, James E, Cloern JE, Lucas LV, Monismith SG. 2002. A comment on the use of flushing time, residence time, and age as transport time scales. *Limnology and Oceanography*, **47**: 1545–1553.

Naiman RJ, Decamps H. 1997. The ecology of interfaces: riparian zones. *Annual Review of Ecological Systems*, **28**: 621–658.

Palmer ME, Ricciardi, A. 2004. Physical factors affecting the relative abundance of native and invasive amphipods in the St. Lawrence River. *Canadian Journal of Zoology*, **82**: 1886–1893.

Pringle CM, Naiman RJ, Bretschko G, Karr JR, Osgood MW, Webster JR, Welcomme RL, Winterbourn MJ. 1988. Patch dynamics in lotic systems: the stream as a mosaic. *Journal of the North American Benthological Society*, **7**: 503–524.

Poff NL, 1997. Landscape filters and species traits: towards mechanistic understanding and prediction in stream ecology. *Journal of the North American Benthological Society*, **16**: 391–409.

Reckendorfer W, Keckeis H, Winkler G, Schiemer F. 1999. Zooplankton abundance in the River Danube, Austria: the significance of inshore retention. *Freshwater Biology*, **41**: 583–591.

Reckendorfer W, Schmalfuss R, Baumgartner C, Habersack H, Hohensinner S, Jungwirth M, Schiemer F. 2005. The integrated river engineering project for the free-flowing Danube in the Austrian Alluvial Zone National Park: contradictionary goals and mutual solutions. *Archiv für Hydrobiologie Supplement*, **155**: 613–630.

Reynolds CS, Descy J-P. 1996. The production, biomass and structure of phytoplankton in large rivers. *Archiv für Hydrobiologie Supplement*, **113**: 161–187.

Reynolds CS, Glaister MS. 1993. Spatial and temporal changes in phytoplankton abundance in the upper and middle reaches of the River Severn. *Archiv für Hydrobiologie Supplement*, **101**: 1–22.

Reynolds CS, Carling PA, Beven KJ. 1991. Flow in river channels: new insights into hydraulic retention. *Archiv für Hydrobiologie*, **121**: 171–179.

Rosgen DL. 1994. A classification of natural rivers. *Catena*, **22**: 169–199.

Schiemer F. 1999. Conservation of biodiversity in floodplain rivers. *Archiv für Hydrobiologie, Supplement*, **11**: 423–438.

Schiemer F, Waidbacher H. 1992. Strategies for conservation of a Danubian fish fauna. In Boon PJ, Calow P, Petts GE. (eds) *River Conservation and Management*. John Wiley & Sons, Ltd, Chichester, pp. 363–382.

Schiemer F, Zalewski M. 1992. The importance of riparian ecotones for diversity and productivity of riverine fish communities. *Netherlands Journal of Zoology*, **42**: 323–335.

Schiemer F, Spindler T, Wintersberger H, Schneider A, Chovanec A. 1991. Fish fry associations: Important indicators for the ecological status of large rivers. *Verhandlungen Internationale Verein Limnologie*, **24**: 2497–2500.

Schiemer F, Keckeis H, Reckendorfer W, Winkler G. 2001a. The 'inshore retentivity concept' and its significance for large rivers. *Archiv für Hydrobiologie, Supplement*, **135**: 509–516.

Schiemer F, Keckeis H, Winkler H, Flore L. 2001b. Large rivers: the relevance of ecotonal structure and hydrological properties for the fish fauna. *Archiv für Hydrobiologie, Supplement*, **135**, 487–508.

Schlosser IJ. 1995. Critical landscape attributes that influence fish population dynamics in headwater streams. *Hydrobiologia*, **303**: 71–81.

Sedell JR, Reeves GH, Hauer FR, Stanford JA, Hawkins CP. 1990. Role of refugia in recovery from disturbances: modern fragmented and disconnected river systems. *The Journal of Environmental Management*, **14**: 711–724.

Stanford JA, Ward JV, Liss WJ, Frissell CA, Williams RN, Lichatowich JA, Coutant CC. 1996. A general protocol for restoration of regulated rivers. *Regulated Rivers. Research & Management*, **12**: 391–413.

Sukhodolov A, Uijttewaal WSJ, Engelhardt C. 2002. On the correspondence between morphological and hydrodynamical patterns of groyne fields. *Earth Surface Processes and Landforms*, **27**: 289–305.

Thorp JH, Delong MD. 1994. The riverine productivity model: an heuristic view of carbon sources and organic processing in large river ecosystems. *Oikos*, **70**: 305–308.

Thorp JH, Delong MD. 2002. Dominance of autochthonous autotrophic carbon in food webs of heterotrophic rivers? *Oikos*, **96**: 543–550.

Thorp JH, Delong MD, Greenwood KS, Casper AF. 1998. Isotopic analysis of three food web theories in constricted and floodplain regions of a large river. *Oecologia*, **117**: 551–563.

Thorp JH, Thoms MC, Delong MD. 2006. The riverine ecosystem synthesis: biocomplexity in river networks across space and time. *River Research and Applications*, **22**: 123–147.

Thoms MC, Parsons M. 2002. Eco-geomorphology: an interdisciplinary approach to river science. *International Association of Hydrological Sciences*, **276**: 113–120.

Tockner K, Pennetzdorfer D, Reiner N, Schiemer F, Ward JV. 1999. Hydrologic connectivity, and the exchange of organic matter and nutrients in a dynamic river-floodplain system. *Freshwater Biology*, **41**: 521–535.

Tockner K, Malard F, Ward JV. 2000. An extension of the flood pulse concept. *Hydrological Processes*, **14**: 2861–2883.

Townsend CR. 1996. Concepts in river ecology: pattern and process in the catchment hierarchy. *Archiv für Hydrobiologie, Supplement*, **113**, 3–21.

Uijttewaal WSJ, Lehmann D, van Mazijk A. 2001. Exchange processes between a river and its groyne fields: Model experiments. *Journal of Hydraulic Engineering-ASCE*, **127**: 928–936.

Vranovsky M. 1995. The effect of current velocity upon the biomass of zooplankton in the River Danube side arms. *Biologia, Bratislava*, **50**: 461–464.

Ward JV, Stanford JA. 1995. Ecological connectivity in alluvial river ecosystems and its disruption by flow regulation. *Regulated Rivers. Research & Management*, **11**: 105–119.

Weitere M, Arndt H. 2002. Top-down effects on pelagic heterotrophic nanoflagellates (HNF) in a large river (River Rhine): do losses to the benthos play a role? *Freshwater Biology*, **47**: 1437–1450.

Weitere M, Scherwass A, Sieben KT, Arndt H. 2005. Planktonic food web structure and potential carbon flow in the lower river Rhine with a focus on the role of protozoans. *River Research and Application*, **21**: 535–549.

Wu J. 1999. Hierarchy and scaling: extrapolating information along a scaling ladder. *Canadian Journal of Remote Sensing*, **25**: 367–380.

22
Hydroecology and Ecohydrology: Challenges and Future Prospects

David M. Hannah, Jonathan P. Sadler and Paul J. Wood

22.1 Introduction

The chapters in this volume clearly demonstrate the need for research at the intersection of hydrology and ecology to advance scientific understanding (Part 1), develop novel methods (Part 2), and manage the balance between growing human demands for water and needs of water-dependent ecosystems (Part 3). It is beyond the scope of this short chapter to attempt a synthesis of the diverse contents of the 21 preceding chapters, although all chapters are cited herein to point toward the present state-of-the-art. Rather than a retrospective, we use this final chapter to present a forward-looking view on the challenges and prospects for the 'emerging discipline' (Bond, 2003) of hydroecology/ ecohydrology and, in doing so, aim to focus attention on issues that require further evaluation and thought. We have suggested that a potential impediment to the development of ecohydrology/hydroecology is the lack of a clear subject definition to serve as a focal point to unite the research community (see Chapter 1). Most importantly, we assert herein that it is not simply the integration of hydrology and ecology that will determine the future prospects for ecohydrology/hydroecology but the way in which this integrative science is conducted. We advocate a truly interdisciplinary (as opposed to multidisciplinary) approach in which ecologists and hydrologists benefit from genuine synergy by embracing advances at the cutting-edge of both sciences and a unified, focused methodological stance. Such an approach should provide more perceptive answers to hydroecological/ ecohydrological problems and management questions. This chapter, and book, end by identifying future challenges and potential research themes that hydroecology/ecohydrology need to address as the discipline moves forward into the future.

Hydroecology and Ecohydrology: Past, Present and Future, Edited by Paul J. Wood, David M. Hannah and Jonathan P. Sadler
© 2007 John Wiley & Sons, Ltd.

22.2 The Need for an Interdisciplinary Approach

The need for, and potential benefits of, research that bridges the gap between traditional discipline boundaries is generally well recognised by scientists (e.g., Palmer and Bernhardt, 2006; Newman *et al.*, 2006), those funding pure, strategic and applied research (e.g., NERC, 2007; Environment Agency, 2004), and organisations that create and implement legislation relating to water (e.g., EU Water Framework Directive; Chapter 13). A single discipline focus may neglect parallel concepts and key process interactions, result in research undertaken at inappropriate (too fine or too coarse) or disconnected (mismatched) scales, and/or fail to make use of the most powerful, cutting-edge investigative tools. This wider realisation of the need for 'hybrid science' and interdisciplinary approaches has certainly stimulated much recent interest in hydroecology.

Although research focusing upon the interaction between hydrology and ecology has existed for some time (Bonell, 2002), true integration (especially between linked and coupled hydrological and ecological processes) remains incomplete. Arguably, this may be related to the way in which such research has been practised. Thus, the logic for adopting a hydroecological approach is persuasive but the way that hydrology and ecology are being integrated requires further careful evaluation. At present, ecologists and biologists appear to be looking at research questions from one perspective and hydrologists (mainly geographers and engineers) from another. Scientists may be seeking to address the same issue or solve the same problem without converging on the most perceptive or robust hydroecological answer(s) due to the absence of a clear theoretical understanding of the key elements in the 'other' discipline (perhaps because they are asking inappropriate question within the 'other' discipline).

If a true paradigm shift is to occur and hydroecology is to flourish, ecologists and hydrologists need to bridge the gap between traditional subject boundaries to build real interdisciplinary teams (e.g., Chapters 12, 18) and so reap benefit from true synergy by embracing advances at the cutting-edge of both sciences. Hence, we see a clear need for a genuine interdisciplinary approach; and we consider this a vital first step for generating new insights into water–ecosystems interactions and, in turn, identifying the future research agendas. To provide fully integrated solutions to issues affecting water-dependent systems, it may be necessary for hydrologists and ecologists to expand their interdisciplinary teams even further to include, for example, atmospheric scientists (e.g., climate change; Chapter 7), geologists (e.g., groundwater–surface water interactions; Chapter 9), biogeochemists (e.g., nutrient availability; Chapter 6), soil scientists (e.g., material properties influencing infiltration and soil biota activity), and social scientists (e.g., socio-economic impacts) to understand the full cascade of processes and their mechanistic interactions.

Making a plea for interdisciplinarity is reasonably straightforward (may be even somewhat trite) but we recognise that building a framework that fosters interdisciplinary collaboration between hydrologists and ecologists is no simple task. There is a need to go beyond mere 'discipline-hopping' (in terms of subject areas) and to meld reductionist Newtonian (often adopted by physical scientists – based upon simplification, universal laws and predictive understanding) and holistic Darwinian (often adopted by ecologists – based upon principles of complex interdependency) philosophies (Chapter 12). This split between Newtonian hydrologists and Darwinian ecologists is probably fuzzier in actuality. Harte (2002, p. 31) identifies 'elements of synthesis' for Newtonian and Darwinian traditions in the context of earth systems science as: (i) simple, falsifiable models, (ii) search for patterns and laws, and (iii) embracing the science of place. Most notably, Harte (2002) highlights the major scientific benefits that could be gained from combining reductionist and holistic systems approaches. Newman et al. (2006) suggest hierarchical scaling theory (e.g., O'Neill, 1986; Blöschl and Sivapalan, 1995) may offer a means of spanning the gap between these two approaches. Within this framework, (reductionist) explanations for phenomena/models can be considered in terms of their significance at different levels of organization (i.e., in a more holistic sense). Newman et al. (2006) propose that by examining adjacent levels in the hierarchy, new discoveries at the higher or lower scale may be made. This nested, coupled approach is not 'new' as such a hierarchical view has already been embraced successfully by some hydroecological/ecohydrological studies (e.g., Chapters 6, 10; Monk et al., 2006; Wood et al., 2001).

Place-based research (or science of place; Harte, 2002) may provide another framework for developing hydroecological collaboration with focus upon: (sub-)system based studies (e.g., Chapters 2, 3, 4, 6, 16, 17, 18, 19, 20, 21), integrated basins studies (Hannah et al., 2007; Zalewski, 2002) or a network of (long term) monitoring sites (e.g., Chapters 8, 14; Patrick et al., 1996; Prowse et al., 2006). Newman et al. (2006) identified three cross-cutting problems that represent barriers to overcoming key challenges in plant–water interaction research but they are applicable more generally to place-based hydroecological research, and also how results can be used to predict response/change and manage the environment. These problems relate back to Harte's (2002) elements of synthesis (above). The first problem relates to *spatial complexity and scaling*, that is the identification of scaling patterns and laws that should lead to improved prediction of cross-scale interactions, but there is a paucity of data at nested scales and poor quantification of spatial ecohydrological interactions. The second problem is connected to *threshold behaviour*, due to the nonlinearity in ecohydrological relationships and, hence, threshold crossing. Newman et al. (2006) suggest that this problem may be overcome by collecting multivariate data sets and field manipulation/experimentation and modelling studies. The third problem stems from *complex feedbacks and interactions*, which may be better understood by long-term, place-based, empirical studies to test models. Hence, there is a clear imperative to collect hydroecological data at nested space and time scales (i.e., at fine resolution for process-understanding, and over the longer-term and across a network of sites to set smaller-scale research in a wider context) and take a multivariate approach, with empirical observations and experimental manipulation conducted in such a way that ecological and hydrological results and models mesh (Chapter 10). To this end, the development of innovative techniques to capture data at appropriate scales for hydroecological research (Chapters 9, 11) and coupled statistical and numerical hydroecological modelling (Chapters 10, 12, 15, 21) have a significant role to play in integrating the science.

22.3 Future Research Themes

The needs for both a unifying definition of hydroecology and a framework to promote true interdisciplinary approach are generic, structural, challenges to the future of hydroecology. In this section, we identify potential research themes (linked to the challenges identified in the preceding section and Chapter 1) that hydroecology needs to address as the discipline moves forward into the future. The examples highlighted from fluvial and riparian ecosystems are indicative, demonstrating the highly interrelated nature of hydroecological research, although we recognise that similar challenges apply to other environments and ecosystems such as drylands (Newman *et al.* 2006) and wetlands (Chapter 14).

22.3.1 Ecosystem Sensitivity to Hydrological Change

An increasing number of ecosystems have been identified as being vulnerable to hydrological change (Angeler, 2007; Blom, 1999; Jasper *et al.*, 2006). However, while this vulnerability is recognised, knowledge is lacking regarding the most ecologically sensitive periods to hydrological extremes (e.g., floods, low flows/droughts and soil moisture deficit), and associated water stress and habitat disturbance (Chapters 3, 5, 7, 14, 17, 18, 19). In addition to focusing upon extremes, there is an evolving awareness of the importance of considering the spectrum of hydrological conditions experienced by habitats, and their linkages to ecosystem structure and functioning (Chapter 10). Long-term data sets have a critical role to play in unravelling these hydroecological associations and setting short-term change/variability in a wider context (Chapter 8).

22.3.2 Disturbance: Water and Ecological Stress

Disturbance is widely accepted as one of the primary driving forces behind ecosystem change (e.g., Dow, 2007; de Vicente *et al.*, 2006; Wilcox *et al.*, 2003; Chapter 5). Most natural disturbances (e.g., floods, drought and fire) occur as a result of a climatological aberration (e.g., Kingsford *et al.*, 2004; Muller *et al.*, 2007), which propagates down the atmosphere–hydrosphere–land surface–biosphere cascade. Superimposed upon natural (climate) disturbances, anthropogenic activities have the potential to enhance or moderate impacts significantly (e.g., Jones *et al.*, 2000; Lake, 2000; Stanford and Ward, 2001). However, the role of predictability and stochasticity in natural disturbance processes has not received the attention warranted (Plachter and Reich, 1998). For example, floodplains are highly dynamic environments that are in a state of constant flux with repeated erosion and sedimentation processes (Chapter 19), and inundation and desiccation events (Junk *et al.*, 1989; Chapter 3). Thus, the riparian habitat is strongly influenced by channel kinetics, determined predominantly by the frequency and magnitude of flood events (Ward *et al.*, 1999; Tockner *et al.*, 2000). This (often seasonal) flood disturbance is thought to maximise biological processes (Naiman and Decamps, 1997), and both in-stream and riparian biodiversity (Naiman *et al.*, 1993; Ward, 1998). In systems where floods are less predictable and 'reset' the physical and biotic environment, they can have catastrophic outcomes for populations of organisms, even though many have adaptations for dealing with inundation (Chapter 3, 17). Urgent research foci include: (i) quantifying variability; (ii) understanding disturbance cause (hydrology) and effect (ecology) and feedbacks; and (iii) predicting disturbance regimes under changing climate (Chapter 5 and 8, respectively) and/or with growing human impact.

22.3.3 Aquatic–Terrestrial Linkages

Aquatic ecosystems are strongly coupled to their terrestrial counterparts as they receive water via overland flow, shallow and deep subsurface flow paths and in some cases direct precipitation (i.e., channel interception). A number of long-standing ecological concepts and theories have been advanced to explain how riverine ecosystems interact with the landscape and affect ecosystem processes. The balance between the supply of autochthonous and allochthonous organic materials from headwaters and floodplains and its importance for riverine energetics, nutrient supply and foodwebs is a fundamental tenet of the three key lotic ecosystem models: the river continuum concept (RCC), the flood pulse concept (FPC) and the riverine productivity model (RPM) (Thorp and Delong, 1994). The FPC (Junk *et al.*, 1989) postulated that in unaltered large river systems, the bulk of riverine animal biomass relates directly or indirectly to production in the floodplains and not downstream transport of material from elsewhere (cf. RCC).

Set within these large-scale theories, there are finer grain linkages characterised by smaller scale 'foodweb subsidies' that move across terrestrial, aquatic and even marine habitats (Drake *et al.*, 2006; Helfield and Naiman, 2006; Polis and Hurd, 1996). A growing number of studies have highlighted the significance of terrestrial arthropods as essential food resources in aquatic stream ecosystems (Kawaguchi and Nakano, 2001) and recent work has shown how aquatic biodiversity can support certain riparian organisms (Paetzold *et al.*, 2005, 2006). The range of spatio-temporal variability in these foodweb subsidies is not fully quantified at present (Baxter *et al.*, 2005) and, hence, represents significant opportunity for new hydroecological research (Chapter 4).

22.3.4 Modern and Palaeo-analogue Studies

The potential value of long-term hydroecological datasets has received extensive comment (e.g., Jackson and Füreder, 2006; Willis *et al.*, 2007). Aside from data generated by routine monitoring, such as annual surveys collected by organisations responsible for managing habitats (e.g., the statutory national environmental monitoring agencies), long-term paired hydrological and ecological datasets that span more than a decade are uncommon (Chapter 10). Long-term data are increasingly regarded as important for understanding larger temporal fluctuations of aquatic communities that result from both anthropogenic activities (e.g., Armitage, 2006) and natural variability (e.g., Bradley and Omerod, 2001; Daufresne *et al.*, 2003; George *et al.*, 2004). In this regard, palaeoecological studies are valuable under-utilised resources. There is a wealth of palaeolimnological research that shows how data derived from subfossil organisms (e.g., diatoms and Cladocera) can be used as baseline information to understand longer-term hydroecological variability (e.g., Battarbee *et al.*, 1989; Quinlan and Smol, 2002; Barker *et al.*, 2005). Although rather too infrequently applied to other aquatic systems (Chapter 8), this palaeoecological approach has demonstrated potential to reveal hydrological change in river basins, and ecological patterns and processes that are associated with such change (e.g., Brayshay and Dinnin, 1999; Davis *et al.*, 2007; Gandouin *et al.*, 2006; Greenwood *et al.*, 2006).

22.3.5 Applied Hydroecology

Balancing anthropogenic demands for water against the water needs of aquatic ecosystems is a pressing global issue (Petts *et al.*, 2006). To achieve this goal requires collaboration between academics (e.g., hydrologists, geomorphologists, engineers and ecologists), practitioners (e.g., water and habitat managers) and stakeholders (e.g., landowners, anglers and recreational users) to balance the multiple and often conflicting pressures associated with the management of the system (e.g., flood alleviation and reservoir management) against the protection, and even enhancement, of ecosystem properties (e.g., conservation and restoration of habitats). River restoration is a good illustration of these issues and arguably represents one of the greatest challenges for hydroecologists (Palmer and Bernhardt, 2006). In many instances, the goals and standards for historic restoration projects have arguably been poorly defined (Palmer *et al.*, 2005). Consequently, there has been a mismatch (even conflict) between what stakeholders would like, what practitioners can deliver with available funds, and scientists' recommendations for environmental and ecological benefit (Gillilan *et al.*, 2005). This has been paralleled by an increase in small scale restoration projects (e.g., at the scale of individual riffles or flow deflectors to enhance local flow velocity) that may be hydraulically and morphologically enhanced but result in minor or insignificant effects on instream communities (e.g., Harrison *et al.*, 2004; Pretty *et al.*, 2003). This may reflect the disconnected nature of the individual projects in relation to the wider river basin context and also the need to identify the appropriate scale at which to undertake restoration. It would seem that clearly defined pre-project aims and standards are required that integrate scientists', practitioners' and users' perspectives; post-project appraisal is also required using multiple indicators (Gillilan *et al.*, 2005; Harrison *et al.*, 2004).

Scientists (hydrologists and ecologists together), environmental managers and stakeholders faced with hydroecological problems require to converge upon a common vision and a unified approach. We hope that this book and the indicative themes suggested above provide additional momentum to propel the future research agenda and the discipline of hydroecology forward.

References

Angeler, D.G. (2007) Resurrection ecology and global climate change research in freshwater ecosystems. *Journal of the North American Benthological Society*, **26**, 12–22.
Armitage, P.D. (2006) Long-term faunal change in a regulated and an unregulated stream – Cow Green thirty years on. *River Research and Applications*, **22**, 947–966.
Barker, P.A., Pates, J.M., Payne, R.J. and Healey, R.M. (2005) Changing nutrient levels in Grassmere, English Lake District, during recent centuries. *Freshwater Biology*, **50**, 1971–1981.
Battarbee, R.W., Stevenson, A.C., Rippey, B., Fletcher, C., Natkanski, J., Wik, M. and Flower, R.J. (1989) Causes of lake acidification in Galloway, south-west Scotland UK: a palaeoecological evaluation of the relative roles of atmospheric contamination and catchment change for two acidified sites with non-afforested catchments. *Journal of Ecology*, 773, 651–672.
Baxter, C.V., Fausch, K.D. and Saunders, C.W. (2005) Tangled webs: reciprocal flows of invertebrate prey link streams and riparian zones. *Freshwater Biology*, **50**, 201–220.
Blom, C.W.P.M. (1999) Adaptations to flooding stress: From plant community to molecule. *Plant Biology*, **1**, 261–273.
Blöschl, G. and Sivapalan, M. (1995) Scale issues in hydrological modeling – a review. *Hydrological Processes*, **9**, 251–290.
Bond, B. (2003) Hydrology and ecology meet – and the meeting is good. *Hydrological Processes*, **17**, 2087–2089.
Bonell, M. (2002) Ecohydrology – a completely new idea? *Hydrological Sciences Journal*, **47**, 809–810.
Bradley, D.C. and Ormerod, S.J. (2001) Community persistence among stream invertebrates tracks the North Atlantic oscillation. *Journal of Animal Ecology*, **70**, 987–996.

Brayshay, B.A. and Dinnin, M. (1999) Integrated palaeoecological evidence for biodiversity at the floodplain-forest margin. *Journal of Biogeography*, **26**, 115–131.

Daufresne, M., Roger, M.C., Capra, H. and Lamouroux, N. (2003) Long-term changes within the invertebrate and fish communities of the Upper Rhônoe River: effects of climatic factors. *Global Change Biology*, **10**, 124–140.

Davis, S.R., Brown, A.G. and Dinnin, M.H. (2007) Floodplain connectivity, disturbance and change: a palaeoentomological investigation of floodplain ecology from south-west England. *Journal of Animal Ecology*, **76**, 276–288.

de Vicente, I., Moreno-Ostos, E., Amores, C., Rueda, F. and Cruz-Pizarro, L. (2006) Low predictability in the dynamics of shallow lakes: Implications for their management. *Wetlands*, **26**, 928–938.

Dow, C.L. (2007) Assessing regional land-use/cover influence on New Jersey pinelands streamflow through hydrograph analysis. *Hydrological Processes*, **21**, 185–197.

Drake, D.C., Naiman, R.J. and Bechtold, J.S. (2006) Fate of nitrogen in riparian forest soils and trees: An N-15 tracer study simulating salmon decay. *Ecology*, **87**, 1256–1266.

Environment Agency (2004) *Hydroecology: Integration for Modern Regulation*. Environment Agency, Bristol.

Gandouin, E., Maasri, A., Van Viliet-Lanoe, B. and Franquet, E. (2006) Chironomid (Insecta: Diptera) assemblages from a gradient of lotic and lentic waterbodies in river floodplains of France: a methodological tool for paleoecological applications. *Journal of Paleolimnology*, **35**, 149–166.

George, D.G., Maberly, S.C. and Hewitt, D.P. (2004) The influence of the North Atlantic oscillation on the physical, chemical and biological characteristitcs of four lakes in the English Lake District. *Freshwater Biology*, **49**, 760–774.

Gillilan, S., Boyd, K., Hoitsma, T. and Kauffman, M. (2005). Challenges in developing and implementing ecological standards for geomorphic river restoration projects: a practitioner's response to Palmer *et al.* (2005). *Journal of Applied Ecology*, **42**: 223–227.

Greenwood, M.T., Wood, P.J. and Monk, W.A. (2006) The use of caddisfly assemblages in the reconstruction of flow environments from floodplain palaeochannels of the River Trent, England. *Journal of Paleolimnology*, **35**, 747–761.

Hannah, D.M., Brown, L.E., Milner, A.M., Gurnell, A.M., McGregor, G.R., Petts, G.E., Smith, B.P.G. and Snook, D.L. (2007). Integrating climate-hydrology-ecology for alpine river systems. *Aquatic Conservation: Marine and Freshwater Ecosystems*, **17**, 636–656.

Hannah, D.M., Wood, P.J. and Sadler, J.P. (2004) Ecohydrology and hydroecology: A 'new paradigm'? *Hydrological Processes*, **18**, 3439–3445.

Harrison, S.S.C., Pretty, J.L., Shepherd, D., Hildrew, A.G., Smith, C. and Hey, R.D. (2004) The effect of instream rehabilitation structures on macroinvertebrates in lowland rivers. *Journal of Applied Ecology*, **41**, 1140–1154.

Harte J. (2002) Towards a synthesis of Newtonian and Darwinian worldviews. *Physics Today*, **55** (10), 29–43

Helfield, J.M. and Naiman, R.J. (2006) Keystone interactions: Salmon and bear in riparian forests of Alaska. *Ecosystems*, **9**, 167–180.

Jackson, J.K. and Füreder, L. (2006) Long-term studies of freshwater macroinvertebrates: a review of the frequency, duration and ecological significance. *Freshwater Biology*, **51**, 591–603.

Jasper, K., Calanca, P. and Fuhrer, J. (2006) Changes in summertime soil water patterns in complex terrain due to climatic change. *Journal of Hydrology*, **327**, 550–563.

Jones, J.A., Swanson, F.J., Wemple, B.C. and Snyder, K.U. (2000). Effects of roads on hydrology, geomorphology, and disturbance patches in stream networks. *Conservation Biology*, **14**, 76–85.

Junk, W.J., Bayley, P.B. and Spinks, R.E. (1989) The flood-pulse concept in river-floodplain systems. *Canadian Special Publication of Fisheries and Aquatic Science*, **106**, 110–127.

Kawaguchi, Y. and Nakano, S. (2001) Contribution of terrestrial invertebrates to the annual resource budget for salmonids in forest and grassland reaches of a headwater stream. *Freshwater Biology*, **46**, 303–316.

Kingsford, R.T., Jenkins, K.M. and Porter, J.L. (2004) Imposed hydrological stability on lakes in arid Australia and effects on waterbirds. *Ecology*, **85**, 2478–2492.

Lake, P.S. (2000) Disturbance, patchiness, and diversity in streams. *Journal of the North American Benthological Society*, **19**, 573–592.

Monk, W.A., Wood, P.J., Hannah, D.M., Wilson, D.A., Extence, C.A. and Chadd, R.P. (2006) Flow variability and macroinvertebrate community response within riverine systems. *River Research and Applications*, **22**, 595–615.

Muller, S.C., Overbeck, G.E., Pfadenhauer, J. and Pillar, V.D. (2007) Plant functional types of woody species related to fire disturbance in forest grassland. *Plant Ecology*, **189**, 1–14.

Naiman, R.J. and Decamps, H. (1997) The ecology of interfaces: Riparian zones. *Annual Review of Ecology and Systematics*, **28**, 621–658.

Naiman, R.J., Decamps, H. and Pollock, M. (1993) The role of riparian corridors in maintaining regional biodiversity. *Ecological Applications*, **3**, 209–212.

NERC (2007) *Next Generation Science for Planet Earth*. Natural Environment Research Council, Swindon.

Newman, B.D., Wilcox, B.P., Archer, A.R., Breshears, D.D., Dahm, C.N., Duffy, C.J., McDowell, N.G., Phillips, F.M., Scanlon, B.R. and Vivoni E.R. (2006) Ecohydrology of water-limited environments: A scientific vision. *Water Resources Research*, **42**, W06302.

O'Neill, R.V., DeAngelis D.L., Allen, T.F.H. and Waide, J.B. (1986), *A Hierarchical Concept of Ecosystems*. Princeton University Press, Princeton

Paetzold, A., Schubert, C.J. and Tockner, K. (2005) Aquatic-terrestrial linkages across a braided river: Riparian arthropods feeding on aquatic insects. *Ecosystems*, **8**, 748–759.

Paetzold, A., Bernet, J.F. and Tockner, K. (2006) Consumer-specific responses to riverine subsidy pulses in a riparian arthropod assemblage. *Freshwater Biology*, **51**, 1103–1115.

Palmer, M.A. and Bernhardt, E.S. (2006) Hydrecology and river restoration: Ripe for research and synthesis. *Water Resources Research*, **42**, WR004354.

Palmer, M.A., Bernhardt, E.S., Allan, J.D., Lake, P.S., Alexander, G., Brooks, S., Carr, J., Clayton, S., Dahm, C.N., Follstad Shah, J., Galat, D.L., Loss, S.G., Goodwin, P., Hart, D.D., Hassett, B., Jenkins, R., Kondolf, G.M., Lave, R., Meyer, J.L., O'Donnell, T.K., Pagano, L. and Sudduth, E. (2005) Standards for ecologically successful river restoration. *Journal of Applied Ecology*, **42**, 208–217.

Patrick, S., Battarbee, RW., Jenkins, A. (1996) Monitoring acid waters in the UK: An overview of the UK acid waters monitoring network and summary of the first interpretative exercise. *Freshwater Biology*, **36**, 131–150.

Petts, G.E., Morales-Chaves, Y. and Sadler, J.P. (2006) Linking hydrology and biology to assess the water needs of river ecosystems. *Hydrological Processes*, **20**, 2247–2251.

Plachter, H. and Reich, M. (1998) The significance of disturbance for population and ecosystems in natural floodplains. In *The International Symposium on River Restoration*, Tokyo, pp. 29–38.

Polis, G.A. and Hurd, S.D. (1996) Linking marine and terrestrial food webs: Allochthonous input from the ocean supports high secondary productivity on small islands and coastal land communities. *American Naturalist*, **147**, 396–423.

Pretty, J.L., Harrison, S.S.C., Shepherd, D.J., Smith, C., Hildrew, A.G. and Hey, R.D. (2003) River rehabilitation and fish populations: assessing the benefit of instream structures. *Journal of Applied Ecology*, **40**, 251–265.

Prowse, T.D., Wrona, F.J., Reist, J.D., Gibson, J.J., Hobbie, J.E., Levesque, L.M.J. and Vincent, W.F. (2006) Climate change effects on hydroecology of Arctic freshwater ecosystems, *Ambio*, **35**, 347–358.

Quinlan, R. and Smol, J.P. (2002) Regional assessment of long-term hypolimnetic oxygen changes in Ontario (Canada) shield lakes using subfossil chironomids. *Journal of Paleolimnology*, **27**, 249–260.

Stanford, J.A. and Ward, J.V. (2001) Revisiting the serial discontinuity concept. *Regulated Rivers: Research and Management*, **17**, 303–310.

Thorp, J.H. and Delong, M.D. (1994) The Riverine Productivity Model – an Heuristic View of Carbon-Sources and Organic-Processing in Large River Ecosystems. *Oikos*, **70**, 305–308.

Tockner, K., Malard, F. and Ward, J.V. (2000) An extension of the flood pulse concept. *Hydrological Processes*, **14**, 2861–2883.

Ward, J.V. (1998) Riverine landscapes: biodiversity patterns, disturbance regimes, and aquatic conservation. *Biological Conservation*, **83**, 269–278.

Ward, J.V., Tockner, K. and Schiemer, F. (1999) Biodiversity of floodplain river ecosystems: Ecotones and connectivity. *Regulated Rivers–Research and Management*, **15**, 125–139.

Wilcox, B.P., Breshears, D.D. and Allen C.D. (2003) Ecohydrology of a resource-conserving semiarid woodland: Effects of scale and disturbance. *Ecological Monographs*, **73**, 223–239.

Willis, K.J., Araujo, M.B., Bennett, K.D., Figueroa-Rangel, B., Froyd, C.A. and Myers, N. (2007) How can knowledge of the past help to conserve the future? Biodiversity conservation and the relevance of long-term ecological studies. *Philosophical Transactions of the Royal Society (Series B)*, **362**, 175–186.

Wood, P.J., Hannah, D.M., Agnew, M.D. and Petts, G.E. (2001) Scales of hydroecological variability within a groundwater-dominated chalk stream. *Regulated Rivers: Research and Management*, **17**, 347–367.

Zalewski, M. (2002) Ecohydrology – the use of ecological and hydrological processes for sustainable management of water resources. *Hydrological Sciences Journal*, **47**, 823–832.

Index

abscisic acid 12–13
accuracy of measurements 149
alblation zones of glaciers 341
alien infestations 329
alpine river hydroecology 339–41
  future directions 353–6
  integrated understanding 351–3
  physicochemical properties 342
    classification 352
    hydrochemistry 346
    stream discharge 342–3
    stream temperature 343–4
    suspended sediment concentration 344–6
  river biota 346
    glacier-fed rivers 346–8
    nonglacier-fed rivers 348–9
    temporal variability 349–50
  water sources 341–2
anisohydric plant species 16–17
aquatic–terrestrial subsidies 57–8, 68
  flow control 58–9
  from land to water 59, 60
  from water to land 59–61
  future research 68
  human impact 67
    riparian deforestation 67
    river channelization and regulation 67–8
  river corridors 61
    braided river reach 63–6
    forrested headwater streams 61–2
    temperate lowland rivers 66–7

average daily flow (ADF) 236
avulsion 376–7

bars 362, 377–80
  evolution 363–6
    avulsion 376–7
    ecological implications 373–4
    migration 374–6
  facies composition 371
  initiation 363
  sedimentology
    ecological implications 362–3
    grain size sorting 363–71
  surface structure and hydraulics 371–3
    cross-sections 372
Behavioral Guidance Structure (BGS) 215
biodiversity in floodplains 41–6
black-water rivers 302
braid bars 363, 364–5
  sedimentology 366–87

canopy conductance 14
carbon dioxide (CO_2), atmospheric 114, 116
  plant response to 117–21
Catchment Abstraction Management Strategies (CAMS) 232
cavitation 15
chemical signaling in transpiration 12–13
chironomids 133–4
Cladocera 134
climate change 113–14, 123–4

Hydroecology and Ecohydrology: Past, Present and Future, Edited by Paul J. Wood, David M. Hannah and Jonathan P. Sadler
© 2007 John Wiley & Sons, Ltd.

differing perspectives of hydrologists and
    ecologists 121–2
  experimental studies 117–21
  future research 122–3
  simulation studies 116–17
  streamflow 114–15
Clivina collaris 45
cluster analysis (CA) 172
cohesion–tension (CT) theory of water
    movement 8, 14
Colorado River basin 100–2
computational fluid dynamics (CFD) 216–17
cross-bar channels 370–1

Danube River 406–7, 409–10
  biota 412–13
Darcy's Law 154–5
decoupling coefficient 13
dendrohydrology 132–3
diatoms 134–5
discrete grain identification
  applications 197
  principles 196–7
dissolved organic carbon (DOC) 103, 106
  in-stream biogeochemical function 106–7
dissolved organic matter (DOM) 82, 412
disturbance, flow-generated 75–80
downstream response to imposed flow
    transformation (DRIFT) 243, 244
drag coefficients of vegetation 277–8
droughts 76, 79–80, 86–7
  impacts 82–4
  recovery 85–6
Dry Weather Flow (DWF) 236
dyes and tracers 153, 159–60
dynamic global vegetation models (DGVM)
    113

ecohydraulics 205–7, 221
  challenges and limits 220–1
  concepts 211–12
  dimensionality 281–3
  drag coefficients of vegetation 277–8
  examples
    drift feeding salmonids 212–15
    fish swim path selection 215–19
  flow–vegetation interactions 269–71
    basic principles 271–2
    nonlinearities 276–7
    roughness properties of vegetation 272–6
  opportunity for engineers and ecologists
    219–20
  origin of concepts 207–8
  reference frameworks 208
    agent 209–11

  Eulerian 209
  Lagrangian 209
  velocity 278–81
ecohydrology
  focus 3–4
  historical perspective 2–3
ecologically acceptable flow regime (EAFR)
    233
El Nino–Southern Oscillation (ENSO) 300–1
electrofishing 391–2
embolisms 15–16
Endrick River 365
ensemble grain size parameter determination
  applications 198–200
  principles 197–8
environmental flows, determining 234–5
  biological response models 235, 238–9
  habitat assessment 235, 237–8
  habitat suitability criteria 240–1
  habitat-inclusive biological models 235,
    241
  hydrological indices 235, 236–7
esturine–coastal floods 78
Eulerian–Lagrangian–Agent method (ELAM)
    211, 218–20
European Union Water Framework Directive
    228, 237
evapotranspiration
  forests
    evaporation and transpiration 21
    understory transpiration 22
  key concepts and processes
    liquid water transport through trees 14–19
    SPAC 8
    transpiration 9–14
    water uptake in roots 19–21
exposed riverine sediments (ERS) 37–8, 52
  flow disturbance 39–41
  flow disturbance, importance to
    invertebrates 41
  lateral and longitudinal connectivity 47–8
  life history patterns and function ecology
    46–7
  physical variability 41–6
  habitats 38
  invertebrate conservation 38–9
  threats to invertebrate biodiversity 50–2
exposure strain rates, vertical and lateral
    213–14

flood pulse concept 131, 361
floodplains 253–4
  see also exposed riverine sediments (ERS)
  adjustment values for vegetation roughness
    274

floods 76, 78–9, 86–7
  constrained streams 80–1
  floodplain rivers 82
flow action, plant response to 320–1
flow regime variability 165–6, 230
  bibliographic analysis 167, 168
  data collection and analysis 172–5
  data integration for hydroecological analysis 175–6, 177, 179
  future directions and challenges 176–8
  hydroecological data requirement 166
  scale 167–70
flow-generated disturbances 75
  definitions 75–6
  droughts 79–80
    impacts 82–4
    recovery 85–7
  floods 78–9
    constrained streams 80–1
    floodplain rivers 82
  future challenges 87–8
  refugia 77–8
  responses 76–7
flow–vegetation interactions 291–2
  dimensionality 281–3
  drag coefficients 277–8
  ecohydraulics
    basic principles 271–2
    need for 269–71
    nonlinearities 276–7
    roughness properties of vegetation 272–6
  empirical illustrations 282
    complex velocity patterns in staggered arrays 286–7
    field measurement of velocity 289–90
    modelling Wienflüss flows 290–1
    velocity profiles in emergent rigid vegetation 284
    velocity profiles in mixture of submerged and emergent rigid vegetation 284–6
    velocity profiles in submerged flexible vegetation 286
    velocity profiles in submerged rigid vegetation 283–4
    velocity variation across partially vegetated channel 287–8
    water surface slope 289
  velocity 278–81
fluvial ecosystems 94
  biogeochemical dynamics 94–5
  biotic communities 95
  material delivery to and within 103–5
    in-stream biogeochemical function 106–7
    riparian interface zone 105–6
forests see trees and forests

frustules 134
future research themes 423–4
  applied hydroecology 425–6
  aquatic–terrestrial linkages 424–5
  disturbance 424
  ecosystem sensitivity to hydrological change 424
  modern and palaeo-analogue studies 425

general circulation models (GCMs) 116
glaciers 341–2
grain size
  determination
    discrete 196–7
    ensemble 197–200
  sorting
    barforms 363–71
groundwater dependent ecosystems (GDEs) 147
groyne fields, flow patterns 408

heterotrophic flagellates (HNF) 413
high resolution remote sensing 185
  depth and morphology 188
    image processing 188–90, 191
    laser scanning 194–6
    photogrammetry 190–4
  future developments 200
  requirements of scale 185–8
  substrate 196
    discrete grain identification 196–7
    ensemble grain size parameter determination 197–200
human impacts on hydroecological patterns 327–9
hydraulic capacitance 18
hydraulic gradient (HG) 96
hydraulic redistribution (HR) 20–1
hydraulic resistance 14, 17
hydraulic retention zones 405–6
  geomorphology and patch dynamics 406–7
  habitat conditions for characteristic biota 409–11
  implications for river management 416–17
  retention and water column processes 411–14
  retention, hydraulics and physiographic conditions 407–8
  significance for river networks 414–16
hydroecology 225–7
  application to water resource problems 242–4, 245–6
    communication and policy development 244–5
  focus 3–4

historical perspective 2–3
water resource management 231–2
  determining environmental flows 234–41
  ecologically acceptable flow regimes 232–4
  riverine system protection 232
hydrogeomorphic template 94
hydrogeomorphical–ecological interactions 295–6, 311–13
  hydrogeomorphical dynamics 296–303
  longitudinal gradients at confluence zones 305
    biological 306–11
    sediment 305–6
  riparian ecosystem 303–5
hyporheic flow 323–5

image processing
  applications 190, 191
  principles 188–9
Indicators of Hydrologic Alteration (IHA) 166, 236
innundation, plant response to 320
interdisciplinary approach 422–3
Intermediate Disturbance Hypothesis (IDH) 48
invertebrates in exposed riverine sediments 37–8, 52
  conservation 38–9
  flow disturbance 41
    life history patterns and function ecology 46–7
    physical variability 41–6
    population variability 47–8
    sustainability 48–50
    threats to biodiversity 50–2
isohydric plant species 16–17

laser scanning
  applications 195–6
  principles 194–5
lateral exposure strain rate 213–14
leaf area index (LAI) 13–14, 22, 23, 25
light sensitivity of stomatal conductance 11–12
Loire River basin 100–2
long-term (palaeo) records 129, 141–2
  key concepts 130–2
  palaeoecohydrology, restoration and enhancement 136–7
    case sudy I – River Culm in SW England 137–8, 139, 140
    case sudy II – Danish Lakes 138–41
  proxies and transfer functions 132

chironomids 133–4
Cladocera 134
Coleoptera 133
dendrohydrology 132–3
diatoms 134–5
pollen and spores 135–6
river–floodplain–lake systems 129–30
Lotic-invertebrate Index for Flow Evaluation (LIFE) 175–6, 179

Manning's n 272
meanders 366–7
Meso-HABSIM 240
meso-scale surveys 238
minimum acceptable flows (MAFS) 231
mini-piezometers and groundwater mapping 151–6
models, mathematical 208
momentum absorbing area (MAA) 272, 275
Murray River 387–9
  velocity profile 393
  fish capture 396–7
  large wood characteristics 394–6

NICOLAS Project 255, 256
nitrate pollution 253–4
  future perspectives 264–5
  landscape perspectives 261
    catchment-scale considerations 263
    nitrogen loading 263–4
    upslope–riparian-zone–channel linkage 262–3
  riparian buffers 254–5
    climatic and hydrological controls 255–8
    effect of vegetation 258–9
    N_2O emissions 260–1
    nitrogen saturation effect 259–60
Numerical Fish Surrogate (NFS) 212, 216–17, 218

Orinoco Basin 295–6, 311–13
  catchment area 297
  characteristics of Orinoco River and tributaries 298
  hydrogeomorphical dynamics 296–303
  riparian ecosystem 303–5
    longitudinal gradients at confluence zones 305
      biological 306–11
      sediment 305–6

palaeo records see long-term records
Palmer Drought Severity Index (PDSI) 300–2
particle imaging velocimetry (PIV) 373

particulate organic matter (POM) 412
perirheic zone 300
permeability of wood 17
PHABSIM (physical habitat simulation) 186, 239
photogrammetry
 applications 193–4
 principles 190–3
photon flux density (PFD) 10
photo-sieving 196
physical–ecological interactions 387–9, 397–400
 Murray River
  fish abundance analysis 397
  fish capture 396–7
  physical and hydraulic characteristcs of large wood 394–6
  study area 389–91
  study methods 391–3
planform bend positions 392
plants, response to atmospheric CO_2 117–21
plant communities, hydroecological patterns of change 317–18
 see also vegetation
 future directions 330–1
 human impacts 327–9
 hydrological–ecological interactions
  ecological drivers 321
  hydrological drivers 319–21
  natural influences 321–2
   lateral dimension 322–3
   longitudinal dimension 325–6
   temporal dimension 326–7
   vertical dimension 323–5
  riverine habitats 318–19
point bars 363
pollen and spores 135–6
pool–riffle sequences 96–8
pools 84
precision of measurements 149
press disturbances 76, 77
principal components analysis (PCA) 172
pulse disturbances 76, 77

ramp disturbances 76, 77
refugia 77–8, 87
rheophylic vegetation 303
riffle–pool sequences 96–8
river continuum concept 130–1, 361
river corridors 57–8, 68
 aquatic–terrestrial subsidies 61
  braided river reach 63–6
  forrested headwater streams 61–2
  temperate lowland rivers 66–7
 flow control 58–9

future research 68
human impact 67
 riparian deforestation 67
 river channelization and regulation 67–8
 subsidies from land to water 59, 60
 subsidies from water to land 59–61
river floods 78, 82
river flow 169–70
river productivity model 414
river regualtion, sustainable 230–1
riverine ecosystems
 barform sedimentology
  ecological implications 362–3
  grain size and sorting 363–71
roots, water uptake 19–21
roughness coefficients for vegetation 273
run-off, plant control of 321

sands and gravel, permeability values 367
sap velocity 18
seasonal droughts 79, 83
sedimentology 361–2, 377–80
 barforms
  ecological implications 362–3
  grain size sorting 363–71
  surface structure and hydraulics 371–3
 barforms, evolution of
  avulsion 376–7
  ecological implications 373–4
  migration 374–6
  permeability values 367
seepage meters 151, 152
slackwater zones 407
Soil–Plant–Atmosphere Continuum (SPAC) 8
South Saskatchewan River 364
specific conductivity 17
specific median discharge (SMED) 176, 177
Squamish River 371
stomatal conductance 10–12
streamflow 114–15
streams
 aquatic–terrestrial subsidies 61–2
 droughts 82–4
  recovery from 85–6
 floods 80–1
suberization 19
supra-seasonal droughts 79, 83
Surface Bypass Collector (SBC) 215
surface water–groundwater (SGW) echange processes 93–4, 107–8
 flow variability and water movements 96
  analysis in basin area 99–102
  flow dynamics 103
  spacial 96–8
  temporal 99

flow variability, implications of
  in-stream biogeochemical function 106–7
  material delivery to and within fluvial
    ecosystems 103–5
  riparian interface zone 105–6
fluvial ecosystems 94
  function 94–5
  structure 95
surface water–groundwater (SGW), field
    monitoring 147–8
  direct hydrological methods 150–1
    mini-piezometers and groundwater
      mapping 151–6
    seepage meters 151, 152
    synoptic surveys 152, 157
  future technical challenges 160–1
  indirect hydrological methods
    dyes and tracers 153, 159–60
    water chemistry and chemical signatures
      153, 158–9
    water temperature and thermal patterns
      153, 157–8
  terminology 148–50
synoptic surveys 152, 157

Tagliamento River 63–6, 363
Terrestrial Ecosystem Model (TEM) 116
total hydraulic strain 217
traffic rule 217
Trannon River 375
transpiration 9–14
trees and forests 7–8
  see also wood
  evapotranspiration in forests
    evaporation and transpiration 21
    understory transpiration 22
  evapotranspiration of trees
    liquid water transport 14–19
    SPAC 8
    transpiration 9–14
    water uptake in roots 19–21
  life cycle hydrolic changes 22–3
    age-related changes in forest composition
      23
    age-related changes in forest stand level
      25–7
    implications for predictive models 27–8
    species composition in aging forests 27
    tree size and stomatal conductance 23–5

structural complexity classes for fragments
    391

UK Biodiversity Action Plan (BAP) 232
unsuberized roots 19

Vegetation Ecosystem Modeling and
    Assessment Program (VEMAP) 116
vegetation *see also* plants responses to
    hydrological drivers 320–1
  roughness and water surface slope 289
vertical exposure strain rate 213–14
vertical hydraulic gradient (VHG) 96, 99
vortex structures 281–2

water, role on Earth 1–2
water chemistry and chemical signatures 153,
    158–9
water column processes 411–14
water processing, plant control of 321
water resource management 226, 245–6
  hydroecology, application of 242–4
    communication and policy development
      244–5
  hydroecology, role of 231–2
    determining environmental flows
      234–41
    ecologically acceptable flow regimes
      232–4
    water allocation 232
  rivers 416–17
  scientific basis 227–30
    principles 230–1
water storage in trees 18
water temperature and thermal patterns 153,
    157–8
wave action, plant response to 320
Weighted Usable Area (WUA) 239
white-water rivers 302
Wienflüss flows 289–90
  modelling 290–1
wood *see also* trees and forests
  fish habitats 389
    fish numbers 395
  structural complexity classes 391

xylem water transport 15